Lean Construction

This book collates the main research developments around Lean Construction over the past 25 years with contributions from many seminal authors in the field. It takes stock of developments since the publication of Koskela's (1992) *Application of the New Production Philosophy to Construction* and, in doing so, challenges current thinking and progress. It also crystallises theoretical conceptualisations and practically situated learning whilst identifying future research challenges, agendas and opportunities for global collaborative actions.

The contributors present the development of Lean Construction as a fundamental part of improving construction productivity, quality and delivery of value to clients and users of built infrastructure. In doing so, the book introduces the reader to the foundational principles and theories that have influenced the way we now understand Lean Construction and has provided very useful insights to students, practitioners and researchers on key junctures over the last 25 years. Highlighting the key contemporary developments and using global case study material the chapters demonstrate good practice but also help introduce new thinking to both lay readers and experienced practitioners alike.

This book is essential reading for undergraduate and postgraduate students, researchers and practitioners with an interest in Lean Construction and construction management, providing a general understanding of the area, current state of the art knowledge as well as providing an insight into areas for future research.

Patricia Tzortzopoulos is Professor of Integrated Design, Associate Dean for Research and Enterprise and Director of the Innovative Design Lab research centre at the School of Art, Design and Architecture, University of Huddersfield. She has developed a number of research projects examining the design process, including front-end design issues, requirements capture and management, evidence-based design, participatory design approaches, Building Information Modelling in the retrofit of existing social housing, adoption of code checking in healthcare design, among others. Patricia is an associate editor of the Brazilian Journal *Ambiente Construido*, and has been an active member of the International Group for Lean Construction since 1998.

Mike Kagioglou is Professor of Process Management and Dean of Art, Design and Architecture at the University of Huddersfield since 2013. Mike's career spans Engineering, Manufacturing, Creativity, Design, Architecture and the Built Environment, engaged in inter and multidisciplinary research at a global level. He was an Academic Director for the £11 million Engineering and Physical Sciences Research Council (EPSRC) funded interdisciplinary Innovative Manufacturing Research Centre (IMRC) in Health and Care

Infrastructures Research and Innovation Centre (HaCIRIC) and was previously the Director of the £8 million EPSRC Salford Centre for Research and Innovation (SCRI) in the built and human environment. Mike has published more than 200 academic referred papers, many industrial reports and two books.

Lauri Koskela is Professor of Construction and Project Management at the University of Huddersfield. Since 1991, Lauri has been involved in research on Lean Construction. His research has focused especially on the theories of production management as well as project management, underlying Lean Construction. Lauri is a founding and continuously active member of the International Group for Lean Construction (IGLC). He was Director and Trustee at Lean Construction Institute – UK from 2010 to 2019. In recent years, he has also focused on the theory of design, and has actively participated in workshops and conferences organised by the Design Society in relation to this theme.

Lean Construction

Core Concepts and New Frontiers

Edited by Patricia Tzortzopoulos, Mike Kagioglou and Lauri Koskela

Routledge
Taylor & Francis Group
LONDON AND NEW YORK

First published 2020
by Routledge
2 Park Square, Milton Park, Abingdon, Oxon OX14 4RN

and by Routledge
52 Vanderbilt Avenue, New York, NY 10017

Routledge is an imprint of the Taylor & Francis Group, an informa business

British Library Cataloguing-in-Publication Data
A catalogue record for this book is available from the British Library

Library of Congress Cataloging-in-Publication Data
A catalog record has been requested for this book

ISBN: 978-0-367-19655-4 (hbk)
ISBN: 978-0-429-20373-2 (ebk)

Typeset in Baskerville
by Swales & Willis, Exeter, Devon, UK

We dedicate this book to the memory of the Danish pioneer and champion of Lean Construction, Sven Bertelsen, who passed away a few days before the book manuscript was completed. Sven has had a major influence on the development and diffusion of Lean Construction, and many of the chapter authors have fond memories of interaction with him. His passion for finding better ways for construction provides a model worth imitating for us, the younger generations.

Contents

Contributors

Thais da C. L. Alves is an Associate Professor at San Diego State University, USA. She has conducted numerous studies in Brazil and the United States, advised students, and written over 100 papers related to implementing Lean in different settings including site and office operations, as well as the supply chain of construction projects. She is an active member of the Lean Construction Institute (LCI) San Diego Community of Practice, and has served as editor, track chair, reviewer and/or conference organiser in multiple editions of the International Group for Lean Construction (IGLC) Conference.

Paz Arroyo is a Quality Leader at DPR Construction, based in San Francisco, California. Paz holds a PhD in civil and environmental engineering (2014), from University of California, Berkeley and a MSc in Engineering (2007) from Catholic University of Chile. She has expertise on both teaching and implementing the Choosing by Advantages (CBA) decision-making system internationally for design and construction decisions involving diverse AEC stakeholders. Paz is also an Adjunct Professor at Catholic University of Chile and co-founder of CollabDecisions, a platform sharing knowledge on collaborative decision making. Her research interests are collaboration, decision making, quality, Lean, and sustainability in construction projects.

Glenn Ballard has been an Adjunct Professor in Civil & Transport Engineering for Norwegian National Technological University since 2014. He has been a Director of Project Production Systems Laboratory at University of California, Berkeley since 2006. His research interests include: project production systems design, contracting strategies, relational contracting, negative and positive incentives, construction work methods design and improvement, including the application of traditional industrial engineering techniques and mimeomotion, decoupling through buffers, and the practice of First Run Studies. He is a Founding Member of IGLC, a co-founder and member of Lean Construction Institute and a member of The Design Society.

Audrey Bascoul is a PhD candidate in the Engineering and Project Management Program at the University of California, Berkeley. Audrey conducted research at Level 10, DPR, and at the University of California, San Francisco (UCSF). She also consulted for Americaribe, a Bouygues' subsidiary located in Florida. She currently does action research at the Lawrence Berkeley National Laboratory. Her research interests include Lean Construction, facility management, structural complexity, visual management and built-in-quality. She holds a BS (Industrial/Mechanical Engineering) and an MS (Industrial/Mechanical Engineering) from the Engineering School Arts et Métiers (France), and an MS (Civil Engineering) from UC Berkeley.

Clarissa Biotto has a PhD from the University of Huddersfield, UK. Clarissa is a Lecturer of the Civil Engineering Department of the Federal University of Sao Carlos (UFSCar), Brazil, where she also develops postdoctoral research on Lean Design Management and BIM. Clarissa's doctoral thesis is entitled "Integration of overlapping design and construction phases through location-based planning tools". Her area of interest is the management of design, off-site production and Lean Construction as the main production management philosophy, as well as location-based scheduling tools and BIM. Clarissa worked as a Lean Consultant in Brazil, where she implemented Lean Construction philosophy and tools in more than 20 construction projects. She holds a master's degree in Construction Management and Economics from the Federal University of Rio Grande do Sul (UFRGS), Brazil. Her dissertation tackled the use of BIM 4D in production management in construction. She is an Architect and Urbanism Designer graduated at the University of São Paulo (USP), Brazil.

Trond Bølviken is Director at Veidekke, one of the leading construction and real estate development companies in Scandinavia, and a part-time professor at the University of Agder, Norway. He has been a member of the International Group for Lean Construction (IGLC) for 15 years and has been first and co-author of several IGLC papers on waste and the nature of production in construction. Trond was local chair of the yearly conference of the International Group for Lean Construction held in Oslo in 2014 (IGLC 22). He is board member of Project Norway, a national program for project process research.

Bhargav Dave is a Senior Researcher at Aalto University with over 18 years of cross research and industry experience in the fields of Information Technology and Construction Management. He has worked on research projects related to information management in construction project lifecycle, with a specific focus on Lean production management systems and their integration with Building Information Modelling. He has authored over 40 scholarly articles and industry reports, and led innovation projects in digitisation of construction processes, including BIMforLean and Intelligent Products (Integrating Internet of Things, Lean and BIM for the construction lifecycle).

Lincoln H. Forbes is a specialist in Lean Project Delivery and Lean management systems. A Registered Professional Engineer, he has over 30 years' experience in quality/performance improvement as well as facilities design and construction management support. His book, *Modern Construction: Lean Project Delivery and Integrated Practices* (Forbes and Ahmed, 2010, CRC Press), is an internationally recognised reference. He is a Fellow of the Institute of Industrial and Systems Engineers (IISE), and is the founder and past president of the IISE's Construction Division. His professional memberships include the Lean Construction Institute (LCI) and the American Society for Quality (ASQ).

Carlos Torres Formoso is a professor in Construction Management at the Federal University of Rio Grande do Sul (UFRGS), Brazil. He has been a member of the International Group for Lean Construction since 1996. He has developed research projects in partnership with contractors, real estate developers and prefabricated systems providers. He has been a visiting scholar in several universities outside Brazil, including the University of California and Norwegian Building Research Institute, among others. He has been a senior advisor of the Brazilian Ministry of the Cities for the National Program for Quality and Productivity in the Habitat (PBQP-H) for 11 years.

Danilo Gomes is a PhD candidate in Architecture at the University of Huddersfield, with a Master Degree in Civil Engineering by the State University of Londrina – Brazil. He has more than 7 years of experience in teaching Design Studios for Architectural Students. He also has more than 10 years of practical experience in delivering Architectural projects in Brazil. His studies on Lean Construction involve especially Lean Design and Collaborative Practices.

Gøril Hannås is associate professor in supply chain management and Head of Department for Working Life and Innovation at the School of Business and Law at University of Agder. She has worked extensively with supply chain management in both private and public sectors, including management of large and complex purchasing contracts. Her research revolves around SCM, contract management and risk management in project-based industries. Gøril Hannås served as director of several boards for enterprises within energy, IT and logistics industries, including the board of directors and Audit Committee for Norway Post, where she was appointed by the Minister of Transportation.

Cynthia dos Santos Hentschke is a PhD student in the Post-Graduation program of Civil Engineering: Construction and Infrastructure from the Federal University of Rio Grande do Sul, Brazil and a Research Exchange Student at the University of Huddersfield. Her main research interests are related to Mass Customisation, New Product development, Design Management and Value Generation. She has over 6 years of experience working with Mass Customisation in construction and Value generation, Additionally, she was a research collaborator at the University of Huddersfield, inspiring her to explore connections between healthcare design, mass customisation and value.

Bo Terje Kalsaas is Professor at the Department of Engineering Sciences, University of Agder. His research and teaching expertise includes supply chain management, Lean production, Lean Construction, project management and engineering management. He has undertaken various research projects, authored numerous publications and guided implementations of process innovations in industry and construction practice for many years. Working with industries including manufacturing, the AEC industry and construction in the offshore oil and gas sector. He was scientific chair of the annual IGLC conference in Oslo in 2014, and he has recently edited a book on Lean Construction.

Sergio Kemmer is a Lean Consultant at Jacobs UK. Previously he worked at the University of Salford as Research Assistant where he contributed to the development of several research projects covering topics including design and construction management, Lean Construction, Building Information Modelling, facilities management and mass customisation. Prior to that he worked for construction companies as production manager and Lean coordinator in Brazil. Since 2002, Sergio has been involved in research and implementation of Lean Construction concepts and tools, especially the Last Planner System™. He has published several papers in the conferences of the International Group for Lean Construction.

Ole Jonny Klakegg, MSc, PhD, has 29 years of experience in research, teaching and consulting within project management. He is Professor in Project Management/Construction Management at the Norwegian University of Science and Technology, Department of Civil and Environmental Engineering. He also holds a part-time position as Special Advisor at WSP Norway. Klakegg has experience from a large number of major projects in Norway in private and public sectors, including building, transport, health, and defence projects. He is program manager for the Construction program in Project Norway. His research interests include project governance, project management and risk/uncertainty management.

William Lichtig is Executive Vice-President, The Boldt Company, USA. Will is an internationally recognised leader in the Lean and Integrated Project Delivery movements. He has authored multiple articles and routinely lectures on these topics. He has coached teams in implementing Lean design and construction methods, including Target Value Design and Built-in Quality. Lichtig developed the Integrated Agreement for Lean Project Delivery, and recently co-chaired the effort to update and improve the Consensus Doc 300. He was recognised by the Lean Construction Institute in 2012 with its "Pioneer Award" and was elected in 2016 to the US National Academy of Construction.

Luciana Miron is Associate Professor at the Faculty of Architecture (since 2006) and the Graduate Program in Urban and Regional Planning (since 2011) of the Federal University of Rio Grande do Sul. Since 2003 she has published papers and participated in the International Group for Lean Construction (IGLC) conferences, having as her main research focus the value generation concept. Field of expertise: Design Process Management, Customer/client Requirements Management, Evaluations of Social Housing Projects, Management and Development of Urban and Regional Projects.

John Rooke studied sociology and ethnomethodology at the University of Manchester and engineering management at the University of Birmingham. Research projects in construction management include: contract management; site health and safety; quality management; knowledge management; production theory. Other research includes: action learning groups and organisational change; small scale evaluation of public health projects.

He has been associated with the International Group for Lean Construction (IGLC) since IGLC 5 in 1997 and in 2011 was scientific chair of IGLC 19 in Lima, Peru. He has enduring interests in research methodology, production theory and organisational process.

Zofia K. Rybkowski's research experience includes Integrated Project Delivery, productivity analysis and Lean Construction, life cycle cost analysis and sustainable design. She has extensive experience as a construction, architectural and engineering researcher and consultant. She has consulted for firms in Boston, San Francisco, Tokyo and Hong Kong. She holds degrees from Stanford, Brown, Harvard, HKUST and the University of California. She is an Associate Professor of Construction Science, teaching Lean Construction to advanced construction science students at Texas A&M University, and a Fellow at the Center for Health Systems and Design. She is a LEED AP.

Rafael Sacks is a Professor of Construction Engineering and Management at the Technion – Israel Institute of Technology. He is the Vice-Dean for Graduate Studies and Research in the Faculty of Civil and Environmental Engineering, and served as Head of Structural Engineering and Construction Management from 2012 to 2015. Rafael's research focuses on the synergies of Building Information Modeling (BIM) and Lean Construction. He has a rich record of publications (some 75 journal articles, 90 peer-reviewed conference papers and numerous research reports). He was awarded the ASCE's Thomas Fitch Rowland Prize for research in construction management in 2016.

Tarcisio Abreu Saurin is an associate professor at the Industrial Engineering Department, at the Federal University of Rio Grande do Sul (UFRGS), Brazil. His main research interests are related to Lean production, Lean Construction, safety management, resilience engineering, modelling and management of complex socio-technical systems. On these topics, he has carried out applied research mainly in construction, healthcare and manufacturing industries.

Olli Seppänen is a Professor of Practice at the Aalto University School of Engineering, Department of Civil Engineering. His research interests include Lean Construction, real-time production control, Lean design management and digitalised construction operations. Before joining academia, Olli co-founded three construction management software start-ups: Dynamic System Solutions LLC, DynaRoad Ltd and Vico Software Inc. Olli worked as Director of VDC Services and managed a services organisation delivering Virtual Design & Construction to companies worldwide. He has been an active member of the International Group of Lean Construction, and was the chair of the Academic Committee of the conference in 2015.

Patricia A. Tillmann works at the University of California, San Francisco (UCSF) as a Senior Lean Manager, supporting the adoption of Lean principles and techniques for design and construction in collaborative project arrangements. She was a visiting scholar at the University of Salford, and at the University of California, Berkeley. She worked as a Lean specialist for a mechanical contractor for 2 years, before joining UCSF in 2016. Patricia is an active participant of the International Group of Lean Construction and has authored several papers on: value generation, benefits realisation, target value design and integrated project delivery.

Iris D. Tommelein is Professor of Engineering and Project Management at the University of California in Berkeley, USA. She teaches and conducts research to develop and advance the theory and principles of project-based production management. She is an expert on Lean Construction, site logistics, layout, operations and methods design, materials management and supply-chain management. She directs the Project Production Systems Laboratory (P2SL – p2sl.berkeley.edu) at UC Berkeley. P2SL is a research laboratory dedicated to developing and deploying knowledge and tools for project management, as well as a learning lab for the Northern California construction industry.

Olav Torp is associate Professor at Norwegian University of Science and Technology. He is deputy head for research at the department and head of research group Building and Construction Engineering. Olav has for over 20 years been working with teaching and research within Construction Project Management. One key research area has been uncertainty analysis and management in large infrastructure projects, and how to go from uncertainty analysis to uncertainty management. He has been first and co-author of a series of publications within the topic of Uncertainty Management, and in recent years some publications on the IGLC-conferences.

Cynthia C. Y. Tsao has worked with educators and researchers in academia and practice to help the Owner-Architect-Engineer-Contractor (OAEC) industry understand and leverage Lean Project Delivery. Cynthia has organised and presented at over 50 trade association and university events. She has worked on multi-project engagements with general contractors, personally coaching over 80 field superintendents. From 2008 to 2011, Cynthia developed an operational structure for the Lean Construction Institute (LCI) and oversaw the unprecedented growth of LCI Communities of Practice from 3 to 17. As a Lean coach, she guides companies in Project Team Alignment, Decision Making Strategy and Design Management.

Stanislaus Tuholski (https://orcid.org/0000-0002-5646-9436), is the Department Head for the Project and Construction Office at the Lawrence Berkeley National Laboratory in Berkeley, California, USA. In his early career, he practised structural engineering (SE) at

Forell/Elsesser Engineers in San Francisco. He worked on several challenging projects including the San Francisco City Hall Retrofit, the Stanford University Science and Engineering Quad, the Asian Art Museum Retrofit and the Pasadena City Hall Seismic Retrofit. In 2004, he joined the Lawrence Livermore National Laboratory. He transferred to the Lawrence Berkeley National Laboratory in 2012.

Daniela Dietz Viana is a Lecturer at the Federal University of Rio Grande do Sul, Brazil, focusing on disciplines from Services and Industrial Engineering. Daniela has a PhD in Civil Engineering (2015), from the Federal University of Rio Grande do Sul. Her Master's degree is also in Civil Engineering from the Federal University of Rio Grande do Sul (2011) and she has a degree in Architecture and Urbanism from the same university (2008). Her research experience is in Construction Management and Economics, having researched the following topics: Last Planner System, Lean Construction, Prefabrication, Construction Industrialisation, engineer-to-order and Language-Action Perspective.

Ruben Vrijhoef is a Senior Researcher at the Department of Management in the Built Environment at the Delft University of Technology, Netherlands. In addition he is a Professor at the Utrecht University of Applied Sciences. His expertise includes supply chain management, Lean Construction, construction logistics, sustainable development, circular economy and strategic procurement in construction. He has undertaken various research projects, authored numerous publications and guided implementations of process innovations in construction practice for many years. This also includes digital coordination and control, and innovative procurement schemes aimed at improved logistics and resource efficiency, and reduced environmental impact of construction transport.

Foreword

This book is about the latest thoughts on Lean Construction and some of the key areas of development since the concept was introduced over 25 years ago.

Since publication of the Egan Report Rethinking Construction, which promoted the introduction of an adaptation of the Toyota Production System, there has been considerable development of Lean Construction worldwide. The Egan Report gave five key drivers: committed leadership, a focus on the customer, integrated processes and teams, a quality driven agenda and a commitment to people, which are as appropriate today as they were in 1997 when the report was published.

This book covers all five key drivers proposed in the Egan Report and more, especially with the introduction of digital engineering.

The automotive industry has seen a huge improvement in productivity thanks to the wide adoption internationally of the Toyota Production System principles. In comparison, the UK Construction industry's productivity has 'flat lined' compared to other industries with frequent delays and budget over-runs. In particular, the high-profile failure of a major UK contractor with a number of others having severe financial difficulties has led many to believe that the Construction Industry's business model, which is highly fragmented in conjunction with low profitability, is not fit for purpose and that change is essential.

The Project 13 initiative is led by the Institution of Civil Engineers and aims to transform the Construction Industry (see www.p13.org.uk). If broadly adopted it will help to tackle the key problems facing the construction industry. However, if this project is to succeed, it must be in conjunction with the strategic deployment of Lean Construction. The construction industry is vital to our economy by improving the built environment and giving employment to millions of citizens. Lean Construction provides a tried and tested method of ensuring construction activities are completed on time, within budget and meet the customer's requirements. In particular, research clearly shows that Lean Construction promotes a safer workplace.

To deploy Lean, a 'top down' strategic approach is essential and the senior team must be committed to giving consistent leadership over the long term. However, Lean Construction deployment often begins 'bottom up' with the use of the many Lean Construction tools available. The use of the Last Planner System (also known as Collaborative Planning) within a Lean Construction deployment is without doubt a hugely effective tool. It has been found to be so effective that Highways England mandate its use across all of their projects. The chapter on the Last Planner System brings the reader the latest developments.

I am optimistic that the Construction Industry is in transition to a much improved delivery model with the more widespread introduction of digital construction. This can only be achieved by considering construction as a production process. The chapter discussing construction production systems at the nexus of Lean and BIM provides an excellent framework to support the integration of Lean thinking and digital construction.

When I was responsible for deploying Lean across Highways England, the support and guidance given by many of the authors was invaluable. At the time of writing, Highways England continue to use Lean thinking across their business and within their supply chain with recorded efficiencies to date approaching £250 million.

The Lean team in Highways England developed a diagnostic tool to measure the level of adoption of Lean Construction both at enterprise level and within projects across their supply chain known as the Highways England Lean Maturity tool (HELMA). However, in many of the HELMA assessments, organisations frequently did not have a systematic method of making decisions. The chapter on Choosing by Advantages explains a highly effective method, which I would recommend to the reader.

Hopefully, other clients who read this book will be inspired, as I was, to adopt a strategic approach to Lean Construction principles and embark on the journey of Lean thinking. In common with all continuous improvement methods it is an endless journey, hence the importance of this book in giving an update on the latest thought by experienced and insightful authors.

I am sure, whether you are new to Lean Construction or you're an experienced practitioner, you will find all of the chapters of this book essential reading. Enjoy.

Derek Drysdale
Director and Trustee
Lean Construction Institute – UK

Acknowledgements

This book would not have materialised without the innovative and hard work of all the authors, the reviewers who peer reviewed all chapters and the publisher's support and patience in completing this work. A special thanks goes to all organisations who are striving to continuously improve by implementing diverse aspects of Lean Construction, and are courageous to work with academia and experiment on improved processes and new paradigms through the co-production of knowledge.

General introduction: Lean Construction: core concepts and new frontiers

Patricia Tzortzopoulos, Mike Kagioglou and Lauri Koskela

The 1990s provided a fertile ground for rethinking construction engineering and management across the spectrum of technology, Information Technology (IT), process reconceptualisation and people interactions in different forms of organising and management. Great strides were made in procurement and partnering practices, integrated databases and whole life costing models, off-site construction, new methods of construction and new financial and costing models. The 1990s were also the era that quality management practices were considered and there was a push to think of "construction as a manufacturing process." What was stark in its absence was any sense of a theoretical framework that could be used for sense-making and articulation of the fundamentals of the organising and making part of construction.

That was true until Koskela introduced his Transformation-Flow-Value (TFV) theory as the new production philosophy to construction. His seminal publication: *Application of the New Production Philosophy to Construction*, in 1992, whilst a visiting scholar at Stanford University, challenged the status quo of theory and practice and called for a re-conceptualisation of production in construction and, since then, it has proven the catalyst for a global movement. The international Group of Lean Construction was established in 1993 and a few hundred publications, PhD theses and practice exemplars have been documented. Significant developments have taken place and new understandings and conceptualisations have been introduced.

This book aims to take stock of progress made during the quarter of a century since the original publication by Koskela, which triggered the Lean Construction movement. In doing so, the book aims to challenge current thinking, advance and crystallise theoretical conceptualisations and practical learning and identify future research challenges and opportunities for global collaborative actions.

The book covers the main thematic areas of knowledge development across the Lean Construction community over the past 25 years, and it has been organised in two parts. Part 1 includes chapters describing some of the Lean Construction themes as developed over time, whilst Part 2 describes Lean Construction approaches, with some examples of applications in practice. Figure I.1 describes in a very schematic form when some of the main Lean themes, concepts and approaches first appeared over publications from the International Group for Lean Construction. The Figure is not precise nor comprehensive, but it aims to provide an overview of developments over time, many of which are discussed across the chapters of this book.

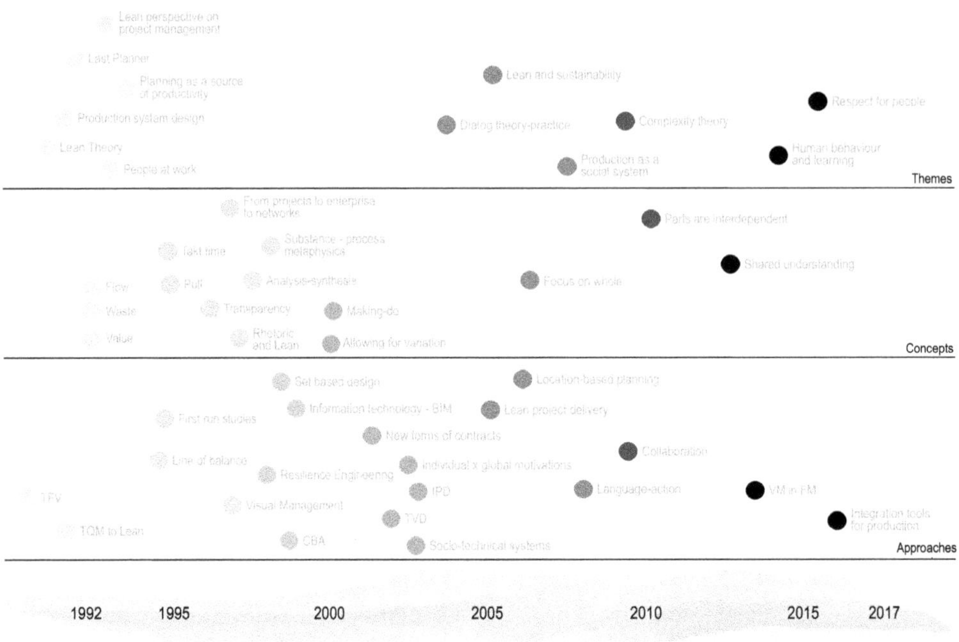

Figure I.1 Overview of some Lean Construction developments

The books brings a compilation of diverse research and practical examples of Lean Construction. We hope the book serves as a useful reference for academics and practitioners who seek to understand and advance production management in construction. The book will be essential reading for undergraduate and postgraduate students, researchers and practitioners with an interest in construction management, providing a general understanding of the Lean Construction area, current state of the art knowledge as well as providing an insight to areas for future research – the new frontiers.

Part 1: Lean Construction themes

Part 1 presents and discusses some of the main Lean Construction themes, as they have been addressed over time by the International Group for Lean Construction. These are not meant to be exhaustive, but pulled together to provide the holistic picture of Lean Construction and to provide insights into its constituent parts. As you are reading these lines, new themes are emerging, existing ones are enriched and many practices become obsolete as new ways of thinking and new technologies, in the broader sense of the word, become available.

Chapter 1, by Lauri Koskela, endeavours to present a consolidated discussion on the theoretical and philosophical foundation of Lean and its instantiation in a particular context, such as construction. It is proposed that at its essence, Lean is a theoretical innovation: the theories and philosophical commitments underpinning Lean allow wider action options in comparison to mainstream operations management. Lean Construction is argued to have evolved in two ways. Originally, Lean Construction started as local problem solving in production management, and gradually developed into a more comprehensive template, covering also contractual and organisational areas. Somewhat later, principles, methods and tools originating in the Toyota Production System have also increasingly been applied in or adapted to the context of construction.

Chapter 2, by Patricia Tzortzopoulos, Cynthia dos Santos Hentschke and Mike Kagioglou discusses New Product Development in manufacturing and how it has influenced Lean design and construction. The chapter discusses design management in the context of Lean. This chapter also presents an overview of developments presented in the International Group of Lean Construction (IGLC), with brief descriptions of the main topics related to Lean design, some of which are linked to other chapters of the book. Finally, areas for future research are proposed.

Chapter 3, by Glenn Ballard, discusses one of the most developed areas of research in Lean, i.e. production planning and control. The chapter presents the Last Planner System, widely acknowledged and implemented worldwide as a means to improve construction planning and control in a context of collaboration and commitment. The chapter presents some of the seminal ideas related to the Last Planner, including a brief history of its development, its principles, and areas of current research.

Chapter 4, by Bhargav Dave and Rafael Sacks, looks at the links between Lean and Building Information Modelling (BIM). The chapter describes an important area that has developed mostly over the past 15 years, discussing the theoretical links between Lean principles and BIM, as well as requirements for production system controls that integrate both process improvement and the use of BIM tools. The chapter also describes Kanbim and VisiLean as existing systems addressing Lean.

Chapter 5, by John Rooke, discusses the importance of people, knowledge and organisation in the context of Lean in general, and production specifically. Rooke contends that understanding the knowledge that people possess is a key to understanding how they organise production and thus how the process of production can be improved. The management of meaning and value chains linking the concepts of flow and commitment are discussed. The chapter concludes by discussing the importance of learning Lean.

Chapter 6, by Patricia A. Tillmann and Luciana Miron, centres around value generation for the customer. The chapter discusses difficulties in defining value, and points to theories adopted by Lean Construction researchers in an attempt to define value, and develop approaches to support value generation in design. Different efforts towards managing value in construction are discussed, and recommendations for future research in the area presented.

Chapter 7, by Carlos Torres Formoso, Trond Bølviken and Daniela Dietz Viana, discusses waste as one of the key concepts in Lean Construction. The chapter presents an overview on existing research on the understanding of waste in the construction industry, including concepts, cause–effect relationships, limitations of existing waste metrics and a discussion on the main types or categories of waste. The chapter also proposes ideas for further research.

Part 2: Lean Construction approaches

In Part 2, insights around innovative Lean approaches including Integrated Project Delivery (IPD), Target Value Delivery (TVD), Choosing by Advantages, among others, are presented. It also includes discussions focusing on managing design and collaboration, and on the management of specific types of projects, e.g. refurbishments, as well as chapters discussing planning, and how the issue of complexity has been addressed across different Lean Construction areas.

Chapter 8, by Glenn Ballard, presents the concept of Target Value Delivery (TVD), which focuses on providing acceptable net benefits throughout the life of an asset to be built. The chapter presents the background and origins of TVD, covering the practical application of the concept, as well as its relationship with other approaches, e.g. Integrated Project Delivery.

Chapter 9, by Thais da C. L. Alves and Will Lichtig, presents Integrated Project Delivery (IPD) as a delivery method embracing collaborative governance and supporting multidisciplinary teamwork, whilst sharing risks and rewards in projects. The chapter discusses the concept, principles and tools applied to IPD, highlighting how it helps achieve benefits such as reduction of cycle times and reduction of the share of non-value adding activities. It also discusses how IPD differs from other delivery methods commonly used in the industry, presenting guidelines for its implementation.

Chapter 10, by Paz Arroyo, focuses around Choosing by Advantages (CBA) as a collaborative and transparent decision-making method to guide the design decision-making process. The method seeks to support collaboration and transparency in decision making. The chapter describes the method, discusses its advantages and presents a case study and recommendations for CBA implementation.

Chapter 11, by Bo Terje Kalsaas, proposes a model to manage interdependencies in detailed design. The chapter presents considerations on design management in general, discussing some of the complexities related to it. A case study is shown to exemplify the need for the model in a design and built context.

Chapter 12, by Sergio Kemmer and Lauri Koskela, centres around refurbishment projects, and discusses how Lean approaches are applicable to this specific type of setting. The chapter presents issues with current approaches used in refurbishment projects, and introduces the theoretical background for production management in refurbishment projects.

Chapter 13, by Ruben Vrijhoef on construction supply chain management, focuses on advancing construction logistics. It highlights the origins and issues around supply chain management, emphasising logistics performance. A case study is included, and the chapter concludes by discussing four roles of construction supply chain management, emphasising improved logistics and environmental performance.

Chapter 14, by Olli Seppänen, discusses location-based management in the context of Lean production. It presents a short history of location-based planning, discussing the system and how to optimise schedules. The chapter also shows a proposed process to combine location-based planning with the Last Planner System, pointing to areas for further research.

Chapter 15, by Clarissa Biotto and Mike Kagioglou, further addresses the issue of location-based planning from a perspective of the integration with production system design. Three tools are described, namely Line of Balance, Flowline and Takt-Time Planning. The chapter concludes by addressing the application of each into different contexts.

Chapter 16, by Tarcisio Saurin and John Rooke, presents construction projects as complex socio-technical systems addressing the implications of complexity theory to Lean Construction, which are still largely unexplored. The chapter discusses how the guidelines for coping with complexity can be used to understand the Last Planner System, and presents recommendations to better handle project complexity through the Last Planner.

Chapter 17, by Audrey Bascoul, Iris D. Tommelein and Stanislaus Tuholski, continues the idea that Lean thinking offers a comprehensive approach to manage project complexity, presenting a framework for that. The framework integrates the Lean Project Delivery System™ with a Multi-Domain-Matrix (MDM) as a means to visualise and model structural complexity on projects. The chapter presents an application of the framework and areas for further research.

Chapter 18, by Ole Jonny Klakegg, Olav Torp, Bo Terje Kalsaas, Trond Bølviken and Gøril Hannås, discusses uncertainty management as a development area for Lean Construction. The chapter proposes the integration of uncertainty management with the Last Planner System (LPS), considering its application to overall project performance. The chapter discusses two fundamental challenges in developing and executing a construction project: making decisions based on scant information and controlling processes characterised by variability.

Chapter 19, by Zofia K. Rybkowski, Lincoln H. Forbes and Cynthia C. Y. Tsao, addresses the important issue of effectively transferring Lean knowledge and skills to owners, architects, engineers and constructors. The chapter summarises the content of Lean Construction courses from 12 US-based universities, including how the content of Lean curricula is evolving. The chapter documents existing developments and raises a question about the potential impact of curriculum standardisation on future continuous improvement initiatives with respect to Lean Construction philosophy, methods, and tools.

Finally, Chapter 20, by Danilo Gomes and Patricia Tzortzopoulos on the concept of collaboration at early project stages provides an analysis of how collaboration has been addressed in construction research. The chapter includes issues around shared understanding, and discusses methods and tools for collaboration, including some of the limitations of existing approaches to collaboration, and questions for further research.

New frontiers

Although most chapters in this book have addressed the question of frontiers of knowledge, it is useful to briefly discuss the new (and partially already existing) frontiers for Lean Construction here. Lean Construction emerged as an innovation based on new theoretical ideas on production. Soon it turned out that for fully implementing and facilitating these innovations in production, changes in the surrounding systems and functions were needed. Inevitably, tensions and contradictions emerged when extending the reach of Lean from production outwards. Especially, frontiers regarding information technology, behaviour, design and operation have arisen. On the other hand, accumulated understanding on the starting point of Lean, namely waste, has revealed that the ultimate frontier is provided by the grand challenges facing humanity.

The Lean frontier on *information technology* (IT) is characterised by the longstanding situation where IT solutions are developed and forwarded as solutions on their own. To some extent this view can be seen to be valid, but mostly digital solutions are there to provide tools for management. For example, some years ago practitioners realised that Building Information Modelling (BIM) is an excellent tool for Lean Construction, and vice versa, as the case can be made that the structured conditions provided by Lean facilitate the implementation of BIM. This can be generalised: How can we frame digitalisation as an enabler of Lean? And how can we make digital process Leaner?

Regarding the frontier on *human behaviour*, the defining contradiction is caused by the refusal of the mainstream academic disciplines for management to theoretically embrace production. In the practical implementation of Lean, there is much emphasis on leadership and behavioural issues; unfortunately, this is hardly supported by the considerable amount of social science oriented research into management. Among the many questions deserving to be studied, the following accentuate: Which kind of behaviours are compatible with Lean systems? How can the quality of working life be promoted through Lean?

The frontier on *design* is subtle. The approaches of Lean product development and Lean design have emerged and are effective. The contradiction here is that they predominantly base on the flow theory (of production), which is powerful and very actionable, whereas the theory of design in itself is underdeveloped and less actionable. Of course, any theory of design should be generic, and thus it is not the responsibility of Lean Construction scholars only to develop it. However, the bottleneck caused by the lack of a powerful theory of design is so significant that this can be defined as a frontier, characterised by the following questions: How is design creativity enabled? How best to arrange collaboration in design? How can the technical and social understandings of design be integrated?

The *operation* phase of a built facility provides for another frontier. Up to now, Lean Construction has been addressing projects and their outputs, buildings or other facilities. The outcomes of projects (after their completion), in the sense of benefits to users, environment and society have not been systematically addressed, either in theory or in practice. A related shortcoming is that continuous improvement of facilities management of existing buildings and structures has hardly been dealt with.

As is well known, humanity faces a number of *grand challenges*, mostly arising from prior human activity. These include climate change, food and water shortage, plastic contamination of oceans, and many others. Many, if not most grand challenges are caused by wasteful production/consumption: it is well-known that a third of all food produced is wasted. Although we lack more precise figures, it is safe to assume the energy waste is of the same magnitude. Obviously, technological solutions are urgently needed in view of grand challenges, and the principles of Lean product development can advantageously be used to accelerate these developments. However, much could – and should – also be done by applying Lean principles for waste reduction. The question arises how Lean can be better positioned to address grand challenges.

Lean gives the impression of being a faddish, somehow sectarian and theoretically unfounded approach to management; this is a wrong impression caused by Lean starting from production, a black hole in management thinking. Rather, Lean should be seen and used as a catalyst for change in the management of human affairs, both at the practical and the theoretical level. In such an endeavour, the new frontiers discussed invite attention and work.

Part 1

Lean Construction themes

1 Theory of Lean Construction

Lauri Koskela

1.1 Introduction

Lean Construction is the counterpart to Lean production (or just Lean as a noun), as it has evolved in the context of construction. Lean is mentioned as one of the world's most influential management ideas (Buchanan and Huczynski, 2016), and as one of the most influential manufacturing paradigms of recent times (Holweg, 2007). What then explains the superiority of Lean? Strangely, the literature has little to say on this question, and the scattered remarks that can be found are often contradictory.

Against this backcloth, and as one of the major functions of a theory is to explain, this chapter endeavours to present a consolidated discussion on the theoretical (and philosophical) foundation of Lean and its instantiation in a particular context, such as construction. The treatment is predominantly based on the author's research into this topic since 1991. Due to space limitations, only the most important, high level concepts are treated here.

For several reasons, the topic to be discussed is complex and multifaceted, and thus a fair number of questions have to be discussed: Is there a theory of Lean, in the first place? Is a theory of Lean needed? If we are looking for the theory of Lean, what are we after? What is the mainstream theory of production management? What then is the theory of Lean? How is Lean Construction different from Lean production?

1.2 Is there a theory of Lean, in the first place?

According to the Japanese scholar Fujimoto (2007), the Toyota Production System (TPS) 'emerged as the unplanned and unexpected result of ... seemingly unrelated innovations, improvements, and initiatives'. This emergence is compatible with the popular view of Lean as a set of practices, which can be bundled into four areas (Shah and Ward, 2003): just-in-time, total quality management, total productive maintenance and human resources management. In a similar vein, Fujimoto (2012) states that the TPS consists of 400 organisational routines. This would indicate that there is no or little theory underlying Lean, which would seem to be just a collection of unrelated practices or routines. However, as the author of this chapter has repeatedly argued (Koskela, 2000; Koskela et al., 2018; Koskela, Tezel, and Patel, 2019), the theoretical and philosophical foundations of Lean can be pinpointed and also the origins of the different theoretical ideas can mostly be determined.

Why, then, has the theory of Lean failed to surface? There are two main reasons; one is related to the Japanese origin of Lean, and the other to the dominant understanding of management research. The Japanese culture tends to emphasise direct personal experience

(Nonaka and Takeuchi, 1995) rather than abstract theories. In alignment with this orientation, the Toyota Production System has not been based on explicit theoretical knowledge, and it is taught to new employees through socialisation (Yadav et al., 2017), rather than through theory-based education. Thus, no or little theory has emerged from Toyota.

One of the central tasks of management research is to decode the theoretical foundations of a management invention created in companies (David and Hatchuel, 2007). It is indeed surprising that the community of management scholars has not succeeded in creating a settled view on the theory of Lean. Here, the explanation is historical. Management research has been decisively influenced by two books published in 1959 (Gordon and Howell, 1959; Pierson, 1959), outlining the contents and objectives of such research. In this conception, production was – just – an application context for managerial methods rather than an object of research (Koskela, 2017). This meant that production, as a phenomenon to theoretically embrace, was thrown out of the domain of management. As Lean arguably is a theoretical innovation in the field of production, the discipline of management has not had access to concepts and principles which are at the heart of this innovation.

1.3 Is a theory of Lean needed?

But even if the theory of Lean could be determined, the question can be raised whether we need such a theory. Namely, Lean production is a generalisation of the Toyota Production System (TPS), which was developed and is still successfully applied without any underlying theory (as discussed above). In turn, Lean production has been applied in countless new contexts, again without the support of a theory.

What indeed would be the role of theory if Lean has been successfully developed and applied without it? Here, general arguments about the many useful functions of a theory, both in research and practice, could be forwarded: explanation, prediction, giving direction, possibility of testing, etc. (Koskela, 2000). However, perhaps a more persuasive argument can be found just by looking at how Lean production is understood by neighbouring disciplines to management.

Let's first consider recent textbooks on organisational behaviour (Buchanan and Huczynski, 2016; Knights and Willmott, 2016; Martin and Fellenz, 2017). As Lean extends to organisational behaviour (for example Liker, 2004), it is of interest to find what they say about Lean. The two first books discuss Lean, even with some enthusiasm, but the third only mentions it in a few sentences without any substantial discussion. As mentioned above, Buchanan and Huczynski (2016, p. 467) state that Lean has become one of the world's most influential management ideas, and characterise it as follows:

> TPS's aim is constant improvement, and reduction in costs through the systematic elimination of waste using the assembly line system.

In turn, Knights and Willmott (2016, p. 540) define Lean production as follows:

> It is associated with just-in-time services and stock inventories where companies do not retain excess labour or stocks of goods but use information technology to ensure recruitment or reordering of stocks when actually needed.

Of course, both the assembly line system and information technology are incidental to Lean. It turns out that the two textbooks discussing Lean cannot explain it.

Let's then turn to current textbooks on economics. According to Samuelson and Nordhaus (2005), economics is the study of how people and society choose to employ scarce resources that could have alternative uses. Thus, decisions under scarcity is the focus of economics. Lean would seem to offer a radically new idea in this context: scarcity can be alleviated through reduction of waste. It is well-known that in all economic activities there is waste. For example, regarding the global food provision, it is estimated that a third of all food produced is lost or wasted (Gustavson et al., 2011). Thus, arguably, economics should either theoretically embrace waste reduction, when economising is discussed, or robustly reject this idea. However, the popular textbooks by McEachern (2017), Parkin (2015) as well as Mankiw and Taylor (2017) do not discuss or mention Lean or waste reduction at all.

Why this failure to explain or silence regarding Lean? There is no common conceptual ground between Lean and established managerial disciplines, such as organisational behaviour and economics, as a theoretical explanation of Lean is missing. A theory of Lean is badly needed for other related disciplines to understand Lean, to ensure a complete coverage of the phenomena they address.

1.4 If we are looking for the theory of Lean, what are we after?

The concept of theory implies as such a certain structure to a theory. According to the well-known account by Whetten (1989), a theory consists of four parts, responding to questions on (1) What, (2) How, (3) Why and (4) When, Where, Who. Regarding the What question, a theory will be based on concepts, factors or variables that refer to its subject. For responding to the How question, the theory will include relationships or principles presented through the selected concepts, factors or variables. The Why question is about the conceptual assumptions underpinning the theory. Finally, the questions When, Where, Who are responded by specifying the contextual assumptions of the theory: in which conditions is the theory valid.

Now, it is possible to outline what we are after when looking for a theory of Lean. Regarding the What question: Theory of Lean is a theory of production (or operations) management. Thus, a conceptualisation of production is needed, along with associated concepts. This focus also determines the major ends to be covered: to get something produced, for minimal costs and for maximal value to the customer (Koskela, 2000).

Regarding the How question: Theory of Lean is a managerial theory. According to Argyris (1996), the task of a theory of management is to produce generalisations that are actionable by managers in everyday life. Further, according to Argyris, managerial theories purport to define activities through which intended consequences can be achieved; thus, the possible means to the ends mentioned above.

Regarding the Why question: There are at least two consequential questions needing to be responded. In producing, we have to interact with the world, and thus we need to have a view on what is there in the world. And we need information and knowledge for producing: where does it come from? These are philosophical questions, the former question falling into ontology (or metaphysics), and the latter into epistemology. Note that the assumptions subscribed regarding these questions are actionable, in the sense of influencing actions. However, customarily, philosophical questions are not discussed in production management, and thus ontological and epistemological assumptions are extremely rarely discussed explicitly.

Lastly, regarding the questions When, Where, Who: The relevant responses concern the empirical conditions where the discussed theory is valid. Although this sounds straightforward, these conditions are often given incompletely or vaguely, if at all.

In practice, it is usually seen that responses to the What and How questions are enough to characterise a theory; this convention is followed here. In turn, the mentioned responses to the Why question occur as general orientations in thinking, and therefore they are covered separately in the following. The When, Where and Who questions, although important, will not be covered.

1.5 What is the mainstream theory of production management?

To fully understand and appreciate the theory and philosophy of Lean, it is necessary to compare it to the old counterpart.

1.5.1 Theory of production

The concept of transformation (Starr, 1989; Holweg et al., 2018) has provided the mainstream theory of production management. It refers to the transformation of inputs into outputs (Starr, 1989). From a practical viewpoint, the related procedure called decomposition provides the power of this theory, although it is rarely explicitly presented. In decomposition, the total productive task is successively broken down into smaller tasks, until they can be assigned to operatives or companies in the supply chain. Decomposition is based on two related assumptions. First, that the tasks emerging from decomposition are by their nature similar input–output transformation as the original total task. Second, that the decomposed tasks are mutually independent. These assumptions make managing easy: all the attention can be turned into optimally executing the decomposed tasks, as in this way the total optimum is (assumed to be) reached.

However, these assumptions are an idealisation: of course, tasks are usually dependent on other tasks. For solving this discrepancy between the idealisation and the reality, a smart move offers itself: the world is changed to correspond to the idealisation. Especially, this happens through buffering, providing a stock of input materials so that different tasks or work stations would seemingly be independent from each other.

The ideas discussed can be presented as a particular theory of production:

- Conceptualisation of production: transformation.
- Central principles: decomposition; optimisation of decomposed sub-transformations with the intention of achieving the global optimum.
- Further principles: buffering for ensuring the (relative) independence of the tasks.

1.5.2 Ontology

Regarding ontology, the mainstream view in the Western world, since Antiquity, has been to see the world as consisting of things and substances (Koskela and Kagioglou, 2005). There are two basic ideas associated with this ontological view. First, that it is possible to separate these things from other things. Second, that the things and substances are basically more or less stable. Based on these tacit ideas, the traditional approach in natural science has been to divide a thing into its constituent parts and to search explanation at the lowest possible level. Note that in production, this sequence is reversed: the product is created from its constituent parts. However, in the preparation of production, the product is divided into its parts in Product Breakdown Structure. Also in the transformation theory of production itself, the description is through stable things, inputs and outputs – the transformation itself remains a black box.

1.5.3 Epistemology

Starting from Antiquity, an influential view on epistemology is that knowledge is stable (like an abstract thing), created and held in the mind, and pushed through deduction towards the world when using it. This is Platonic epistemology. In the mainstream view on production management, this epistemology is underpinning the command and control approach: knowledge is held by management, and pushed towards the world through plans.

1.5.4 Outcomes

Mainstream management in the West has thus been underpinned by internally coherent but implicitly theoretical and philosophical ideas. The problematic nature of these ideas began to surface in the 1970s (Schonberger, 1996), especially in the form continuously growing buffers needed in production as well as in quality problems. These issues became especially visible in comparisons with the performance of the Japanese industries.

1.6 What is the Lean theory?

1.6.1 Theory of production

Lean implies extending the theoretical understanding of production to two other theories, namely the flow theory and value generation theory, while also using the transformation theory when appropriate (Koskela, 2000). The current Japanese explanation of *monozukuri* (a term for a broad understanding of manufacturing) is in essence similar to this position (Fujimoto, 2019):

> We can therefore adopt a broad definition of manufacturing seen as creating and maintaining effective flows of value-carrying design information to the customers or, more simply, "creating good flows of good design information to customers." This is the basic concept of *monozukuri* as it is taught in Japan and the theoretical interpretation of the Toyota Production System.

In the quoted passage, 'good flows' refer to material and information flows, while 'good design information' refers to both design information and its embodiment in material products from the viewpoint of customer needs, thus representing value generation.

A key ingredient in the flow model theory of production is time: what is happening to objects of production in the time line and correspondingly, what do the subjects of production, men and machines, do. When monitoring this, it is realised that not all time is used for transformation. Objects of production are stored, or they are transferred, or inspected. Correspondingly, subjects of production may be waiting, moving or redoing. Such non-transformation stages are called waste, as such stages do not add value for the customer.

It is noteworthy that this theory of production powerfully extends the prescription for action in comparison to the transformation theory. Namely, now it is obvious that we can improve production through waste reduction.[1] Good flows, referred to by Fujimoto, are those with few interruptions and stops. Looking at a typical lead-time, the biggest waste is made up by waiting, in the form of inventory. This waste can thus be reduced by compressing the lead time. These principles were foundational for just-in-time production and further for the Toyota Production System. Subsequently, Hopp and Spearman (1996) showed, using queueing theory, that one common cause for waste is temporal variability, in terms of unpredictability regarding when a job will arrive at a workstation, and how long it's completion will take.

The third theory of production, related to value generation, was seminally presented by Shewhart (1931). Production is seen as an interaction between the customer and the supplier. The customer has requirements and wishes, and the supplier converts these into products that fulfil those. The production process comprises design, which converts requirements and wishes into a product specification, and production, which creates the product, as close to the specification as possible. This theory of production again extends the prescription for action, focusing especially on quality management (Koskela, Tezel, and Patel, 2019) but also on design. Regarding the latter, while theoretical progress has been slow (Pikas, 2019), there have been definite methodical advancements springing from practice.

1.6.2 Ontology

A counterpart to thing ontology existed already in Antiquity, namely process ontology (or metaphysics) (Rescher, 1996). Here, the phenomena in the world are viewed to be temporal processes. The suitable scientific approach is then to identify and address the dependences that this phenomenon has on other processes. It is also natural to pay attention to incremental change produced through those dependences.

Lean subscribes to process ontology in two ways. The approach of continuous improvement can be interpreted as a systematic managerial action focusing on incremental change. The considerable focus on collaboration in Lean is underpinned by process ontology: the many dependences, often poorly visible in advance, between tasks in design and production, are tackled through collaboration.

1.6.3 Epistemology

Aristotle developed an epistemology that is a counterpart to that of Plato. According to Aristotle, theoretical knowledge can be pulled out from the world (Losee, 2001). That knowledge, gained through induction, may then be put into practical use through deduction.

The Aristotelian epistemology is a hallmark of Lean, in several ways (Koskela et al., 2018). The central issue of Lean, waste, typically exists in the material world, and the analysis of its root causes represents induction; knowledge about the root causes can then be used through deduction for the elimination of the waste in question. The Plan-Do-Check-Act cycle, seminally proposed by Shewhart (1931), renders this epistemology operational.

Note that epistemological action occurs at the most down-to-earth level in production. Pushing and pulling are well-known practical concepts in the management of production. Pushing can be characterised as Platonic: deduction from ideas through a plan to action. Pulling, in turn, is Aristotelian: basing action on the state of the world (that is, the production system). In Lean, both pulling and pushing are used.

1.6.4 Outcomes

The significance of the discussed theoretical and philosophical starting points is that they widen the action space in production in critical ways, enabling effective action towards waste reduction, and value increase, through collaboration, continuous improvement and learning from doing. Arguably, these factors have ensured the superior results of advanced Lean production systems.

1.7 How is Lean Construction different from Lean production?

An often-heard question is: How is Lean Construction different from Lean production? Actually, the question is somewhat vague as it is not sure whether Lean production is meant to refer (1) to the embodiment of Lean especially in car manufacturing and perhaps in other broadly similar instances or (2) to generic principles, methods and knowledge about Lean.

At the outset, it is worthwhile to remember the differences between industrial contexts. Construction has its peculiarities: one-of-a-kind/craft production, site production, temporary organisation (Koskela, 2000), which contrast to mass production in a factory by a permanent organisation, and of course there are many other varieties.

Let us then tackle the first interpretation of the title question. All embodiments[2] of Lean, in car manufacturing, healthcare, construction and other fields are arguably contextual, in the sense that the characteristics and specific problems of each industry are mirrored in the production model, methods and tools. Thus, in practice, Lean can be expected to be different, more or less, in different industries.[3] Also, the Toyota Production System must be seen as a contextual application of general principles, rather than a generic model to emulate when addressing other contexts (see Lillrank, 1995). Thus, Lean Construction and Lean car manufacturing are different but comparable: outcomes of application of the (broadly) same principles and methods in different situations.

The answer to the second interpretation of the question could be simple: Lean Construction could be defined as the contextual application of what is seen as generic principles, methods and tools of Lean (see Gao and Low, 2014). However, this is a Platonic understanding (Koskela et al., 2018) of the situation: there is a static body of knowledge that can be applied in practical situations through design. An alternative understanding would be Aristotelian: the body of knowledge about Lean is dynamic, it is accumulating through inventions and discoveries done in practical application. Indeed, this is the way the Toyota Production System has developed, according to Fujimoto (2007). Thus, given that construction has rather different characteristics in comparison with car manufacturing, another way of creating Lean Construction would be to start from the specific problems and situation as perceived in construction.

Actually, this is largely the way Lean Construction emerged. Construction had encountered chronic problems over many decades, and against that backcloth, Ballard (2000) developed the Last Planner System, that actually for many years was more or less equated with Lean Construction (see Chapter 3 by Ballard). Other original inventions made in the framework of Lean Construction are Integrated Project Delivery (Matthews and Howell, 2005 – see Chapter 9 by Alves and Lichtig) and its underlying contractual forms (Lichtig, 2006), as well as Target Value Delivery (Ballard, 2012 – see Chapter 8 by Ballard). Indeed, similarly to the emergence of the Toyota Production System, in construction one started from local problem solving, and over time, a new production model emerged.[4]

However, over time it has become evident that many individual methods and tools in the Toyota Production System are generic and can very well be applied in construction. On the other hand, there are popular methods used only in Lean Construction (and thus in the wider arena of Lean production), exemplified by the Last Planner System and Choosing by Advantages (Arroyo, Tommelein, and Ballard, 2015 – see Chapter 10 by Arroyo), that are generic and can be applied across industries in situations for which they have been developed.

It is now possible to summarise the response to the question examined. Lean Construction is an amalgamation of a contextual production model emerging from attempts to solve industry-specific problems in construction, and the application of generic Lean production principles, methods and tools, originated in car manufacturing.

1.8 Concluding discussion

In this chapter, it has been proposed that at its essence, Lean is a theoretical innovation: the theories and philosophical commitments underpinning Lean allow wider action options in comparison to the mainstream operations management (an overview is given in Table 1.1). The innovative features include the flow and value generation theories of production, process ontology and Aristotelian epistemology. In contrast, the traditional approach to production has been based on the transformation theory of production, thing ontology and Platonic epistemology. All these foundations (both for Lean and the traditional approach) have largely been invisible due to varied reasons, mainly the denial in management research of the need to conceptualise production, and the mistaken view that philosophical commitments are outside practical affairs. It is this invisibility that has rendered Lean so challenging to explain and understand.

Table 1.1 Summary of the theoretical and philosophical underpinnings of mainstream production management and Lean production management (based on Koskela, Vrijhoef and Broft, 2020)

Element of theoretical and philosophical foundation	*Mainstream management of production*	*Lean management of production*
Production theory	Transformation: Production is conceptualised as transformation, and associated principles are used for managing it.	Transformation, flow and value generation: Production is conceptualised from all these three angles, and associated principles are used in a balanced manner for managing it.
Epistemology	Platonic: Knowledge is held by managers and engineers, and it is pushed towards the world deductively through plans and designs and their implementation.	Aristotelian: The deductive push of knowledge is acknowledged; however, knowledge is also inductively created, in terms of using production as a scientific laboratory.
Ontology	Thing: Management addresses clear-cut, stable things. Productive processes are decomposed into independent tasks for division of work. Changes are abrupt, realized by introducing new things into the setting.	Process: Management concerns interrelated, fluid processes (embracing also things). Productive processes are decomposed into individual tasks, but their dependencies are tackled through collaboration. Changes may be abrupt or continuous.

Lean production, emerging from car manufacturing, is biased towards mass manufacturing in stable factory conditions by a permanent organisation. Construction shows very different characteristics: one-of-a-kind products, site production and temporary organisation. This difference in context has caused Lean Construction to evolve in two ways. Originally, Lean Construction started as local problem solving in production management, and gradually evolved into a more comprehensive template, covering also contractual and organisational areas. Somewhat later, principles, methods and tools originating in the Toyota Production System have also increasingly been applied in or adapted to the context of construction. One consequence of this difference of context is that the Lean Construction community is largely distinct from the wider Lean production community.

Clarification of theoretical and philosophical starting points of Lean in general and specifically Lean Construction is necessary both for diffusing these effective approaches and advancing them further. It can only be hoped that the ideas proposed here trigger wider research and discussion on these matters.

Notes

1 More on the concept of waste is presented in Chapter 7.
2 In the edited book *The Routledge Companion to Lean Management* (Netland and Powell, 2016), there are 17 chapters on the application of Lean in industries or contexts different from the original focus on car manufacturing.
3 There has been research trying to validate the hypothesis that there exists a 'best practice Leanness' transcending national borders and industries. Fukuzawa (2019) provides a critical review on such research.
4 Another example of local problem solving extending towards a new production model is provided by agile software engineering (Sutherland and Sutherland, 2014).

References

Argyris, C. (1996) 'Actionable knowledge: Design causality in the service of consequential theory', *The Journal of Applied Behavioral Science*, 32(4), pp. 390–406.

Arroyo, P., Tommelein, I. D. and Ballard, G. (2015) 'Selecting globally sustainable materials: A case study using choosing by advantages', *Journal of Construction Engineering and Management*, 142(2).

Ballard, G. (2000) *The Last Planner System of Production Control.* PhD Thesis, The University of Birmingham, Birmingham, UK.. Available at http://leanconstruction.org.uk/media/docs/ballard2000-dissertation.pdf.

Ballard, G. (2012) 'Target value design', *Proc. DESIGN 2012, the 12th International Design Conference*, Dubrovnik, Croatia.

Buchanan, D. A. and Huczynski, A. A. (2016) *Organizational Behaviour.* 9th edition, London, UK: Pearson.

David, A. and Hatchuel, A. (2007) 'From actionable knowledge to universal theory in management research', in A. B. Shani, S. A. Mohrman, W. A. Pasmore, B. Stymne & N. Adler (eds). *Handbook of Collaborative Management Research.* Thousand Oaks, CA: Sage Publications, pp. 33–47.

Fujimoto, T. (2007) *Competing to Be Really, Really Good: The Behind-the-Scenes Drama of Capability-Building Competition in the Automobile Industry.* Vol. 22, Tokyo: International House of Japan.

Fujimoto, T. (2012) 'The evolution of production systems', *Annals of Business Administrative Science*, 11, pp. 25–44.

Fujimoto, T. (2019) 'Key concepts for improving supply chain robustness', *Discussion Paper Series*, Number 511, Manufacturing Management Research Center (MMRC). University of Tokyo.

Fukuzawa, M. (2019) 'Critique on the lean production system research', *Annals of Business Administrative Science*, 18(3), pp. 85–101.

Gao, S. and Low, S.P. (2014) 'The Toyota way model: An alternative framework for lean construction', *Total Quality Management & Business Excellence*, 25(5–6), pp. 664–682.

Gordon, R.A. and Howell, J.E. (1959) *Higher Education for Business*. New York, NY, USA: Columbia University Press.

Gustavson, J., Cederberg, C., Sonesson, U., van Otterdijk, R. and Meybeck, A. (2011) *Global Food Losses and Food Waste*. Gothenburg, Sweden: Swedish Institute for Food and Biotechnology (SIK), pp. 1–38.

Holweg, M. (2007) 'The genealogy of lean production', *Journal of Operations Management*, 25(2), pp. 420–437.

Hopp, W. J. and Spearman, M. L. (1996) *Factory Physics Foundations of Manufacturing Systems*. Boston, MA: Irwin, Times Mirror.

Knights, D. and Willmott, H. (2016) *Introducing Organizational Behaviour and Management*. 3rd edition, Andover, UK: Cengage Learning EMEA.

Koskela, L. (2000) *An Exploration towards a Production Theory and its Application to Construction*. PhD thesis, VTT: Technical Research Centre of Finland.

Koskela, L. (2017) 'Why is management research irrelevant?', *Construction Management and Economics*, 35(1–2), pp. 4–23.

Koskela, L., Ferrantelli, A., Niiranen, J., Pikas, E. and Dave, B. (2018) 'Epistemological explanation of lean construction', *Journal of Construction Engineering and Management*, 145, p. 2.

Koskela, L., Tezel, A. and Patel, V. (2019) 'Theory of quality management: Its origins and history', *Proc. 27th Annual Conference of the International Group for Lean Construction*, Dublin, Ireland. Available at iglc.net.

Koskela, L., Vrijhoef, R., Broft, R. (2020) 'Construction supply chain management through a lean lens', in S. Pryke (ed.), *Construction Supply Chain Management*, 2nd edition, Abingdon: Routledge.

Koskela, L. J. and Kagioglou, M. (2005) 'On the metaphysics of production', *Proc. 13th Annual Conference of the International Group for Lean Construction*, Sydney, Australia. Available at iglc.net.

Lichtig, W. A. (2006) 'The integrated agreement for lean project delivery', *The Construction Lawyer*, 26(3), pp. 25–32.

Liker, J. K. (2004) *The Toyota Way*. New York: McGraw-Hill.

Lillrank, P. (1995) 'The transfer of management innovations from Japan', *Organization Studies*, 16(6), pp. 971–989.

Losee, J. (2001) *A historical Introduction to the Philosophy of Science*. Oxford, UK: Oxford University Press.

Mankiw, N.G. and Taylor, M.P. (2017) *Economics*. 4th edition, Andover, UK: Cengage Learning EMEA.

Martin, J. and Fellenz, M. (2017) *Organizational Behaviour and Management*. 5th edition. Andover, UK: Cengage Learning EMEA.

Matthews, O., & Howell, G. A. (2005) 'Integrated project delivery an example of relational contracting', *Lean Construction Journal*, 2(1), pp. 46–61.

McEachern, W. A. (2017) *Economics: A Contemporary Introduction*. 11th edition. Andover, UK: Cengage Learning EMEA.

Netland, T. H. and Powell, D. J. (2016) *The Routledge Companion to Lean Management*. Abingdon, UK: Routledge.

Nonaka, I. and Takeuchi, H. (1995) *The knowledge-Creating Company: How Japanese Companies Create the Dynamics of Innovation*. Oxford, UK: Oxford University Press.

Parkin, M. (2015) *Economics*. 12th edition. Harlow, UK: Pearson Education.

Pierson, F.C. (1959) *The Education of American Businessmen*. New York, NY, USA: John Wiley.

Pikas, E. (2019) *Causality and Interpretation: Integrating the Technical and Social Aspects of Design*. PhD Thesis, Aalto University, Espoo, Finland.

Rescher, N. (1996) *Process Metaphysics: An Introduction to Process Philosophy*. Albany, NY: Suny Press.

Samuelson, P. and Nordhaus, W. (2005) Economics. 18th edition. Boston, MA: McGraw-Hill.

Schonberger, Richard J. (1996) *World Class Manufacturing: The Next Decade*. New York, NY, USA: The Free Press, p. 275.

Shah, R. and Ward, P. T. (2003) 'Lean manufacturing: Context, practice bundles, and performance', *Journal of Operations Management*, 21(2), pp. 129–149.

Shewhart, W. A. (1931) *Economic Control of Quality of Manufactured Product*. New York, NY, USA: van Nostrand.

Starr, M. K. (1989) *Managing Production and Operations*. Englewood Cliffs, NJ: Prentice Hall.

Sutherland, J. and Sutherland, J. J. (2014) *Scrum: The Art of doing Twice the Work in Half the Time*. Currency. London: Random House Business.

Whetten, D. A. (1989) 'What constitutes a theoretical contribution?' *Academy of Management Review*, 14(4), pp. 490–495.

Yadav, O. P., Nepal, B. P., Rahaman, M. M. and Lal, V. (2017) 'Lean implementation and organizational transformation: A literature review', *Engineering Management Journal*, 29(1), pp. 2–16.

2 Lean product development and design management

Patricia Tzortzopoulos, Cynthia dos Santos Hentschke and Mike Kagioglou

2.1 Introduction

For over 30 years, researchers and practitioners from diverse fields of knowledge aimed to improve the efficiency and effectiveness of the New Product Development process (NPD), addressing complex design and production issues. At the end of the 1990s Kagioglou et al. (1998) stated that the manufacturing industry was under transformation due to increased globalised competition, which reflected a need for companies to search for fast responses to changing customers' requirements. The ability of companies to respond to those requirements relies on communication with customers and the capacity to translate their requirements into finished products and services, which is the main concern of NPD. This has also been the main focus of Lean design and construction initiatives and remains as current today as it was more than 25 years ago.

Lean Construction has focused on reducing waste in construction projects, and diverse initiatives have been successfully implemented in the industry to manage production, e.g. the Last Planner System of production planning and control (see Chapter 3 by Ballard). Lean design addresses design as a production process, in which information is transformed, value added to it continually informed by clients, as opposed to the transformation of physical materials. The importance of design has been highlighted in Lean as the main means to generate value to clients. However, the adoption of Lean thinking is still modest in design. Attempts to implement Lean design highlighted the need to integrate design and production, emphasising the adoption of Lean thinking from briefing and conceptual design throughout the whole development process (Emmitt, Sander, and Kirk-Christoffersen, 2004).

The complexity in identifying and managing requirements, the multiple and sometimes competing requirements for efficiency, sustainability and cost, handovers between professionals and organisations, the inherently temporal nature of construction teams and the continuously evolving technological agenda make the undertaking of projects increasingly challenging. As such, recent developments have promoted improvements in various aspects of the process, often bringing about significant savings and increases in quality. This book presents many of these within the context of Lean Construction.

This chapter starts by briefly discussing NPD in other sectors with a main focus on manufacturing. It then discusses how developments in manufacturing have influenced Lean design and construction. Following, the chapter presents issues around design management. An overview of Lean product development and design management research by the International Group for Lean Construction (IGLC) is presented to identify the progress made over time, as well as highlight the evolution of the subject area in that period outside IGLC. Within this overview, subsections present brief descriptions of the main topics related to Lean design, some of which are discussed in subsequent chapters in detail. Finally, areas for future research are identified.

2.2 New Product Development process

Cooper (1994, p. 3) defines the New Product Development process as: 'a formal blue print, road map, template or thought process for driving a new product project from the idea stage through to market launch and beyond'. According to Ulrich and Eppinger (2000), product development is the process through which a product is conceived, designed and launched in the market, and includes feedback from both production and product use. New Product Development begins with the perception of a market opportunity and typically involves the capture and management of customer requirements, concept development, product design, market launch and collection and dissemination of feedback data (Cooper et al., 1998; Yazdani and Holmes, 1999). Hence, it includes both product and production process design activities. This concept makes explicit the importance of design, but also the interfaces between design and production, through the links between information and physical production (Tzortzopoulos, 2004).

A number of process models describing stages, activities and stakeholders were proposed aiming to support the management of the NPD (Kagioglou et al., 2000; Wang et al., 2011). Examples of influential models include: Clark and Fujimoto (1991) focusing on execution, manufacturing and ramp-up; Cooper (2001), who proposed the stage-gate management approach, which has been largely influential and applied in industry (Kagioglou et al., 1998); and Rozenfeld et al. (2006) describing process, and including after-launch activities, follow-up and plans for product discontinuity. The purpose of those process models is to be a reference or a generic template of the process, providing a set of tools and techniques to support the development of different activities, including the use of Information Technology (Tzortzopoulos, 2004). These models also aimed to ensure that the definition of the NPD is customer and market oriented (Kagioglou et al., 1998; Wang et al., 2011).

Kagioglou et al. (1998) argues that there is no single solution to NPD; thus, companies should adopt different combinations of available tools and approaches according to their particular market context, product and process complexity, duration of projects and other factors. When considering the available approaches, companies should find a balance in the combination of the success drivers pursued, as each approach offers advantages and disadvantages that must be considered (Kagioglou et al., 1998). The same authors also state that the drivers for success relate to flexibility and speed, control and focus, which provide guidelines to the NPD.

In manufacturing, there has also been research looking at Lean product development specifically, in an attempt to both identify Toyota's NPD strategies (as the leading NPD company at the time), and define strategies to improve NPD through Lean approaches. As discussed in Allen Ward's study of the Toyota Production System, Lean product development addresses the following issues (Ward and Sobek, 2014):

- Reduction of long development cycle times;
- Reducing the high costs of development;
- The need for more innovative solutions;
- Reduction of production costs;
- Redevelopment cycles.

Lean NPD was defined by Wang et al. (2011, emphasis added) as '*the application of Lean principles to the product development process to eliminate waste*', considering adaptations to materials and information flows, which supports the development of the company's value stream pulled by customers' orders. Morgan and Liker (2006) state that waste in product development can be regarded as the issues that negatively affect the overall performance of the process, affecting product cost, quality, development time and production capability.

Griffin, Langerak, and Eling (2019) further state that there is a need for the establishment of an overarching theory of NPD to fully understand its benefits. A similar lack of a holistic view of NPD occurs in construction. NPD process models and elements of the Lean NPD have influenced research in design and construction. Some of the main influences are described as follows.

2.2.1 Product development process models in design and construction

In the late 1990s, developments from manufacturing inspired construction researchers to develop process models adapted to the characteristics of the industry (Ballard and Koskela, 1998; Kagioglou et al., 1998, 2000; Tzortzopoulos, 2004). Ballard and Koskela (1998) stated that the main lesson to be learned concerns value generation, specially the identification and capture of customers' requirements and their translation into product specifications (Ballard and Koskela, 1998).

Tzortzopoulos (2004) defines the product development process in the built environment as 'the set of activities needed for the conception and design of a built product, from the identification of a market opportunity to its delivery to the client'. As such, it includes the design front-end, design development, design-related activities that occur during production (and the interfaces between design and production processes), as well as post occupancy evaluations.

Process models present phases, each of which provides deliverables, increasing progressively the level of design detail through a series of creative activities (Ballard and Koskela, 1998). Process models must portray relevant information e.g.: (i) the content of main activities; (ii) precedence relationship between activities; (iii) main inputs, outputs and how information flows along the process; (iv) the roles, interests and responsibilities of the stakeholders (Kagioglou et al., 2000; Tzortzopoulos and Formoso, 1999).

The process protocol, proposed by Kagioglou et al. (2000), is one example of such models. It defines project phases focusing particularly on the early stages, defined as the fuzzy front-end. Kagioglou et al. (2000) based the development of the process protocol around six key principles: (i) holistic project view, encompassing design, construction and demolition of projects and embracing customers, business and technical perspectives of construction; (ii) consistent process, and adoption of a standard approach for performance measurement and control, reducing the ambiguity and uncertainty in the design process, assisting continuous improvement; (iii) progressive design fixity, an approach based on hard and soft gates for design that enables the approval of evolving solutions throughout the process; (iv) co-ordination, managing activity zones; (v) stakeholder involvement and team work, especially at early stages of design, by promoting a collaborative environment, communication at the right time and supporting decision making; and (vi) feedback offering lessons learnt through a legacy archive.

The RIBA plan of works is the most widely used process model in the UK. Figure 2.1 illustrates a generic model for construction projects, adapted from the RIBA plan of works (2013), outlining the scope of design management within this process.

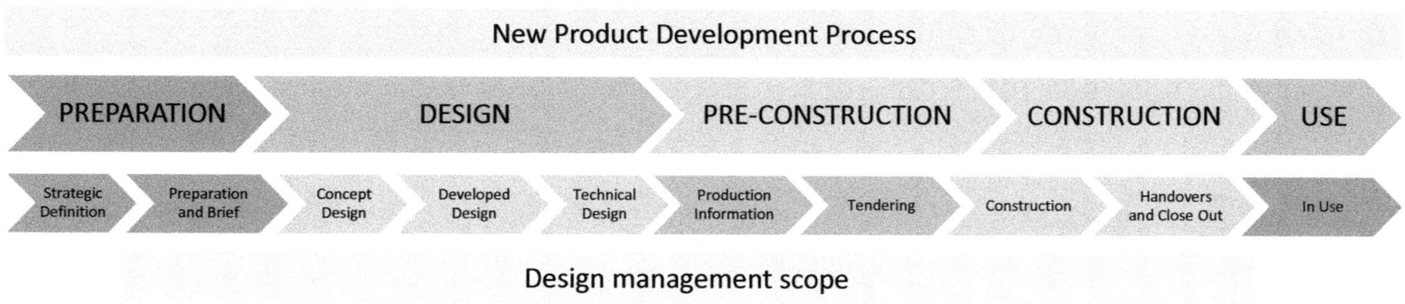

Figure 2.1 General New Product Development process in construction and design management scope

Although process models have been widely discussed, there are limitations in practice in the adoption of such models, specifically at more detailed activity levels, and hence limited empirical data about their success. Some of the issues include:

- Tailoring a generic process model to the specific situation – at the company level, but also at the project level, which involves diverse company's processes;
- Having enough detail for medium- and short-term management;
- Process models generally do not provide inputs to how communication and collaboration is organised and how decisions are made.

As the structure of the industry is fragmented, there is no ownership of such models or a collective agreement, and there is no knowledge maturity to adopt such models consistently across all parties in the sector. The RIBA plan of works comes closer to being more widely adopted – also making it harder to change and adapt in incorporating current thinking and embedding current innovation practices.

2.3 Design management in the context of Lean

Design management is an element of the product development process focused on organising the design team and understanding its nature, stages and activities, aiming to support communication, coordination and improve the integration of information flows. According to the Design Management Institute (DMI), design management includes the processes, decisions and strategies that enable the creation and innovation of products and services that provide organisational success and improve quality of life (DMI, 2019). Furthermore, it aims to provide practices and methods to design, including decision making, controlling processes, maintaining costs within the expected and ensuring profitability (Cooper, Junginger, and Lockwood, 2011). Accordingly, DMI (2019) states that design management fosters the collaboration and synergy between design and business perspectives, aiming to improve the effectiveness of the process.

As such, 'design management endeavours to establish managerial practices focused on improving the design process, thus creating opportunities for the development of high-quality innovative products through effective processes' (Tzortzopoulos and Cooper, 2007, p. 18). This definition highlights the emphasis on the process, linked to the basic ideas of Lean Construction.

In Lean, design has been examined as a production process (Ballard and Koskela, 1998). Hence, following Koskela's Transformation-Flow-Value (TFV) theory, three perspectives are considered: (i) transformation, i.e. the conversion of product requirements into the design of a product; (ii) flow, i.e. the flow of information across design activities and stakeholders; and (iii) value generation, i.e. design consists of the process of adding value through the fulfilment of customers' requirements. Some early developments include Huovila, Koskela, and Lautanala (1997), who developed a conceptual framework for design management considering TFV, and Tzortzopoulos and Formoso (1999), who discussed the application of the TFV principles through the development of a design process model.

Understanding design as a flow of information can lead to reducing waste by minimising non-value adding activities, and the potential to increase efficiency (Ballard and Koskela, 1998). This enables the reduction of time, for example: (i) waiting for available information; (ii) inspection of design solutions; (iii) rework, i.e. reviewing the design solutions according to the requirements (Ballard and Koskela, 1998). The same authors state that viewing design as a flow supports the identification and coordination of interdependencies and integration between design and construction.

The design process is also the means to fulfil costumer requirements, which involves capturing clients' requirements, translating them into design solutions, and producing goods as specified in the design, generating value (Tzortzopoulos, 2004). The value generation process of fulfilling customers' needs occurs in cycles of information capture and conversion to deliver a product or service (Koskela, 2000).

Lean Construction research has proposed the use of diverse approaches to support product development and design, in recognition of its essential role in providing value to customers. The following section highlights some of the main issues in design as discussed in the Lean literature. These have been the levers for a number of Lean design developments, which are briefly presented in the section after that.

2.3.1 Design management problems

Design problems have been described as ill-defined, or wicked problems. Wicked problems have particular characteristics, e.g. there is no definitive definition of a solution to a problem – understanding the problem and building its solution occurs in cycles of comprehension and formulation of the solution (Buchanan, 1992; Whelton and Ballard, 2002). Hence, the information captured to understand the problem is dependent on the possible solutions. As there is no definitive solution, problem solving is framed by time and resource limitations. As such, structured processes and group management skills are needed to support the problem identification and the development of the best solution possible to satisfy multiple perspectives, design criteria and project needs (Buchanan, 1992; Coyne, 2005). Furthermore, the definition of problems results from the combination of the complexity of the project context variables and stakeholders value sets.

The nature of design problems influences the management of the process. The Lean literature has highlighted design management issues, which can lead to poor design quality, and have a strong impact on the effectiveness of the production stage. These issues are summarised in Table 2.1, and briefly discussed as follows.

Design managers tend to focus on managing projects, tasks, resources and contracts (Howell, Windahl, and Seidel, 2010) and less on managing people, production processes, environment and technology (Koskela et al., 2002; Pikas et al., 2015). This has led to disjointed management and contradicting methods; management of deliverables focused on producing models and drawings, while needs, requirements and alternatives are poorly specified or studied; misalignment between design flows, between others (Pikas, Koskela, and Seppänen, 2017). These are issues which have proved difficult to tackle over time.

According to Kagioglou et al. (2000) the lack of coordination, high variability and poor communication in construction projects are partially caused by fragmentation. For instance, the briefing process is performed by a team which is different from the one involved in production, who should deliver a project according to predefined design criteria (Ballard and Koskela, 1998). Tauriainen et al. (2016) add that within the current unstructured design management practices, where different teams adopt different design practices and lack standards for collaboration, the chances for error and conflict increase.

Multiple issues have been discussed around managing requirements and generating value. During design, there are multiple trade-offs between conflicting requirements, and

Table 2.1 Summary of design management issues

	Issues	*Examples of references mentioning issues*
Collaboration and shared understanding	Poor communication and coordination between stakeholders	(Koskela, Ballard, and Tanhuanpää, 1997); (Tzortzopoulos and Formoso, 1999); (Whelton and Ballard, 2002).
	Poor collaboration and lack of integration between design teams	(Whelton and Ballard, 2002); (Arayici et al., 2011).
	Lack of shared understanding between teams	(Tillmann et al., 2013); (Gomes, Tzortzopoulos, and Kagioglou, 2016); (Schöttle and Tillmann, 2018); (Gomes and Tzortzopoulos, 2018).
Requirements and value	Poor requirements processing	(Whelton and Ballard, 2002); (Lima, Formoso, and Echeveste, 2008).
	Difficulties in the identification of customers' needs and its translation into requirements and product specifications	(Ballard and Koskela, 1998); (Whelton and Ballard, 2002); (Santos, Kistmann, and Fischer, 2004); (Pikas, Koskela, and Seppänen, 2017).
	Limited time and effort for identifying customers' needs	(Lima et al., 2009); (Hentschke et al., 2014).
	Difficulties in considering and balancing multiple stakeholders' needs and goals, which are often conflicting	(Ballard and Koskela, 1998); (Whelton and Ballard, 2002); (Tillmann et al., 2013).
Process	Design errors identified late in the process, and rework Lack of understanding of what constitutes waste in design	(Ballard and Koskela, 1998); (Tzortzopoulos and Formoso, 1999); (Mryyian˙ and Tzortzopoulos, 2013); (Svalestuen, Lædre, and Lohne, 2014).
	Large amount of non-value adding activities during the design process, causing delays and reducing the time to improve design	(Ballard and Koskela, 1998)
	Lack of adequate design documentation; deficient and missing information	(Koskela, Ballard, and Tanhuanpää, 1997); (Tzortzopoulos and Formoso, 1999); (Tauriainen et al., 2016).
	Lack of planning and control, especially at early design stages	(Ballard and Koskela, 1998); (Tzortzopoulos and Formoso, 1999); (Whelton and Ballard, 2002); (Khan, 2016).
	Poor consideration of interdependencies between design activities	(Svalestuen, Lædre, and Lohne, 2014); (Khan, 2016).
	Poor adaptation of design flows to cope with BIM tools	(Tauriainen et al., 2016); (Arayici et al., 2011).

divergent design criteria which need to be considered from different perspectives (Ballard and Koskela, 1998). However, such trade-offs commonly happen based on inadequate information within a context of elevated budget and schedule pressure.

The Lean literature also highlights a number of issues around the design process itself. One important issue is the relationship between design errors and waste. It is known that complexity causes design iteration which can be value-adding or wasteful. Waste as defined by Womack and Jones (1996) is 'any activity, which absorbs resources but creates no value'. It is also known that waste in design arises from delays, waiting, design errors, over processing and negative iteration (Ballard, 2000). Such waste can have major impacts by undermining efforts to complete construction on time. Therefore, elimination of waste in both design and construction requires an emphasis on controlling and eliminating errors in design, as errors are main contributors of waste and value loss.

An important design management issue is the lack of effective planning and control to reduce the influence of complexity and uncertainty, and to ensure that the information flow is adequate and consistent to support design decision making. According to Ballard and Koskela (1998), early design stages are especially hard to evaluate and control due to the lack of clear deliverables hindering the estimation of the amount of work done and remaining. Moreover, there should be formal ways to control information flows, and the use of soft and hard gates have been described to support this (Emmitt, Sander, and Kirk-Christoffersen, 2004).

However, many managers and designers believe that the design process is not suitable for planning due to its creative and iterative nature and that excessively detailed design plans would hinder creativity (Khan, 2016). According to the same author, design planning and control still is one neglected area in construction resulting in a chaotic environment and the large use of improvisation to solve design issues. Furthermore, many problems in design can be a consequence of the lack of consideration of interrelationship and precedence relations between design activities. Particularly, many activities are reciprocally interdependent, which adds complexity to design development and variability in the workflow.

Finally, the use of technology to support design has been at the forefront of many initiatives, and Building Information Modelling/Management (BIM) has been posed as a means through which information flows can be better managed, increasing productivity, efficiency and quality (Arayici et al., 2011; Dave et al., 2013; Eastman et al., 2011) – see Chapter 4 by Dave and Sacks for a detailed discussion around Lean and BIM.

BIM provides great potential for improvement, but at the same time creates challenges related to its implementation. According to Arayici et al. (2011), such challenges include: (i) overcoming resistance to change; (ii) adapting existing design workflows and training staff; (iii) collaboration and integration among different design teams and clear understanding among stakeholders; (iv) developing new managers' capabilities for BIM to improve the definition, schedule and evaluation of design activities; (v) improving the information flow to avoid disruption. Hence, one of the greater issues in BIM implementation relates to the need for changing the design process, working within predefined and agreed structures to produce and share information. Ballard and Koskela (1998) highlighted the need for a solid understanding of engineering and construction to be embedded in any specific applications to support design and production, which applies for BIM.

The following section presents some prominent Lean design approaches, which were developed aiming to resolve some of the issues described above. The section starts by presenting an analysis of research efforts reported throughout the past Lean Construction conferences, describing the main areas of work, and offering references for further details on each area.

2.4 Overview of product development and design management research at the International Group for Lean Construction (IGLC)

This section aims to provide a broad overview of Lean product development and design management developments over time, as discussed in papers published in the Annual Proceedings of the Conference International Group for Lean Construction (IGLC). An analysis of the 306 papers published from 1997 to 2018 was developed to provide a broad overview, as well as insights about future research needs in the area.

Firstly, the analysis aimed to portray how the topic evolved over time in a quantitative form. Figure 2.2 illustrates that the number of design management related papers has increased over time in the conference. In general, from 1997 until 2018, the relative frequency of papers related to design in relation to other areas represents an average of 17.62% of the conference papers, which is relatively low considering the importance of the field. However, it is noteworthy that recently the proportion of design management papers in relation to the total has increased, reaching a peak in 2016 with 33% (39/120) of the publications.

Figure 2.3 presents an analysis of the purpose of the product development and Design Management research, which was categorised into: (i) theoretical, and (ii) empirical. The first category includes papers focused on theory and predominantly based on the literature. Systematic literature reviews, literature review papers to develop a new solution to a problem and the analysis of secondary data from previous studies were classified as theoretical. Empirical papers are those focussed on applying or adapting solutions in a specific context, e.g.: (i) the prescription of a solution to a practical problem; (ii) implementation of Lean product development models, approaches, tools and techniques; (iii) assessments of the application of Lean and its approaches, methods, tools and techniques; (iv) use of IT or BIM tools in design; among others.

The analysis shows that the focus of the research has been mainly empirical, rather than theoretical, as shown in Figure 2.3. Interestingly, early on, from 1996 to 2000, the topic was mostly addressed from a theoretical perspective. The analysis illustrated in Figure 2.3 indicates that the proportion of empirical and theoretical papers has oscillated over the years, and theoretical research represents around a third of all design related publications. The predominance of empirical papers portrays a peculiar characteristic, as there is a wide range of solution developments, adaptations and implementations. Furthermore, many solutions are mentioned once or there are partial implementations of tools, reflecting some fragmentation of the body of design management knowledge by the IGLC community. Additionally, this highlights that theory is not evolving at the same rate as practical implementations, and there are limited successful examples, and hence poor support to new Lean design implementations.

Noteworthy is the fact that most papers adopt qualitative methods, many employing case studies. However, from 2010 the applications of Design Science Research (DSR), a prescriptive research approach, started to emerge. Furthermore, many of the solutions proposed are briefly implemented or tested, in a qualitative way, which contributes to the fragmentation of knowledge in the area.

The topics discussed and frequency of papers on Lean New Product Development and design management has increased in the last 10 years in IGLC, as shown in Figure 2.4. Early papers discussed the three views of design (TFV) as presented earlier in this

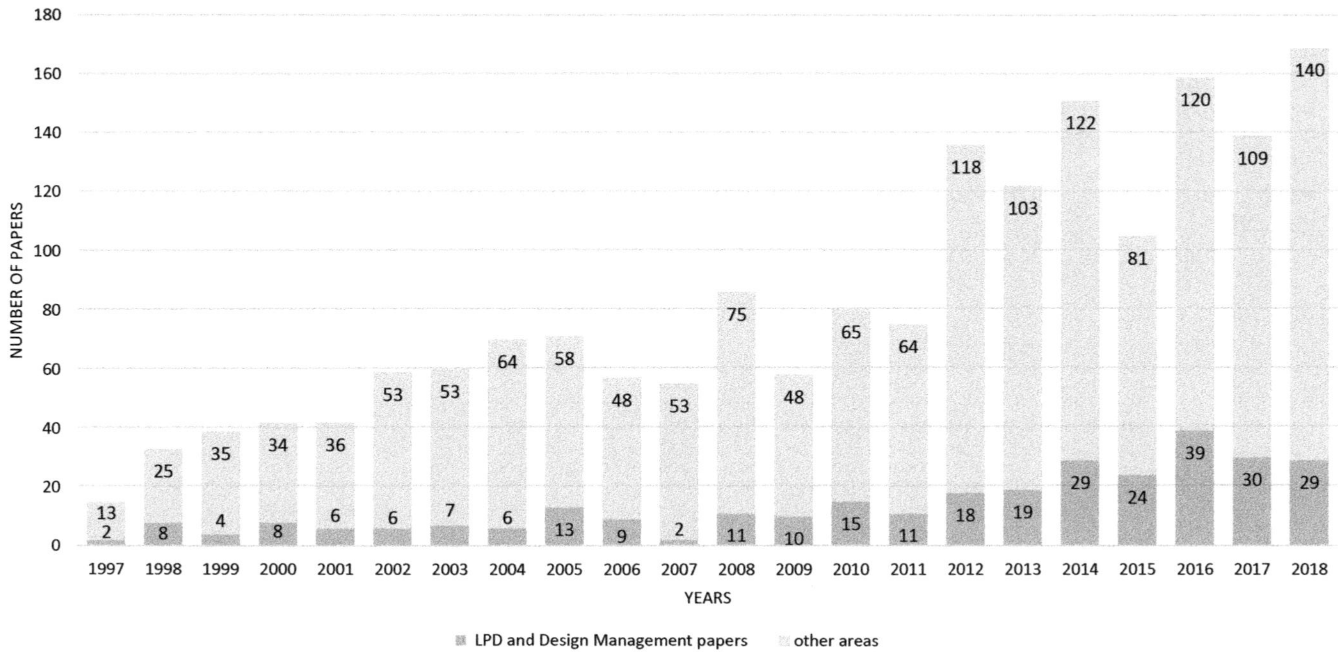

Figure 2.2 Frequency of papers related to LPD and Design Management in the IGLC from 1997 to 2018

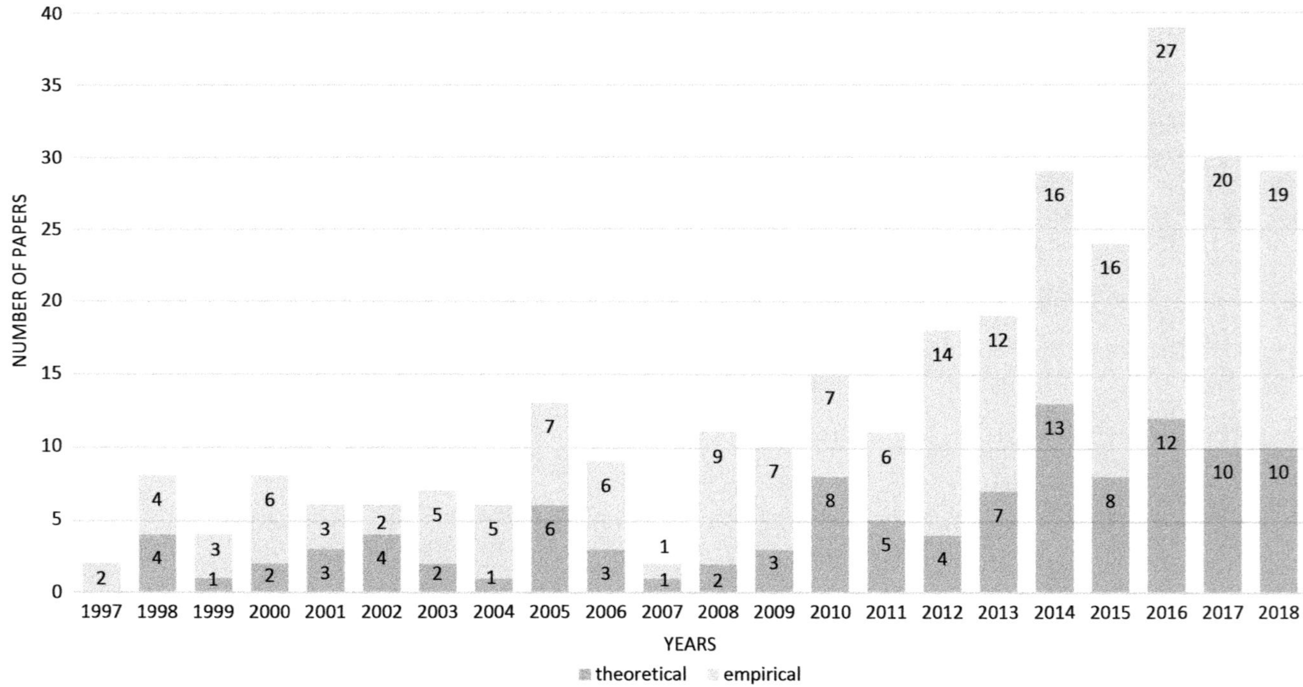

Figure 2.3 Purpose of the research

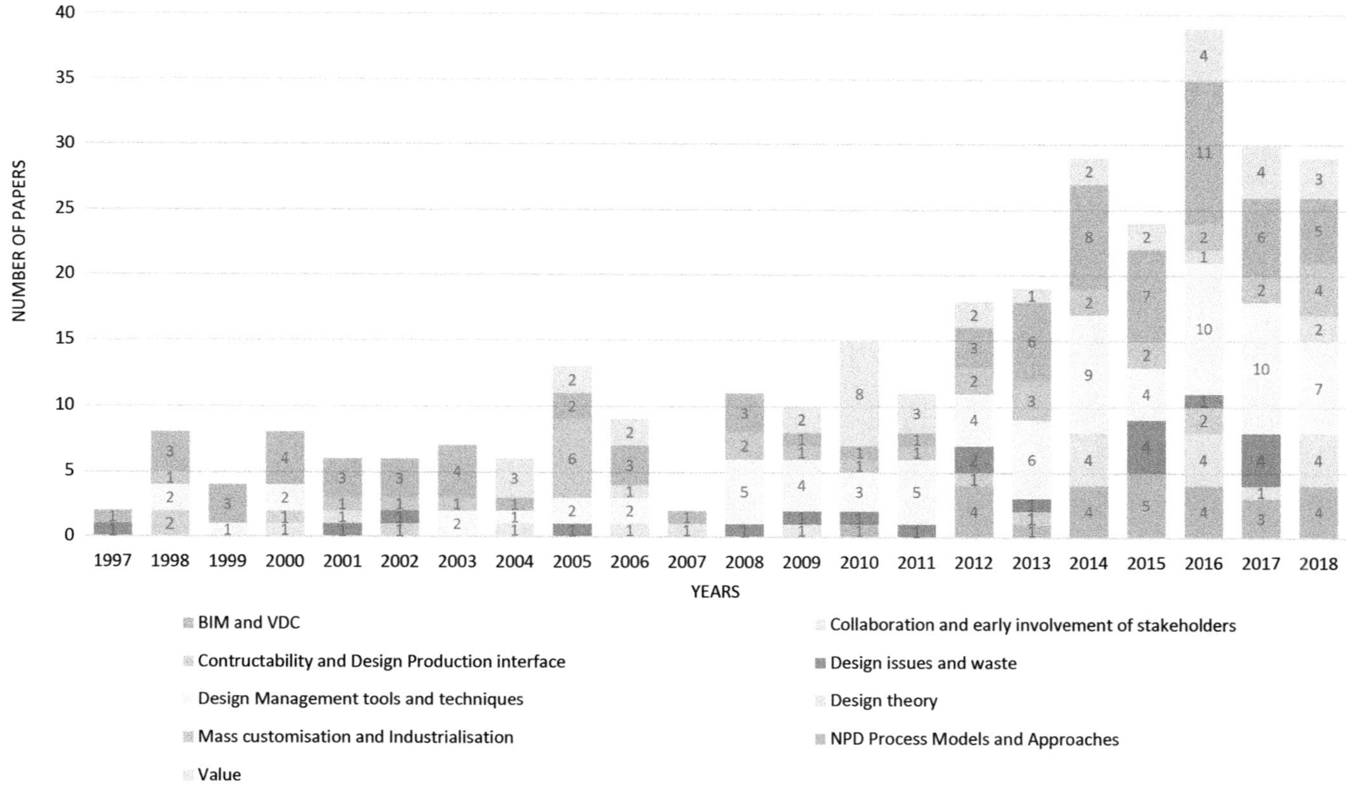

Figure 2.4 Development areas along the years

chapter. Many of these discuss the adaptation of Lean Construction principles from manufacturing to design in the construction industry.

From 2013 to 2016 the volume of NPD process models and approaches as well as the design management tools and techniques increased, possibly due to the transformations in the industry regarding BIM. There have been four or more BIM related papers each year in the last five years presented at the conference, with further technology papers not directly related to design. This reinforces the relevance of recent technological developments in construction.

Other areas have been brought to light in the conference, such as value, design issues and waste and mass customisation and industrialisation. The least frequently discussed topics analysed were collaboration and early involvement of stakeholders (17), constructability and design construction interface (8), and design theory (6) respectively. Those less discussed development areas may offer fruitful opportunities of further research.

Many topics have been discussed under the umbrella of Lean New Product Development and design management. Considering this, a timeline presenting when some influential developments first appeared in IGLC is shown in Figure 2.5.

Table 2.2 presents a categorisation of Lean product development and design management themes, which are briefly discussed as follows.

2.4.1 Design theory

Koskela has promoted the view that design theorising started in Antiquity, with the method of analysis (from geometry) and rhetoric as the first models of design by analogy. It is proposed that these two models still provide a superior understanding on design.

The method of analysis has been discussed as the proto-theory of design. Koskela et al. (2014) discussed how the method of geometrical analysis, proposed by Aristotle, provides a theoretical underpinning for design, which has been forgotten over time. The authors highlighted that this compels researchers to see the evolution of design under a new light – and although scholars of design have endeavoured to discover the core

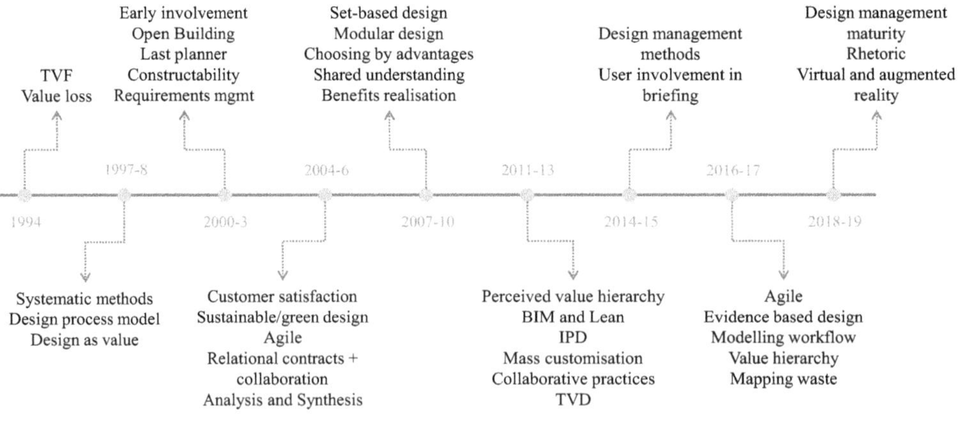

Figure 2.5 Timeline of the emergence of the main topics in Lean product development and design management

Table 2.2 Categorisation of Lean product development and design management themes

Themes	*Description*	*Examples of papers*
Design theory	Concepts underpinning design and product development	(Codinhoto et al., 2006); (Koskela, 2015); (Koskela, Arroyo, and Ballard, 2018); (Pikas et al., 2018).
Value	Value generation, assessment of clients' perception of value and satisfaction, the concept of value, requirements capture and management, between others	(Emmitt, Sander, and Kirk-Christoffersen, 2004); (Salvatierra, Pasquire, and Miron, 2012); (de Brito, Formoso and Rocha, 2012); (Drevland and Tillmann, 2018).
New Product Development approaches	Research on the conceptualisation, adaptation or implementation of approaches such as: Concurrent Engineering (CE), Target Value Delivery (TVD), Integrated Project Delivery (IPD), Benefits Realisation, Building Information Management (BIM) and the links between Lean and BIM	(Kamara, 2003); (Ballard, 2006); (Parrish et al., 2008a); (Tillmann et al., 2012); (Hamdi and Leite, 2012); (Ahuja, Sawhney, and Arif, 2014); (Franco and Picchi, 2016); (Harris, 2016); (Soliman Junior et al., 2018).
Design management tools and techniques	tools and techniques used to support a specific design activity or part of the product development process based on Lean thinking, promoting Lean principles, collaboration, efficiency, reduction of waste, etc. For example: Choosing By advantages (CBA); Set Based Design (SBD); application of the Last Planner System to design; Design Structure Matrix (DSM)	(Furtmeier and Tommelein, 2010); (Arroyo, Ballard, and Tommelein, 2014); (Khalife et al., 2018).
Mass customisation and industrialisation	Research around mass customisation strategies, prefabrication, industrialisation and modularisation	(Barlow, 1998); (Hook and Stehn, 2005); (Höök, 2006); (Tillmann and Formoso, 2008); (Maxwell and Aitchison, 2017); (Martinez, Tommelein, and Alvear, 2017).
Collaboration and early involvement of stakeholders	Papers focussed on improving collaboration to establish a shared understanding and early involvement of stakeholders enhancing the holistic view of design and construction	(Jara, Alarcón and Mourgues, 2009); (Gomes, Tzortzopoulos, and Kagioglou, 2016); (Schöttle, Arroyo, and Christensen, 2018).
Design issues and waste	Papers describing design management problems and waste, such as: rework, waiting times, fragmentation, poor documentation, material waste and inadequate design	(Whelton and Ballard, 2002); (Lima, Maia, and Neto, 2012); (Pikas, Koskela, and Seppänen, 2017).

theory, progress has been slow and results fragmented. Koskela et al. (2014) further argue that this missing of the core theory has arguably contributed to the maintenance of disciplinary fragmentation around design, calling for a unified design science.

Koskela et al. (2014) also highlighted important terminological problems. A starting point to understand the nature of design has been the analysis-synthesis-evaluation model, with initial proposals developed around the 1960s, then influential models e.g. Markus and Arch (1973), which have been refined into many different variants, including Pahl and Beitz (2013) model.

However, there have been discussions around the concept of analysis and synthesis. According to Koskela et al. (2014), the two types of analysis include finding (a solution) and proving (an assertion, say, on the validity of a proposed solution). The authors further argue that the main difference of these is that in the former, one endeavours to create a chain of inferences from the problem towards a solution, whereas in the latter a solution is first guessed and then analysed for its validity. In the design literature, analogous approaches have been called problem-oriented and solution-oriented strategies, and it is recognised that completing a design requires the application of both (Emmitt and Ruikar, 2013; Wynn and Clarkson, 2005).

From the existing models, some state that analysis precedes synthesis (Asimow, 1962), some that synthesis precedes analysis (Hall, 1962); such models have tried to accommodate and clarify both types, with varying success (it is important to note that in these models the meanings of analysis and synthesis have drastically drifted from the original sense of these terms, as used in Greek geometry) (Koskela et al., 2014).

These descriptive models have a common feature in the assumption of selections and decisions between different ideas, or sub-solutions, being made in satisfactory manner; however, the models are silent on how the decisions are or should be made.

The terms analysis and synthesis have maintained a longstanding prestige in design, and as the understanding of their original meaning has been corrupted, new meanings have been given to them in different knowledge domains. This has led to a fundamental confusion of the role and meaning of analysis and synthesis in design. The current popular understanding in the design literature of analysis as a rational stage and synthesis as a creative stage is in direct contradiction with the ancient understanding.

Another important issue is rich connections between rhetoric and design, which have been analysed in prior research (Buchanan, 1985, 2001; Ballard and Koskela, 2013; Koskela, 2015). It is known that rhetoric originated from the need of citizens in ancient Greece to make speeches in a court of law. Hence, the discipline of rhetoric was then built around the unit of a speech.

Koskela (2015) contends that legal proceedings, as they have evolved from Antiquity, embrace important and effective principles for the collaborative – and simultaneously competitive – pursuit of a common goal. Hence, these principles can also be useful in the context of collaborative design. Koskela (2015) presented the seven principles as: (i) hear both parties; (ii) reasoned judgment; (iii) right to appeal; (iv) use of both logical and rhetorical arguments and reasoning; (v) standardised proceedings and documents; (vi) public nature of proceedings; as well as (vii) dedicated and structured space. The main idea is that, up to now in design theorising, the angle of competition of ideas has either mostly been abstracted away or only some specific topic has been examined; however, theoretical and practical gains have been modest. Consequently, the question on how this competition should be arranged has been left disregarded, and provides fertile grounds for new research (Koskela, 2015).

Furthermore, design theorising has covered several narrower topics (partly based on Kroes, 2002), such as creativity in design thinking, design education, design effort, conceptual design as a process, design progress, communication of design knowledge, managing design information, the role of computers in design, design as a cognitive activity, decision making in design, design intent/rationale, collaborative design, and team cognition. For a further discussion on Lean theory, please refer to Chapter 1 by Koskela.

2.4.2 Value

Value generation has been discussed as a main goal for Lean Construction and a focus for design management. The concept of value is complex and ambiguous, and as such its development has been informed by theories from different domains.

Value generation is considered by identifying design criteria and customer needs to be used as input to the process (Ballard and Koskela, 1998; Tzortzopoulos and Formoso, 1999). Furthermore, the consideration of customers' requirements involves gathering information about demand and feedback from surveys, post occupancy evaluation and facilities management (Emmitt, Sander and Kirk-Christoffersen, 2004). Value can be improved through a combination of better requirements management, collaborative interactions, communication (knowledge sharing), and supporting creativity through adequate management of design activities (Dave et al., 2013).

The value gained at early design through the efficient capture and management of requirements and intent can be significant, which is one of the core tenets of Lean. Buildings are built to serve a purpose, and the construction industry should focus on that purpose, as illustrated in Figure 2.6.

Creating value should be more important than minimising design costs, as the impact of any investment in design is likely to be greatly outweighed by having a building that supports its users' needs and hence supports business profitability. Consequently, a shift of focus from the delivery of products to the generation of value and benefits to clients is needed.

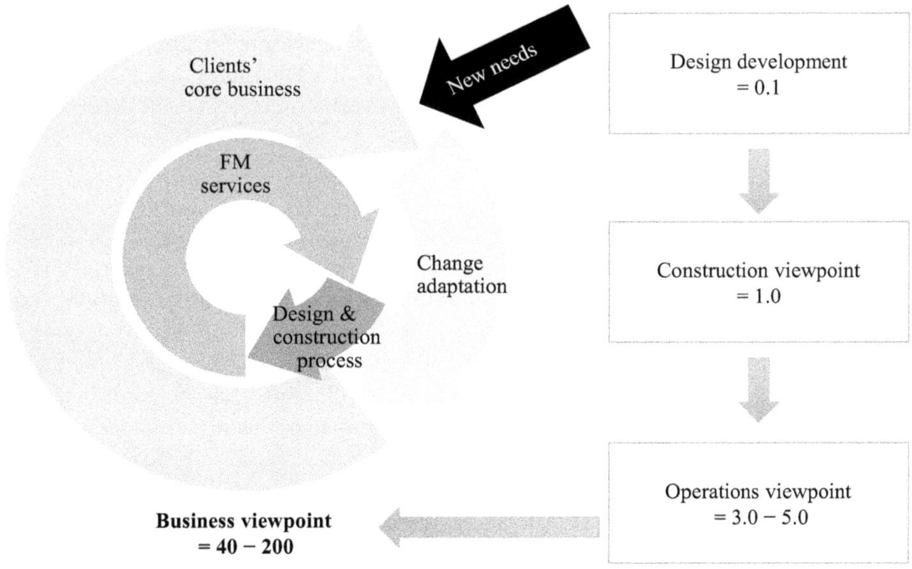

Figure 2.6 Project life cycle value (adapted from Dave et al., 2013)

The main concern is no longer the capital asset in its own right, but linking clients' business strategy to projects, maximising revenue generation and managing the delivery of benefits in relation to different stakeholder groups (Winter et al., 2006).

The traditional approaches to 'value' have been focused on defining a set of requirements in the project brief, and designing and delivering a building through a generally linear process. There is not much emphasis on evaluating and refining the design early against requirements through an iterative process (Ballard and Koskela, 1998). However, requirements evolve over time as ideas are tested and alternatives evaluated, but this notion is not sufficiently considered or supported by traditional processes or tools. 'Value' is more typically associated with value engineering, which is generally more aligned to construction cost reduction.

In Lean, generating value implies understanding the client's purposes and helping them to fulfil those purposes with minimum resources (e.g. costs, time). Integrated Lean and BIM concepts can help in this, as it enables the industry to better focus on life cycle values (Dave et al., 2013).

At the IGLC conferences, reflexions about the concept of Value itself emerged in 2009 with Salvatierra-Garrido, Pasquire and Thorpe's paper, and discussions on how value perception is established in a hierarchical manner for customers, which was also highlighted by Lima et al. (2009). From 2011, some research focused on adapting the means–end chain theory to understand how project attributes and consequences in use can lead to the achievement of goals and values desired by customers (Bonatto, Miron, and Formoso, 2011; de Brito and Formoso, 2014; Hentschke et al., 2014; Nascimento da Silva and Miron, 2017). Overall, research around value as a concept, and how to better support its achievement in design is still needed. Further discussions around value generation can be found in Chapter 6, by Tillmann and Miron.

2.4.3 New Product Development approaches

This broad title is used here to present some developments on the conceptualisation, adaptation or implementation of overarching product development approaches, such as: Concurrent Engineering (CE); Target Value Delivery (TVD); Integrated Project Delivery (IPD) and Benefits Realisation. Each of these approaches is very briefly described, and references made to sources for further information.

2.4.3.1 Concurrent Engineering (CE)

One of the first New Product Development approaches to be discussed in IGLC was CE, in the early 2000s. CE involves the simultaneous consideration of multiple requirements from diverse stakeholders (Kamara, 2003). It introduced the concept of overlapping activities, where two or more activities should be performed simultaneously. Kagioglou et al. (1998) and Ballard and Koskela (1998) state that its combination with cross-functional integration and teamwork could bring benefits to construction projects. The concept has influenced much research in construction, and brought to light the need to integrate the multidisciplinary project team into a unified development process, enabling downstream aspects of design and construction and non-value adding activities (waste) to be considered throughout design (Love and Gunasekaran, 1997). Some of the more recent ideas around overlapping design and construction are discussed in Chapter 15 by Biotto and Kagioglou.

2.4.3.2 Target Value Delivery (TVD)

Target Value Delivery (TVD), also referred to in the literature as Target Value Design, is the practice of steering design to targets, using costs as a driver to design, and not as a consequence of it, constraining the design and building of a facility to a maximum cost (Ballard, 2008). Collaboration is an essential element of the approach, which originates in the area of value engineering. Glenn Ballard is the main proponent of TVD, and details can be found in Chapter 8.

According to Zimina, Ballard, and Pasquire (2012), TVD is an adaptation of target costing to construction, that is, an effective management technique for achieving cost predictability, focusing on attaining market defined price and accomplishing financial returns. The approach is mentioned here as its implementation impinges changes in the way design is developed. A main change is that both cost and value are considered as drivers to design development; hence there is a push for designing to targets, instead of calculating the cost of a complete design (Zimina, Ballard, and Pasquire, 2012). Such approach needs to be supported by the use of BIM, e.g. enabling the fast production of design alternatives, and fast cost estimation of alternatives. Successful examples of TVD implementation have been reported, where the approach is used in conjunction with IPD, briefly discussed as follows.

2.4.3.3 Integrated Project Delivery (IPD)

The American Institute of Architects (AIA) defines Integrated Project Delivery (IPD) as: 'a project delivery approach that integrates people, systems, business structures and practices into a process that collaboratively harnesses the talents and insights of all participants to reduce waste and optimise efficiency through all phases of design, fabrication and construction'.

IPD focusses on collaboration, waste minimisation and integration of key stakeholders. It enables key features of Lean, such as early contractor involvement, integrated design and the whole life cycle approach. IPD leverages early contributions of knowledge and expertise through use of new technologies, allowing all team members to better realise their best potential while expanding the value they provide throughout the project life cycle. It is important to note that the collaborative structures inherent to an IPD legal agreement break down barriers and enable benefits to be achieved (Fischer et al., 2014). IPD is discussed in detail in Chapter 9 by Alvez and Lichtig.

2.4.3.4 Benefits realisation

Benefits realisation emerged in the information systems sector during the 1990s. It was motivated by the low success of technology implementation on generating the expected business benefits of investments, related to the organisations' strategic aims (Ward and Daniel, 2006). The same authors explain that the focus on benefits also brought a different perspective to business justification, complementing the traditional justification based only on monetary gains.

In the UK, the Office of Government Commerce (OGC), a governmental agency that provides policy standards and guidance on good practices of project management, recommends that projects should clarify the benefits they intend to achieve in their business case, track them during implementation and evaluate their realisation after delivery (OGC, 2007).

A benefits realisation model, BeReal, was developed based on the ideas of benefits management suggested by the OGC and others (Sapountzis et al., 2010). The model focuses on the development of healthcare infrastructure. The BeReal model includes a series of workshops for the definition of the expected benefits of projects, and a full description of such benefits with the definition of how and when their realisation will be measured (Sapountzis et al., 2010). Then, a design process in which key stakeholders evaluate the options against benefit criteria follows. Finally, an assessment phase is carried out, in which benefits are reviewed based on interviews, questionnaires, post occupancy evaluation and other techniques. Hence, the BeReal model is based on the understanding of value as the achievement of a purpose, giving emphasis to the advantages that can be generated over the life cycle of a project, which is the reason for its implementation and justification for the investment.

In the IGLC, benefits realisation has been addressed suggesting the active management of expected intent throughout project delivery, starting from capabilities to the realisation of expected benefits (Rooke et al., 2010; Tillmann, Tzortzopoulos, and Formoso, 2010; Tillmann et al., 2012; Kagioglou and Tzortzopoulos, 2016). Chapter 6 by Tillmann and Miron briefly discusses the concept further.

2.4.3.5 Building Information Management (BIM)

BIM has become a ubiquitous term across the industry and there is widely documented research on the theme. A host of benefits from the adoption of BIM in design have been highlighted, including, e.g.: reduced design development life cycle; effective capture and flow down of intent; reduced rework; increased iteration for value improvement; improved predictability of investment and life cycle costs; enhanced ability to engage with stakeholders (Kiviniemi, 2005).

The links between Lean and BIM have attracted much interest since 2011 with, e.g. the UK government BIM strategy requirement that all suitable government construction projects use BIM by 2016. It has been recognised that BIM contributes directly to Lean goals, enabling Lean processes. Furthermore, Lean processes can also facilitate the adoption of BIM, hence there are diverse positive synergies between the two (Sacks et al., 2010; Tauriainen et al., 2016).

Dave et al. (2013) have highlighted some characteristics of a Lean and BIM project, specifically for the design stage, as those that include: (a) collaborative development of design and detailing; (b) co-location of the design team; (c) involvement of downstream stakeholders in design; (d) using last planner in design; (e) detailing BIM models for construction use.

In the IGLC conferences, BIM papers first appeared in 2010. Following, many papers integrating the implementation of BIM with other product development approaches and tools as a way to support design have been presented. Some discuss how to solve specific design problems through the use of BIM based tools and management, such as clash detection, design coordination, information management, construction and demolition waste management. Chapter 4, by Dave and Sacks, presents a detailed discussion around Lean and BIM.

2.4.4 Design management tools and techniques

Design management in Lean Construction realises its principles through the application of tools and techniques to support a specific design activity or part of the product development process, promoting Lean principles such as collaboration, efficiency, and reduction of waste. This section briefly presents some of the existing tools, i.e. Choosing by Advantages (CBA); Set Based Design (SBD); Last Planner System (LPS) applied to design, and Design Structure Matrix (DSM).

2.4.4.1 Choosing by Advantages (CBA)

Suhr (1999) defined Choosing by Advantages as a decision-making system in which the significance of the advantages guide design decisions. Parrish and Tommelein (2009) describe CBA as 'a system that considers advantages of alternatives and makes comparisons based on these advantages.'

CBA includes decision-making methods, from very simple to extremely complex (Suhr, 1999). The overarching idea is that by using sound methods, practitioners can make better decisions that will guide their actions, and produce improved outcomes. The main CBA principles as presented by Suhr (1999) are shown in Table 2.3.

CBA includes the phases: (i) stage-setting, an innovation stage; (ii) the decision making; and (iii) implementation. The first appearance of choosing by advantages in the IGLC was in 2009 with Parrish and Tommelein (2009)'s paper. For detailed information on CBA, see Chapter 10 by Arroyo.

2.4.4.2 Set-based design

The concept of set-based design involves a simple idea – that designers need to consider sets of alternative design solutions from the start of the process, instead of developing one alternative in detail. Hence, designers consider sets of possible solutions and gradually narrow the set of possibilities to converge on a final solution. Such narrowing down happens based on additional information from design development, testing, the customer etc. A wide net from the start, and gradual elimination of weaker solutions, makes finding the better solution more likely (Sobek II, Ward, and Liker, 1999), as designers reason about, develop and communicate sets of solutions in parallel and relatively independently. According to Ward et al. (1995) Toyota's engineers and managers seek to delay decisions until the last responsible moment, so the approach is viewed as a funnelling process.

In Lean Construction, Parrish et al. (2008b) sates that this approach seeks to maintain the design space as open as possible for as long as possible. The final solution is defined based on discussions about design alternatives and their trade-offs considering different stakeholders' perception of value (Parrish et al., 2008b). Ballard (2008) summarises set-based design, arguing that: 'the basic idea is to apply all relevant criteria in producing,

Table 2.3 Choosing by Advantages (CBA) main principles (based on Suhr, 1999)

The cornerstone principle	To consistently make sound decisions, decision makers must learn and skilfully use sound methods of decision making
The fundamental rule of sound decision making	Decisions must be based on the importance of advantages
The principle of anchoring	Decisions must be anchored to the relevant facts
The methods principle	Different types of decisions call for different sound methods of decision making

evaluating and choosing from design alternatives from the beginning of design, rather than introducing new criteria as new stakeholders come onto the team'.

According to Ballard and Koskela (1998), set-based design integrates the perspectives of conversion, flow and value. Parrish et al. (2008b) discuss that set-based design promotes communication, avoids rework and supports the development of solutions focused on value within target cost boundaries. As a method, set-based design requires a shift in design management thinking, as it allows more of the design effort to proceed concurrently and defers detailed specifications until trade-offs are more fully understood. As such, it keeps design options flexible for as long as possible during the process (Ward and Sobek II, 2014).

Finally, it should be mentioned that, although facing some cultural barriers and organisational issues, the implementations of set-based design in building projects has presented sparse but good results, as well as promising opportunities for improvements (Ballard, 2008).

2.4.4.3 Last Planner System (LPS) applied to design

There have been attempts at implementing the last planner system of production planning and control to design, with initial attempts reported on the IGLC conference as early as 1997 by Koskela, Ballard, and Tanhuanpää (1997).

According to Fosse and Ballard (2016), the last planner can support key aspects of planning and control during the design process. It is argued that it can improve the transparency of the design process through scheduling and provide metrics to control progress (PPC – Percent Plan Complete) and establish benchmarking opportunities across projects. Benefits pointed out in the literature also include improvements on collaboration and commitment, by supporting and promoting the understanding of the role of other designers in the process. According to Khan (2016), although LPS improves construction planning and control reliability combining collaboration and coordination, a limited number of studies to date have successfully applied LPS in the design process.

There have been reported benefits in implementing for instance weekly work plans in design; however, the implementation of medium-term planning in design has proven challenging. This is mainly because of the high variability in the process, in which one piece of new information may change the priority of design activities quite dramatically. The literature also states that the implementation of LPS combined with BIM could significantly improve design workflow (Khan, 2016), but this is currently an area for further research.

2.4.4.4 Design Structure Matrix

Design Structure Matrix is a method for information flow representation, where tasks are organised according to their expected chronological order in rows and columns as a matrix (Koskela, Ballard, and Tanhuanpää, 1997). A precedence matrix is developed to show design dependencies, where the information flow between activities can be visualised.

Hence, the tool can be used to represent and manage design flows (Ballard and Koskela, 1998). The marks on the diagonal of the matrix represent that a task provides input for an earlier task, which represents the loops on the design process (Ballard and

Koskela, 1998). Khan (2016) states that DSM can improve design planning by considering the precedence between tasks. Furthermore, Hassan (1996) proposes that DSM can be regarded as a technique to optimise the order of design tasks, identify design loops and plan design based on the needed iterations.

Koskela, Ballard, and Tanhuanpää (1997) describe an experiment in which the optimal sequence of design tasks is analysed by means of the Design Structure Matrix: a schedule is prepared on the basis of the ordered tasks, and the execution of tasks is controlled through the last planner system. The authors state that resulting design process was more disciplined in comparison to a process managed in a conventional style. However, there are difficulties in implementing the DSM in design, as the information needed to produce the matrix needs to be produced at a detailed level, which is very difficult to predict in advance in design. There are clear opportunities to further investigate its application to support design planning in construction.

2.4.5 Mass customisation and industrialisation

Mass customisation (MC) is a strategy that aims to offer products that fulfil customers' requirements, potentially adding value, through flexible process and structure, with costs and delivery time similar to mass production (Fogliatto, da Silveira, and Borenstein, 2012; Hart, 1995; Jiao, Ma, and Tseng, 2003). The main challenges in the application of MC are capturing customers' requirements (Barlow and Ozaki, 2003; Martinez, Tommelein, and Alvear, 2017; Tillmann and Formoso, 2008) and keeping the balance between offering variety and maintaining affordability (Martinez, Tommelein, and Alvear, 2017) without affecting the efficiency of the production process (Nahmens and Bindroo, 2011). Several MC practices have been associated to Lean Construction concepts, e.g. reducing lead time, increasing value by systematically capturing customer requirements, and increasing output flexibility.

At the IGLC, initial discussions about mass customisation and industrialisation began with concepts of MC and open building, by Barlow (1998) and Cuperus (2001) respectively. Research shifted to identify connections between related topics such as modularisation, open building and prefabrication around the year 2005. These papers are also concerned with design and production process of industrialised and prefabricated systems, and how to improve productivity, reducing on-site activities and approximating construction to manufacturing and its benefits. The discussions about mass customisation were re-established in 2008 by Tillmann and Formoso, and little research was published in the area until more recently. In 2017, the concept of platform design started to be discussed. The area provides fertile grounds for further research, which could also address issues around construction productivity.

2.4.6 Collaboration and early involvement of stakeholders

Collaboration is an essential concept across Lean Construction, and a diversity of research efforts have identified benefits from its adoption, proposing it could help solve many of the deeply rooted problems of design. The term collaboration has been used widely, carrying a variety of meanings. As a consequence of the lack of conceptual clarity, the concept of collaboration remains amorphous (Poirier, Forgues, and Staub-French, 2016).

Attempts at understanding collaboration have resulted in normative definitions and checklists of antecedents and criteria for successful collaboration (D'Amour et al., 2005).

However, researchers have argued that normative definitions fail to account for the evolutionary and dynamic nature of collaborative interactions, because they tend to consider collaboration as an endpoint, rather than an evolving feature of the activity (Gomes and Tzortzopoulos, 2018; Poirier, Forgues, and Staub-French, 2016).

Nonetheless, procedures for collaborative working are needed to support designers concurrently working on the same project. In this context, the following issues need to be considered:

- how to ensure that all stakeholders have up-to-date information;
- how to make sure that several people are not editing the same information at the same time;
- how to notify about ongoing work and changes.

Also, other issues have to be addressed when coordinating multidisciplinary design, including (Dave et al., 2013):

- what information is needed by designers for critical decision making, and when is it needed?
- there are many temporary changes in trying to find solutions, so when is it adequate to exchange information?
- is there a need to 'freeze' design solutions so that other disciplines can start design?

A value stream (process) mapping exercise can help understand these issues including identifying critical design handover points, the level of detail required from each discipline at handovers, and general scheduling of the design process. Hence, effective information management through collaboration is a prerequisite for process improvement. Collaboration in early project stages is discussed in Chapter 20 by Gomes and Tzortzopoulos.

2.4.7 Design issues and waste

Regarding the concept of waste in design, as discussed earlier in this chapter, design errors are waste. Activities that are developed but not taken forward, could be classed as wasteful. However, in practice, it is very challenging to identify which design activities may lead to waste, and which interactions are needed to generate value – e.g. the development of design alternatives is needed to enable teams and clients to better understand what is possible within the project constraints – and, indeed, the concept of waste in design is still poorly understood (Mryyian and Tzortzopoulos, 2013). Further discussions around design management issues have been presented earlier in this chapter.

2.5 Future research

Improving the design process is still seen as a bottleneck for the improvement of the construction industry. Although there have been attempts to adapt Lean Construction principles and develop Lean design tools over the last two and a half decades, the impact of Lean thinking is still much stronger in the production process (Svalestuen, Lædre, and Lohne, 2014).

Long ago, Ballard and Koskela (1998) proposed a research agenda for design management, pointing out: (i) a need to understand design management as a facilitator to value generation and waste reduction, including the relationship among stakeholders, and (ii) a need for the identification of tools and techniques to integrate product and process design and to plan and control early design stages. Complementarily, Tzortzopoulos and Formoso (1999) indicated that research in design management should consolidate the value generation view of design, further investigating cycle times, waiting activities and variability, aiming to reduce waste without compromising creativity. The above issues are still relevant today. Some potential topics for further research in Lean product development and design management include:

- Further consideration of the concepts of analysis and synthesis and how clarifying these can support better practices in design management – this missing of the core theory has arguably contributed to the maintenance of disciplinary fragmentation around design, calling for a unified design science;
- Rhetoric and how its principles could be adapted to design and used to support design management;
- Theorising on collective action has developed slowly and has remained fragmented – existing concepts such as common ground, shared understanding, boundary objects present a multitude of fertile ideas that can be investigated further in support of collaboration; also, there is a need to better understand how theories link to practices like e.g. the big room;
- How to support creativity in design; how a better understanding of creative processes could support better design outputs;
- How simplicity concepts could be used to help manage design (e.g. Maeda, 2006);
- Communication and action – while systematic processes provide structure, the communication practices by which action occurs require further attention. To understand how communication and action can be improved, it is vital to achieve an understanding of stakeholder decision framing (Whelton and Ballard, 2002);
- NPD urges for new holistic approaches which cope with the current needs for innovation, speed and adding value for customers, this means that existing practices and solutions should be adapted to support companies in the competitive environment of start-ups, open innovation and so on (Griffin, Langerak, and Eling, 2019) in manufacture as well as in construction;
- The need for a systemic view also is reflected in design management, which according to Uusitalo et al. (2017) could be improved by the combination of different Lean design management methods, tools and techniques;
- Research around value as a concept, and how to better support its achievement in design is still needed;
- Value generation/early design/optioneering – links to client value; links to benefits; links to BIM through fast development of alternatives;
- BIM and LPS implementation in design – how they complement each other at detailed activity levels and how complexity theory could help in building resilience in design management;
- Other technological developments e.g. robotics, 3D printing – and how they affect design. There has been nearly no consideration of such technological developments by the IGLC community, and these provide somehow untapped rich grounds for research;

- Artificial Intelligence – its developments in IT and there are fertile opportunities to further investigate its use for example in supporting project planning or in the evaluation of the design of existing buildings. For example, a rules-based and case-based optimisation of design and production planning with automatic order processing and progress checking could eliminate the need for planners! AI-based location-based tools could also eliminate crew clashing on building spaces and crews standing whilst other work gets completed;
- Improved levels of automation in design – code checking, standardised products, platforms, mass customisation;
- Platform design and mass customisation – the area provides fertile grounds for further research, such as exploring the balance of flexibility–productivity trade-offs and further connections with Lean Construction concepts as ways to improve productivity, value generation and potential of practices to be adapted to new contexts;
- Better understanding the concept of waste in design can provide insights into effective tools for managing the process and its intricated information flows.

The relentless pursuit of better NPD and design management processes is never ending. Developments on the above will certainly help bring about significant improvements. Furthermore, the interface between academia, industry and professional institutions needs to become more agile, responsive and welcoming of new innovations (both process and technological). The need for holistic seeing-the-whole understandings has never been more relevant. As the boundaries between disciplines, project phases and actors become more permeable there will be a need to embrace an approach where processes influence and are influenced by human and technological agency in an iterative and evolving manner whilst focusing on delivering projects and wider space and social benefits addressing societies' grand challenges.

References

Ahuja, R., Sawhney, A. and Arif, M. (2014) 'BIM based conceptual framework for lean and green integration.' *Proc. 22nd Annual Conference of the International Group for Lean Construction*, Oslo, Norway. Available at: www.iglc.net/Papers.

AIA – American Institute of Architects. (2007) *Integrated Project Delivery: A Guide*. AIA California Council. Available at: info.aia.org/siteobjects/files/ipd_guide_2007.pdf.

Arayici, Y., Coates, P., Koskela, L., Kagioglou, M., Usher, C. and O'Reilly, K. (2011) 'Technology adoption in the BIM implementation for lean architectural practice', *Automation in Construction*, 20(2), pp. 189–195. Elsevier B.V. doi: 10.1016/j.autcon.2010.09.016.

Arroyo, P., Ballard, G. and Tommelein, I. D. (2014) 'Choosing By Advantages and Rhetoric in Building Design: Relationship and Potential Synergies', *Proc. 22nd Annual Conference of the International Group for Lean Construction*, Oslo, Norway. Available at: www.iglc.net/papers/details/989.

Asimow, M. (1962) *Introduction to Design: Fundamentals of Engineering Design*. Englewood Cliffs, Prentice-Hall.

Ballard, G. (2000) *The Last Planner System of Production Control*. PhD Thesis, The University of Birmingham, Birmingham, UK. Available at: http://leanconstruction.org.uk/media/docs/ballard2000-dissertation.pdf.

Ballard, G. (2006) 'Rethinking Project Definition in Terms of Target Costing', *Proc. 14th Annual Conference of the International Group for Lean Construction*, Santiago, Chile. Available at: www.iglc.net/Papers.

Ballard, G. (2008) 'The lean project delivery system: An update', *Lean Construction Journal*, 1–19.

Ballard, G. and Koskela, L. (1998) 'On the agenda of design management research', *Proc. 6th Annual Conference of the International Group for Lean Construction*, Guarujá, Brazil. Available at: www.iglc.net/Papers.

Ballard, G. and Koskela, L. (2013) 'Rhetoric and design', *Proc. 19th International Conference on Engineering Design ICED 13*, Seoul, Korea. Available at: http://eprints.hud.ac.uk/id/eprint/25190/1/Koskelaiced.pdf

Barlow, J. (1998) 'From craft production to mass customisation? Customer-focused approaches to housebuilding', *Proc. 6th Annual Conference of the International Group for Lean Construction*, Guarujá, Brazil. Available at: www.iglc.net/Papers.

Barlow, J. and Ozaki, R. (2003) 'Achieving "customer focus" in private housebuilding: Current practice and lessons from other industries', *Housing Studies*, 18(1), pp. 87–101. doi: 10.1080/0267303032000076858.

Bonatto, F. S., Miron, L. I. G. and Formoso, C. T. (2011) 'Evaluation of social housing projects based on user perceived value hierarchy', *Proc. 19th Annual Conference of the International Group for Lean Construction*, Lima, Peru. Available at: www.iglc.net/Papers.

de Brito, J. N. S. and Formoso, C. T. (2014) 'Using the means-end approach to understand perceived value by users of social housing projects', *Proc. 22nd Annual Conference of the International Group for Lean Construction*, Oslo, Norway. Available at: www.iglc.net/Papers.

de Brito, J. N. S., Formoso, C. T. and Rocha, G. S. (2012) 'Estudo da Formação de Valor de Usuários Finais de Empreendimentos Habitacionais de Interesse Social', *Encontro Nacional de Tecnologia do Ambiente Construído 2012*, p. 14.

Buchanan, R. (1985) 'Declaration by design: Rhetoric, argument, and demonstration in design practice', *Design Issues*, 2(1), pp. 4–22.

Buchanan, R. (1992) 'Wicked problems in design thinking', *Design Issues*, 8(2), pp. 5–21.

Buchanan, R. (2001) 'Design and the new rhetoric: Productive arts in the philosophy of culture', *Philosophy & Rhetoric*, 34(3), pp. 183–206.

Clark, K.B. and Fujimoto, T. (1991). *Product Development Performance*. Boston, USA: Harvard Business School Press.

Codinhoto, R., Koskela, L., Tzortzopoulos, P. and Kagioglou, M. (2006) 'How analysis and synthesis have been understood in design', *Proc. 14th Annual Conference of the International Group for Lean Construction*, Santiago, Chile. Available at: http://usir.salford.ac.uk/9372/.

Cooper, R. (1994) 'Third-generation new product processes', *Journal of Product Innovation Management: An International Publication of the Product Development & Management Association*, 11(1), pp. 3–14.

Cooper, R., Junginger, S. and Lockwood, T. eds. (2011) *The Handbook of Design Management*. Bloosbury, UK: A&C Black.

Cooper, R., Kagioglou, M., Aouad, G., Hinks, J., Sexton, M. and Sheath, D. (1998) 'The development of a generic design and construction process', *European Conference, Product Data Technology (PDT) Days*, pp. 1–10.

Cooper, R.G. (2001) *Winning at New Products: Accelerating the Process from Idea to Launch*. 3rd edition, Cambridge, MA, Perseus Books Group.

Coyne, R. (2005) 'Wicked problems revisited', *Design Studies*, 26(1), pp. 5–17.

Cuperus, Y. (2001) 'An introduction to open building', *Proc. 9th Annual Conference of the International Group for Lean Construction*, Singapore, Singapore. Available at: www.iglc.net/Papers.

D'Amour, D., Ferrada-Videla, M., San Martin Rodriguez, L., and Beaulieu, M. D. (2005) 'The conceptual basis for interprofessional collaboration: Core concepts and theoretical frameworks', *Journal of Interprofessional Care*, 19(1), pp. 116–131.

Dave, B. Koskela, L. Kiviniemi, A. and Tzortzopoulos, P. (2013) *Implementing Lean in Construction: Lean Construction and BIM*. London: CIRIA Guide C725.

DMI – Design Management Institute (2019) *What is Design Management*. Available at: www.dmi.org/page/What_is_Design_Manag.

Drevland, F. and Tillmann, P. A. (2018) 'Value for Whom?', *Proc. 26th Annual Conference of the International Group for Lean Construction*, Chennai, India. Available at: www.iglc.net/Papers.

Eastman, C., Teicholz, P., Sacks, R. and Liston, K. (2011) *BIM Handbook: A Guide to Building Information Modeling for Owners, Managers, Designers, Engineers and Contractors.* Hoboken, NJ: John Wiley & Sons.

Emmitt, S. and Ruikar, K. (2013) *Collaborative Design Management.* London and New York: Routledge.

Emmitt, S., Sander, D. and Kirk-Christoffersen, A. (2004) 'Implementing value through lean design management', *Proc. 12th Annual Conference of the International Group of Lean Construction*, Helsingør, Denmark. Available at: www.iglc.net/Papers.

Fischer, M., Reed, D., Khanzode, D. A. and Ashcraft, H. (2014) 'A Simple Framework for Integrated Project Delivery', *Proc. 22nd Annual Conference of the International Group for Lean Construction*, Oslo, Norway. Available at: www.iglc.net/Papers.

Fogliatto, F. S., da Silveira, G. J. C. and Borenstein, D. (2012) 'The mass customization decade: An updated review of the literature', *International Journal of Production Economics*, 138(1), pp. 14–25. Elsevier. Available at:: 10.1016/j.ijpe.2012.03.002.

Fosse, R. and Ballard, G. (2016) Lean design management in practice with the last planner system', *Proc. 24th Annual Conference of the International Group for Lean Construction*, Boston, USA. Available at: www.iglc.net/Papers.

Franco, J. V. and Picchi, F. A. (2016) 'Lean design in building projects: Guiding principles and exploratory collection of good practices', *Proc. 24th Annual Conference of the International Group for Lean Construction*, Boston, USA. Available at: www.iglc.net/Papers.

Furtmeier, F. and Tommelein, I.D. (2010) 'Explorative application of the multi-domain matrix methodology in lean design', *Proc. 18th Annual Conference of the International Group of Lean Construction*, Haifa, Israel. Available at: www.iglc.net/Papers.

Gomes, D. and Tzortzopoulos, P. (2018) 'Building shared understanding during early design', *Proc. 26th Annual Conference of the International Group of Lean Construction*, Chennai, India. Available at: www.iglc.net/Papers.

Gomes, D., Tzortzopoulos, P. and Kagioglou, M. (2016) 'Collaboration through shared understanding in early design stage', *Proc. 24th Annual Conference of the International Group of Lean Construction*, Boston, USA. Available at: www.iglc.net/Papers.

Griffin, A., Langerak, F. and Eling, K. (2019) 'The evolution, status, and research agenda for the future of research in NPD cycle time', *Journal of Product Innovation Management*, 36(2), pp. 263–280. Available at: 10.1111/jpim.12484.

Hall, A.D. (1962) *A Methodology for Systems Engineering*. Princeton, NJ: Van Nostrand.

Hamdi, O. and Leite, F. (2012) 'BIM and lean interactions from the BIM capability maturity model perspective: A case study', *Proc. 20th Conference of the International Group for Lean Construction*, San Diego, USA. Available at: www.iglc.net/Papers.

Harris, B. N. (2016) 'Building information modeling: A report from the field', *Proc. 24th Annual Conference of the International Group for Lean Construction*, Boston, USA. Available at: www.iglc.net/Papers.

Hart, C. W. L. (1995) 'Mass customization: Conceptual underpinnings, opportunities and limits', *International Journal of Service Industry Management*, 6(2), pp. 36–45. Available at: 10.1108/09564239510084932.

Hassan, T. (1996) *Simulating Information Flow to Assist Building Design Management.* PhD Thesis, Loughborough University, UK.

Hentschke, C., Formoso, C., Rocha, C., and Echeveste, M. (2014) 'A method for proposing valued-adding attributes in customized housing', *Sustainability (Switzerland)*, 6(12), pp. 9244–9267. Available at: 10.3390/su6129244.

Höök, M. (2006) 'Customer value in lean prefabrication of housing considering both construction and manufacturing', *Proc. 14th Annual Conference of the International Group for Lean Construction*, Santiago, Chile. Available at: www.iglc.net/Papers.

Hook, M. and Stehn, L. (2005) 'Connecting lean construction to pre-fabrication complexity in Swedish volume element housing', *Proc. 13th Annual Conference of the International Group for Lean Construction*, Sydney, Australia. Available at: www.iglc.net/Papers.

Howell, D., Windahl, C. and Seidel, R. (2010) 'A project contingency framework based on uncertainty and its consequences', *International Journal of Project Management*, 28(3), pp. 256–264.

Huovila, P., Koskela, L. and Lautanala, M. (1997) 'Fast or concurrent: The art of getting construction improved', *Lean Construction*, 143, pp. 159.

Jara, C., Alarcón, L.F. & Mourgues, C. (2009) 'Accelerating interactions in project design through extreme collaboration and commitment management–A case study', *Proc. 17th Annual Conference of the International Group for Lean Construction*, Taipei, Taiwan. Available at: www.iglc.net/Papers.

Jiao, J., Ma, Q. and Tseng, M. M. (2003) 'Towards high value-added products and services: mass customization and beyond', *Technovation*, 23(10), pp. 809–821. Available at: 10.1016/S0166-4972(02)00023-8.

Kagioglou, M. Cooper, R. Aouad, G. Hinks, J. Sexton, M. and Sheath, D. (1998) *Final Report: Process Control.* University of Salford, Salford, UK.

Kagioglou, M., Cooper, R., Aouad, G., and Sexton, M. (2000) 'Rethinking construction : The generic design and construction process protocol', *Engineering, Construction and Architectural Management*, 7(2), pp. 141–153.

Kagioglou, M. and Tzortzopoulos, P. (2016) 'Benefits realisation: An investigation of structure and agency', *Proc. 24th Annual Conference of the International Group for Lean Construction*, Boston, USA. Available at: www.iglc.net/Papers.

Kamara, J. M. (2003) 'Enablers for concurrent engineering in construction', *Proc. 11th Annual Conference of the International Group for Lean Construction*, Virginia, USA. Available at: https://pdfs.semanticscholar.org/83d6/85e50cb121931ad5e510215ca8ec98872a18.pdf.

Khalife, S., Mneymneh, B.E., Tawbe, A., Chatila, M.H. and Hamzeh, F. (2018) 'Employing simulation to study the role of design structure matrix in reducing waste in design', *Proc. 26th Annual Conference of the International Group for Lean Construction*, Chennai, India. Available at: www.iglc.net/Papers.

Khan, S. (2016) *Using the Last Planner System to Improve Workflow in BIM-Based Building Design Projects.* PhD Thesis, University of Huddersfield, UK.

Kiviniemi, A. (2005) *Requirements Management Interface to Building Product Model.* PhD Thesis, Stanford University.

Koskela, L. (2000) *An Exploration towards a Production Theory and Its Application to Construction.* PhD thesis, Technical Research Centre of Finland, VTT.

Koskela, L. (2015) 'Where rhetoric and lean meet', *Proc. 23rd Annual Conference of the International Group for Lean Construction*, Perth, Australia. Available at: www.iglc.net/Papers.

Koskela, L., Arroyo, P. and Ballard, G. (2018) 'Designing as a court of law', *Proc. 26th Annual Conference of the International Group for Lean Construction*, Chennai, India. Available at: www.iglc.net/Papers.

Koskela, L., Ballard, G. and Tanhuanpää, V.P. (1997) 'Towards lean design management', *Proc. 5th Annual Conference of the International Group for Lean Construction*, Gold Coast, Australia. Available at: www.iglc.net/Papers.

Koskela, L., Codinhoto, R., Tzortzopoulos, P., and Kagioglou, M. (2014) 'The Aristotelian proto-theory of design', In A. Chakrabarti & L.T.M. Blessing (eds), *An Anthology of Theories and Models of Design: Philosophy, Approaches and Empirical Explorations*, Springer, London, pp. 285–303.

Koskela, L., Howell, G., Ballard, G. and Tommelein, I. (2002) 'The foundations of lean construction', *Design and Construction: Building in Value*, 291, pp. 211–226.

Kroes, P. (2002) 'Design methodology and the nature of technical artefacts', *Design Studies*, 23(3), pp. 287–302.

Lima, L., Miron, L., Leite, F. and Formoso, C. (2009) 'Perceived value in social housing projects', *Proc. 17th Annual Conference of the International Group for Lean Construction*, Taipei, Taiwan. Available at: www.iglc.net/Papers.

Lima, L. P., Formoso, C. T. and Echeveste, M. E. S. (2008) 'Client requirements processing in low-income house-building using visual displays and the house of quality', *Proc. 16th Annual Conference of the International Group for Lean Construction*, Manchester, UK. Available at: www.iglc.net/Papers.

Lima, M.M., Maia, S.C. and Neto, J.D.P.B. (2012) 'Design inadequacies analysis in low income housing service areas', *Proc. 20th Annual Conference of the International Group of Lean Construction*, San Diego, USA. Available at: www.iglc.net/Papers.

Love, P. and Gunasekaran, A. (1997) 'Concurrent engineering in the construction industry', *Concurrent Engineering*, 5(2), pp. 155–162. Available at: 10.1177/1063293X9700500207.

Maeda, J. (2006) *The Laws of Simplicity*. Cambridge, MA: MIT press.

Markus, T. and Arch, M. (1973) *Optimisation by Evaluation in the Appraisal of Buildings*. Value in Building, London, Ed. Hutton e Devonald, Applied Science, 82–111.

Martinez, E., Tommelein, I. D. and Alvear, A. (2017) 'Integration of lean and information technology to enable a customization strategy in affordable housing', *Proc. 25th Annual Conference of the International Group for Lean Construction*, Heraklion, Greece. Available at: www.iglc.net/Papers.

Maxwell, D. W. and Aitchison, M. (2017) 'Design-value in the platform approach', *Proc. 25th Annual Conference of the International Group for Lean Construction*, Heraklion, Greece. Available at: www.iglc.net/Papers.

Morgan, J.M. and Liker, J.K. (2006) *The Toyota Product Development System – Integrating People, Process, and Technology*. New York, Productivity Press.

Mryyian, M. and Tzortzopoulos, P. (2013) 'Identifying source of design error in the design of residential buildings', *Proc. 21st Annual Conference of the International Group for Lean Construction*, Fortaleza, Brazil. Available at: www.iglc.net/Papers.

Nahmens, I. and Bindroo, V. (2011) 'Is customization fruitful in industrialized homebuilding industry?', *Journal of Construction Engineering and Management*, 137(12), pp. 1027–1035. doi: 10.1061/(ASCE)CO.1943-7862.0000396.

Nascimento da Silva, M. and Miron, L. I. G. (2017) 'Hierarchy of value perceived by groups of users about their neighbourhood' *Proc. 25th Annual Conference of the International Group for Lean Construction*, Heraklion, Greece. Available at: www.iglc.net/Papers.

OGC - OFFICE of GOVERNMENT COMMERCE. (2007) *Managing Successful Programmes (MSP)*. London, The Stationery Office.

Pahl, G. and Beitz, W. (2013) *Engineering Design: A Systematic Approach*. London: Springer Science & Business Media.

Parrish, K. and Tommelein, I.D. (2009) 'Making design decisions using choosing by advantages', *Proc. 17th Annual Conference of the International Group for Lean Construction*, Taipei, Taiwan. Available at: www.iglc.net/Papers.

Parrish, K., Wong, J.M., Tommelein, I.D. and Stojadinovic, B. (2008a) 'Set-based design: Case study on innovative hospital design', *Proc. 16th Annual Conference of the International Group for Lean Construction*, Manchester, UK. Available at: www.iglc.net/Papers.

Parrish, K., Wong, J. M., Tommelein, I. D., and Stojadinovic, B. (2008b) 'Set-Based Design: Case Study on Innovative Hospital Design', *Proc. 16th Annual Conference of the International Group for Lean Construction*, Manchester, UK. Available at: www.iglc.net/Papers.

Pikas, E., Koskela, L., Dave, B. and Liias, R. (2015) 'Case study on design management: Inefficiencies and possible remedies', *Proc. 23rd Annual Conference of the International Group for Lean Construction*, Perth, Australia. Available at: http://eprints.hud.ac.uk/id/eprint/25402/1/iglcKosCase.pdf.

Pikas, E., Koskela, L. and Seppänen, O. (2017) 'Design Management in a Design Office: Solving the Problem of Relevance', *Proc. 25th Annual Conference of the International Group for Lean Construction*, Heraklion, Greece. Available at: 10.24928/2017/0315.

Pikas, E., Koskela, L., Tredal, N., Knotten, V., and Bølviken, T. (2018) 'The Dual Nature of Design Management', *Proc. 26th Annual Conference of the International Group for Lean Construction*, Chennai, India. Available at: 10.24928/2018/0537.

Poirier, E., Forgues, D., and Staub-French, S. (2016) 'Collaboration through innovation: Implications for expertise in the AEC sector', *Construction Management and Economics*, 34(11), pp. 769–789.

RIBA – Royal Institute of British Architects (2013) 'RIBA Plan of work 2013"'. Available at: www.ribaplanofwork.com/Download.aspx

Rooke, J.A., Sapountzis, S., Koskela, L.J., Codinhoto, R. and Kagioglou, M. (2010) 'Lean knowledge management: The problem of value', *Proc. 18th Annual Conference of the International Group for Lean Construction*, Haifa, Israel. Available at: www.iglc.net/Papers.

Rozenfeld, H., Forcellini, F.A., Amaral, D.C., Toledo, J.C. and Silva, S.L. (2006) *Gestão de Desenvolvimento de Produtos*. São Paulo, Editora Saraiva.

Sacks, R. Koskela, L. Dave, B. Owen, R. (2010) 'The interaction of lean and building information modeling', *Construction Journal of Construction Engineering and Management*, 136(9), pp. 968–980.

Salvatierra, J., Pasquire, C. and Miron, L. (2012) 'Exploring value concept through the IGLC community: Nineteen years of experience', *Proc. 20th Annual Conference of the International Group for Lean Construction*. San Diego, USA. Available at: www.iglc.net/Papers.

Salvatierra-Garrido, J., Pasquire, C. and Thorpe, T. (2009) "Value in Construction from a Lean Thinking Perspective: Current State and Future Development," *Proc. 17th Annual Conference on Lean Construction*, Taipei, Taiwan. Available at: www.iglc.net/Papers.

Santos, A., Kistmann, V. B. and Fischer, S. (2004) 'Assessing total value requirements on low income houses in Brazil', *Proceedings 12th Annual Conference of the International Group for Lean Construction*, Helsingør, Denmark. Available at: www.iglc.net/Papers.

Sapountzis, S., Yates, K., Lima J. B. and Kagioglou, M. (2010) 'Benefits realisation: Planning and evaluating healthcare infrastructures and services', In M. Kagioglou and P. Tzortzopoulos (eds), *Improving Healthcare through Built Environment Infrastructure*, Oxford, Blackwell Publishing, pp. 166–195.

Schöttle, A., Arroyo, P. and Christensen, R. (2018) 'Demonstrating the value of an effective collaborative decision-making process in the design phase', *Proc. 26th Annual Conference of the International Group for Lean Construction*, Chennai, India. Available at: 10.24928/2018/0500.

Schöttle, A., and Tillmann, P.A. (2018) 'Explaining the benefits of team goals to support collaboration', *Proc. 26th Annual Conference of the International Group for Lean Construction*, Chennai, India. Available at: 10.24928/2018/0490.

Sobek II, D.K., Ward, A.C. and Liker, J.K. (1999) 'Toyota's principles of set-based concurrent engineering', *Sloan Management Review*, Winter 1999, 40(2); ABI/INFORM Global, pp. 67.

Soliman Junior, J., Baldauf, J., Formoso, C. T., and Tzortzopoulos, P. (2018) 'Using BIM and lean for modelling requirements in the design of healthcare projects', *Proc. 26th Annual Conference of the International Group for Lean Construction*, Chennai, India. Available at: 10.24928/2018/0455.

Suhr, J. (1999) 'The choosing by advantages decision making system', *Quorum*, Westport, USA (ISBN: 978-1-56720-217-5). Available at: www.decisioninnovations.com/

Svalestuen, F., Lædre, O. and Lohne, J. (2014) 'Research work on the design process within the IGLC conference', *Proc. 22nd Annual Conference of the International Group for Lean Construction*, Oslo, Norway. Available at: www.scopus.com/inward/record.url?eid=2-s2.0-84923365390&partnerID=tZOtx3y1.

Tauriainen, M., Marttinen, P., Dave, B., and Koskela, L. (2016) 'The Effects of BIM and Lean Construction on Design Management Practices', *Procedia Engineering*, 164, pp. 567–574. Available at: 10.1016/j.proeng.2016.11.659.

Tillmann, P., Tzortzopoulos, P., Formoso, C. T., and Ballard, G. (2013) 'Contributions of existing practices to pursuing value in construction projects', *Proc. 21st Annual Conference of the International Group for Lean Construction*, Fortaleza, Brazil. Available at: www.iglc.net/Papers.

Tillmann, P., Tzortzopoulos, P., Sapountzis, S., Formoso, C., and Kagioglou, M. (2012) 'A case study on benefits realisation and its contributions for achieving project outcomes', *Proc. 20th Annual Conference of the International Group for Lean Construction*, San Diego, USA. Available at: www.iglc.net/Papers.

Tillmann, P. A. and Formoso, C. T. (2008) 'Opportunities to adopt mass customisation – a case study in the Brazilian house building sector', *Proc. 16th Annual Conference of the International Group for Lean Construction*, Manchester, UK. Available at: www.iglc.net/Papers.

Tillmann, P.A., Tzortzopoulos, P. and Formoso, C.T. (2010) 'Analysing benefits realisation from a theoretical perspective and its contribution to value generation', *Proc. 18th Annual Conference of the International Group for Lean Construction*, Haifa, Israel. Available at: www.iglc.net/Papers.

Tzortzopoulos, P. (2004) *The Design and Implementation of Product Development Process Models in Construction Companies*. Ph.D. Thesis, University of Salford, UK.

Tzortzopoulos, P. and Cooper, R. (2007) 'Design management from a contractor's perspective: The need for clarity', *Architectural Engineering and Design Management*, 3(1), pp. 17–28. doi: 10.1080/17452007.2007.9684626.

Tzortzopoulos, P. and Formoso, C. (1999) 'Considerations on application of lean construction principles to design management', *Proc. 7th Annual Conference of the International Group for Lean Construction*, Berkeley, USA. Available at: http://iglc.net/Papers/Details/88.

Ulrich, K.T., Eppinger, S.D. (2000). *Product Design and Development.* 2nd edition, USA: McGraw-Hill, p. 358.

Uusitalo, P., Olivieri, H., Seppänen, O., Pikas, E., and Peltokorpi, A. (2017) 'Review of lean design management: Processes, methods and technologies', *Proc. 25th Annual Conference of the International Group for Lean Construction*, Heraklion, Greece. Available at: www.iglc.net/Papers.

Wang, L., Ming, X. G., Kong, F. B., Li, D., and Wang, P. P. (2011) 'Focus on implementation: A framework for lean product development', *Journal of Manufacturing Technology Management*, 23(1), pp. 4–24. doi: 10.1108/17410381211196267.

Ward, A., Liker, J.K., Cristiano, J.J. and Sobek, D.K. (1995) 'The second Toyota paradox: How delaying decisions can make better cars faster', *Sloan Management Review*, 36, pp. 43.

Ward, A., and Sobek II, D. (2014) *Lean Product and Process Development.* Cambridge, MA: 2nd Lean Enterprise Institute.

Ward, J. and Daniel, E. (2006) *Benefits Management.* John Wiley & Sons, Chichester, UK.

Whelton, M. and Ballard, G. (2002) 'Project definition and wicked problems', *Proc. 10th Annual Conference of the International Group for Lean Construction*, Gramado, Brazil. Available at: www.iglc.net/Papers.

Winter, M., Smith, C., Morris, P. and Cicmil, S. (2006) 'Directions for future research in project management: The main findings of a UK government-funded research network', *International Journal of Project Management*, 24(8), pp. 638–649.

Womack, J.P. and Jones, D.T. (1996) 'Beyond Toyota: how to root out waste and pursue perfection', *Harvard Business Review*, 74(5), pp. 140–158.

Wynn, D. and Clarkson, J. (2005) 'Models of designing', In Clarkson J, Eckert C (eds.) *Design process improvement: a review of current practice*, London, Springer, pp. 34–59.

Yazdani, B. and Holmes, C. (1999) 'Four models of design definition: Sequential design centered, concurrent and dynamic', *Journal of Engineering Design*, 10(1), pp. 25–37.

Zimina, D., Ballard, G. and Pasquire, C. (2012) 'Target value design: Using collaboration and a lean approach to reduce construction cost', *Construction Management and Economics*, 30(5), pp. 383–398. doi: 10.1080/01446193.2012.676658.

3 The Last Planner System

Glenn Ballard

3.1 Introduction

The Last Planner System (LPS) can usefully be understood as a specification of the social component in a socio-technical system. Technical systems involving different work structures (activity based, location based, takt based) are presented elsewhere in the book containing this chapter (see Chapters 14 and 15). All work with LPS as the social planning system.

Project planning has traditionally been understood as though the plans alone were sufficient to achieve desired outcomes, with no discretion for adjustment in execution. Thus, perfect plans exactly describe how to act in order to achieve the desired outcomes, and a deviation of actions from plans indicates either imperfect plans or incompetence in plan execution.

This perspective was neatly captured in Laufer and Howell (1993), which critiques Critical Path Method (CPM), the dominant approach to construction project planning. The second of the 10 'ground rules' for that approach is: *A plan's time horizon should be maximal and a plan should be comprehensive, detailed and complete.*

This assertion is contradicted by the widely held belief that 'All plans are forecasts, and all forecasts are wrong. Forecast error increases with forecast length and level of detail' (Nahmias, 1997). If what Nahmias says is true, perfect plans are impossible, hence the traditional understanding is incorrect. What, then, is the relationship between planning, plans, and plan execution?

Suchman (1987) offers an answer with the example of manoeuvring a canoe through river rapids:

> The purpose of the plan in this case is not to get your canoe through the rapids, but rather to orient you in such a way that you can obtain the best possible position from which to use those embodied skills on which, in the final analysis, your success depends.

Suchman's description suggests that plans are produced as a preparation for a future action in a situation that cannot be fully described in advance. This virtually erases the distinction between planning and preparation. Planning is a type of preparation similar to arranging for materials to be delivered to a job site, or practising ahead of playing a football match. This understanding resonates with a quote attributed to General Dwight D. Eisenhower: 'Plans are nothing. Planning is everything'.

Suchman (1987) also suggests that planning should be considered as something in which a number of different people participate, and thus is incorrectly conceived as the job only of that person with the title 'planner' or 'manager'.

Here are some of the important implications from the works of Laufer and Howell (1993), Nahmias (1997) and Suchman (1987):

- The circumstances in which tasks are executed cannot be fully captured in plans created prior to the action being planned;
- Skills are needed to accomplish planned tasks, no matter how 'perfect' the task plan, and those skills include adapting to situations as found during plan execution;
- Planning is done at different levels of detail and forecast by different people in a project organisation;
- Planning extends through task completion, with the circle being completed back to the original plan, revising it in response to new information relevant to either or both project objectives and the path to follow in order to achieve them.

The LPS was created as a reaction to the Critical Path Method (CPM) approach and can be understood as a response to Laufer and Howell's (1993) call for a new project planning paradigm. The strategy was to shield project production from the baneful influence of CPM by limiting project master schedules to the milestone level of detail, and by having those directly responsible for doing the work to also decide how to do the same within each phase. As such, it anticipated Koskela and Howell's later critique of project management theory (Koskela and Howell, 2002).

The defensive strategy was founded on these presuppositions (Ballard and Tommelein, 2016):

A. All plans are forecasts and all forecasts are wrong. Forecast error varies with forecast length and level of detail;
B. Planning is dynamic and does not end until the project is completed;
C. Involving those who will directly supervise or perform the work being planned results in better plans and greater ability to adapt plans when needed;
D. Actors within a project production system can make choices that help or hinder achieving project objectives, i.e. actors have discretion;
E. Understanding project objectives and the current and future state of the project helps actors make better choices. [In other words, in order for project team members to help, they must know what we're trying to do, the current state of the project, and possible future states.]

The LPS itself has changed over time, through extension of functions and through improvement of methods for performing functions. The intent is to deliberately improve it continuously into the future. The remainder of this chapter is devoted to the following topics: (a) A brief history of the LPS; (b) LPS functions; (c) LPS principles; and (d) Further development of the LPS.

3.2 A brief history of the Last Planner System

In 1990, while working as a consultant to Brown and Root on a paper machine project during the engineering phase, this author stumbled upon the phenomenon of low workflow reliability. The first indication was the low overlap between what tasks should have been started in a given week and the tasks actually started. A meeting was held with the discipline leads (e.g. structural, mechanical, electrical engineering) for that project in order to better understand this mismatch between SHOULD and DID[1]. The conversation revealed a connected anomaly. When asked, the discipline leads estimated that they were able to complete between 50% and 70% of planned tasks each week. This anecdotal evidence was subsequently strengthened by measuring the percentage of planned tasks completed on seven industrial projects by seven different construction companies. The average percentage of weekly work plan tasks completed each week (PPC) ranged between 33% and 70%, with the grand average = 54% (Howell and Ballard, 1995).

3.2.1 Matching DID and WILL

The current functions of the LPS are shown in Figure 3.1, and these evolved over time.

In Figure 3.1, Tasks Made Ready (TMR) is the same measurement as Percent Plan Complete (PPC), but compares the weekly work plan in the last week of the lookahead to the weekly work plan actually committed, looking for tasks that were not made ready. TMR measures the extent to which scheduled tasks are being made ready, i.e. the effectiveness of constraints analysis and removal. Tasks Anticipated (TA) also compares the weekly work plan in the last week of the lookahead to the weekly work plan actually committed, but looking for tasks in the latter that were not in the former. TA measures the extent to which tasks that should be done are anticipated, i.e. the task breakdown process.

At this early stage, the focus was on increasing the match between WILL and DID, i.e. getting better at doing what we say we're going to do. The key concept was quality, specifically the quality of assignments. This quality checklist was developed for tasks in weekly work plans:

- Definition;
- Soundness;
- Sequence;

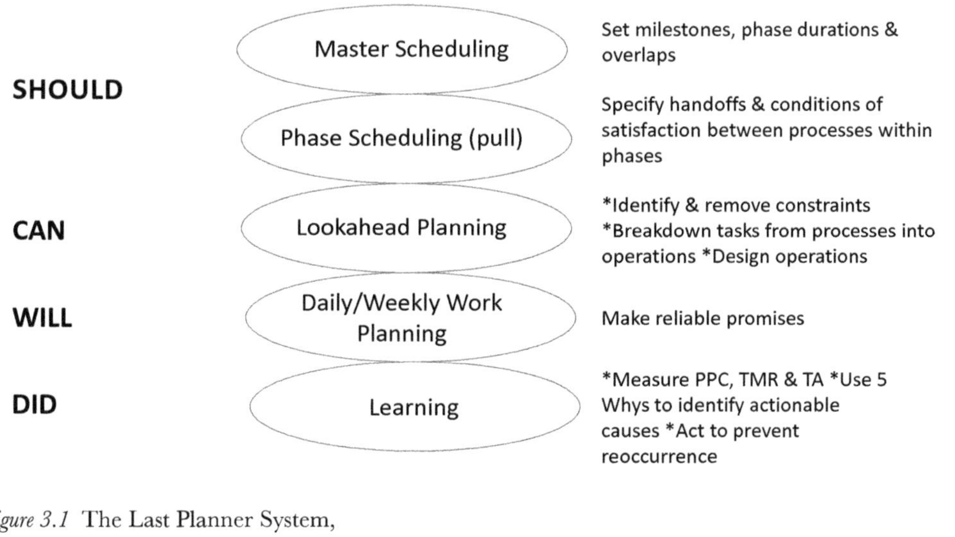

Figure 3.1 The Last Planner System, adapted from Ballard et al. (2007)

- Size;
- Learning – perfect planning may be impossible, but we can aspire to never making the same mistakes twice.

A definition was considered adequate if the workers performing the task could understand what was to be done, where it was to be done, and could determine what materials, information and resources were needed.

A task was considered to be sound if all constraints had been removed, materials were in hand, and work space was available, etc. A task was considered to be in the proper sequence if it could be done now without paying a penalty later in the project. A task was considered properly sized when the target task duration/subcrew productivity was adjusted for the capabilities of performers. All tasks assigned to a work group (design squad or construction crew) were considered properly sized when they matched the available capacity at budget unit rates 'adjusted for work mix and conditions' (Ballard and Howell, 1994b).

Although a characteristic of the planning and control process rather than the task itself, learning was also a requirement for increasing the match of DID with WILL. Failures to complete planned tasks had to be identified so that actionable causes could be found and countermeasures taken.

The quality checklist was tested and refined on the Koch Mid-Plants Project in Corpus Christi, Texas during 1993–4. Piping foremen were divided into two groups, those with PPP <50% and those with PPC>50%. 'The productivity of the 2nd group was 30 points higher than the 1st group … if the first group was performing at 15% over budget, the second group was performing at 15% under budget' (Ballard and Howell, 1994b).

3.2.2 Matching WILL and CAN

In four papers authored by Howell and Ballard in the 1994 Lean Construction conference in Santiago, Chile (Ballard and Howell, 1994a, 1994b; Howell and Ballard, 1994a, 1994b), we introduced the idea of creating a buffer of ready work from which front line supervisors could select tasks and place them on weekly work plans as commitments. We also shared what had been done on the Koch MidPlants project to ensure that materials were available. What we did not do was provide a more detailed description of how to create the buffer of ready work.

At the end of the Koch Mid-Plants Project, the refinery manager asked this author to review what they were doing to improve planning and control of refinery maintenance. One of the key concepts found when doing that review was 'make ready', emphasised by the consultants as a necessity prior to releasing maintenance tasks for commitment each day.

Reflecting on the concept, it became apparent that PPC could be 100%, but a project could fall behind schedule if tasks were not being made ready in the right sequence and rate (see Rother, 2009, who describes this is a natural progression in learning). To avoid that possibility, a process was needed to make scheduled tasks ready, i.e. to make them sound, to express them at the level of granularity appropriate to weekly work plans,[2] and to develop operations designs (work methods) to be tested in the first run of each type of operation (for a description of methods used to perform LPS functions, see Ballard and Tommelein, 2016, pp. 12–21). The primary output of such a lookahead planning process is a buffer of ready work available for commitment on weekly work plans. This differentiates the lookahead planning within the LPS from the commonly practised lookahead planning, which is devoted to focusing attention on what SHOULD happen in the near term – an early warning of mobilisation. The first publication specifically on LPS lookahead planning was *Lookahead Planning: The Missing Link in Production Control* (Ballard, 1997).

3.2.3 Phase schedules: improving specification of SHOULD

The next development in the LPS was a partial extension to specification of SHOULD. No guidance on planning (as distinct from control) had been provided apart from the advice to keep project master schedules at milestone level. In 1997, the Next Stage project was initiated by Linbeck Construction, with the objective of constructing a series of 7000-seat enclosed amphitheatres around the United States. The first of those was to be located between Dallas and Fort Worth, and a team was selected to design and build it. In the planning of that project, Neenan's Mike Daley introduced a process for collaboratively producing a schedule, using sticky notes on a wall, working backward from the target milestone. This was designed to encourage the reliable promising conversation between the people playing different, interdependent roles.

The collaborative planning of work to be done in project phases was developed to fill the need for planning across work scopes, as distinct from planning only or primarily within individual work scopes. Documentation of the pull planning method was first published in a Lean Construction Institute white paper (Ballard, 2000), and further developed in Ballard and Howell (2003).

3.3 Last Planner System functions

An explicit description of LPS functions first appeared in Ballard, Hammond, and Nickerson (2009). The following list is taken from Ballard and Tommelein (2016):

1. Specifying what tasks should be done when and by whom, from milestones to phases between milestones, to processes within phases, to operations within processes, to steps within operations;
2. Making scheduled tasks ready to be performed;
3. Replanning/planning to complete, to achieve project objectives;
4. Selecting tasks for daily and weekly work plans – deciding what work to do next;
5. Making release of work between specialists reliable;
6. Making visible the current and future state of the project;
7. Measuring planning system performance;
8. Learning from plan failures.

The innovations LPS brought in comparison to traditional practice were primarily in numbers 2, 5, 7 and 8 above. An explicit process was prescribed for making scheduled tasks ready to be performed, workflow reliability was an explicit objective, planning system performance was measured versus exclusive focus on project progress and cost, and learning was integrated into practice. These are new components, but arguably the more radical innovations concerned the social dimension; deployment of planning and control responsibilities from top to bottom of the project organisation, and focus on reliable promising between interdependent players as the key to successful execution.

3.4 Last Planner System principles

Principles are guides to action and are central in the LPS, which is properly understood, consistent with its orientation toward the social system of planning, as the complex of functions to be performed and principles to follow in their performance. Principles are derived from presuppositions about the nature of the relevant reality, and methods are invented to implement the principles. Methods are improved systematically, including by borrowing and adapting from other applications, e.g. agile methods. A method is 'Lean' and belongs to the LPS if it is fit for purpose in performing LPS functions, regardless where the method originated.

LPS principles were first articulated in Ballard, Hammond, and Nickerson (2009). The list here is taken from Ballard and Tommelein (2016):

1. Keep all plans, at every level of detail, in public view at all times within the project;
2. Keep master schedules at milestone level of detail;
3. Plan in greater detail as the start date for planned tasks approaches;
4. Produce plans collaboratively with those who are to do the work being planned;
5. Re-plan as necessary to adjust plans to the realities of the unfolding future;
6. Reveal and remove constraints on planned tasks as a team;
7. Improve workflow reliability in order to improve operational performance;
8. Don't start tasks that you should not or cannot complete. Commit to perform only those tasks that are properly defined, sound, sequenced and sized;
9. Make and secure reliable promises, and speak up immediately should you lose confidence, so that you can keep your promises (as opposed to waiting as long as possible and hoping someone else speaks up first);
10. Learn from breakdowns (unintended consequences of actions taken);
11. Underload resources to increase reliability of work release;
12. Maintain workable backlog; a backlog of ready work (tasks ready to be executed) to buffer against capacity and time loss.

Only #3 can reasonably be said to be shared with traditional project planning, and even then the more detailed levels of planning tend to be left to performers and to be invisible to traditional planning and control systems. A contract management perspective naturally includes shifting responsibility for detailed planning along with the performance risk.

3.5 Industrial and academic reception of the Last Planner System

There has been considerable international effort to pilot the LPS, evaluate it through research, and to understand it conceptually and theoretically. Some 241 out of 1,433 papers from annual conferences of the International Group for Lean Construction have 'Last Planner' in their titles, and more papers concern the LPS or some facets of it. Three particularly important streams of research and development relevant to the LPS are those on visual controls (e.g. Tezel, 2011), work structuring (e.g. Seppänen, Ballard, and Pesonen, 2010; Tsao et al., 2000) and simulation (e.g. Tommelein, 1997).

With regard to industry practitioners, one indicator of take-up is the number of affiliates (formal and informal) of the United States' Lean Construction Institute, which have tended to start their Lean journeys with the LPS: Argentina, Australia, Austria, Brazil, Canada, Chile, Denmark, Finland, France, India, Ireland, Mexico, Norway, Peru, Russia, Spain, Switzerland, and the UK.

3.6 Conclusions and further development of the Last Planner System

Invented to fill a hole in the project management tool kit (Ballard, 2000), LPS has grown over time with the recognition that traditional project management is fundamentally broken and needs a complete overhaul. That is happening now on multiple fronts. Classical theories of contracts as transactions are being confronted with the theory of contracts as blends of transactional and relational. Commercial terms that tie compensation to fixed scopes of work are being confronted by shared risk and reward, which promotes working together as a team to produce value for the project's customers. Technological developments in information management are facilitating collaboration across distance and across organisational boundaries. 3D printing looms as a near-future threat to segments of the construction supply system; many components that are now purchased fully formed can be designed and fabricated where and when needed. Global warming has prompted application of Lean thinking to both environmental and social sustainability. The true nature of projects as complex adaptive systems is confronting and replacing traditional notions of management.

As stated at the beginning of this chapter, the LPS is by no means exempt from change. The author of this chapter is currently working with a group of construction industry practitioners (owners, architects, engineers, general contractors, specialty contractors, management consultants) to further develop and improve the LPS. Teams are working on five tasks, with results scheduled for publication in the first quarter of 2019:

1. Extending LPS to the planning and control of the entire project;
2. Specifying the chunks in which work is to be done and released between trades in each construction phase;
3. Learning from breakdowns;
4. Metrics;
5. Alternative methods for design.

In the 2016 Current Process Benchmark for the LPS (Ballard and Tommelein, 2016), the only recommendation regarding business case validation and project execution strategies was to limit project master schedules to milestones, and subsequently to collaboratively plan how to do the work in the phases between milestones with those who are to execute the plans. Task Team #1 is developing process guidelines for business case validation and for development and assessment of project execution strategies, including the incorporation of stochastic planning methods in project master scheduling – principally, postponement and hedging. Postponement involves deferring a decision or action until better information is available for deciding or acting. Hedging is creating opportunities for future action. For example, when a decision to use a specific component has not yet been made or may change, alternative pathways can be built into project schedules, enabling the facility design to move forward regardless of which machine is ultimately selected.

Previously, LPS was unbiased as to what work structures were applied, but over time it has become evident that having some form of work structure other than the CPM 'activity' approach is much needed, and that some forms are more appropriate than others. Task Team #2 is developing the guidelines for creating those work structures.

Learning from breakdowns has been among the LPS functions that is least well implemented. Task Team #3 is seeking reasons and countermeasures to remedy that situation. LPS is designed to be a learning system, and is not achieving its full potential without systematic learning from experience.

Extension to the entire project brings with it the responsibility to provide metrics that not only inform planning and control system improvement, as has been the case with LPS until now, but also to provide metrics that inform how to improve project performance. Task Team #3 is developing guidelines for those metrics.

Designing and making are essentially connected, but very different types of work. It is no surprise that different methods might be more effective. Task Team #5 is looking for instances of those methods that have been successfully implemented in practice, and for methods from other domains that can be adapted to built environment design.

Notes

1 Should, Can, Will and Did were proposed as functions of project planning and control in Ballard and Howell (1994a). Traditional project control relies on an after-the-fact comparison of what DID happen with what SHOULD have happened, neglecting the need and opportunity for proactively steering toward SHOULD, which includes lookahead planning to make ready what SHOULD be done so it CAN be done when needed.

2 Terminology for task breakdown was developed later: projects are composed of phases, phases are composed of processes, processes are composed of operations, operations are composed of steps, and steps are composed of elemental motions (Ballard and Tommelein, 2016, p. 9).

References

Ballard, G. (1997) 'Lookahead planning: the missing link in production control', *Proc. 5th Annual Conference of the International Group for Lean Construction*, Gold Coast, Australia, Available at: www.iglc.net.

Ballard, G. (2000) *The Last Planner System of Production Control.* PhD Thesis, The University of Birmingham, Birmingham, UK. Available at: http://leanconstruction.org.uk/media/docs/ballard2000-dissertation.pdf.

Ballard, G., Hammond, J. and Nickerson, R. (2009) 'Production control principles', *Proc. 17th Annual Conference of the International Group for Lean Construction*, Taipei, Taiwan, Available at: www.iglc.net.

Ballard, G. and Howell, G. (1994a) 'Implementing lean construction: stabilizing work flow', *Presented at the 2nd Annual Conference of the International Group for Lean Construction*, Santiago, Chile. Available in Alarcon, L. (ed.) (1997), Rotterdam, The Netherlands, Lean Construction, A.A. Balkema.

Ballard, G. and Howell, G. (1994b) 'Implementing lean construction: improving downstream performance', *Presented at the 2nd Annual Conference of the International Group for Lean Construction*, Santiago, Chile. Available in Alarcon, L. (ed.) (1997), Rotterdam, The Netherlands, Lean Construction, A.A. Balkema.

Ballard, G. and Howell, G. (2003) 'An update on last planner', *Proc. 11th Annual Conference of the International Group for Lean Construction*, Blacksburg, VA. Available at: www.iglc.net.

Ballard, G. and Tommelein, I. (2016) 'Current process benchmark for the Last Planner® System', *Lean Construction Journal*, pp. 57–89.

Ballard, G., Tommelein, I., Koskela, L. and Howell, G. (2007). 'Lean construction tools and techniques', pp 251–279 in Best, R. and de Valence, G. (eds.) *Design and Construction: Building in Value*, London and New York: Routledge.

Howell, G. and Ballard, G. (1994a) 'Beyond can-do', *Presented at the 2nd Annual Conference of the International Group for Lean Construction*, Santiago, Chile. Available in Alarcon, L. (ed.) (1997), Rotterdam, The Netherlands, Lean Construction, A.A. Balkema.

Howell, G. and Ballard, G. (1994b) 'Reducing inflow variation', *Presented at the 2nd annual conference of the International Group for Lean Construction*, Santiago, Chile. Available in Alarcon, L. (ed.) (1997), Rotterdam, The Netherlands, Lean Construction, A.A. Balkema.

Howell, G. and Ballard, G. (1995) 'Factors affecting project success in the piping function', *Proc. 2nd Annual Conference of the International Group for Lean Construction*, Albuquerque, New Mexico, Available at: www.iglc.net.

Koskela, L.J. and Howell, G. (2002) 'The underlying theory of project management is obsolete', *Proc. PMI Research Conference*, pp. 293–302.

Laufer, A. and Howell, G.A. (1993) 'Construction planning: revising the paradigm', *Project Management Journal, 24*(3), pp. 23–33.

Nahmias, S. (1997) *Production and Operations Management*. Chicago, IL, USA: Irwin.

Rother, M. (2009) *Toyota Kata*. New York: McGraw-Hill Professional Publishing.

Seppänen, O., Ballard, G. and Pesonen, S. (2010) 'The combination of last planner system and location-based management system', *Proc. 18th Annual Conference of the International Group for Lean Construction*, Haifa, Israel. Available at: www.iglc.net.

Suchman, L. (1987) *Plans and Situated Actions*. Cambridge University Press, Cambridge, UK.

Tezel, B.A. (2011) *Visual Management: An Exploration of the Concept and Its Implementation in Construction*. PhD Thesis, University of Salford, Salford, UK.

Tommelein, I.D. (1997) 'Discrete-event simulation of lean construction processes', *Proc. 5th Annual Conference of the International Group for Lean Construction*, Gold Coast, Australia. Available at: www.iglc.net.

Tsao, C.C., Tommelein, I.D., Swanlund, E. and Howell, G.A. (2000) 'Case study for work structuring: installation of metal door frames', *Proc. 8th Annual Conference of the International Group for Lean Construction*, Brighton, UK. Available at: www.iglc.net.

4 Production control systems for construction at the nexus of Lean and BIM

Bhargav Dave and Rafael Sacks

4.1 Introduction

Production management in construction is directly linked to the successful completion of construction projects, yet it is an area that remains relatively ad hoc and dependent on manual processes (Zhang, 2005). A meta-analysis of research into production in construction covering the period from 1970 to 2000 (Horman and Kenley, 2005) reported that almost 49.6% of time was wasted in non-value adding activities. Similarly, research in Sweden showed that only 15%–20% of workers' time was spent on direct value-adding work (Jongeling and Olofsson, 2007). Many other studies around the world have similarly reported sub-optimal performance of construction projects in terms of productivity and efficiency (Ramaswamy and Kalidindi, 2009; Teicholz, Goodrum, and Haas, 2001).

These observations are explained in large part by sub-optimal performance of the production management systems in construction (Ballard and Howell, 1998; Howell, Laufer, and Ballard, 1993). In a study of the root causes of failures in construction activities, Koch (2005) reported production planning and control, communications and cooperation, and design activities to be the main factors. In construction, traditional planning and control processes with weak communication and coordination between stakeholders have been identified (Navon and Sacks, 2007; Sacks, Radosavljevic, and Barak, 2010b; Sarhan and Fox, 2012). Koskela and Howell (2001) reported that production management was a relatively neglected area and that there had been a decline in the attention given to it by general contractors and project managers up to that time.

Much of the growth in the theory and practice of Lean Construction can be seen as a response to this state of affairs. Koskela's Transformation-Flow-Value (TFV) theory pinpointed the construction management profession's narrow focus on transformation as the governing paradigm, to the neglect of the flow and value views (Koskela, 1992). The Last Planner System (LPS) (Ballard, 2000, also described in Chapter 3) grew from a recognition of the need to cope with the inherent uncertainty in construction planning. The LPS partly filled the gap left by the failure of the Critical Path Method (CPM) to model the instability of the various flows at the level of detail necessary for effective production control.

Information systems based on CPM became increasingly sophisticated, but no more effective. Chua, Nguyen, and Yeoh (2013) emphasised the role of information technology in planning and production management, especially from a constraints management perspective; however, they concluded that the current information management systems did not offer effective solutions. In a survey of a large-scale project management information and control system, Futcher (2001) highlighted the need for data entry at the project level as one of the most significant obstacles to successful implementation. Similarly, Koskela and Kazi (2003) drew attention to the ineffectiveness of construction information systems that did not support the core production processes and, in many cases, proved to be counterproductive.

Nevertheless, the quantity of information required to build a modern building demands the use of information technology. The new, Lean view of flow and value in construction opened the door to the development of new, quite different information systems for production management on-site. The 'OurPlan' system, developed by DPR Construction Inc., was one of the first commercial systems to provide IT support for the LPS (Ourplan, 2013). vPlanner, from Ghafari Associates, was another (Emdanat, Linnik, and Christian, 2016).

These developments took place roughly in parallel with the growth and maturation of Building Information Modelling (BIM). BIM was a logical step in the application of computing to management of product information in construction, changing the paradigm of representation of buildings using drawings and documents (paper or digital) to one of modelling the buildings with digital prototypes (Sacks et al., 2018). The impacts of Lean Construction and BIM on the construction industry are each profound in their own right; however, they not only arose in parallel, but they also had synergistic effects where they were applied together (Sacks et al., 2010a).

Given that the information needed for construction encompasses not only the product information, which is the main focus of BIM, but also the process information, the possibility of devising production control systems that exploit models became apparent. Research of the basic requirements for such systems identified the need for support negotiation and commitment between trade crews, Lean production planning with sophisticated pull flow control, and effective communication and visualisation of flow (Sacks, Radosavljevic, and Barak, 2010b). The KanBIM system was an early prototype of BIM-based production support systems.

In this chapter, we review the state of the art of the theory and of the resulting production control processes and information system tools that lie at the nexus of Lean Construction and BIM. The next section sets out the background to the issues, describing the nature of production on construction sites, and the inherent split between general contractors and subcontracted trade crews in terms of economic incentives, access to product and to process information, and production control systems. We present the theoretical relationships between Lean Construction and BIM, before proceeding to outline the principles that have been defined for information systems that exploit the synergies between Lean and BIM. The fourth section brings evidence for the interactions and the operation of production control systems based on BIM, from both experiments conducted in the laboratory and experience with the use of a commercial system in the field. Finally, we compare the results and draw conclusions.

4.2 Background

4.2.1 The disconnect between the product and the process

Production management in construction is information intensive. To succeed, people at various operational levels, in both office and site organisations, must communicate continuously and seamlessly. Without appropriate information systems, they need to be in constant interaction with each other. The information that flows consists of two main types: product information and process information (Sacks, Radosavljevic, and Barak,

2010b). The product information defines design-related knowledge, i.e. *what* is to be built/constructed, whereas the process information defines the process of construction itself, i.e. the *how*, *when*, *where* and *who* of the construction process.

Traditionally, drawings and specifications have been the primary tools for communicating product information, and they have well-established, albeit seriously flawed, procedures. The shortcomings include design errors, missing information, contradictions in the way the product or the work is defined across different documents, details that are not constructible and multiple versions of the same information. BIM has been developed in large part as a response to those flaws, and it is rapidly replacing drawings as the medium of communication of product information (Sacks et al., 2018).

Process information is commonly generated using digital planning tools and communicated in text documents and charts (such as large-format Gantt schedules). However, because it changes dynamically with the status of the work on-site, it is often communicated primarily by word of mouth, with updates to more formal documents lagging production itself. The charts are difficult to comprehend; they hide spatial conflicts, and they do not show the resource allocations or the planned pace of work. Where drawings and documents are the primary means of communicating product and process information, there is a fundamental disconnect between the two, which often renders it extremely difficult for workers to do the right work at the right time, with the right materials and the right equipment in the right place.

4.2.2 Connecting the last mile

While there is a general trend of increased Information and Communication Technology (ICT) spend in construction organisations, there is little evidence that the core construction processes are being improved or addressed (Dave et al., 2008). Research has shown over the years that production on-site, i.e. the core construction processes, has remained relatively untouched despite the availability of improved ICT tools elsewhere in the industry, and indeed productivity has remained largely stagnant (Teicholz, Goodrum, and Haas, 2001). The industry has failed to connect the 'last mile' of information flow, between the office and the site, effectively. As a result, even where companies have sophisticated Enterprise Resource Planning (ERP) systems, most construction crews still end up devising their own solutions to coordinate production in construction.

4.2.3 Distinction between operations and process

The need to spread risk in an industry that experiences frequent and large fluctuations in demand has led to high levels of fragmentation in the construction industry in most parts of the world. Most projects are managed by general contractors who employ multiple subcontractors to perform most of the work in each project. More than 90% of European construction companies employ between 1 and 9 workers (Census, 2016). As a result, production planning and control is split among general contractors, subcontractors and construction crews. Overall responsibility for coordination still lies with the general contractor, who receives the designs and specifications along with the overall project plan, and then divides these into smaller work packages and allocates them to subcontractors. However, as a result, subcontractors and site crews have a far narrower view, restricted mostly to their own operations and tasks. This creates further disconnects within the already isolated construction site.

The information disconnect is exacerbated by the conflicting economic incentives of general contractors and subcontractors under the standard commercial terms of construction contracts, in which general contractors purchase products rather than services from subcontractors (i.e. they measure and pay for the results of the work, not for the work resource). General contractors prioritise the flow of the product, whereas subcontractors prioritise their productivity, and these economic imperatives are often at odds with one another. In particular, where construction plans are unreliable, subcontractors protect themselves by allowing large buffers of work to accumulate before assigning their resources.

From a Lean thinking point of view, this means that the *processes* defined by Shingo (the flow of products) and the *operations* (the flow of work crews in the construction context) (Shingo and Dillon, 1989) are managed by separate organisations with conflicting motives and unequal access to information. Under these conditions, it is extremely difficult to achieve good workflow in construction.[1]

4.2.4 Synergies of Lean and BIM

Given the development of the Last Planner® System (Ballard, 2000) and other tools rooted in Lean thinking as a response to the neglect of production management in the field, on the one hand, and the development of BIM in response to the shortcomings of drawings for communicating product information on the other hand, researchers began to identify significant synergies between these two hitherto independent developments.

Sacks et al. (2010a) first laid out the interactions between them, identifying 56 distinct points of intersection between Lean Construction principles and BIM functionalities. Among those, some 45 are specifically relevant for production control, which are shown in Figure 4.1.

Researchers have also carried out analyses of construction projects in which they identified synergistic effects of parallel BIM and Lean Construction implementation. In one such study, Gerber, Becerik-Gerber, and Kunz (2010) identified a number of positive interactions, and concluded that the lack of prior intent on the part of the project team to merge Lean and BIM did not diminish the value they gained. From a research perspective, they felt that this underscored the validity of their findings.

4.3 Requirements for Lean and BIM production control systems

Koskela's proposition of the TFV theory of production in construction (Koskela, 1992, 2000) and Ballard's development of the LPS (Ballard, 2000) form the foundation of Lean Construction practice. Guiding principles for production management systems in construction can be distilled from these contributions. These are some of the key ideas:

- Support smooth workflow through coordination;
- Reduce variability through sound assignments;
- Support pull techniques;
- Measure and monitor to support continuous process improvement;

	Lean Principles	Reduce Variability		Reduce cycle times		Reduce batch sizes	Increase flexibility		Select an appropriate production control approach		Standardise	Institute continuous improvement	Use visual management		Design the production system for flow and value				Verify and Validate	Go and see for yourself	Decide by consensus consider all options
BIM Functionality		Improve quality (reduce product variability)	Improve flow (reduce production variability)	Reduce production and supply cycle durations	Reduce inventory	Strive for single piece flow	Reduce changeover/ setup times	Use multi-skilled teams	Use pull flow systems	Level production			Visualize production methods	Visualize production process	Simplify	Use parallel processing	Use reliable technology	Ensure capability			
Visualization of form	Aesthetic and functional evaluation	1,2													3				5	6	4
Automated generation of drawings and documents		11		22	(-52)	53											54	54			
Collaboration in design and construction	Multi-user simultaneous editing of a single discipline model			23						36						36					
	Multi-user simultaneous viewing of merged or separate multi-discipline models	2,13		24				33											46		49
Rapid generation and evaluation of multiple construction plan alternatives	Automated generation of construction tasks	14		25	(-29)		31								(-41)						
	Discrete event simulation		15	25	(-29)					37					(-41)				47		
	4D visualization of construction schedules	2	40	25	(-29)						17		40	40		40			47		49
Online/electronic object-based communication	Visualizations of process status		29	26	30	30			34					34			(-42)		47	48	
	Online communication of product and process information	18		26	30	30			34		38		38	34			(-42)				49
	Computer-controlled fabrication	19		27			32										(-42)				
	Integration with project partner (supply chain) databases		20	28					35								(-42)				
	Provision of context for status data collection on site/off site		21		30	30			34			39					(-42)		47	48	

Figure 4.1 Interactions between Lean Construction and BIM with relevance to production control. Numbers in parentheses indicate potentially negative interactions. Numbers in parentheses indicate potentially negative interactions.

- Maintain a buffer of unassigned tasks;
- Support distributed decision making through collaboration.

Table 4.1 provides an overview of the aspects of construction planning and control at four levels (it extends an earlier definition from Sacks, Radosavljevic, and Barak, 2010b), which outlined aspects for the first three levels. To be effective and useful, a production planning and control IT system must cater for all four levels.

In addition, a Lean and BIM production control system must address the two major information strands of production systems:

- The process – planning, simulation, visualisation, monitoring and control of production system;
- The product – definition, representation and visualisation of the built product.

Addressing the product information aspects requires delivery of information in a BIM model at LOD 400, i.e. with complete construction details, after resolution of all system

Table 4.1 Aspects of construction planning and control at four levels of resolution

Planning/ control level *Aspects*	*Master planning*	*Lookahead planning and make-ready process*	*Weekly planning and control*	*Daily production control*
Resolution	Milestones	Work Packages	Tasks	Tasks
Goal	Work structuring; subcontracting	Make ready, constraint release: long lead-time material and information supply	Short-term resource, space and material coordination	Task go/no go decisions; pull of materials
Responsible function	Construction manager	Construction manager in consultation with trade managers	Trade team leaders and managers	Trade team leaders and managers
Approach to define planning units	Top-down division: high-level work breakdown structure	Top-down division of milestones into work packages	Bottom-up aggregation of parts into tasks	n/a
Tools	Contract terms, CPM and process optimisation	Location-based planning; constraint release; and line of balance scheduling	Pull priority; negotiation	Production status Andon displays; mobile device feedback
Measures	Earned-value measures	Work package maturity index (degree of constraint release)	Maturity index; construction flow index (CFI); percent plan complete	Percent task complete; crew waiting time
Task dependencies	Contractual, CPM technological constraints	Technological constraints (such as FS); resource levelling and space conflict resolution	Flexible working relationships	Current conditions: resource and shared equipment coordination

coordination issues (LOD refers to the Level of Development of the construction details in a BIM model; LOD 400 is the level at which elements are modelled with sufficient detail and accuracy for fabrication, see BIMFORUM, 2017). These capabilities are well understood and are fairly well developed in many field BIM software applications (Sacks et al., 2018). Addressing the process information aspects encompasses support for production system design, integration of resources (such as procurement, inventory, personnel, etc.), communication between personnel at all levels and delivery of accurate process status information to facilitate better decision making.

The following paragraphs detail some of the key requirements for Lean planning and control, and discuss information delivery requirements and opportunities for them. The list is based in part on Sacks, Treckmann, and Rozenfeld (2009).

Plan for stable work – plan project activities effectively, predicting problems and safety issues. Activities and their constraints (space, materials, equipment, crews, information, safety, etc.) must be modelled and associated with building product objects, so that they can be used for filtering in the make-ready process. 4D CAD is a useful technique for planning tasks with consideration of space resources and temporary facilities (Akinci et al., 2002; McKinney and Fischer, 1998).

Communicate standardised processes to workers. Standardisation of construction methods is key to achieving predictability and stability of plans and is crucial for the continuous improvement efforts. Production details modelled in BIM can be animated and displayed in the field.

Monitor production and record performance benchmarks for improvement experiments. Visual interfaces for workers to input production data, and automated interfaces for integrating monitoring data from sensors of various types are essential.

Make process transparent to all. Electronic status boards and displays on mobile devices can show the current status of tasks and of constraints (such as material stores and equipment locations).

Filter work packages for maturity to ensure stability. The LPS (Ballard, 2000) can be supported by visual status charts that show the readiness of equipment, materials, space, information, etc. BIM can support dynamic safety-conscious work filtering using the CHASTE model (Rozenfeld, Sacks, and Rosenfeld, 2010).

Pull product information for work packages when needed. Provide workers with the most up-to-date product information by ensuring that it is pulled from the BIM model online only when needed. This should obviate the need for version control for product information (Sacks, Korb, and Barak, 2017, Chapter 11).

Pull detailing and fabrication/assembly of building system components according to short-term planning to match production flow. Signalling upstream that product information for subsequent production stages will be needed within some time step can be automated in a system where the progress of work is monitored continuously. In this way, the product objects can be detailed to LOD 400 only as and when needed. Where fabrication off-site is needed, this can also be pulled using signals from the production control system.

Provide pull flow signals to regulate workflow. Work in progress in construction is not visible, as it might be on a factory floor in a manufacturing facility, so explicit directives are needed to pull work. Online pull flow signals are needed and must be communicated to teams. Visual labels like those implemented in the KanBIM prototype (Gurevich and Sacks, 2013) can provide such signals.

Just-in-time delivery of material and parts. BIM can provide accurate and automated preparation of bills of materials for JIT delivery (Chau, Anson, and Zhang, 2004) and colour-coded model view interfaces for communicating clear pull signals.

Pull management attention to where it is needed, to release bottlenecks or facilitate flow. Visual production flow monitors and safety risk levels can be used to attract management attention to nodes of instability or danger, e.g. use of Andon lights (Pereira, 1998).

Empower managers and crew leaders to respond flexibly to change. Design or process changes can be disruptive. Visual planning interfaces can enable managers to adapt construction plans/material and resource orders/work assignments flexibly and responsibly, by providing situational awareness and the ability to explore 'what-if' planning scenarios through simulation.

From the above discussion, it emerges that there is a clear opportunity, and indeed the need, for software systems that support the full Lean production management workflow. Such a system should not only support the lookahead/make-ready process and the weekly work planning function (the Last Planners), but it should also support production management on the job site itself, i.e. it should support the trade crews in their day-to-day work, as they make workflow decisions.

4.4 BIM software with construction management functionality

There are several commercial BIM systems that support construction management/production management activities, including, but not limited to, the following:

a) Integration of the various discipline models (such as architectural, structural and MEP) in a federated BIM model to gain a better understanding of the design;
b) Clash detection – analysing clashes between the different building systems for constructability analysis;
c) Construction sequencing – linking of the project plan (mostly at master plan level) to the model and creating a simulation of the project (also known as 4D CAD);
d) Design review and communication, including marking-up of design for clarification;
e) Visualisation of design during construction (i.e. viewing product information in the field);
f) Quantity and cost take-off;
g) Constraints analysis;
h) Evaluation of 'what-if' scenarios;
i) Visual tracking of construction progress.

To determine to what degree existing BIM tools offer construction management functionality, six commercially available tools were selected and assessed. The six were *Autodesk Navisworks, Tekla BIMSight, Solibri Model Checker, Synchro, Vico Office* and *Tekla Construction Management.* All of these systems offer the first four features listed above (a to d), and some offer additional features.

Table 4.2 provides a comparison of the systems from the viewpoint of their construction management capabilities. It is clear that, in terms of detailed production planning, scheduling and control capabilities, the capabilities of these tools remain limited. Barring Vico and Synchro, none of them offer detailed production planning and resource linkage

Table 4.2 Assessment of construction management functions of BIM tools

BIM Tools Features	*Autodesk Navisworks*	*Synchro*	*Vico Office*	*Tekla Construction Manager*	*Tekla BIMSight*	*Solibri Model Checker*	*Autodesk 360 Plan*
Visualisation of federated model	✓	✓	✓	✓	✓	✓	
4D simulation	✓	✓	✓	✓		✓	
Visualisation of temporary works		✓	✓				
Clash detection	✓	✓	✓	✓	✓	✓	
Resource loading		✓					
Constraints analysis			✓				✓
Visual task information							✓

to the production plans, and most offer 4D planning functionality only at the master plan level. Both Synchro and Vico demonstrate some capability to provide detailed production planning by enabling the development of detailed plans and links to the model. However, neither of them supports 'pull' production management. Many users find Vico a difficult and complex tool to master and use, requiring extensive training and experience, particularly where they do not have previous training in location-based planning. It is, therefore, considered appropriate for the lookahead planning phase, and less so for day-to-day production control.

Furthermore, none of the systems provide detailed constraints analysis at task level or assignment of responsibility and real-time task status updates to enable accurate production planning. One of the cornerstone principles of pull planning is to carry out detailed constraints analysis at the lookahead and weekly planning level and to assign responsibility to task/trade leaders/managers to ensure the constraints are removed before the tasks are considered for production (Ballard, 2000). However, such a level of production planning is not yet available in any of the commercially available systems.

As discussed, a range of information sources must be integrated in the production management systems in real time to enable accurate decision-making. Up-to-date status information concerning material procurement and delivery, equipment status and availability, labour availability, space occupation, etc., must be made available to users in Andon-type display interfaces. None of the BIM systems reviewed support dynamic integration of such information sources, nor the display of the production system status. Thus, whereas these commercial software tools offer significantly better construction production planning and control functionality than traditional 2D CAD or even BIM platforms, there is still a significant gap in functionality and strong potential for the development of production planning and control systems that better fulfil the requirements.

4.5 Research and development of Lean and BIM production control systems

Following the realisation of the potential for synergistic implementation of Lean Construction and BIM, first examples emerged around 2005 (Eastman et al., 2008; Khanzode et al., 2006), and experimentation began in academia and in industry to develop and test new software tools. At Stanford University, the work focused on developing a 'Virtual Design and Construction' approach in which 3D models and 4D CAD technology were used and enhanced to support Lean Construction implementation (Khanzode et al., 2006). At Technion, the value of visual task planning and control using BIM models as the vehicle for the information was explored in interface mock-ups, such as the one shown in Figure 4.2 (Sacks, Treckmann, and Rozenfeld, 2009). This was the seed for the KanBIM prototype, which was designed in collaboration with Tekla Oy (a Finnish BIM software company) and the University of Reading and later implemented and tested at the Technion. Shortly thereafter, research on the VisiLean research prototype began at Salford University (Dave, Boddy, and Koskela, 2011).

The following sections outline the development of Lean and BIM production management tools for construction, using KanBIM and VisiLean as examples.

4.5.1 Experimental prototypes

KanBIM was a prototype production workflow management software developed to facilitate experimentation on construction sites which aimed to explore the impact of up-to-date process status information on the workflow itself. Figure 4.3 illustrates the software system architecture (full details can be found in Sacks et al., 2013). The primary database records the product and the process information. At its heart is a federated BIM model with the building elements from all the fabrication and construction disciplines at a construction level of detail. The process aspects are instantiated by applying construction methods, aggregating objects for association with work packages and generating model objects to represent temporary equipment. The work status information is updated as users in the field (primarily trade crew managers) report the as-made status of work packages using the interface provided. Although the system architecture defined interfaces for planning at different levels of detail of the LPS, in the experimental prototype only the team leaders' interface was implemented (Figure 4.4).

The novelty of the KanBIM concept was that the process status was presented together with the product information to provide an accurate, up-to-date and accessible picture of the project's status. Project personnel could filter the information on production status, supply chains, quality control, equipment availability, space occupation and design information maturity to enhance their situational awareness with minimal effort. The guiding ideas were that:

- improved situational awareness would lead to decision-making regarding work to be done that would avoid the waste of waiting and of rework;
- materials, design information, equipment and work itself could be pulled (rather than pushed) by the people using the KanBIM interface directly from the field.

KanBIM also offered two novel workflow measurement indices that were designed to encapsulate data from various streams and present it in a format that would communicate the production status to trade crew leaders and construction engineers at a glance. The maturity index (MI) was compiled from the status measures of the various constraints that apply to any pending task. When all of the constraints for a given task had been resolved, a traffic light symbol on the KanBIM task label changed from 'no entry' to green. When a trade crew leader reported start, interruption or completion of a task, the labels showed 'people at work', 'stop' and 'tick' traffic sign symbols, respectively. The pull

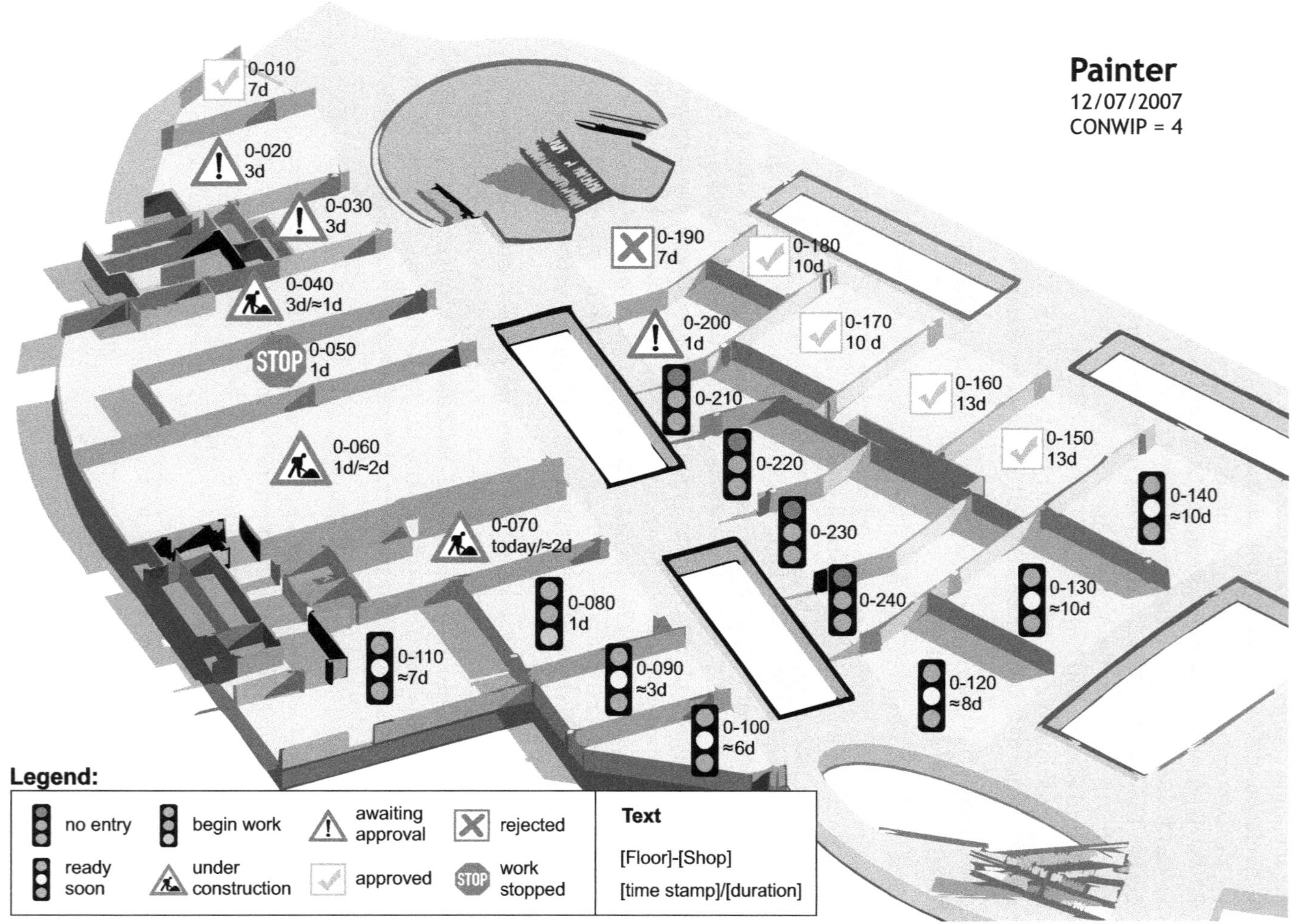

Figure 4.2 Proposed 3D visualisation of past, present and future work status for a trade (Sacks, Treckmann, and Rozenfeld, 2009)

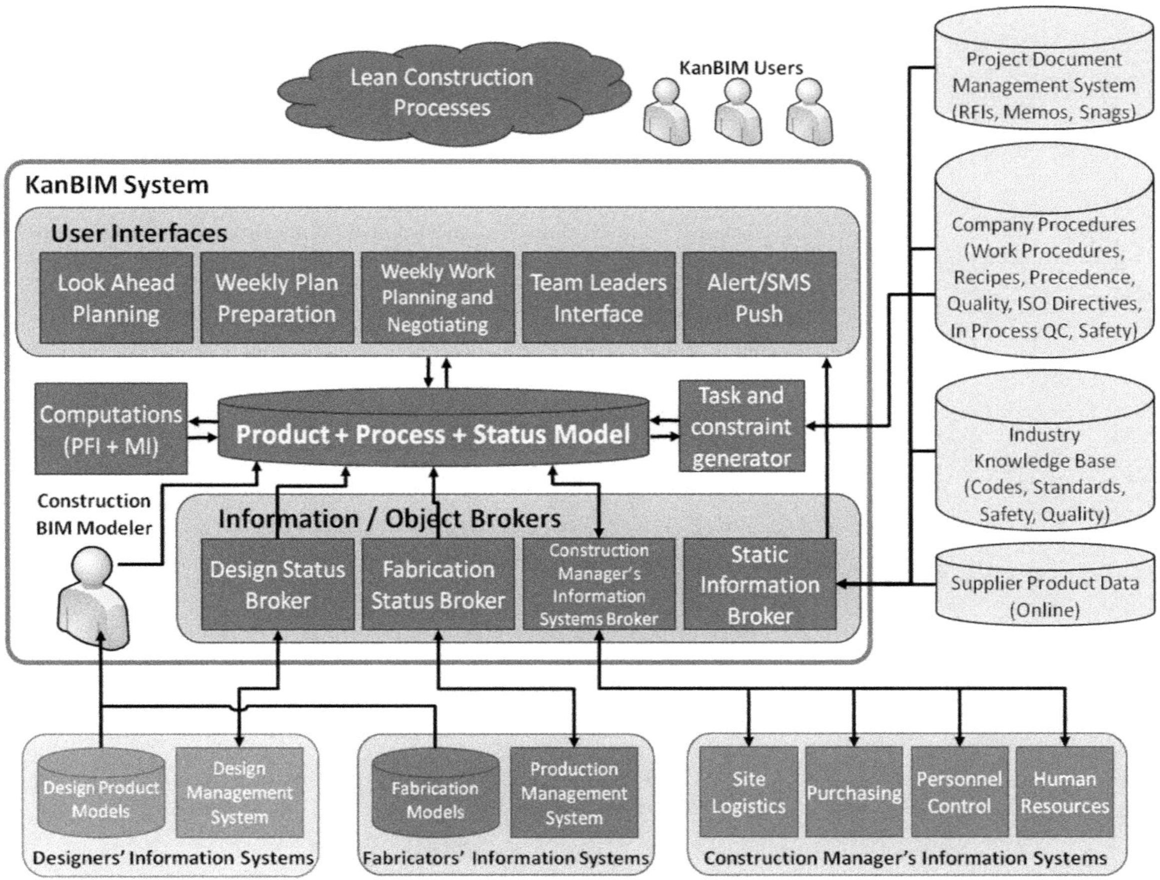

Figure 4.3 KanBIM system architecture chart (Sacks et al., 2013)

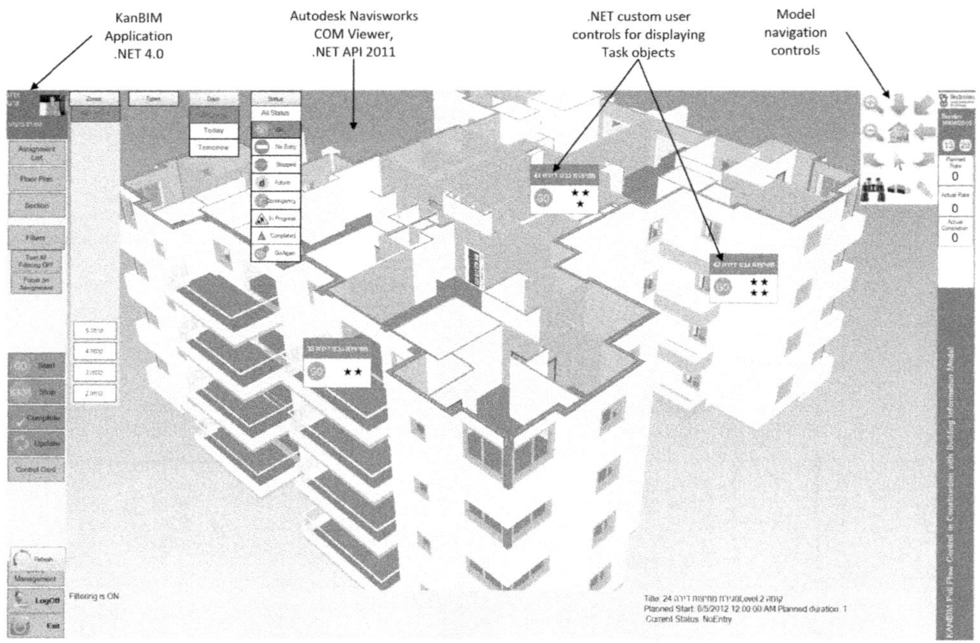

Figure 4.4 KanBIM large-format touchscreen interface, showing the .NET 4.0 application, the embedded Navisworks COM viewer, the custom task controls overlaid on the model and the model navigation controls

flow index represented the priority with which a task should be selected for execution from the set of mature tasks. This was a measure of the degree to which the constraints of subsequent tasks had been resolved. A task with a higher pull priority index carried a greater probability that the product it worked on could be completed, thus reducing WIP.

The KanBIM prototype was tested in the field (Sacks et al., 2013) and in the laboratory, using a virtual construction site experimental setup (Gurevich and Sacks, 2013). The observations pointed to positive potential effects of the system on the ability of site personnel to visualise the process itself, with a reduction of wasted time spent 'looking' for work. The site superintendent in the field tests summarised his views with the claim that the system would enable him to 'essentially double the scope of work that he could reasonably supervise'. The experiments also identified specific shortcomings, all of which pointed to valuable guidance for development of commercial tools. Among the recommendations:

- the reliability and completeness of the data in the system is a critical determinant of its usefulness. Information must be drawn in real time from supply chain monitoring systems, the BIM model must be continuously updated as design progresses, and the status of the work itself must be fed back into the system continuously;
- the need for continuous real-time update means that users must obtain sufficient benefit from the system to motivate them to also report status. Considering the potential unreliability of human reporting, it also suggests that tracking design, supply chain and production status should be automated as far as possible;
- the system must be easily accessible to all who use it at all times. This requires provision of the system on mobile devices with appropriate interfaces, including use in the field.

4.5.2 Commercial software development

The first commercial software to offer Last Planner-style workflow control was called 'Ourplan' (Ourplan, 2013). Ourplan was developed by a team at the DPR Construction company in the US before being spun off to a dedicated IT company. It was used in numerous DPR projects and the projects of other construction companies. The company was later acquired by Autodesk, after which the software was entirely rewritten and rebranded as 'BIM 360 Plan'. There are also several software applications that facilitate LPS-style planning to some degree or another, but neither they nor Ourplan use BIM models.

Most of the early LPS tools suffered a rather prosaic barrier to introduction into the market: lack of knowledge and skills in Lean Construction practices in general, and of the LPS in particular. As a result, company representatives were forced to invest disproportionate time in coaching new users.

VisiLean is a 'software as a service' cloud solution for production planning and control in construction projects (VisiLean, n.d.). VisiLean uses BIM models directly, for visualising both product and process information. VisiLean research was contemporaneous to KanBIM research and there were influences from KanBIM research interface mock-ups; as such, its design and implementation were informed by the principles derived from theory (outlined in the section: Requirements for Lean and BIM production control systems, above), by the rich experience gained from experimentation with research prototypes (outlined in the section: Experimental prototypes), and by feedback from practitioners who tested it during its alpha phase. Table 4.3 lists and discusses some of the key requirements that guide its ongoing development.

Table 4.3 VisiLean system requirements

Requirement	*Discussion*
Process and product visualisation at the 'coal face'	This is one of the most significant and overlapping requirements that emerges from past research (and from prior case studies and exploration of the problem area). This means that the information regarding the planning and scheduling, along with the relevant design information (that forms the product model), should be made available to workers (construction teams) on-site.
Supporting constraints analysis and management	This is an essential requirement for a production management system and an aspect that is absent from existing tools (see section 4 above). The system should allow the teams to identify, analyse and assign constraints. Once assigned, the system should also help users track the status of the constraints linked to the tasks. Finally, the efficiency of constraint removal actions should be assessed and reported.
Supporting collaboration, work negotiation and communication between the trade crews	Addressing people issues, building trust, improving coordination and communication and securing commitment to the production plan are some of the critical issues that a new system must address. All previous research initiatives recognise this aspect and proposed collaboration as a key requirement for a production planning system. Collaboration spans the entire life cycle of the project, starting from Lean work structuring and continuing to Lookahead planning, weekly commitment planning and daily execution, feedback and coordination (including start-stop signals). Another key aspect to recognise is that the production management process is highly people centric and the goal of the computer system should be to support the collaboration rather than to automate the process. Many previous attempts have failed where the users have been alienated from the system due to high levels of automation or the complicated nature of the system.
Enable 'pull' flow control and plan stability	From a Lean perspective, 'pull' production management is a key to reducing variation, which is a major cause of waste and uncertainty in construction. All previous researchers have identified the need for the system to support a combination of 'push' and 'pull' techniques to maintain plan stability and reduce variation. As specified in section 3, the system should support the workflow starting from Master Planning and Phase Planning, going on to Lookahead planning, weekly work planning and, finally, to day-to-day execution. Throughout the latter three levels of resolution, the system enables users to generate a workable backlog of constraint-free tasks that can be selected based on their priorities to improve flow and reduce work in progress. At the same time, the system supports coordination of labour allocations by signalling to downstream crews when preceding tasks are completed.

Koskela's TFV conceptualisations of production comprise the theoretical foundation for the software. All three are necessary and a good system design should apply them in concert. Table 4.4 describes some of the specific features within VisiLean that address the requirements that stem from TFV theory.

VisiLean supports different levels of planning resolutions such as Phase, Lookahead and Weekly planning (in line with the LPS) and the tracking and execution support with the help of mobile apps. The direct integration with BIM helps to filter activities and visualise the plan and the model at the time of collaborative planning sessions. Figure 4.5 shows the VisiLean system with a filtered model and plan, in this case showing all completed activities.

Figure 4.6 shows the collaborative planning features of VisiLean. Each row represents either an individual worker/supervisor or an organisation (subcontractor, designer, main contractor, etc.). The tasks are shown as movable sticky notes with their different shades of grey showing the current status of the tasks. It is possible to model constraints and perform detailed constraints analysis on the tasks for lookahead and weekly planning. The tasks become ready for execution only when constraints have been tackled. It is also possible to import quantities or attach actual and planned quantity data with tasks. Also, it is possible to link any type of file such as images, documents or drawings with the tasks, which can then be pushed to the mobile apps in the field.

Table 4.4 VisiLean features addressing TFV concepts

VisiLean Features	*Transformation*	*Flow*	*Value*
Planning and scheduling	Task planning and scheduling in Phase, Lookahead and Weekly Planning (software) interfaces help to maintain consistency in task specification.	Managing the flow of resources through constraints analysis and management. Assigning constraints to tasks and the responsibility to manage them to workers and teams.	Reduction in making-do through improved performance of constraints removal process, leading to better performing planning and scheduling system leading to sounder tasks, less rework and better quality.
Task management	Assigning task completion responsibility at the Last Planner level to workers.	To make input flows visible through linking of constraints to tasks (and their status).	
Simultaneous visualisation of process and product	Visualisation of task information in process and model views, i.e. where and when the task is supposed to be executed.	Managing the flow of work between long-term, medium-term and short-term planning processes.	Reduction of confusion through joint appraisal of production plan in both process and product views, improving the quality of work and reducing risk of rework and delays due to misunderstandings.
Production control	Production control features of starting, stopping and completing tasks and their visualisation in both the model and process views.	Managing and visualising in-process flow between production tasks by visualisation of task statuses in both product and process views.	Visibility of upstream task completion to downstream stakeholders.

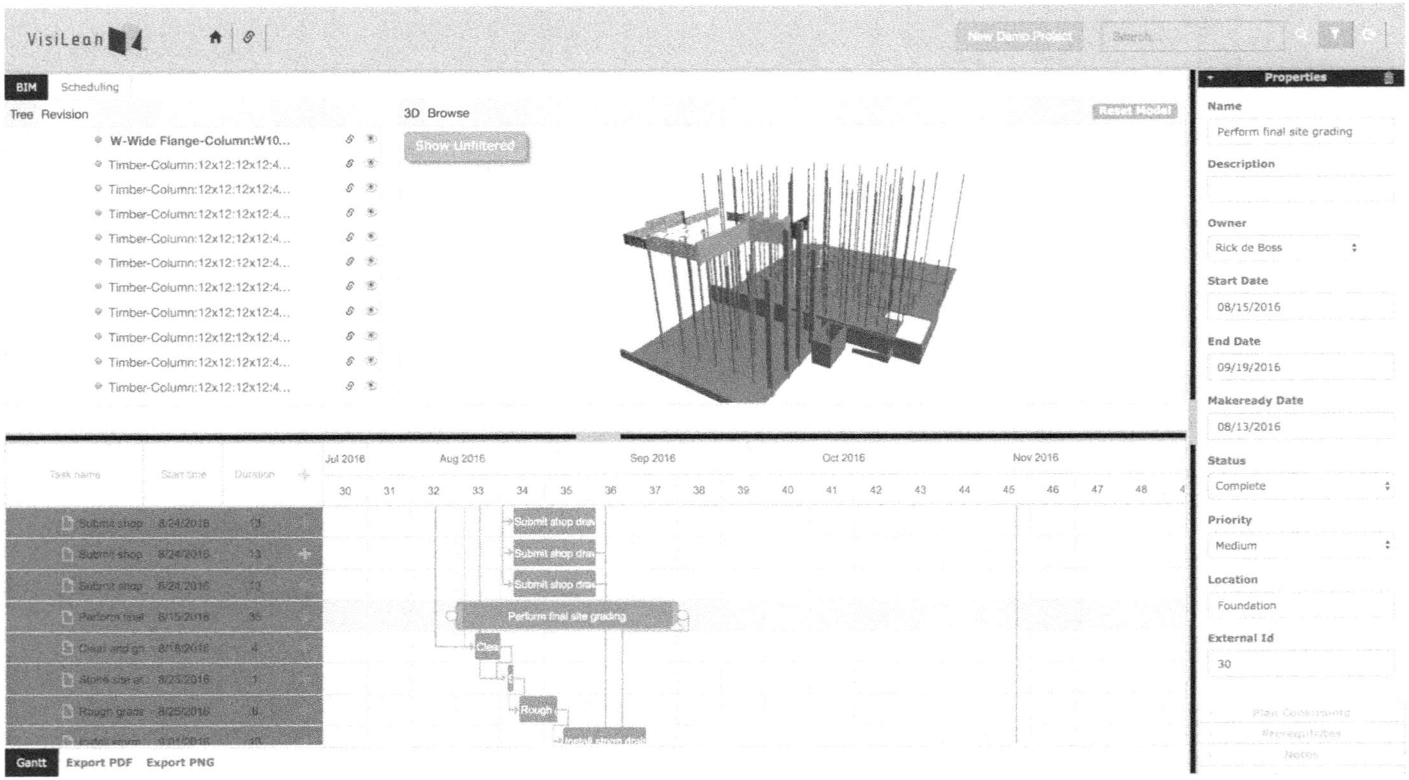

Figure 4.5 VisiLean model showing filtered activities in Plan and Model views

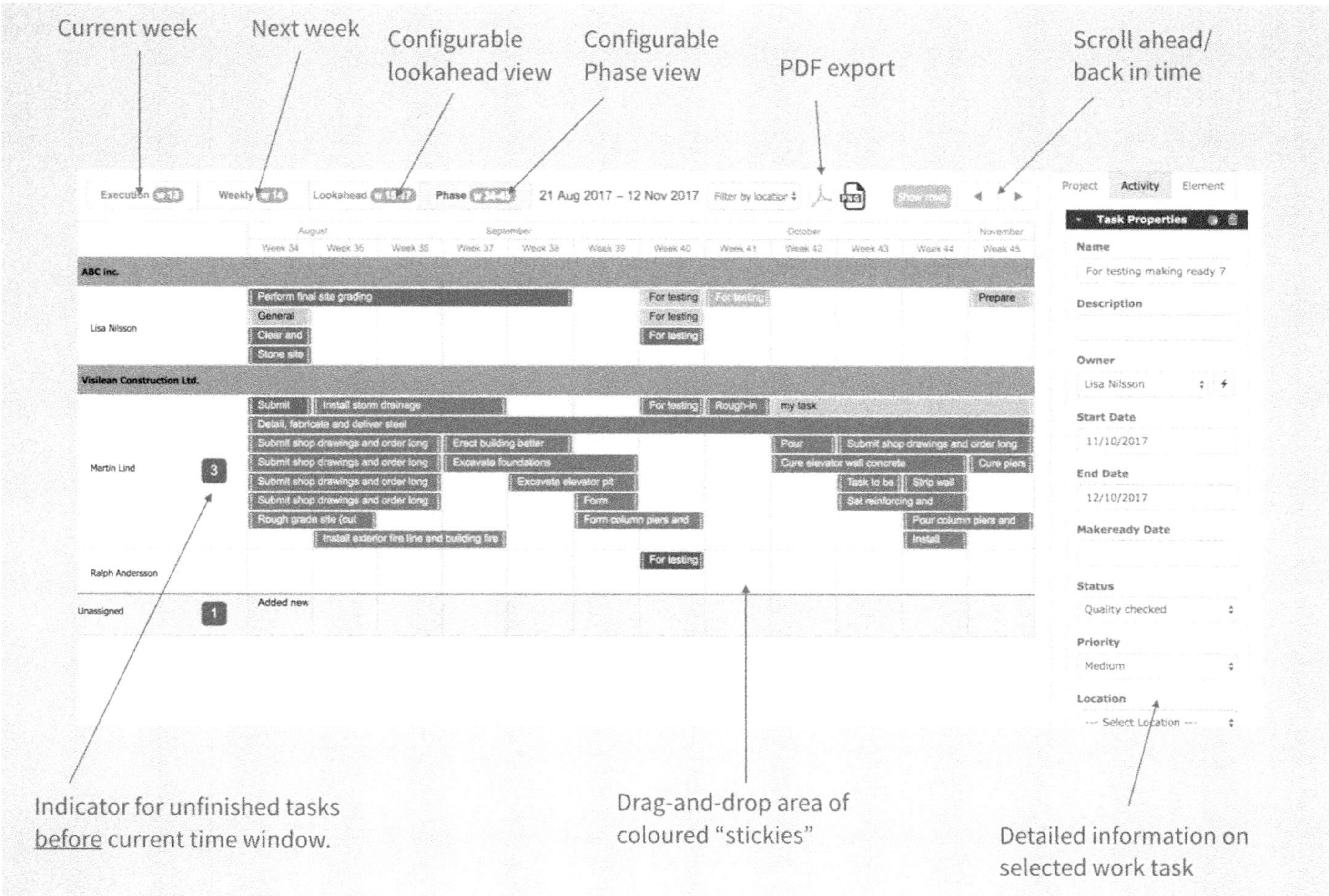

Figure 4.6 Collaborative planning features of VisiLean

4.6 Case studies of integrated Lean/BIM production planning and control

VisiLean has been implemented on a wide number of projects ranging from infrastructure and commercial to residential and in countries such as the UK, USA, Brazil, Finland and Sweden. The three case studies listed in Table 4.5, and detailed below, range from a small commercial project to a large mixed-use project. They illustrate the capabilities of a typical Lean and BIM production control system, highlighting advantages and disadvantages. Each case study was evaluated in a plus/delta analysis with project participants.

4.6.1 Project A

Project A was a supermarket building located in central Finland. The project was managed by a medium-sized construction company. The company had implemented BIM solutions prior to the case project and the main motivation for adopting VisiLean was to integrate their production planning and control workflow with the BIM models. The company had a sister company that produced precast concrete elements and acted as a supplier on most of its projects. A key requirement was to track the precast element supply chain visually from factory to installation.

Plus (advantages): The personnel responsible for VisiLean implementation were the site engineer and BIM manager, who looked after model management, plan updates, linking project plan to model, user management and training. Setup and training on the cloud-based software were straightforward.

The project used prefabricated components from the sister company, the premises of which were located approximately one hour's journey from the construction site. With VisiLean, it became possible to track individual shipments both in the task view as well as in the model view. Timely reporting of production status became much easier and the latency between planned and actual reports was reduced by a significant margin due to the app-based workflow. As a result, the effectiveness of weekly and lookahead planning meetings increased greatly.

Another major improvement was that the weekly collaborative planning meetings became more interactive as the model and up-to-date status information of all planned and completed tasks were readily available (see Figure 4.7).

Delta (areas for improvement): Users on the project recommended a few user interface improvements:

Table 4.5 VisiLean Case Studies

	Project A	*Project B*	*Project C*
Location	Pieksämäki, Finland	Espoo, Finland	Stockholm, Sweden
Nature of Project	Commercial/ Supermarket	Commercial/Retail	Residential
Size	€1.9 m	€220m	107 MSEK = €10m
Project Duration	8 months	2017–2019	05/2016–10/2017
Responsible VisiLean Coordinator	Site Engineer/BIM Manager	Project Manager, BIM Consultant	Site Manager, BIM Consultant

Figure 4.7 VisiLean's main interface with a model and a work programme from Project A

- A static guide/legend to show which task status corresponds to the respective colour would help the crew in visualising the information quickly;
- An audible notification in the app or a vibration signal would help notify workers when a task was assigned to him/her in VisiLean;
- Linking and visualising several project plans (e.g. groundworks, structural, finishing activities, etc.) and several BIM models in the software would enable more effective integration;
- Further, a 4D simulation at the weekly and lookahead level would help the crew plan better.

4.6.2 Project B

Project B was a shopping mall located in the city of Espoo in Finland. The project consisted of the complete redevelopment of the old shopping district and construction of a new mixed-use area including 36,000 square metres of retail premises, 550 new apartments and 1,400 parking spaces. The project began in 2017 and completion is planned for 2020. The project value is €220 million.

The main contractor began a process of thorough digitisation within the organisation. The company adopted a BIM workflow in which 4D scheduling and quantity take-off are supported by software such as Synchro and Vico Office. However, the company identified a gap in their solution stack for a site production management system where Lean processes and BIM can be integrated. VisiLean was selected to fill this gap and was implemented.

The team started with a small user group pilot of VisiLean with one subcontractor and five users. The project plan from Primavera P6 was imported into VisiLean, and an IFC model was imported. The BIM consultant played a key role as project coordinator during the initial pilot period, and this ensured a smooth rollout. The project had installed smartboards for planning purposes and all weekly coordination meetings were carried out using them.

The first subcontractor to be invited to use the system was the earthworks contractor. Once the master plan was imported, training was provided in how to create lookahead and weekly plans using the VisiLean system. The mobile app was also introduced to ensure constraint management and accurate tracking of tasks from the field (see representative Figures 4.8 and 4.9). Following a three-month pilot period, the company decided to fully roll out the VisiLean system across this project and three other subcontractors were invited to the system. At the time of writing, five organisations and 18 users were registered in the system.

The team used VisiLean during the weekly planning meetings, where each team would prepare their weekly plan using VisiLean and commit to the activities. The teams would also use VisiLean to continuously track the performance planning by reviewing progress on the previous week's tasks. The team members updated their tasks using the mobile apps. The BIM manager regularly used UAVs (Unmanned Air Vehicle) to take images of the construction and overlaid project drawings to evaluate progress side-by-side with VisiLean displays (Figure 4.10). Occasionally, 4D animation from other BIM systems was also used to compare progress against plans.

Plus (advantages): The team found the production tracking features supported by the mobile app to be useful and simple to operate. The app provided the workers continuous

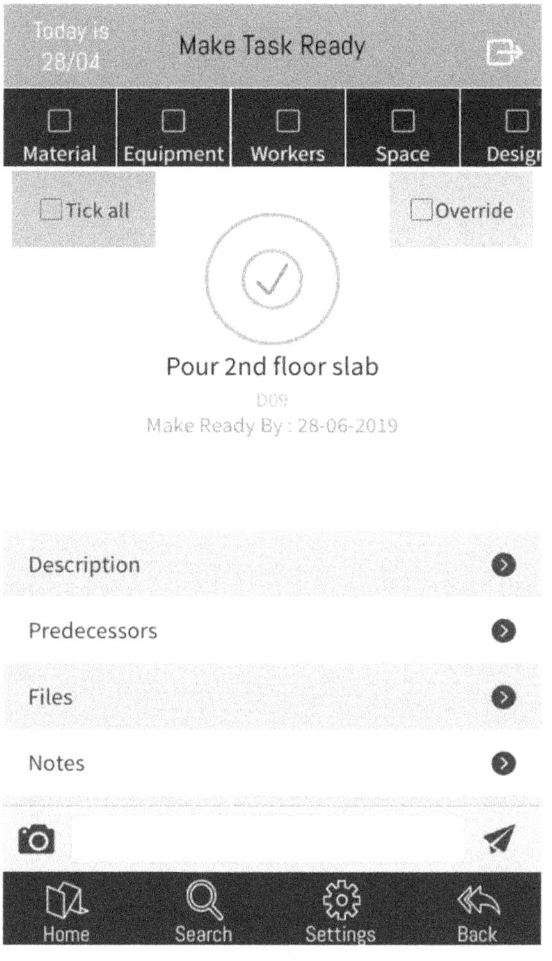

Figure 4.8 Constraint management in the mobile app

access to up-to-date information that defined the status of work. This made coordination between site crews more effective as they received notifications when their respective tasks became ready, reducing the need for informal/traditional communication methods. The latency between plans and actual progress tracking was reduced from four to six weeks to less than eight hours. The mobile app was perceived as a major advantage.

Delta (areas for improvement): The project used a 4D system for high-level project planning and visualisation using BIM. The BIM coordinator had already linked project plans with the model using this tool. However, it was not possible to import this information (model linked with project plan) to VisiLean. This meant that the BIM coordinator had to recreate these links. The 4D tool can export this information in the IFC format, and the capability to import this information in VisiLean is now planned.

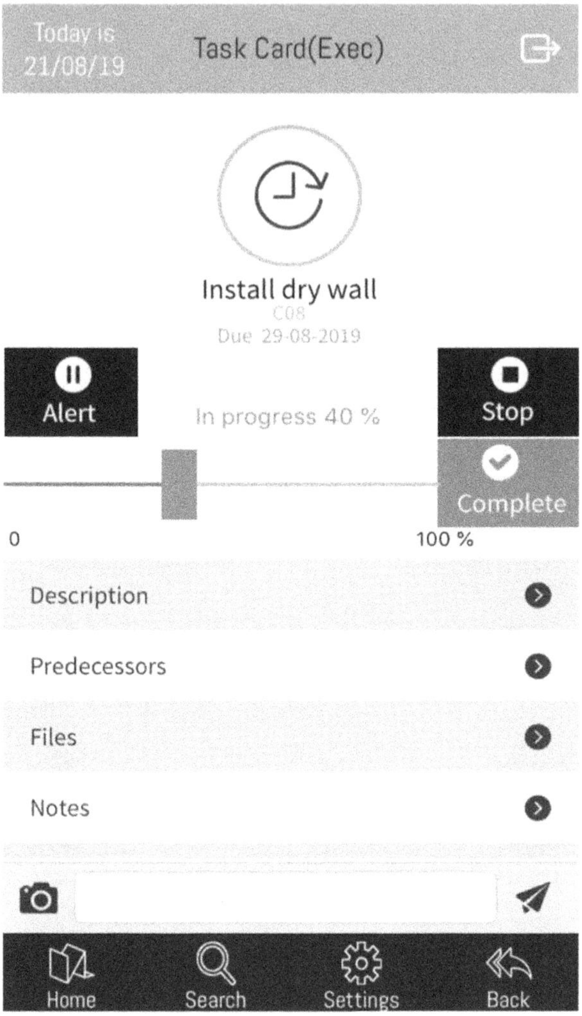

Figure 4.9 Mobile phone app for reporting task status

4.6.3 Project C

Project C was a residential construction project, a single building with 80 apartments in Vårberg, Stockholm, Sweden. VisiLean was implemented first in a four-week pilot study involving three students and the construction project team. Following the pilot period, the company decided to purchase the VisiLean software and continued to use the software for the duration of the project.

The first step was to import the construction plan and the BIM model, after which the BIM consultant linked model objects to plan activities. In the next step, the project manager was briefed about the process. Before VisiLean adoption, the project used Excel to track its plans, and this was the first functionality to be replicated in VisiLean. The project team members were then invited to the VisiLean system and briefed about the

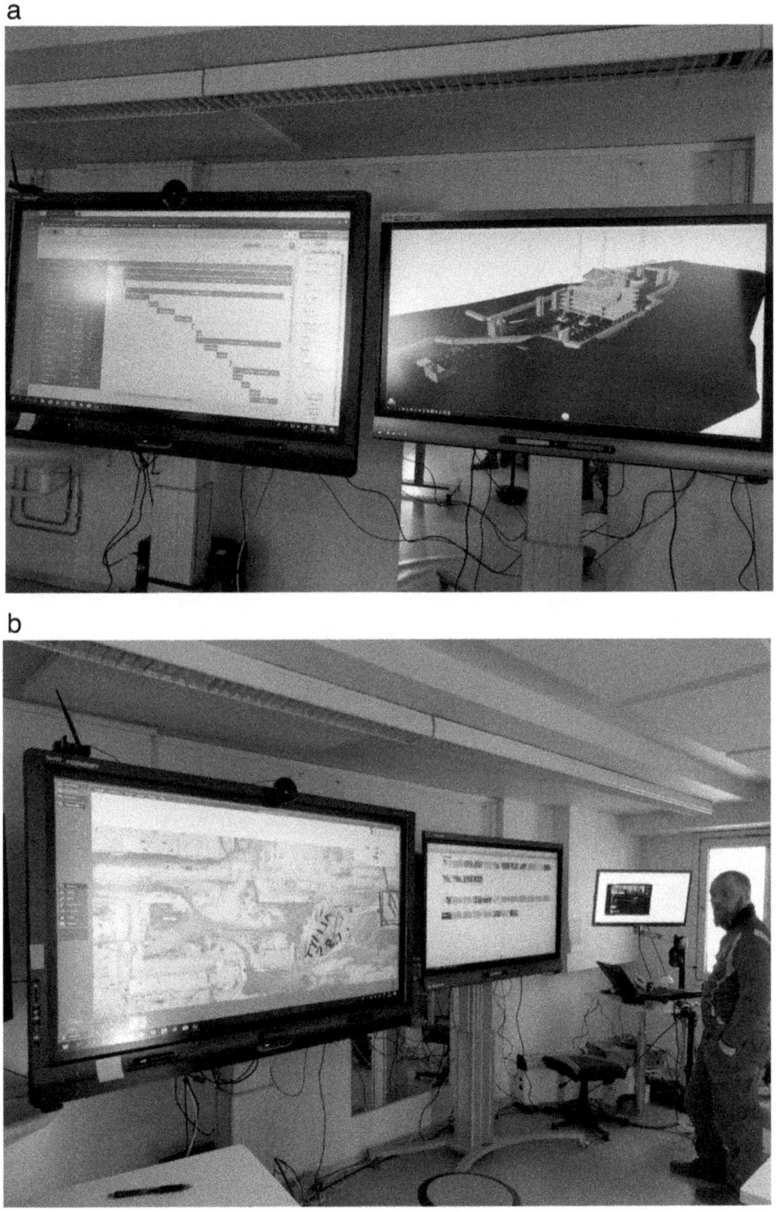

Figure 4.10 a) VisiLean displayed on a smartboard alongside a 4D animation system display; b) Drone images are reviewed alongside planning and coordination information displayed with VisiLean

process to be followed. In the following production planning sessions, the team used Excel and VisiLean side by side. After this handover period, Excel was entirely replaced by VisiLean as a collaborative planning tool, and all the subsequent planning sessions were organised within VisiLean.

In each week's planning session, the team would prepare their lookahead and weekly plans and then track them using the mobile apps. A dedicated screen was set up in the site office to provide 24/7 updates from the site. This can be likened to the visual Andon boards frequently used in Lean manufacturing. In this project, five organisations with 22 users collaborated using the system.

Unlike projects A and B, Project C had been using Excel for its lookahead and weekly planning (Figure 4.11). The Excel sheet was also used to track the project, with the site coordinator collecting all information through personal communication and feeding it

a

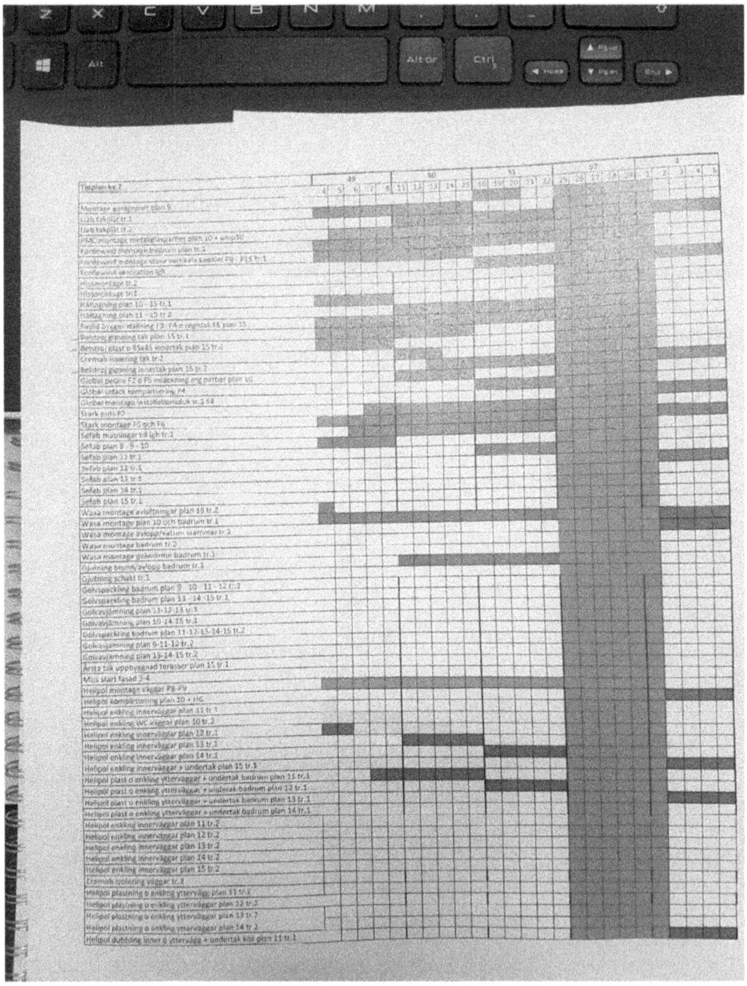

Figure 4.11 a) Before: Site work planned in a five-week lookahead; b) During adoption: Excel and VisiLean used in parallel, introduction of mobile app; c) After: VisiLean alone is used in planning sessions. Work updates are made through the mobile app, with project status always up to date

b

c

Figure 4.11 (Continued)

into the system. With VisiLean, the site team performed both lookahead and weekly planning directly using the collaborative planning features. No external systems were used to import the plan and all the tasks were created directly in VisiLean.

To accommodate the team's familiarity with Excel-style reports, a tabular report with a similar look and feel was developed within VisiLean. Using this report template, information was collated automatically, and printed hand-outs were provided to subcontractors and posted on the wall. Once the teams were familiar with the reporting app, the project management team decided to use VisiLean as an electronic 'Andon' tool where task statuses from each team were clearly visible 24/7. A new read-only view showing process status was made available to all users in different formats, on-site or remote. At a glance, any team member could visualise and respond to a changing production status in the field (Figure 4.12).

Plus (advantages): VisiLean proved to be particularly useful as a visual tool for the production phase. People on-site found it easy to use, providing the information they needed. The functionality for sending both the right production plans and the right design drawings to the right people at the right time was found to be very useful. Implementation required some training of the people involved and tying the service to the

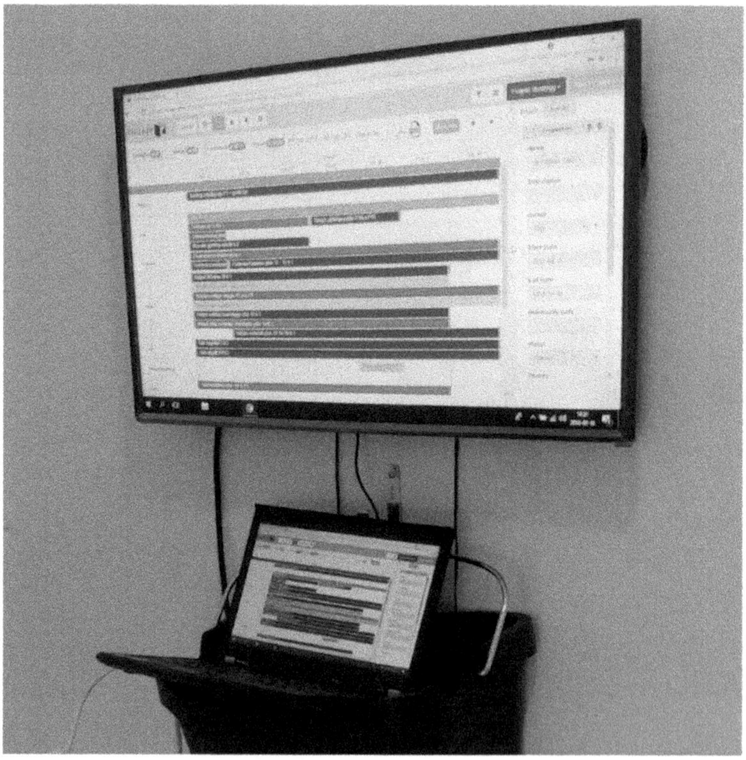

Figure 4.12 Andon display mode in the site office

process. The most significant difference between this case study and Projects A and B was the presence of a full-time coordinator who managed the information with Visi-Lean and supported the site crew whenever needed. This made the implementation smoother and more effective.

Delta (areas for improvement): The team requested additional functions for time planning and model visualisation. Specifically, the team asked for better focus of the information content in the BIM model to provide only the information needed by the workers and supervisors on-site. Instead of presenting the whole model, the workers should be provided with only the information required to complete the task at hand.

4.7 Conclusion

Subcontracting and the consequent fragmentation of construction project organisations can result in the neglect of production management. Traditional project planning and control methods focus on the product aspects, implementing the viewpoint of the general contractor, whereas trade subcontractors focus on the process aspects, primarily trying to optimise productivity. Lean Construction tools, such as the LPS, aim to refocus the attention of all partners on production flow.

In this chapter, we have explained how software tools that incorporate BIM models with process flow functionality based on Lean principles can support closer and more effective collaboration between those who focus on product and those who focus on process. The chapter outlined the requirements for a construction management system that supports TFV and the LPS, described the state of the art in Lean and BIM integration, and followed the path of research and development from early prototype applications, through testing in the laboratory and in the field, to the emergence of commercial systems. The three case study projects illustrate how the principles are beginning to be applied in practice.

Today, software applications are available that tackle planning resolutions starting from high-level milestones to detailed daily tasks and everything in between (phase, lookahead and weekly planning). While top-down integration of product definition and production planning has been effectively achieved, bottom-up information flow from the field remains a big challenge. BIM integration in the field is only beginning to emerge, and field-based resource management with feedback from IoT (Internet of Things) sensors is not yet available. Tools that have been developed and tested in research, such as the MI and the CFI, have not yet been implemented, partly as a result of the difficulty of measuring the status of the process. Mobile technologies are still used mostly in 'read-only' mode, and they may be exploited further to monitor process status in the future.

Of the 11 detailed requirements listed at the end of the section above titled Requirements for Lean and BIM production control systems, five are fulfilled (fully or in part) by the state-of-the-art commercial production planning and control software systems that fully integrate Lean and BIM:

- Monitor production and record performance benchmarks for improvement experiments;
- Make process transparent to all;
- Pull product information for work packages when needed;

- Provide pull flow signals to regulate workflow (although, note that regulating workflow is not automated, remaining the responsibility of the crews themselves);
- Pull management attention to where it is needed, to release bottlenecks or facilitate flow.

Other requirements cannot be thoroughly met, given the current lack of effective real-time monitoring of the status of the supply chain and of work in progress, which means that information about all constraints is unavailable. These include:

- Plan for stable work;
- Filter work packages for maturity to ensure stability;
- Pull detailing and fabrication/assembly of building system components according to short-term planning to match production flow;
- Just-in-time delivery of material and parts (although ERP systems deal with this aspect in planning, it is more difficult to control).

Tools exist for 'Communicating standardised processes to workers', but detailed process models and standardised ways of working are not yet tackled by many companies. Finally, 'Empower managers and crew leaders to respond flexibly to change' is a soft requirement, which one could argue is partially achieved by providing accurate information through Andon-type interfaces.

Clearly, then, much still remains to be done to improve Lean/BIM production control systems. Progress in capturing as-built data and providing in-process quality feedback is essential, but it is also at the forefront of Research and Development (R&D) in academic laboratories and in a wide array of construction tech startup companies. The potential of augmented reality interfaces to help communicate work processes and methods and to support decision making has not yet been realised. Applying Artificial Intelligence (AI) methods such as machine-learning to automate product-based planning, planning optimisation and adjustment of short-term plans to optimise production flow, is still in its infancy but holds great promise.

In addition to the limitations posed by the current state of development of the tools, Lean and BIM integration for production planning and control also faces some practical hurdles. Since the software tools support people who execute processes, it is critical to have the right processes in place. Where people are not familiar with procedures such as the Last Planner System and its underlying principles, they may find it difficult to use the tools effectively. Also, any gaps in the information that a system provides may leave crews frustrated. For example, if a model is incomplete or not developed to a reasonable LOD for execution, or if planning information is not updated regularly, the tools can become a burden to use. Further research and development work is needed to tackle these issues to ensure smooth deployment of Lean and BIM in the field.

Note

1 Sacks (2016) defined 15 conditions for good workflow in construction. They include project, process, operations and feeder conditions that cover aspects such as minimisation of cycle time, batch size and work-in-process inventory of projects and of locations, balanced work for crews, minimal re-entry flow and stable production rates.

References

Akinci, B., Fischer, M., Levitt, R., and Carlson, R. (2002) 'Formalization and Automation of Time-Space Conflict Analysis', *Journal of Computing in Civil Engineering*, 16, pp. 124–134.

Ballard, G. (2000) *The Last Planner System of Production Control*. PhD Thesis, The University of Birmingham, Birmingham, UK. Available at: http://leanconstruction.org.uk/media/docs/ballard2000-dissertation.pdf.

Ballard, G. and Howell, G. (1998) 'Shielding Production: Essential Step in Production Control', *Journal of Construction Engineering and Management*, 124(1), pp. 11–17.

BIMFORUM. (2017) *Level of Development Specification Guide*. Reinhardt, J., Bedrick, J., (chairs). BIM Forum, AGC of America, Arlington, VA.

Census. (2016) *Construction: Summary Series: General Summary: Employment Size Class by Subsectors and Industries for U.S. and States: 2012*. US Census Bureau, Washington, DC.

Chau, K.W., Anson, M. and Zhang, J.P. (2004) 'Four-Dimensional Visualization of Construction Scheduling and Site Utilization', *Journal of Construction Engineering and Management*, 130(4), pp. 598–606.

Chua, D.K.H., Nguyen, T.Q. and Yeoh, K.W. (2013) 'Automated Construction Sequencing and Scheduling from Functional Requirements', *Automation in Construction*, 35, pp. 79–88.

Dave, B., Boddy, S. and Koskela, L. (2011) 'Visilean: Designing a Production Management System with Lean and BIM', *Proc. 19th Annual Conference of the International Group for Lean Construction*, Lima, Peru. Available at: www.iglc.net.

Dave, B., Koskela, L., Kagioglou, M. and Bertelsen, S. (2008) 'A Critical Look at Integrating People, Process and Information Systems Within the Construction Sector', *Proc. 16th Annual Conference of the International Group for Lean Construction*, Manchester, UK. Available at: www.iglc.net.

Eastman, C.M., Teicholz, P., Sacks, R. and Liston, K. (2008) *BIM Handbook: A Guide to Building Information Modeling for Owners, Managers, Architects, Engineers, Contractors, and Fabricators*. John Wiley & Sons, Hoboken, NJ, USA.

Emdanat, S., Linnik, M. and Christian, D. (2016) 'A Framework for Integrating Takt Planning, Last Planner System and Labor Tracking', *Proc. 24th Annual Conference of the International Group for Lean Construction*, Boston, MA, USA. Available at: www.iglc.net.

Futcher, K. (2001) 'User Survey on a WAN Portfolio MIS Used for Portfolio/Project management in Hong Kong', *Construction Informatics Digital Library*, pp. 44-1–44-14. Available at: http://itc.scix.net/.

Gerber, D.J., Becerik-Gerber, B. and Kunz, A. (2010) 'Building Information Modeling and Lean Construction: Technology, Methodology and Advances from Practice', *Proc. 18th Annual Conference of the International Group for Lean Construction*, Haifa, Israel. Available at: www.iglc.net.

Gurevich, U. and Sacks, R. (2013) 'Examination of the Effects of a KanBIM Production Control System on Subcontractors' Task Selections in Interior Works', *Automation in Construction*, 37, pp. 81–87.

Horman, M. and Kenley, R. (2005) 'Quantifying Levels of Wasted Time in Construction with Meta-Analysis', *Journal of Construction Engineering and Management*, 131(1), pp. 52–61.

Howell, G., Laufer, A. and Ballard, G. (1993) 'Interaction between Subcycles: One Key to Improved Methods', *Journal of Construction. Engineering and Management*, 119(4), pp. 714–728.

Jongeling, R. and Olofsson, T. (2007) 'A Method for Planning of Work-Flow by Combined use of Location-Based Scheduling and 4D CAD', *Automation in Construction*, 16(2), pp. 189–198.

Khanzode, A., Fischer, M., Reed, D. and Ballard, G. (2006) *A Guide to Applying the Principles of Virtual Design & Construction (VDC) to the Lean Project Delivery Process*. CIFE, Stanford University, Palo Alto, CA, USA.

Koch, C. (2005) 'Failures in combined knowledge and material supply chains', *Proc. International Engineering Management Conference, IEEE*, pp. 818–821.

Koskela, L. (1992) *Application of the New Production Philosophy to Construction*. Center for Integrated Facility Engineering, Department of Civil Engineering, Stanford University. Technical Report number: 72.

Koskela, L. (2000) *An Exploration Towards a Production Theory and its Application to Construction*. PhD thesis, Technical Research Centre of Finland, VTT, Helsinki.

Koskela, L. and Kazi, A.S. (2003) 'Information Technology in Construction: How to Realise the Benefits?', *Socio-Technical and Human Cognition Elements of Information Systems*, E.C. Steve Clarke, ed., Idea Group Inc. (IGI), Hershey, PA, USA, p. 295.

Koskela, L.J. and Howell, G. (2001) 'Reforming Project Management: The Role of Planning, Execution and Controlling', *Proc. 9th Annual Conference of the International Group for Lean Construction*, Singapore, Singapore. Available at: www.iglc.net.

McKinney, K. and Fischer, M. (1998) 'Generating, Evaluating and Visualizing Construction Schedules with Cad Tools', *Automation in Construction*, 7(6), pp. 433–447.

Navon, R. and Sacks, R. (2007) 'Assessing Research Issues in Automated Project Performance Control (APPC)', *Automation in Construction*, 16(4), pp. 474–484.

Ourplan. (2013) 'The Visual Planning Tool for Lean Construction'. Available at: http://our-plan.com/(Accessed: 1st Jan 2013).

Pereira, P. (1998) 'Just in Time and Autonomation: Applying TPS in Vertical Housing', *6th Annual Lean Construction Institute Congress*, Lean Construction Institute, Park City, Utah.

Ramaswamy, K.P. and Kalidindi, S.N. (2009) 'Waste in Indian Building Construction Projects', *Proc. 17th Annual Conference of the International Group for Lean Construction, Tapei, Taiwan*. Available at: www.iglc.net.

Rozenfeld, O., Sacks, R. and Rosenfeld, Y. (2010) 'CHASTE - Construction Hazard Analysis with Spatial and Temporal Exposure', *Construction Management and Economics*, 27(7), pp. 625–638.

Sacks, R. (2016) 'What Constitutes Good Production Flow in Construction?', *Construction Management and Economics*, 34(9), pp. 641–656.

Sacks, R., Barak, R., Belaciano, B., Gurevich, U. and Pikas, E. (2013) 'KanBIM Lean Construction Workflow Management System: Prototype Implementation and Field Testing', *Lean Construction Journal*, 9, pp. 19–34.

Sacks, R., Eastman, C.M., Lee, G. and Teicholz, P. (2018) *BIM Handbook: A Guide to Building Information Modeling for Owners, Designers, Engineers, Contractors and Facility Managers*. John Wiley & Sons, Hoboken, NJ, USA.

Sacks, R., Korb, S. and Barak, R. (2017) *Building Lean, Building BIM: Improving Construction the Tidhar Way*. Routledge, Oxford, UK.

Sacks, R., Koskela, L., Dave, B.A. and Owen, R. (2010a) 'Interaction of Lean and Building Information Modeling in Construction', *Journal of Construction Engineering and Management*, 136(9), pp. 968–980.

Sacks, R., Radosavljevic, M., and Barak, R. (2010b) 'Requirements for Building Information Modeling Based Lean Production Management Systems for Construction', *Automation in Construction*, 19(5), pp. 641–655.

Sacks, R., Treckmann, M. and Rozenfeld, O. (2009) 'Visualization of Work Flow to Support Lean Construction', *Journal of Construction Engineering and Management*, 135(12), pp. 1307–1315.

Sarhan, S. and Fox, A. (2012) 'Trends and Challenges to the Development of a Lean Culture among UK Construction Organisations', *Proc. 20th Annual Conference of the International Group for Lean Construction*, San Diego, CA, USA. Available at: www.iglc.net.

Shingo, S. and Dillon, A.P. (1989) *A Study of the Toyota Production System: From an Industrial Engineering Viewpoint*. Produce What Is Needed, When It's Needed. Taylor & Francis, Abingdon, UK.

Teicholz, P., Goodrum, P. and Haas, C. (2001) 'U.S. Construction Labor Productivity Trends, 1970–1998', *Journal of Construction Engineering and Management*, 127(5), pp. 427–429.

VisiLean. (n.d.). 'VisiLean'. Available at: http://visilean.com/(Accessed: 25th May 2017).

Zhang, X. (2005) 'Critical Success Factors for Public–Private Partnerships in Infrastructure Development', *Journal of Construction Engineering and Management*, 131(1), pp. 3–14.

5 People and knowledge

Lean organisation

John Rooke

5.1 Introduction

This chapter looks at the concept of 'people', 'knowledge' and 'organisation', arguing that they are closely related. Knowledge is always a possession of people, either as individuals, or collectives. Understanding the knowledge that people possess is a key to understanding how they organise production and this how the process of production can be improved. It follows that, since knowledge is both a product and a producer of an organisation, achieving Lean organisation requires that we understand not only a knowledge production theory, but also the state of knowledge in the existing organisation.

Within the context of Transformation, Flow and Value – TFV (Koskela, 2000), the people aspect of production is addressed primarily in the V perspective. The value (or supply) chain is effectively a set of relationships linking and governing organisational subsystems which each add value to the product. These relationships are, above all, human relationships. This insight is central to both the Toyota Production System (TPS) and the Last Planner System (LPS – see description in Chapter 3).

The necessary nature of these relationships and the basis for their improvement is addressed in the concepts of conversation for action (Winograd and Flores, 1986) and improvement cycle (Deming, 1986). The former is built around the concept of commitment and the latter around learning. Both concepts are constructive (or normative) in that they do not set out to simply describe the organisation of production as it is, but to offer prescriptions as to how it should be done. Both emphasise: (i) the principled investigation of and engagement with reality; (ii) clear and co-operative communication.

Communication can also be analysed in terms of flows (F). In the simplest terms, this is merely a matter of mapping the lines of communication in a similar matter to value stream mapping. However, building on Shingo's distinction between process flow and operational flow, it is possible to perceive a conceptual link between the F and V perspectives. Thus, conversations for action and improvement cycles can be identified as communicative (or knowledge) operations. The theoretical consequences of such a move are yet to be fully explored.

Koskela's (2001) theory of Lean management will be examined and found to be adequate to an analysis of these issues. However, it is concluded that the most important challenge currently facing Lean Construction in people management is the initial transformation that orients the organisation towards continual improvement. On this basis, an extension to Koskela's theory is proposed.

The origins of people management in production theory predate Lean thinking and can be found in the Shewhart's (1931) concepts of 'customer satisfaction' and 'improvement cycle'. The development of the TPS began with Toyota's invention of the auto-activated weaving machine, which stopped whenever a thread broke, which Ohno (1988, p. 6) characterises as 'automation with a human touch'. This invention emphasises the importance of gemba, as well as of stopping production, rather than producing defective products. It provides the conceptual basis for Ohno's practice of empowering assembly workers to call a halt to production in order to prevent defects (Ohno, 1988). Combined with a simply specified research practice ('ask why five times') this constitutes total quality management (Ohno, 1988).

The concept of the 'last planner' translates this practice for the organisation of planning, enabling front line managers to halt production when conditions warrant it. Under the rubric 'culture', Seymour explores the centrality of knowledge to these and other production practices (Seymour and Rooke, 1995; Seymour, 1996; Seymour, Shammas-Toma, and Clark, 1997). The unique adequacy requirement is suggested as a set of criteria for evaluating reports on these concerns by Rooke (1997).

V theory (Koskela, 2000) begins the formalisation of this human side of production into Lean theory. Koskela (2001) goes on to propose a novel theory of management, employing the key concepts of organisation, planning, adherence and improvement. Subsequently, the language action perspective is introduced as providing a practical method for improving commitments (Macomber and Howell, 2003; Macomber, Howell, and Reed, 2005) and occasioning further theoretical shifts (Howell et al., 2004; Slivon et al., 2010). Pasquire (2012) introduces the idea of common or shared understanding as one of the flows that must be managed on projects. Rooke (2012) synthesises the unique adequacy requirement into a design science perspective.

5.2 Lean is always about people

If the core of Lean thinking, or production theory, is the understanding and design of production processes (Koskela, 1992, 2000), it has always been recognised that these processes depend on people to make them happen. Of Liker's (2004) 14 principles of the Toyota Production System only five (2–5, 8) directly address the design of effective production processes. Six (1, 6, 7, 12–14) can be said to be about knowledge, while three (9–11) concern the development of people (principles 1, 9–11 and 12–14).

Liker begins his exposition with an insistence on the importance of having 'a long-term philosophy, Even [sic] at the expense of Short-Term Financial Goals' (Liker, 2004, p. 71). Easily overlooked by those eager to get to the nitty-gritty of process design, this principle is concerned with establishing the fundamental organisational values without which building a Lean transformation will be impossible. These values are not controversial, but in most organisations are more honoured in the breach than the practice. Likers' formulation: 'Do the right thing for the company, its employees, the customers, and society as a whole' (Liker, 2004, p.72) almost mirrors that of the European Foundation for Quality Management's Excellence Model, which specifies people, society, customer and business results (EFQM, 2017). One problem with this model is that 'business' results refers to the interests of investors, rather than to the well-being of the company itself. In the Western context, where the interests of shareholders can often be restricted to short-term profit, this creates particular barriers to Lean transformation. The problems of Lean transformation and culture more generally are taken up below in the section: Lean culture.

Table 5.1 Liker's 14 principles according to main focus

No.	*Principle*	*Main Focus*
1	Base your management decisions on a long-term philosophy, even at the expense of short-term financial goals	Knowledge
6	Standardised tasks are the foundation for continuous improvement and employee empowerment	
7	Use visual control so no problems are hidden	
12	Go and see for yourself to thoroughly understand the situation (*Genchi Genbutsu*)	
13	Make decisions slowly by consensus, thoroughly considering all options; implement decisions rapidly	
14	Become a learning organisation through relentless reflection (*Hansei*) and (*Kaizen*)	
9	Grow leaders who thoroughly understand the work, live the philosophy, and teach it to others	People
10	Develop exceptional people and teams who follow your company's philosophy	
11	Respect your extended network of partners and suppliers by challenging them and helping them improve	
2	Create continuous process flow to bring problems to the surface	Production process
3	Use 'pull' systems to avoid overproduction	
4	Level out the workload	
5	Build a culture of stopping to fix problems, to get quality right the first time	
8	Use only reliable, thoroughly tested technology that serves your people and processes	

Three of Liker's principles (2, 6 and 7) are difficult to categorise, being deeply relevant to both process design and knowledge management. Thus, principle 2, labelled 'process' in Table 5.1, is concerned with making problems visible and is thus relevant to the principle of transparency. Equally, principles 6 and 7, labelled 'knowledge' here, are intimately concerned with process design and are categorised as such by Liker.

Three principles directly address people management. The first of these (principle 9) establishes the need for leaders who are engaged in not only the values but the day-to-day work of the organisation. This latter engagement is demanding, requiring that managers have both a deep understanding of work and the humility to learn from their subordinates. In this way, managers become conduits of organisational communication, teaching organisational values to operatives, while learning from them about the productive work of the organisation. Managers are also responsible for encouraging the development of competence and teamwork among those for whom they are responsible (principle 10) and extending these values and practices into the supply chain (principle 11). The other three principles are concerned with learning, beginning with *genchi genbutsu* and the five whys (principle 12) and progressing through decision-making by consensus (principle 13) to 'relentless reflection and continuous improvement' (principle 14; Liker, 2004, p. 250).

The Last Planner System (LPS) carries the balance between people and production process into construction. The switch from stable production line manufacture to dynamic project organisation, involved a change in focus from reducing inventory to increasing planning reliability, but retained the emphasis on time. Early papers stressed the flow aspects of production (Ballard and Howell, 1994a, 1994b; Howell and Ballard, 1997; Ballard and Howell, 1998). However, at the heart of the system was an emphasis on a very human process: the art of making reliable promises, an act of communication and the basic building block of project organisation (Macomber and Howell, 2003; Howell et al., 2004; Macomber, Howell, and Reed, 2005). The wide range of visual management tools, ranging from value stream mapping and location based planning diagrams to kanban and andon, among others, are also designed with the importance of communication at the forefront of concern.

5.3 People management is all about knowledge

Arguably, everything discussed in the previous section can be subsumed under the heading of knowledge. But what is knowledge? Ryle (2009) convincingly argues, contrary to Western philosophical tradition, that it consists of more than information and theory (that which can be stated as facts) and encompasses 'knowing how', the skills or competence with which we act in the world. The argument is coherent with Wittgenstein's (1958) insight that language does not merely describe the world, but is an active engagement with it; linguistic expressions can be regarded as tools which we use to do things in the world. Coupled with Wittgenstein's further insights that language is entirely a product of social interaction and thus intimately tied to particular ways of life, this provides the basis for a Wittgensteinian study of people.

Given this broad definition of knowledge, Winch (1990) proposes that the study of people is best conceived of as the study of knowledge. He proposes that the particular bodies of knowledge held by particular groups of people is the proper topic for epistemology. Thus, rather than the traditional attempt to establish an abstract and objective specification of knowledge, as philosophers attempt to do, philosophical methods should be used to understand the reasons that human beings have for their activities. A key feature of this kind of research is the recognition that it is meaning, rather than effective cause, which explains human behaviour.

Ethnomethodology (EM) is the discipline which has progressed this type of research agenda the furthest. Anderson and Sharrock (2017, p. 3) observe that EM is based on an 'analogy which Garfinkel runs between social life and production processes'. It would not be surprising then to find that it has a natural affinity with Lean thinking. A closer examination confirms this to be the case. Leading features of EM which are relevant to a theory of production are:

- A focus on the local production of organisation (even the biggest organisational systems must be produced through the concerted efforts of individual members interacting with each other);
- The treatment of production processes as having an essentially temporal nature (Sharrock and Anderson, 1986; Koskela, 1992);
- The realisation that the product of these processes is the locally achieved objectivity of an organisation's existence (see Schutz, 1972 on intersubjectivity);
- That, thus, it is the shared information, concepts and competence of an organisations' members involved in this local production that account for organisational reality;
- The consequent methodological prescription that reports of organisational reality should be based in an everyday competence in that local production;
- Further, that such reports should, as a matter of methodological principle, eschew as explanatory devices, concepts and competences that are not involved in that local production;

- A product of EM inquiry is the detailed description of these processes;
- While EM is primarily (or at least initially) constituted as a discipline alternate to sociology and thus primarily sharing sociology's concerns, an additional concern of EM is to produce hybrid studies that bring EM insights to bear on the problems of other disciplines.

The idea of the gemba, the local site of production, is central to such a study. The unique adequacy requirement of methods (Garfinkel and Wieder, 1992) is crucial to it. This is a set of criteria for assessing the validity of research into knowledge. From our perspective here, we can view research as anything from a gemba walk to a major academic research project, from planning a minor kaizen initiative to requirement capture for a complex facility. The requirement has two parts: first, that the researcher is competent in the activities analysed (the weak requirement); second, that the analysis employs only concepts native to those activities (the strong requirement). In layman's terms, the weak requirement is that the researcher knows what they are talking about, with the proviso that knowledge is not simply a grasp of the facts (knowledge that), but the necessary experience to be competent in the activity analysed (knowledge how).

Whereas the weak requirement makes a strong appeal to common sense and is often easily accepted by workers, managers and academics, the strong requirement presents more difficulty in understanding and acceptance. It demands that the analysis is restricted to concepts native to the knowledge analysed. This means that the researcher should attempt to reproduce the knowledge that is being studied, rather than theorise, or explain it. This is necessary because the understandings of members under study are ontologically equivalent to the understandings of researchers. Thus, while management studies draw on a panoply of psychological and sociological approaches, these not only tend to fail in providing sufficient empirical grounding to the study; they also involve importing irrelevant theoretical explanations into the setting. For this reason, EM concerns have long had a place in Lean Construction thinking (see, for instance: Seymour, 1996; Rooke et al., 2008).

Considering Liker's seven 'people principles' in this light, we can see the range of knowledge involved in the Toyota Way: knowledge of basic cultural values; knowledge of the nature of the work; the ability to cope with the changing demands of the work; teamwork; ability to act as an ambassador for the Toyota Way; appreciation of the significance and importance of gemba; rational and consensual decision-making skills; learning skills; improvement skills (Liker, 2004).

5.4 Information physics

Having proposed in the previous section that the people side of production theory is fundamentally about meaning, it is necessary to pause and recognise that there are important aspects of knowledge and communication on construction projects that properly fall under the auspices of construction physics (Hopp and Spearman, 2000; Bertelsen et al., 2007). For convenience, these aspects will be referred to below as information physics.

These aspects can be described using the terms 'repository', 'flow' and 'field'. The term 'repository', denoting the site where knowledge is stored, is familiar enough in information science (e.g. Rotenstreich, 2017). However, the foundational insight in knowledge management is that where knowledge is stored in the capabilities of human beings, the matter becomes much more complex (Davenport and Prusak, 1997). In knowledge management terms, it may be useful to posit a storage view of knowledge, analogous to a T theory view of production: a black box; we know that knowledge is 'in there', but we don't know what it consists of. Thus, we may know that an engineer is necessary to design the formwork for a concrete feature and that skilled workers are necessary to place and fix it, without having much insight into the precise nature of the knowledge that makes these people competent.

We can also speak unproblematically of data flows, where data has an identifiable physical presence (e.g. Clark, 1983). This approach has proved useful in facilitating information 'pull' (Caldas and Soibelman, 2002), as well as mapping information flows in an integrated visual management system (Brady et al., 2017).

While there is some precedent for treating the broader concept of knowledge in a similar way, at least to the extent that it can be understood as 'sticky', or 'leaky'. Both characteristics of knowledge can prove problematic, depending on circumstances. Competence is sticky knowledge and is hard to transfer. Information is leaky, not only is it not sticky, but it can easily find its way into places it was not intended to go, such as competing companies (Brown and Duguid, 2001). This, of course, is good news for enterprises which have spent years developing the Lean skills of their employees and partners. Not only do these skills stay with the individuals concerned, they also inhere in teams (Senge, 1990). There is anecdotal evidence that on the rare occasions when Toyota managers have been successfully lured away to other companies, they have not flourished in the non-Lean cultures in which they have found themselves.

The term 'field', taken from visual management, denotes the physical area from which a communication can be perceived (Greif, 1991). It represents the reciprocal opposite concept of a 'vista', in wayfinding, the sum of what can be seen from any particular point. It should be noted that these terms, like 'storage' and 'flow', can be applied to any kind of information. Most information takes a visual form at some point in its 'career' (Harper, 1997), but even where this is not the case, we can speak of audio fields, for instance. However, they are less applicable to the sticky process of transferring skills. This more complex process will be further explored in the next section.

A further key concept is 'transparency'. This has been described as the ability of a production process to communicate (Formoso, Santos, and Powell, 2002). Stripped of its anthropomorphism, we can render this as: the degree of ease with which an organisation's members can access the information they need about the way the organisation is functioning.

These concepts can provide us with a partial theoretical explanation of the TPS principle of *genchi genbutsu*. This calls for managers to physically situate themselves at the site of production, within the informational fields generated directly by the production process with consequently short information flows involved in the decision-making process. Last Planner takes this principle into the management of complex project organisations. In the LPS, not only is the physical point of decision making brought as close as possible to the physical point of production, but the decision-making capability is delegated to those members who are the principle repositories of knowledge of the process concerned.

5.5 The management of meaning

Information physics is concerned with the delivery of information to the right people at the right time. Achieving the optimum form and content for the information delivered is, like other aspects of organisational reality, an exercise in the management of language. A language may be spoken or written, but it can also take the form of a set of visual cues, as in the visual management perspective. Searle's (1976) taxonomy of illocutionary acts provides a more generic underpinning for Galsworth's (1997) taxonomy of visual management tools (Rooke, 2017). The essential characteristic of language is that it is meaningful.

Language is the way that we analyse and organise the world. This insight comes to us from the philosopher Wittgenstein. Originally adopting the conventional philosophical view that language represents the world, he came to the realisation that language is better understood as a set of tools with which we do things in the world. For Wittgenstein (1958), this had the utility of (dis)-solving a whole range of philosophical problems which stemmed from the notion that words must stand for things. The same insight has practical implications for us in that it informs Searle's taxonomy (Searle, 1976).

The production of organisation and the organisation of production would be impossible without it. Crucial to the study of how language works is the concept of gemba. It is the local production of commitments that produces organisation. It is the articulation of these commitments into production plans that leads to the creation of value. A powerful demonstration of this is provided by a study of the 'Tidhar Lean Boot Camp' (Kerem et al., 2013). Last planner is the organisational form that brings the central focus on gemba into the construction industry. In addition to language skills, there are other skilful ways in which we act in the world. For our purposes here, the most important of these are the construction operatives who perform the physical operations that drive the construction processes.

To attain an adequate understanding of any of these skilled activities (whether linguistic or directly productive) is to understand them from the inside. In Garfinkel's (2002) words, 'the analyst must be *vulgarly* competent to the local production and reflexively natural accountability of the phenomenon of order', or to put it another way, must understand how these things work out in the gemba. Approaches to the study of meaning which attempt to achieve a panoramic objectivity, rather than this localised understanding, will inevitably tend to overlook important features of this phenomenon. The study of people is the study of meaningful activity and meaning is subjective in its essence.

It also follows that the ability to look (Goodwin, 2000; Kerem et al., 2013) and listen (Rooke, 2005) are the crucial skills of both managers and researchers.

5.6 Value chains: linking flow and commitment

Shingo has a unique perspective on flow, which sees it as fundamental, not just to the analysis of processes, but to the operations which drive them. Processes, in the original analysis, are about moving matter (materials and components) around, but this movement is achieved through the operations of people and machines, which in turn consist of centred flows of activity (Shingo, 1985, 1988). Bertelsen (2017) suggests that an operational analysis should supplement Flow and Value analyses.

Moving now from the physics of production to the management of meaning, the 'conversation for action' (Macomber and Howell, 2003) can also be viewed in terms of operations and flows. This is a negotiation that utilises Searle's (1976) taxonomy of illocutionary acts to specify commitment. The importance of this cannot be emphasised too strongly. According to Winograd and Flores (1986), an organisation can be analysed entirely in terms of the commitments made by its members. While there is reason to doubt this, it is nevertheless the case that without those commitments the organisation would cease to exist – change them and you change the organisation. This realisation derives from an insight that belongs to Searle's teacher, Austin, as well as to Wittgenstein, that the main purpose of language is to manage the world we live in. Combining this perspective with that of Shingo, we can see conversations for action as operations that drive, not only the flow of communication, but through the commitments made, the production process itself.

In terms of TFV theory, the concept of a value chain provides a link between the operational cycles of conversations for action, on the one hand, and flows (or value streams) on the other. The value stream map is a widely used Lean tool, employing the flow concept. However, this flow is driven by the effort to satisfy customer needs: requirements and expectations drive value generation (Koskela, 2000). Macomber and Howell (2003) argue that the mechanism through which this is achieved is the conversation for action, which begins with a request and progresses through offer and commitment to declarations of completion and satisfaction. They argue that the achievement of planning reliability through the Last Planner System is facilitated by employing this conversation for action model.

It is also possible to see the whole value chain as a single conversation for action which extends throughout the project, beginning with customer need (request) and ending in post-occupancy survey. Formoso, Tzortzopoulos, and Liedtke (2002) provide a useful overview of the phases that this conversation goes through. The recognition of the importance of this value analysis leads to a focus on the often neglected extreme ends of the project: client activities at the front end (Tzortzopoulos et al., 2006); and post-occupancy evaluation (Codinhoto et al., 2009); as well as the through-life management of value (Tillmann et al., 2012). Ideally, these activities link up in a trans-project learning cycle.

Pasquire (2012) points out that shared, or common, understanding is a key concept for understanding these processes, proposing that it constitutes one of the flows which must be managed on projects. If 'flow' is taken to indicate a necessary precondition for task initiation (Koskela, 2000; Bertelsen et al., 2006), the proposal is unproblematic and useful. Clearly, shared understanding is a necessary precondition if commitments are to be reliable; that is to say, if the author of a request is to be satisfied with the outcome. Pasquire and Court (2013) offer some methods of achieving shared understanding drawn from fieldwork and a new analysis of the concept, to include: structural (information flow); relational (trust); and cognitive (shared language and vision) elements. However, the concept also mutates, such that: 'knowledge is the basic resource and understanding is the output of the transformation of knowledge' (2013, p. 46).

Thus, it is now knowledge (presumably, 'knowledge that'), rather than understanding that flows, while understanding is somehow generated out of the transformation of this knowledge. On the one hand, the conception of a 'knowledge flow' is more in line with the construction physics view outlined above. On the other, the attempt to then fit 'understanding' into the same causal framework is difficult to comprehend. First, it would seem to imply that one can know something without understanding it. Second, it is not clear in Pasquire and Court's account how the transformation of knowledge is performed, or what it is that changes about the knowledge.

Notwithstanding these difficulties, the notion of a shared understanding is useful, not only to the process of commitment management, but to those of improvement and design discussed in the following section.

5.7 Improvement, learning and design

The concept of the improvement cycle as an active and scientific learning process pre-dates the TPS by a number of years, in the form of the Shewhart cycle (Deming, 1986). Toyota's great contribution to this development is the concept of zero defects, or total quality (TQM), which involves stopping production to address the origins of defects as they arise (Ohno, 1988). Last Planner adopts this approach to the improvement of planning reliability.

There is a parallel to be drawn between this approach to learning and Garfinkel's (1984) use of 'breaching procedures' to study social organisation through studying what happens when members experience problems communicating. Despite this, and despite extensive ethnomethodological studies of routine work in a range of industries, there has been little attempt to make detailed studies of the *in vivo* learning processes involved in either TQM or conventional Shewhart cycles. Similarly lacking are detailed studies of the acquisition of skills.

There have been a number of theoretical investigations. For instance, Powell (1999) has suggested an action learning approach, based on the work of Revans. Hirota, Lantelme, and Formoso (1999), on the other hand, base their approach on Vygotsky. It is significant that both these approaches stress the role of activity in learning and its social nature. Revans (1998) in particular stresses the importance of learning through active experience of solving problems in the workplace.

However, more detailed studies of actual real time learning processes in industry have also emerged. In one of the earliest of these, Samudio, Alves, and Chambers (2011) describe a team's experience of employing the practice of *genchi genbutsu*. The following year, Skinnarland and Yndesdal (2012) published a similar study of last planner implementation. Fauchier and Alves (2013) is particularly significant for its emphasis on skill acquisition. More recently, Tillmann and Sargent (2016) have reported on Last Planner and BIM integration, while Alves et al. (2017) have investigated the learning generated by sharing already existing supplier evaluations with the suppliers in question.

There would appear to be two possible ways to approach improvement: the experimental method inherent in Shewhart's improvement cycle; or Ohno's focus on breakdown, a kind of natural experiment. While Last Planner favours the latter, most other initiatives adopt the former. The key difference is that while Shewhart type initiatives are planned, Ohno's are reactive, based on the logical analysis of a naturally occurring situation. The 'five whys' is intended to focus members' inquiries into a deep causal analysis of failures in quality or timely delivery.

One way of looking at continual improvement is as a sub-set of design. In one sense, it is concerned with the design of the production process. 'Construction completes design', as Glenn Ballard says. In another sense then, improvement initiatives can contribute directly to the product design. Crucially, learning plays a central role in both processes; they can both be seen as research processes (though not academic research processes); they both seek to add value. Drawing on Simon (1996), Koskela (2008) makes these connections explicit, characterising the research process involved as 'design science'.

Tzortzopoulos et al. (2006) offers a comprehensive view of activities at the 'design front end', which helps to highlight both the similarities and the intense interconnection between the two. The decision-making process at this early stage is concerned with both the design of the facility and the (re-)design of the company's business processes, the distinction often being insufficiently made. An implication is that the analysis of client requirements is essential to the design process and this analysis is a production process analysis. Two suggestions can be made for guiding this analysis. First, that it should be based in the competencies and theories of client organisation members (unique adequacy requirement). Second, that Ohno's five whys technique can be adapted to address final rather than effective cause. Aristotle identifies four types of cause, the effective cause of the cause–effect relationship being that usually focused on in modern analysis. The final cause of something is its purpose, making this kind of analysis particularly appropriate to many design issues. In asking 'why' five times of the purpose of a proposed design feature it is often possible to understand how that purpose can be achieved through more economical or satisfactory means.

Rocha et al. (2012) apply the design science approach in two projects, developing conceptual schemas to aid decision making in production management and design respectively. Rooke (2012) integrates the unique adequacy requirement into the approach, using participant observation to test and improve the effectiveness of a wayfinding system.

5.8 A new theory of management

In an important, but neglected paper Koskela (2001) suggests that in addition to a better theory of production, the TPS implies a better theory of management. Koskela bases his argument on a comparison of Johnston's (1995) 'management as organisation' thesis, with Spear and Bowen's (1999) identification of a 'scientific method' approach, and the resolution of an apparent contradiction between the two. Further evidence for Koskela's proposal can be found in the analysis of Liker's (2004) 14 principles above. Clearly, these principles are equally balanced between a theory of production and a theory of management (or perhaps, to put it more prosaically, between a theory of production management and a theory of people management). Thus, Koskela's new theory brings together many of the issues considered in this chapter. He proposes four aspects of management: organising; planning; adherence; and improvement. Viewed from a knowledge perspective, each of these aspects presents a major learning challenge for the industry.

Organising is perhaps the key aspect, necessary to the successful accomplishment of the other three. The Last Planner System, based as it is on the importance of gemba, represents a radical innovation in organisation, which distributes responsibility for planning to where it can be most effectively performed. The most obvious challenge for industry members is to learn how to operate this system. A second challenge is to ensure that those responsible for planning have the skills necessary for the kind of collaborative planning that the system requires (Seed, 2014). While the system restricts planners to making plans only about matters in which they can be assumed to have competence, this does not in itself guarantee effective planning. For instance, Kemmer et al. (2007) note the difficulty of adequately specifying constraints in medium-term planning, leading to failures in task completion on site.

Liker's (2004, p. 237) principle 13 is the one most pertinent to the planning process: make decisions slowly by consensus, thoroughly considering all options. The Last Planner process of look-ahead planning and the forming of weekly work packages takes this principle forward into construction. Last Planner also addresses the question of adherence, with its emphasis on the formation of reliable commitments. However, the question then becomes 'how do we get people to adhere to the Last Planner?' Thus, adherence is perhaps the aspect in which Lean Construction is currently weakest.

Percentage plan complete is probably still regarded as the key measure of conformance, and thus of the effectiveness of the planning system, but many difficulties have been identified in implementation (Samad, Hamzeh, and Emdanat, 2017). However, it is necessary to draw a distinction between adherence to commitments within a Lean Construction organisation and adherence to the organisational principles of a lean construction system itself. Much work has been done from the earliest days of IGLC to develop methods of ensuring reliable commitments and this work continues (see section on value chains above). However, the major challenge facing Lean Construction is to find more effective ways of advocating, teaching and learning these and other Lean methods. This implies a fifth aspect of management: organisational change.

As noted in the previous section, continual improvement predates the TPS in its origins. Continual improvement implies organisational change through the gradual effects of the changes in work practices that it introduces. However, the kind of change required to institute a culture of continual improvement in the first place is of a different order. It is this kind of change that will be discussed in the following section.

5.9 Learning Lean

There is a large literature on the difficulties of making the change to Lean, particularly Last Planner (Brady, Tzortzopoulos, and Rooke, 2011). Rooke, Seymour, and Fellows (2003) found structural patterns of resistance to change in the industry. Skinnarland and Yndesdal (2014) found inadequate training, resistance to train and lack of contractor engagement among four companies attempting to go Lean. Salvatierra et al. (2015) found that uncertainty, inadequate training and an inadequate range of available Lean tools, presented barriers to Lean implementation, despite 90% of managers seeing Lean as valuable.

Freeman and Seppänen (2014) identified lack of trust to be an obstacle to implementing location based planning, but found it possible to overcome this over time by persuading members that their own interests, as well as the effectiveness of the project, are enhanced by using the tool. Pekuri et al. (2012) also identified trust as an important factor. Koskenvesa and Koskela (2012) found that failure to understand the Lean Construction management model acted as a barrier to Last Planner implementation. Furthermore, apparent gains can be lost (Leino, Heinonen, and Kiurula, 2014).

There has been a great deal of effort put into developing and evaluating teaching methods. Some of these focus on developing simulations (for instance: Smith and Rybkowski, 2013; Heyl, 2015), or other computer supported learning (Feliz et al., 2014). It is noticeable that Feliz et al. (2014) also stress learning in action. A workshop approach has also been suggested (Izquierdo, Cerf, and Gomez, 2011). Several courses and programmes have been developed (Raghavan et al., 2014); see also Chapter 19 by Rybkowski, Forbes and Tsao in this book.

However, there has always been a strong tendency to favour workplace situated learning closer to the spirit of the Shewhart Cycle. Action learning was an early candidate for Lean implementation, with some degree of success (Davey et al., 2000; Hirota and Fomoso, 2001) and the development of the Last Planner System itself followed this type of approach (Ballard, 2000). The advantage of this approach is that it takes learning out of the classroom and workshop and encourages managers to learn on the job.

Developing Lean leadership is another approach that has been widely explored (Orr, 2005; Pekuri et al., 2012; Bettler and Lightner, 2013; Hämäläinen, Ballard, and Elfving, 2014) along with the related approach of mentoring (Tillmann, Ballard, and Tommelein, 2014). Leadership begins to phase into a more political approach, which has also proven important. Miles (1998) suggested, for instance, that Lean change in projects could best be achieved from below through informal alliances between small committed companies.

Addressing the issue of conflicting interests more directly, Integrated Project Delivery (IPD) has sought, with some success, to give a formal contractual basis to Lean initiatives (Lichtig, 2010; Mauck et al., 2009; Suttie, 2013). Seymour, Shammas-Toma, and Clark

(1997) found that institutionalised conflict in the industry, formalised into contracts, is mirrored in the way non-conformance to specification is measured, explained and managed. Lean construction initiatives inevitably contradict the requirements of such contracts. IPD replaces the former informal practice of merely ignoring the contract (see Chapter 9 by Alves and Lichtig in this book for further reading on IPD).

5.10 Conclusion

It has been argued that a focus on people is central to Lean and that a focus on knowledge is central to the study and management of people. Knowledge encompasses not just the facts and theories that people hold, but their competence. The latter includes the physical and mental skills necessary for the performance of work tasks, the social skills necessary for successful collaboration and the values that underlie the organisational ethos. It has been suggested that EM provides the most developed approach to researching these forms of knowledge and that there is a natural analogy between EM research and Lean management.

While some part of the management of knowledge can be seen as information physics, the larger part concerns the management of meaning. A key concept is the gemba: the local site where the production of both consumer value and of the project organisation itself take place. The basic building block of the project organisation is the achievement of reliable commitments. Chains of commitments constitute the value chain which channels the production flow. When these value chains are supplemented by learning cycles oriented towards the promotion of flow and the elimination of waste, the Lean organisation emerges.

Koskela (2001) proposes four aspects of management: organisation; planning; adherence; improvement. These cover most of the matters discussed above, but not the question of fundamental organisational change – how companies and projects can be transformed into Lean organisations. This issue has been much discussed in IGLC conferences in various guises and provides a major research challenge for the years ahead.

References

Alves, T.D.C.L., Desai, P., Needy, K.L., Hegwood, A. and Musick, S. (2017) 'Impact of Supplier Evaluation on Product Quality', *Proc. 25th Annual Conference of the International Group for Lean Construction*, Heraklion, Greece. Available at: www.iglc.net.

Anderson, R. and Sharrock, W. (2017) Has Ethnomethodology Run It's Course? Unpublished Manuscript. Draft 7.1. Available at: www.sharrockandanderson.co.uk/wp-content/uploads/2017/11/Run-its-Course-VII.pdf (Accessed: 24th Dec 2017).

Ballard, G. (2000) *The Last Planner System of Production Control*. PhD Thesis, The University of Birmingham, Birmingham, UK Available at: http://leanconstruction.org.uk/media/docs/ballard2000-dissertation.pdf.

Ballard, G. and Howell, G. (1994a) 'Implementing Lean Construction: Stabilizing Work Flow', *Presented at the 2nd Annual Conference of the International Group for Lean Construction*, Santiago, Chile. Available in Alarcon, L. (ed.) (1997), Lean Construction, A.A. Balkema, Rotterdam, The Netherlands.

Ballard, G. and Howell, G. (1994b) 'Implementing Lean Construction: Improving Downstream Performance', *Presented at the 2nd Annual Conference of the International Group for Lean Construction*, Santiago, Chile. Available in Alarcon, L. (ed.) (1997), Lean Construction, A.A. Balkema, Rotterdam, The Netherlands.

Ballard, G. and Howell, G. (1998) 'Shielding production: essential step in production control', *Journal of Construction Engineering and Management*, 124(1), pp. 11–17.

Bertelsen, S. (2017) *The Unruly Project: A New Understanding of its Nature and Management*, Holte, Denmark, Sven Bertelsen aps.

Bertelsen, S., Henrich, G., Koskela, L. and Rooke, J. (2007) 'Construction Physics', *Proc. of 15th Annual Conference of the International Group for Lean Construction*, East Lansing, Michigan, USA. Available at: www.iglc.net.

Bertelsen, S., Koskela, L., Henrich, G. and Rooke, J. (2006) 'Critical Flow - Towards a Construction Flow Theory', *Proc. 14th Annual Conference of the International Group for Lean Construction*, Santiago de Chile, Santiago. Available at: www.iglc.net.

Bettler, R. and Lightner, B. (2013) 'Applied Leadership Model for Lean Construction: A New Conversation', *Proc. of 21st Annual Conference of the International Group for Lean Construction*, Fortaleza, Brazil. Available at: www.iglc.net.

Brady, D., Tzortzopolous, P., Rooke, J., Formoso, C. and Tezel, A. (2017) 'Improving Transparency in Construction Management: A Visual Planning and Control Model', *Engineering, Construction and Architectural Management*, ISSN 0969-9988 (In Press). Available at: http://eprints.hud.ac.uk/id/eprint/33542/.

Brady, D.A., Tzortzopoulos, P. and Rooke, J. (2011) 'An Examination of the Barriers to Last Planner Implementation', *Proc. 19th Annual Conference of the International Group for Lean Construction*, Lima, Peru. Available at: www.iglc.net.

Brown, J.S. and Duguid, P. (2001) 'Knowledge and organization, a social practice perspective', *Organization Science*, 12(2), pp. 198–213.

Caldas, C.H. and Soibelman, L. (2002) 'Automated Classification Methods: Supporting the Implementation of Pull Techniques for Information Flow Management', *Proc. 10th Annual Conference of the International Group for Lean Construction*, Gramado, Brazil. Available at: www.iglc.net.

Clark, A.P. (1983) *Principles of Digital Data Transmission*. 2nd edition, London, UK, Pentech Press.

Codinhoto, R., Tzortzopoulos, P., Kagioglou, M., Aouad, G. and Cooper, R. (2009) 'The impacts of the built environment on health outcomes', *Facilities*, 27(3/4), pp. 138–151. doi: 10.1108/02632770910933152.

Davenport, T.H. and Prusak, L. (1997) *Working Knowledge: How Organisations Manage What They Know*. Boston, MA, USA, Harvard Business School Press.

Davey, C., Powell, J., Cooper, I. and Hirota, E. (2000) 'Innovation and Culture Change Within a Medium-Sized Construction Company: Success Through the Process of Action Learning', *Proc. 8th Annual Conference of the International Group for Lean Construction*, Brigthon, UK. Available at: www.iglc.net.

Deming, W.E. (1986) *Out of the Crisis: Quality, Productivity and Competitive Position*. Cambridge, UK, Cambridge University Press.

EFQM (2017) 'The EFQM Excellence Model'. Available at: www.efqm.org/efqm-model/criteria/results (Accessed: 19th Nov 2017).

Fauchier, D. and Alves, T.D.C.L. (2013) 'Last Planner® System Is the Gateway to Lean Behaviors', *Proc. 21st Annual Conference of the International Group for Lean Construction*, Fortaleza, Brazil. Available at: www.iglc.net.

Feliz, T., Reed, D., Draper, J. and Macomber, H. (2014) 'Leveraging Software for Learning-in-Action Using Commitment-Based Planning', *Proc. 22nd Annual Conference of the International Group for Lean Construction*, Oslo, Norway. Available at: www.iglc.net.

Formoso, C.T., Santos, A.D. and Powell, J. (2002) 'An exploratory study on the applicability of process transparency in construction sites', *Journal of Construction Research*, 3(1), pp. 35–54.

Formoso, C.T., Tzortzopoulos, P. and Liedtke, R. (2002) 'A model for managing the product development process in house building', *Engineering, Construction and Architectural Management*, 9(5/6), pp. 419–432. ISSN 0969-9988.

Freeman, C. and Seppänen, O. (2014) 'Social Aspects Related to LBMS Implementation – A Case Study', *Proc. 22nd Annual Conference of the International Group for Lean Construction*, Oslo, Norway. Available at: www.iglc.net.

Galsworth, G.D. (1997) *Visual Systems: Harnessing the Power of a Visual Workplace*. New York, USA, AMACOM, American Management Association.

Garfinkel, H. (1984) *Studies in Ethnomethodology*. Cambridge, UK, Polity Press.

Garfinkel, H. (2002) *Ethnomethodology's Program: Working out Durkheim's Aphorism*. Lanham, MD, USA, Rowman & Littleford.

Garfinkel, H. and Wieder, D.L. (1992) 'Two incommensurable, asymmetrically alternate technologies of social analysis', in G. Watson and R.M. Seiler (eds.), *Text in Context*, Sage, London, UK, pp. 175–206.

Goodwin, C. (2000) ''Practices of Seeing: Visual Analysis: An Ethnomethodological Approach', in T. van Leeuwen and C. Jewitt (eds.), *Handbook of Visual Analsysis*, Sage, London, UK, pp. 157–182.

Greif, M. (1991) *The Visual Factory: Building Participation Through Shared Information*. Cambridge, UK, Productivity Press.

Hämäläinen, J., Ballard, G. and Elfving, J. (2014) 'Are Tools and Training Enough - An Argument for Leadership', *Proc. 22nd Annual Conference of the International Group for Lean Construction*, Oslo, Norway. Available at: www.iglc.net.

Harper, R.H.R. (1997) *Inside the IMF: Documents, Technology and Organizational Action*. London, UK, Academic.

Heyl, J.V. (2015) 'Lean Simulation in Road Construction: Teaching of Basic Lean Principals', *Proc. 23rd Annual Conference of the International Group for Lean Construction*, Perth, Australia. Available at: www.iglc.net.

Hirota, E.H. and Fomoso, C.T. (2001) 'Barriers to Management Innovations: Communicating Meanings', *Proc. 9th Annual Conference of the International Group for Lean Construction*, Singapore, Singapore. Available at: www.iglc.net.

Hirota, E.H., Lantelme, E.M.V. and Formoso, C.T. (1999) 'Learning How to Learn Lean Construction Concepts and Principles', *Proc. 7th Annual Conference of the International Group for Lean Construction*, Berkeley, USA. Available at: www.iglc.net.

Hopp, W. and Spearman, M. (2000) *Factory Physics: Foundations of Manufacturing Management*. 2nd edition, Boston, MA, USA, Irwin McGraw-Hill.

Howell, G. and Ballard, G. (1997) 'Implementing Lean Construction: Reducing Inflow Variation', *Presented at the 2nd Annual Conference of the International Group for Lean Construction*, Santiago, Chile. Available in Alarcon, L. (ed.) (1997), Lean Construction, A.A. Balkema, Rotterdam, The Netherlands.

Howell, G., Macomber, H., Koskela, L. and Draper, J. (2004) 'Leadership and Project Management: Time for a Shift from Fayol to Flores', *Proc. 12nd Annual Conference of the International Group for Lean Construction*, Elsinor, Denmark. Available at: http://iglc.net/Papers/Details/302.

Izquierdo, J.L., Cerf, M. and Gomez, S.A. (2011) 'Lean Construction Education: Basic Management Functions Workshop', *Proc. 19th Annual Conference of the International Group for Lean Construction*, Lima, Peru. Available at: www.iglc.net.

Johnston, R.B. (1995) 'Making manufacturing practices tacit: a case study of computer aided production management and lean production', *Journal of the Operational Research Society*, 46(10), pp. 1174–1183.

Kemmer, S.L., Heineck, L.F., Novaes, M.D.V., Mourao, C.A.M. and Alves, T.D.C. (2007) 'Medium-Term Planning: Constributions Based on Field Application', *Proc. 15th Annual Conference of the International Group for Lean Construction*, East Lansing, Michigan, USA, pp. 509–518.

Kerem, C., Barak, R., Sacks, R. and Priven, V. (2013) 'Learning to See – Managers Working in the Gemba as Part of the Tidhar Way Training Program', *Proc. 21st Annual Conference of the International Group for Lean Construction*, Fortaleza, Brazil. Available at: www.iglc.net.

Koskela, L. (1992) *Application of the New Production Philosophy to Construction*. Center for Integrated Facility Engineering, (CIFE), Department of Civil Engineering, Stanford University. Technical Report number: 72.

Koskela, L. (2000) *An Exploration Towards a Production Theory and Its Application to Construction*. PhD Thesis, VTT Technical Research Centre of Finland, VTT Publication 408. Available at: www.vtt.fi/inf/pdf/publications/2000/P408.pdf.

Koskela, L. (2001) 'On New Footnotes to Shingo', *Proc. 9th Annual Conference of the International Group for Lean Construction*, Singapore, Singapore.

Koskela, L.J. (2008) 'Which Kind of Science is Construction Management?', *Proc. 16th Annual Conference of the International Group for Lean Construction*, Manchester, UK. Available at: www.iglc.net.

Koskenvesa, A. and Koskela, L. (2012) 'Ten Years of Last Planner in Finland–Where are we', *Proc. 20th Annual Conference of the International Group for Lean Construction*, San Diego, CA, USA. Available at: iglc.net.

Leino, A., Heinonen, R. and Kiurula, M. (2014) 'Improving Safety Performance Through 5S Program', *Proc. 22nd Annual Conference of the International Group for Lean Construction*, Oslo, Norway. Available at: iglc.net.

Lichtig, W. (2010) Presentation at *12th Annual Lean Construction Congress*. Boulder, CO.

Liker, J.K. (2004) *The Toyota Way*. New York, USA, McGraw Hill.

Macomber, H. and Howell, G (2003) 'Linguistic action: contributing to the theory of lean construction', *Proc. 11th Annual Conference of the International Group for Lean Construction*, Blacksburg, Virginia. Available at: www.researchgate.net/publication/228918270_Linguistic_Action_Contributing_to_the_Theory_of_Lean_Construction

Macomber, H., Howell, G. and Reed, D. (2005) 'Managing Promises with the Last Planner System: closing in on uninterrupted flow', *Proc. 13th Annual Conference of the International Group for Lean Construction*, Sydney, Australia. Available at: iglc.net.

Mauck, R., Lichtig, W., Digby, C. and Darrington, D. (2009) 'Integrated Project Delivery: Different Outcomes, Different Rules', Presented to *the 48th Annual Meeting of Invited Attorneys*, pp. 105–125.

Miles, R.S. (1998) 'Alliance Lean Design/Construct on a Small High Tech Project', *Proc. 6th Annual Conference of the International Group for Lean Construction*, Guarujá, Brazil. Available at: iglc.net.

Ohno, T. (1988) *Toyota Production System; Beyond Large Scale Production*. Portland, OR, USA, Productivity Press.

Orr, C. (2005) 'Lean Leadership in Construction', *Proc. 13th Annual Conference of the International Group for Lean Construction*, Sydney, Australia. Available at: iglc.net.

Pasquire, C. (2012) 'The 8th Flow: Common Understanding', *Proc. 20th Annual Conference of the International Group for Lean Construction*, San Diego, CA, USA. Available at: iglc.net.

Pasquire, C. and Court, P. (2013) 'An Exploration of Knowledge and Understanding - the Eighth Flow', *Proc. 21st Annual Conference of the International Group for Lean Construction*, Fortaleza, Brazil. Available at: www.iglc.net.

Pekuri, A., Herrala, M., Aapaoja, A. and Haapasalo, H. (2012) 'Applying Lean in Construction - Cornerstones for Implementation', *Proc. 20th Annual Conference of the International Group for Lean Construction*, San Diego, CA, USA. Available at: www.iglc.net.

Powell, J.A. (1999) 'Action Learning for Continuous Improvement and Enhanced Innovation in Construction', *Proc. 7th Annual Conference of the International Group for Lean Construction*. Berkeley, USA. Available at: iglc.net.

Raghavan, N., Kalidindi, S., Mahalingam, A., Varghese, K. and Ayesha, A. (2014) 'Implementing Lean Concepts on Indian Construction Sites - Organisational Aspects and Lessons Learned', *Proc. 22nd Annual Conference of the International Group for Lean Construction*, Oslo, Norway. Available at: www.iglc.net.

Revans, R.W. (1998) *ABC of Action Learning; Empowering Managers to Act and to Learn From Action*. London, UK, Lemos & Crane.

Rocha, C.G.D., Formoso, C.T., Tzortzopoulos-Fazenda, P., Koskela, L. and Tezel, A. (2012) 'Design Science Research in Lean Construction: Process and Outcomes', *Proc. 20th Annual Conference of the International Group for Lean Construction*, San Diego, CA, USA. Available at: www.iglc.net.

Rooke, C.N. (2012) *Improving Wayfinding in Old and Complex Hospital Environments*. Thesis submitted in partial fulfilment of the requirements of the degree of doctor of philosophy, School of the Built Environment, The University of Salford. Available at: http://usir.salford.ac.uk/27358/ (Accessed: 11th Dec 2013).

Rooke, J. (1997) 'Developing a more empirical approach to culture, attitude and motivation in construction management research: a critique and a proposal', *Journal of Construction Procurement*, 3(2), pp. 45–55.

Rooke, J. (2005) 'Listening is at the Heart of Lean Construction (and every other worthwhile change initiative): A Review of SPEAR'S Paper', *IPHRP Working Paper*, University of Salford.

Rooke, J. (2017) 'Language Action Perspective and Visual Communication', *Wittgenstein's Builders*. Available at: http://jalfro.altervista.org/language-action-perspective-visual-communication/ (Accessed: 24th Dec 2017).

Rooke, J., Seymour, D. and Fellows, R. (2003) 'The claims culture; A taxonomy of industry attitudes', *Construction Management and Economics*, 21(2), pp. 167–174.

Rooke, J., Seymour, D., Koskela, L., Bertelsen, S., Owen, B. and Cleary, M. (2008) 'Inquiry Pull Research: an ethnomethodological approach to lean construction research or a lean approach to ethnomethodological research?', *Proc. 16th Annual Conference of the International Group for Lean Construction*, Manchester, UK. Available at: www.iglc.net.

Rotenstreich, S. (2017) 'The Difference between Electronic and Paper Documents', Computer Science Department, The George Washington University (Accessed: 22nd Nov 2017).

Ryle, G. (2009) *The Concept of Mind*. Abingdon, UK, Routledge.

Salvatierra, J.L., Alarcón, L.F., López, A. and Velásquez, X. (2015) 'Lean Diagnosis for Chilean Construction Industry: Towards More Sustainable Lean Practices and Tools', *Proc. 23rd Annual Conference of the International Group for Lean Construction*, Perth, Australia. Available at: www.iglc.net.

Samad, G.E., Hamzeh, F.R. and Emdanat, S. (2017) 'Last Planner System – the Need for New Metrics', *Proc. 25th Annual Conference of the International Group for Lean Construction*, Heraklion, Greece. Available at: www.iglc.net.

Samudio, M., Alves, T.D.C. and Chambers, D. (2011) 'Employing the Principle of "Going and Seeing" to Construction', *Proc. 19th Annual Conference of the International Group for Lean Construction*, Lima, Peru. Available at: www.iglc.net.

Schutz, A. (1972) *The Phenomenology of the Social World*. G. Walsh and F. Lehnert (trans.), Evanston, IL, Northwestern University Press.

Searle, J.R. (1976) 'A Taxonomy of Illocutionary Acts' In *Language and Society*, 5(1), pp. 1–23.

Seed, W.R. (2014) 'Integrated Project Delivery Requires a New Project Manager', *Proc. 22nd Annual Conference of the International Group for Lean Construction*, Oslo, Norway. Available at: iglc.net.

Senge, P.M. (1990) *The Fifth Discipline: The Art and Practice of the Learning Organisation*. London, UK, Century Business.

Seymour, D. (1996) 'Developing Theory in Lean Construction', *Proc. 4th Annual Conference of the International Group for Lean Construction*, Birmingham, UK. Available at: www.iglc.net.

Seymour, D. and Rooke, J. (1995) 'The culture of the industry and culture of research', *Construction Management and Economics*, 13(6), pp. 511–523.

Seymour, D., Shammas-Toma, M. and Clark, L. (1997) 'Limitations of the use of tolerances for communicating design requirements to site', *Engineering, Construction & Architectural Management*, 4(1), pp. 3–22.

Sharrock, W.W. and Anderson, R.J. (1986) *The Ethnomethodologists*. Chichester, UK, Ellis Horwood.

Shewhart, W.A. (1931) *Economic Control of Quality of Manufactured Product*. London, UK, Macmillan.

Shingo, S. (1985) *A Revolution in Manufacturing: The SMED System*. A.P. Dillon (trans), Portland, OR, USA, Productivity Press.

Shingo, S. (1988) *Non-Stock Production: The Shingo System for Continuous Improvement*. Cambridge, MA, USA, Productivity Press.

Simon, H.A. (1996) *The Sciences of the Artificial*. 3rd edition, Cambridge, MA, USA, MIT Press.

Skinnarland, S. and Yndesdal, S. (2012) 'The Last Planner System as a Driver for Knowledge Creation', *Proc. 20th Annual Conference of the International Group for Lean Construction*, San Diego, CA, USA. Available at: www.iglc.net.

Skinnarland, S. and Yndesdal, S. (2014) 'Barriers to a Continuous Learning Process in Construction', *Proc. 22nd Annual Conference of the International Group for Lean Construction*, Oslo, Norway. Available at: iglc.net.

Slivon, C.A., Howell, G.A., Koskela, L. and Rooke, J. (2010) 'Social construction: Understanding construction in a human context', *Lean Construction Journal*, pp. 2–11. Available at: www.leanconstruction.org/lcj/paper_2010_issue.htm.

Smith, J.P. and Rybkowski, Z.K. (2013) 'The Maroon-White Game a Simulation of Trust and Long-Term Gains and Losses', *Proc. 21st Annual Conference of the International Group for Lean Construction*, Fortaleza, Brazil. Available at: iglc.net.

Spear, S. and Bowen, H.K. (1999) 'Decoding the DNA of the Toyota production system', *Harvard Business Review*, September-October, 77(5), pp. 96–106.

Suttie, J.B. (2013) 'The Impacts and Effects of Integrated Project Delivery on Participating Organisations With a Focus on Organisational Culture', *Proc. 21st Annual Conference of the International Group for Lean Construction*, Fortaleza, Brazil. Available at: iglc.net.

Tillmann, P., Ballard, G. and Tommelein, I. (2014) 'A Mentoring Approach to Implement Lean Construction', *Proc. 22nd Annual Conference of the International Group for Lean Construction*, Oslo, Norway. Available at: www.iglc.net.

Tillmann, P. and Sargent, Z. (2016) 'Last Planner & BIM Integration: Lessons From a Continuous Improvement Effort', *Proc. 24th Annual Conference of the International Group for Lean Construction*, Boston, MA, USA. Available at: iglc.net.

Tillmann, P., Tzortzopolous, P., Sapountzis, S., Formoso, C., and Kagioglou, M. (2012) 'A case study on benefits realisation and its contributions for achieving project outcomes', *Proc. 20th Annual Conference of the International Group for Lean Construction*, San Diego, CA, USA. Available at: www.iglc.net.

Tzortzopoulos, P., Cooper, R., Chan, P. and Kagioglou, M. (2006) 'Clients' activities at the design front-end', *Design Studies*, 27(6), pp. 657–683.

Winch, P. (1990) *The Idea of a Social Science and its Relation to Philosophy*. 2nd edition, London, UK, Routledge and Kegan Paul.

Winograd, T. and Flores, F. (1986) *Understanding Computers and Cognition: A New Understanding for Design.* New Jersey, USA, Ablex Publishing Corporation.

Wittgenstein, L. (1958) *Philosophical Investigations*. 2nd edition, G.E.M. Anscombe (Trans.), Oxford, UK, Blackwell.

6 Value generation

Bringing the customer into perspective

Patricia A. Tillmann and Luciana Miron

6.1 The starting point: bringing the customer into perspective

One of the most significant contributions that Lean thinking brought to construction was a greater focus on customers.[1] In production, the rationale of the value generation concept is that the value of a product can be determined only in reference to customers, and the goal of production is to satisfy customers' needs (Levitt, 1960). In this context, generating value means delivering *what the customers want, when they want and in the amount that they want.*

The technical report produced by Lauri Koskela in 1992 is one of the first publications to mention the importance of considering the customer's perspective in construction. The report focuses on the application of the Lean production philosophy to construction and describes the early 1990s paradigm shift in manufacturing. At that time, companies were moving from an internal focus on process improvement towards understanding customer demand as the new basis of competition. Increasingly, businesses were adopting a value-based strategy, becoming customer-oriented, in contrast to competitor-oriented. One of the heuristic production principles suggested during that period reflects this shift:

> *Increase output value through systematic consideration of customer requirements.*[2]

The applicability of the value concept of production to the construction industry was further explained in 2000 by Koskela, in his thesis, *An Exploration towards a Production Theory and its Application to Construction.* There, he contrasts the internally oriented managerial model based on the transformation concept to the emerging model in which the customer is the reference to determine the value of a product (Koskela, 2000). The conceptual roots of value generation (from a production perspective) emerged in the quality movement. In his seminal work, Shewhart (1931) presented the relationship between customers, who have requirements, wishes and expectations, and suppliers, who deliver value to these customers by providing products and services that fulfil their needs. Koskela (2000) uses Shewhart's model to propose five principles to explain value generation:

- P1. Ensuring that all customer requirements, both explicit and latent, have been captured;
- P2. Ensuring that relevant customer requirements are available in all phases of production and that they are not lost when progressively transformed into design solutions, production plans, and products;
- P3. Ensuring that customer requirements have a bearing on all deliverables for all roles of the customer;

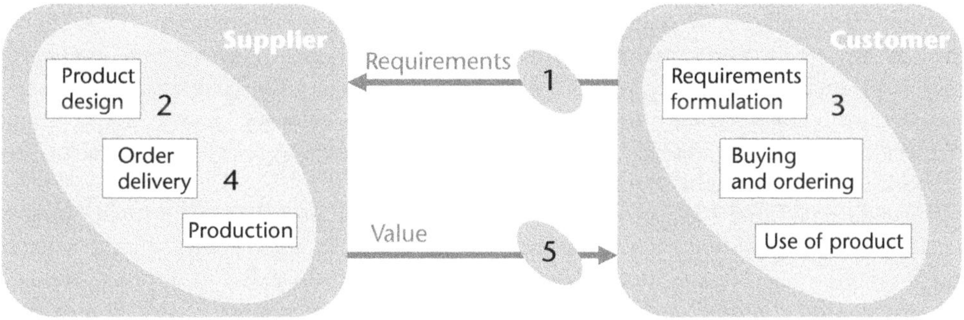

The numbers refer to principles as follows: 1. Requirements capture. 2. Requirements flowdown. 3. Comprehensiveness of requirements. 4. Capability of production subsystems. 5. Measurement of value.

Figure 6.1 Five principles of value generation proposed by Koskela (2000)

- P4. Ensuring the capability of the production system to produce products as required;
- P5. Ensuring by measurement that value is generated for the customer.

By highlighting these principles, Koskela's framework (Figure 6.1) provided a greater understanding of Shewhart's initial customer–supplier model. While Shewhart's model pinpointed an existing customer–supplier relationship, Koskela added elements that further characterise the interaction between these two.

The idea of designing products to satisfy customer's needs can be traced back to the early writings of influential scholars in neoclassical economics: 'the economic measure of satisfaction is that which a person would be just willing to pay for any satisfaction rather than go without it' (Marshall, 1879). Arguably, even the first distinctions of *use-value* and *exchange-value* by Aristotle already had implicit the existence of an individual who has needs to be satisfied.

Throughout the years, production-oriented disciplines developed creative approaches that further consider the customer in product design, i.e. human-centred design (ISO 9241-210, 2010; ISO 13407, 1999), user-centred design (Norman, 1988) and design thinking (Brown, 2008; Dorst, 2011). These approaches focus on enabling greater attention to customer requirements, i.e. requirements management, customer participation, collaborative design approaches, among others.

From the early discussions around the concept of value generation, much has evolved to support the application of the concept in different industries. However, bringing the concept to Lean Construction in a meaningful way is still an ongoing effort. Since Koskela's early publications advocating the inclusion of the customer perspective to the scope of project management capabilities, 25 years have passed and many scholars have engaged with the task of further understanding the concept and its applicability to the construction industry. Some specific questions that scholars have focused on include:

- What is value? How do customers perceive value in the construction industry?
- Who are the customers in construction? Whose perspective should be considered?
- How to manage value in construction? How can we improve project management practices in construction to better focus on the customer's perspective?

In an effort to answer these questions, the Lean Construction community has discussed different ideas and tested approaches to support a customer-orientation. This chapter aims to review and discuss some of these contributions. The starting point was a literature review of papers published in the last 25 years in the International Group of Lean Construction (IGLC) conferences and the *Lean Construction Journal* (LCJ). Papers were selected based on the keywords 'value', 'value generation', 'customer perspective' and 'customer requirements management'.

It was observed that the word 'value' is utilised in a variety of ways and with different meanings.[3] This variation is perhaps due to the multiple applicability of the concept. One clear distinction in the literature, for instance, is the possibility of applying the concept of value for the internal analysis of production, and not only to understand external demand. As the quality movement focused on conformance to specification, the customer–supplier relationship has also been used to improve hand-offs internally in production. This created a distinction between generating value based on the demand of external customers (or buyers) and configuring the production system to meet internal demands, or fulfil conditions of satisfaction between hand-offs.

Although both perspectives contribute to project management in construction, the literature review revealed that, while the internal customer perspective has been widely discussed, understanding value from an external customer perspective and supporting its practical application is a much less explored topic in Lean Construction. Thus, the external customer perspective is the focus of this chapter.

In this chapter, generating value is understood as the *ability to deliver a built environment that is fit for purpose, i.e. fulfils intent and considers the needs of multiple customers (stakeholders) involved, within the limitations of resources (i.e. time and money) imposed by the circumstances of the project.*

The chapter recognises the predominant role of design in generating value to external customers, over physical production. As suggested in Koskela (2000, p. 111),

> In design, the customer requirements are translated into a design solution. In production, this design solution is realised. Thus the functional performance, the primary attribute of customer value, is determined in design, barring defective production. Thus the value aspect in design is much more significant, and by nature different in comparison to production.

In other words, the design process plays a fundamental role in revealing the customers' perspective and helping them to envision and anticipate the value expected from a project. During construction, such vision starts to be realised, and it is fully achieved during the use of the building.

The literature review included not only a selection of major contributions in the Lean Construction literature but also cited related publications from other knowledge domains. The reason for that was to understand how the concept of value has evolved. The result of this exercise is summarised in Figure 6.2. The dots represent key ideas that originated in different fields and have influenced the current understanding of value and value management in Lean Construction. The literature review suggests that the conceptual evolution of value generation has borrowed and benefited from ideas that originated not only in the construction industry but also in different disciplines, such as manufacturing, economics, sociology and marketing. Throughout this chapter, these different perspectives are discussed.

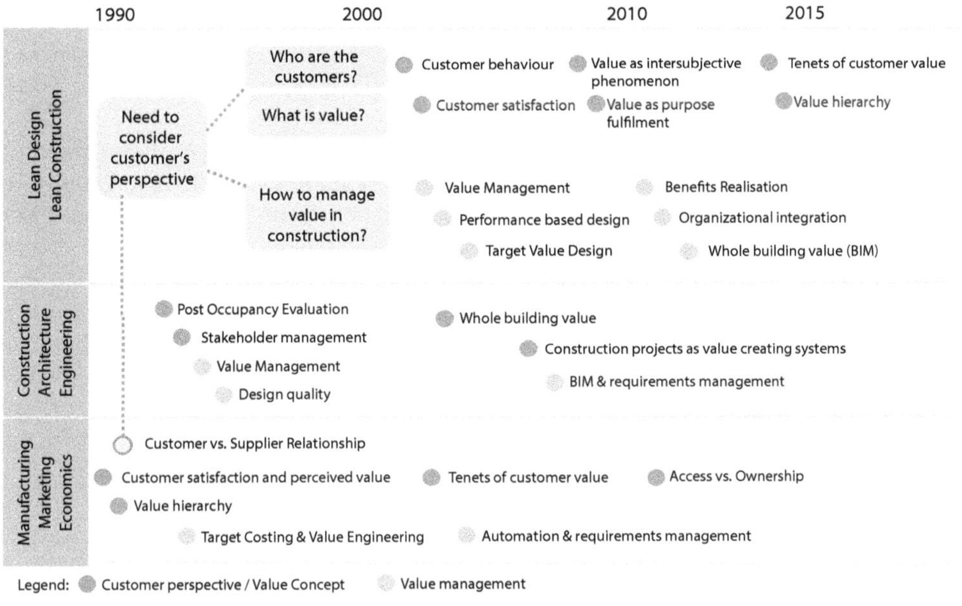

Figure 6.2 Main influences on the value concept in Lean Construction

6.2 Understanding the customers' perspective

6.2.1 Value for whom?

Project definition is the first phase in the delivery of a construction project, and it involves capturing and understanding customer requirements (needs and desires) and translating those into criteria for both product and process design. Several publications within the Lean community have contributed to a better understanding of this process: from understanding what value means and who the customers are, to defining techniques to manage requirements of different stakeholders throughout the project delivery process.

One peculiarity of the built environment as a product is that it is constituted by two elements: the building and the land. The complexity that this brings to value generation is that both land and the building will play a role in the value equation. On the one hand, the building should accommodate a specific use and conform to a wide range of performance requirements, such as safety, habitability, durability, maintainability and economic aspects related to its operation. On the other hand, the land will influence the building characteristics and also establish a relationship with its surroundings (Topalov, 1979). As a result, the built environment creates a variety of social relations with different customer groups, which will have individual expectations and may be affected by the building in different ways. Because of such complexity, a fundamental question was raised to bring the customers' voice into construction, which was: Who are the customers? Whose voice should be considered for generating value?

The notion that construction projects are complex, characterised by a multi-customer (stakeholder) environment, and that multiple perspectives should be considered when generating value, has been a major topic of discussion (i.e. Pearce, 2003; Bertelsen and Emmitt, 2005; Emmitt, Sander, and Christoffersen, 2005; Tzortzopoulos et al., 2006; Tillmann et al., 2011; Drevland and Klakegg, 2017; Formoso and Miron, 2017).

One of the main contributions of these studies is the recognition that managing requirements in a multi-customer environment can be complex and often different customer requirements are conflicting (Tzortzopoulos, Cooper, Chan, and Kagioglou, 2006; Formoso and Miron, 2017). Along with the recognition of such complexity, the debate on whose perception of value counts is still under discussion (i.e. Drevland and Klakegg, 2017; Drevland and Tillmann, 2018). While some researchers recognise the perspective of a paying customer (i.e. Drevland and Klakegg, 2017), others highlight the intricate relationship between different customers (or stakeholder) groups and their often different ways of perceiving value (Drevland and Tillmann, 2018). These observations perhaps help explain the challenge for understanding not only who should be involved in the value definition, but also how to manage multiple (often conflicting) requirements throughout the delivery of a built facility.

6.2.2 Concepts that have helped to understand the customer's perspective

Since the introduction of the customer perspective in the Lean Construction literature, scholars and practitioners have relied on concepts originated in other fields, i.e. economics, sociology and marketing, to explain what generating value for customers means. In particular, the area of marketing is pointed out as fertile for the role it plays in providing information about customer perceptions to new product development (Griffin and Hauser, 1996). Some of the proposed concepts from marketing and their application to construction have been analysed, i.e. (Miron and Formoso, 2003, Santos, Kristmann, and Fischer, 2004, Miron and Formoso, 2010, Rooke et al., 2010, Bonatto, Miron, and Formoso, 2011, Brito and Formoso, 2014, Drevland and Lohne, 2015, Nascimento da Silva and Miron, 2017, Drevland, Lohne, and Klakegg, 2018). These authors highlight important contributions to further understand what value from a customer perspective is. Some of the concepts discussed in these studies include: customer-perceived value, customer personal values, customer satisfaction, value as an intersubjective phenomenon and value as purpose fulfilment.

6.2.2.1 Customer perceived value

Extensive research in marketing has focused on the concept of perceived value. From an overview of major publications, Sánchez-Fernández and Iniesta-Bonillo (2007) defined two ways to approach customer-perceived value: as a uni-dimensional or a multi-dimensional construct. The first represents value as a single concept based on a utilitarian perspective. Although this perspective considers that a uni-dimensional construct might be produced by the effects of multiple antecedents, it does not ponder value as an aggregate concept formed from several components (Sánchez-Fernández and Iniesta-Bonillo, 2007). Based on this perspective, Monroe (1990) argues that customers' perception of value represents a trade-off between the benefits they perceive in the product and the sacrifice they perceive by paying its price (Monroe, 1990).

$$\text{Value} = \text{benefits}/\text{sacrifices}$$

Benefit is what *one gets* and sacrifice is what *one gives* in order to get the benefits. The perceived benefits are related to customers' judgements about product quality (Monroe, 1990) and a combination of physical and service attributes (Ravald and Gronroos, 1996). By contrast, the total sacrifice includes the sum of the purchase price, repairs and maintenance, risk of failure or poor performance (Ravald and Gronroos, 1996). Therefore, according to this perspective, customer-perceived value involves a comparison between the *utility* provided by a product (positive aspects, benefits) and its *disutility* as a result of the payment (negative aspects, sacrifices). Although this uni-dimensional approach possesses the merit of simplicity, it does not explore the complexity of consumers' perceptions of value (intangible, intrinsic and emotional factors/dimensions) (Sánchez-Fernández and Iniesta-Bonillo, 2007).

Conversely, the multi-dimensional approach understands perceived value from several interrelated dimensions (e.g. utilitarian, hedonistic, social and altruistic) that form a holistic representation of a complex phenomenon (Sánchez-Fernández and Iniesta-Bonillo, 2007). This approach is represented in the synthesis proposed by Sánchez-Fernández and Iniesta-Bonillo (2007):

(i) The nature of perceived value is complex and multi-dimensional;
(ii) The concept of perceived value implies an interaction between a customer/user and a product;
(iii) Value is relative by virtue of its comparative, personal and situational nature; and
(iv) Value is preferential, perceptual and cognitive-affective by nature.

In addition, Holbrook (2006) explains that the 'customer value entails subjective hierarchical preferences based on an individual's situation-specific comparisons of one object (product) with another' (Holbrook, 2006, p. 715). Holbrook (2006, p. 716) also categorises customer value as extrinsic (or utilitarian value) and intrinsic (or hedonic value):

(i) Extrinsic: economic value (product or consumption experience serves as a means to a customer's objectives); social value (status-enhancing favourable impression or esteem-evoking material possessions);
(ii) Intrinsic: hedonic value (pleasure in consumption experiences); altruistic value (consumption behaviour affects others).

These concepts, borrowed from marketing, have influenced the understanding of value within the Lean Construction community. One example is the nine tenets on the nature of value published by Drevland and Lohne (2015) and Drevland, Lohne, and Klakegg (2018), based on the seminal classification offered by Holbrook (2006). From these tenets, a comprehensive definition of value was proposed. However, according to the authors, this definition has limitations because it is too voluminous to be practical. Although Holbrook's definition helps to understand the complex nature of perceived value, its complicated structure makes it difficult to support its practical application (Sánchez-Fernández and Iniesta-Bonillo, 2007; Drevland and Lohne, 2015; Drevland, Lohne, and Klakegg, 2018).

6.2.2.2 Customer (personal) values

Value, as perceived by the customers, is different from *values*. The latter can also be viewed as sociological values (Rooke et al., 2010) or personal values (Rokeach, 1973; Woodruff, 1997). Personal values are intrinsic beliefs that consist of mental representations of need, and which are used by subjects as a base for equating decision and conflict processes (Woodruff, 1997). Thus, customers' personal *values* will influence their perception of *value* (Gutman, 1982; Woodruff, 1997; Woodruff, Schumann, and Gardial, 1993), but those two terms do not have the same meaning. While *values* are related to core beliefs, morals and ideas (Thomson et al., 2003), *perceived value* is related to a judgement of an object by a subject.

From social psychology, Rokeach (1973) is the most quoted scholar in the topic of personal values. He distinguishes personal values into instrumental (means) and terminal (ends) values when considering the existence of a functional relationship between both. The instrumental values mean the favourite social conduct and behaviour modes to achieve personal goals, while terminal values represent the favourite final states of existence (Rokeach, 1981). Rokeach's studies had a great influence on the development of the Hierarchical Value Maps (HVM), a technique used to understand customer-perceived value. HVM was built based on a means–end chain (Gutman, 1982), which connects the customer's perception about concrete attributes of a product/project (objective and tangible attributes) to their personal values (subjective and intangible objectives). Some scholars have analysed the applicability of the HVM to understand the customer's value in housing studies, i.e. Brito and Formoso (2014), Hentschke et al. (2014).

6.2.2.3 Customer satisfaction

Customer satisfaction is another concept extensively studied in marketing research. Based on literature review and empirical data collection, Giese and Cote (2000) conclude that 'satisfaction is a summary affective response of varying intensity; with a time-specific point of determination and limited duration; directed toward focal aspects of product acquisition and/or consumption'.

Although measuring satisfaction may be an important indicator, it is not sufficient to understand the complexity of customer perception. Some marketing scholars have identified that the perceived value can be understood as an antecedent of customer satisfaction (Woodruff, 1997; Woodruff, Schumann, and Gardial, 1993). Both concepts, namely 'customer satisfaction' and 'customer-perceived value', describe evaluative judgements about a product in use situation (Woodruff, 1997).

Figure 6.3 shows an interpretation where perceived value is an antecedent of disconfirmation that results in satisfaction. Value is perceived in two different moments: before the acquisition (desired value) and after the use of the product (received value). Oliver, Balakrishnan, and Barry (1994) explain customer satisfaction as an output of the disconfirmation-type satisfaction model: they compare the desired value to received value and confirm or disconfirm their expectations, resulting in satisfaction or dissatisfaction.

The relationship between satisfaction and perceived value have been analysed by different scholars in the context of low-income housing (Lima et al., 2009; Miron and Formoso, 2010; Bonatto, Miron, and Formoso, 2011; Morais, Carneiro and Neto, 2014). Lima et al. (2009) and Miron and Formoso (2010) proposed conceptual models that incorporated: the ratio between benefits and sacrifices, and the value hierarchy within the set of benefits. Thus, the percentage of satisfaction could be analysed for the different

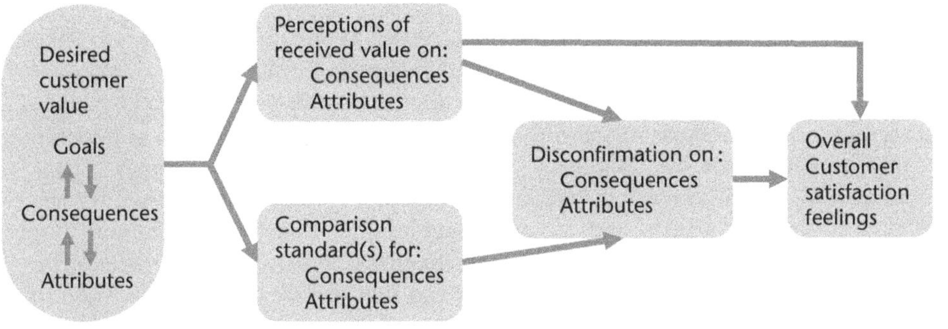

Figure 6.3 The relationship between customer value and customer satisfaction (adapted from Woodruff, 1997)

levels of the hierarchy of benefits. Bonatto, Miron, and Formoso (2011) proposed a method for evaluating value generation based on the hierarchical perspective. The evaluation of customers' satisfaction was focused on aspects beyond the attributes of the products, such as the consequences of use and project objectives.

6.2.2.4 *Perceived value as an intersubjective phenomenon*

There has been reasonable debate about understanding value as a subjective or an objective phenomenon. According to Thomson et al. (2003):

(i) Value can be *subjective* – when remaining internalised within an individual or organisation, or when framed against a customer's values; or
(ii) Value can be *objective* – when expressed and negotiated in a common language by individuals and organisations within a project, or when the relationship between benefits and sacrifices is compared with appropriate units of measurement (unidimensional, utilitarian perspective).

Within the Lean community, Rooke et al. (2010) argue that 'objective' and 'subjective', rather than being mutually exclusive categories, are more like points on a continuum in which objectivity is socially established from the stream of our perceptions, the so-called intersubjective control. According to these authors, perceived value is an intersubjective phenomenon. The perceived value can be governed by sociological values (principles, guidelines for living, or personal values), but it is a social construct.

Understanding value as an intersubjective phenomenon presupposes a collective perception of the value of an object. If an individual changes his/her perception of value over an object, that would be just the opinion of one individual. In order to change the collective perception, that individual would have to convince groups of people to develop the same opinion. For the perceived value is not a subjective perception existing in the imagination of one individual – neither does an objective phenomenon exist independently of human consciousness and human beliefs (personal values). Perceived value is rather an intersubjective perception, existing in the shared imagination of a representative number of people.

6.2.2.5 *Value as purpose fulfilment*

Past research (i.e., Saxon, 2005; Winter et al., 2006; Winter and Szczepanek, 2008) has criticised the predominant understanding of value generation in the construction industry. They argue that project management practices are traditionally focused on the creation and delivery of physical products to specification, cost and time without much consideration of how these physical assets will ultimately contribute to fulfil a purpose, or intent.

Winter et al. (2006) argue that buildings are often expected to contribute for the realisation of business strategies and for the generation of benefits to different customers groups, i.e. stakeholder, users, investors, etc. Therefore, generating value is ensuring that the built facility provides the right capabilities and that those capabilities are used to realise expected benefits.

Such a perspective inspired the exploration of different concepts within the Lean Construction community, such as through life management (Koskela, Siriwardena, and Rooke, 2008) and benefits realisation (Rooke et al., 2010; Tillmann, Tzortzopoulos, and Formoso, 2010; Tillmann et al., 2012a; Kagioglou and Tzortzopoulos, 2016). While the former proposes a more holistic view of value as purpose fulfilment throughout the entire building life cycle, the latter suggests the active management of expected intent throughout project delivery – from generation of capabilities to the realisation of expected benefits.

In practice, this view is aligned with project delivery models that include not only design and construction but also the responsibility over the facilities' operation – shifting the focus from the delivery of a physical product to the provision of a solution. Conceptually, it approaches an Aristotelian idea about the value of an object: 'true value is intrinsic to the natural end the item serves'. This perspective of value brings a distinction between the intrinsic (valuable in and of itself) and instrumental value (valuable as means to achieve an end) of a built facility.

6.2.3 Considerations when addressing customers' perspective

Based on the literature review for this session, the following characteristics have been identified as relevant when addressing the value concept in construction:

- Value for whom
    - Projects are complex, characterised by a multi-customer environment: paying customers, user customers, project stakeholders, suppliers and the society as a whole. Often, these different groups have requirements that are conflicting and need to be properly managed.
- Concepts that have helped in understanding customers' perspective
    - Customer-perceived value: uni-dimensional construct (utilitarian) or multi-dimensional construct (e.g. utilitarian, hedonistic, social, altruistic). The cognitive trade-off between benefits and sacrifices has been the most quoted and operationalised concept of perceived value as a uni-dimensional construct. The HVM is an approach that relates multi-dimensional components inside customer-perceived value.
    - Customer values: personal values are intrinsic beliefs that influence customer perception of value. Thus, 'value', as perceived by the customers, is different from 'values';

- Customer satisfaction: satisfaction and perceived value are important indicators of products/built environment;
- Value as an intersubjective phenomenon: perceived value is an intersubjective phenomenon, resulting from a negotiation among a representative number of customers; and
- Value as the fulfilment of a purpose: perceived value is also related to how well an object fulfils the intent it serves for.

6.3 Efforts to manage value in construction projects

At the beginning of the 2000s, Koskela, Huovila, and Leinonem (2002) explained that the main principle of value generation is to eliminate value loss by analysing requirements rigorously, systematically managing their flow-down and seeking optimisation. The same authors suggested the adoption of 'value management' as a name for the practical application of the concept and recommended the use of some methods adopted in the Japanese manufacturing industry.

Since then, different scholars have made contributions to support value management in construction: (i) developing strategies to engage stakeholders in defining requirements; (ii) proposing requirements in management guidelines, methods and techniques; (iii) testing the advantages and disadvantages of these different strategies; and (iv) analysing the applicability of technology to support requirements' management. This session discusses some of these key contributions organised in the following items: understanding customers' requirements; from customers' requirements to product specification; working within the boundaries of economically feasible solutions; the role of project integration in managing value; and considerations for managing value in construction projects.

6.3.1 Understanding customers' requirements

6.3.1.1 Actively engaging stakeholders in defining the value proposition

Efforts to support the active participation of the customer in the design of built facilities have been underway for a long time. In the UK, research to develop and test value management approaches – which would later influence the work of other researchers within the Lean community – date back to the 1980s and 1990s. One example was a benchmark comparison of different methods to support value management in construction (Kelly and Male, 1993). This effort resulted in the proposition of frameworks (i.e. Kelly and Male, 1993; Green, 1994) and a diversity of definitions, procedures and standards for supporting value management in the construction industry (i.e. Male et al., 2007).

A key contribution of these efforts was to better understand the role of collective decision making during the design process as a key component to value generation. Such contribution is reflected in early definitions of the concept:

> value management is concerned with developing a common understanding of the design process and identifying explicitly and agreed statements of design objectives. It focuses on establishing a common decision framework in project settings, around which participants can think and communicate.
>
> (Green, 1994, p. 51)

Another influential concept developed in the late 1990s was the Design Quality Indicator (DQI). The DQI is still used to assist project customers (stakeholders) in defining priorities and assessing the quality of design based on the three fundamental Vitruvian Principles: *durability*, *utility* and *aesthetics* (Gann, Salter, and Whyte, 2003; Thomson et al., 2003). The proposed framework evaluates project design based on the ability to meet a variety of physical, aspirational and emotional needs of customers. Gann, Salter, and Whyte (2003) explain that these three elements provide a framework by which different customers can organise their needs and priorities to support the design process.

These contributions had an influence on early discussions about value management in the Lean Construction community. One example was the work of Emmitt, Sander, and Christoffersen (2004, 2005), also described in Thyssen et al. (2010). They presented a value management approach introduced in a Danish construction company in which customers discuss and agree on 'value parameters' to drive design efforts. The approach involves a series of workshops to define requirements and priorities and track design against set criteria (Figure 6.4). It also brings a conceptual framework to assess design quality. The authors used such approach as a way to discuss different customers' expectations and priorities, by means of using a common framework to steer the design process.

The key benefit of these early studies was dealing with a subject that is still currently discussed within the Lean community: value as an intersubjective phenomenon and the need to promote dialogue among the different customers to define and steer project delivery towards an agreed set of goals. Similar techniques are still utilised to support value generation in construction and researchers report on their benefits for customers' alignment and increased collaboration to achieve an agreed purpose (Schöttle and Tillmann, 2018).

A more recent approach that has also influenced studies within the Lean community is Benefits Realisation Management (BRM) (Kagioglou and Tzortzopoulos, 2016). It emphasises the need to actively manage the realisation of benefits expected from investments, suggesting that the relationship between project outputs and expected outcomes must be fused, and seen as part of an integrated whole. BRM has its roots on the Benefits Realisation Approach (BRA), developed in the Information Systems and Technology (IS/IT) sector in the 1990s. BRA is concerned with the systemic understanding of interdependent elements to generate the desired outcome (Thorp, 1998). It suggests aligning customer (stakeholder) interests and expected benefits through discussion and agreement and establishing project governance as a way to monitor the realisation of benefits (Ward and Daniel, 2006). The approach also touches on the dynamicity of customer requirements management, suggesting that a project's objective and means for its achievement might change over time or require further refinement; thus, monitoring the realisation of benefits also includes a learning and adaptation component (Farbey, Land, and Targett, 1999).

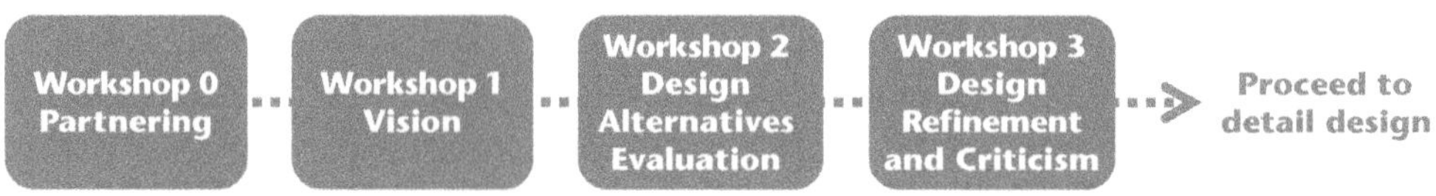

Figure 6.4 Steps of a value management technique proposed in Thyssen et al. (2010)

The BRA inspired the development of the BeReal model, a model that was tested in the context of healthcare facilities by Sapountzis, Harris, and Kagioglou (2008) and Sapountzis et al. (2010). It recommends a continuous cycle to support value generation: (a) envisioning outcomes; (b) implementing the necessary actions to achieve the outputs that will contribute to expected outcomes; (c) checking intermediate results; and (d) dynamically adjusting the path leading from investment to outcomes (Sapountzis, Harris, and Kagioglou, 2008).

The BeReal model follows a similar structure of more traditional value management techniques (i.e. Thyssen et al., 2010). The main difference, however, is the shift from a focus on design quality to making more explicit the overarching expected benefits of a construction project, while planning for and tracking their realisation (Tillmann et al., 2012a; Tillmann, Tzortzopoulos, and Formoso, 2010). In the BeReal model, customers (key stakeholders) get together to assess multiple design options based on their opinion on how likely those options are able to fulfil the expected outcomes of investment. Similarly to the benefits of other value management approaches, one of its major contributions is to tackle the intersubjective component of value generation (Rooke et al., 2010).

Several other techniques were developed and tested within the Lean community to promote the active engagement of customers and to capture their preferences during the design process. Examples that became popular and are widely used in the US include 'set-based design' and 'choosing-by-advantages' (CBA) (Parrish et al., 2008; Arroyo, Tommelein, and Ballard, 2012, 2013; Kpamma et al., 2014a, 2017; Arroyo et al., 2016; Arroyo and Long, 2018; Schöttle, Arroyo, and Christensen, 2018). While set-based design allows for decisions to be delayed and multiple alternatives to be developed (Parrish et al., 2008), CBA supports sound decision making using specific comparisons of advantages of alternatives and generating an environment that integrates multiple perspectives, creates transparency, separates 'value' from cost, and documents the decision-making rationale (Arroyo et al., 2016; see also Chapter 10).

In addition to these different strategies of engaging customers to define and make more explicit the purpose of projects, some authors have also analysed the benefits of simply engaging customers in the design process, regardless of whether a specific technique to make the value proposition explicit is used (i.e. Sfandyarifard and Tzortzopoulos, 2011; Kpamma et al., 2014b; Bascoul et al., 2018). These studies indicate benefits such as: (a) opportunities for customers to voice their views; (b) ensuring requirements are not lost when translated into design solutions; and (c) bringing expertise upfront from downstream players to allow for better solution development.

6.3.1.2 Using available data to understand the customers' requirements

Other means to understand customers' requirements in the early stages of design without necessarily actively engaging them is the analysis of data collected from the performance of existing buildings. Post Occupancy Evaluation (POE) has historically (since the 1960s) been an important approach for measuring whether customer value has been delivered by obtaining feedback about a building's performance in use, including energy performance, Indoor Environment Quality (IEQ), occupants' satisfaction, productivity, etc. (Li, Froeseb, and Bragerc, 2018).

POE methods can broadly include energy and water assessment, indoor environmental quality physical measurements, occupant survey questionnaires, focus group meetings, structured interviews, visual records, walkthroughs, and technical measurement of building structure, services and systems (Sanni-Anibire, Hassanain, and Al-Hammad, 2016). However, Li, Froeseb, and Bragerc (2018) observe that the application of questionnaires to get building occupants' feedback is still the predominant method used for POE. Questionnaires generally aim to understand satisfaction levels with several aspects of the building: aesthetics, functionality, comfort, etc. Such information is used by designers to understand how building attributes can better support customer (user, occupant) activities, informing the design of new facilities. Current research is also exploring the applicability of Building Information Modelling (BIM), Geographic Information Systems (GIS) and other software to facilitate the visualisation of data in space (Li, Froeseb, and Bragerc, 2018).

The application of POE to inform the design of new buildings has been an extensive topic of research within the Lean community since the early 2000s (i.e. Santos, Kristmann, and Fischer, 2004; Lima et al., 2009; Miron and Formoso, 2010; Bonatto, Miron, and Formoso, 2011; Morais, Carneiro, and Neto, 2014; Hatoum et al., 2018). More recently, established frameworks (i.e. HVM) have also been used for collecting, analysing and displaying data from POEs (i.e. Bonatto, Miron, and Formoso, 2011; Brito and Formoso, 2014; Hentschke et al., 2014; Nascimento da Silva and Miron, 2017). The application of such an approach has been explored to fulfil different purposes: understanding the outcomes of building assessments (Bonatto, Miron, and Formoso, 2011; Hentschke et al., 2014), supporting data collection and processing (Brito and Formoso, 2014), overlapping qualitative and quantitative data (Bonatto, Miron, and Formoso, 2011; Hentschke et al., 2014), comparing the perception of different customers (Bonatto, Miron, and Formoso, 2011; Nascimento da Silva and Miron, 2017) and as a visual device to support decision making.

In addition to the extensive use of POE, recent research has been done to evaluate the possibility of using Evidence-Based Design (EBD) to inform the design of newly built facilities, especially in healthcare (Rybkowski and Ballard, 2008; Zhang, Tzortzopoulos, and Kagioglou, 2016, 2017). EBD is 'the conscientious and judicious use of current best evidence, and its critical interpretation, to make significant design decisions for each unique project. These design decisions should be based on sound hypotheses related to measurable outcomes' (Hamilton, 2006). Rybkowski and Ballard (2008) list some examples of health benefits associated with EBD decisions. Those include: reduced length of stay thanks to views of foliage and sunlight (Walch et al., 2005), reduced patient falls thanks to rubberised flooring, reduced hospital-acquired infections thanks to single patient rooms, reduced drug costs thanks to patient stress reduction from quieter rooms, reduced nursing turnover thanks to a less stressful work environment, increased market share, and increased philanthropy thanks to more patient-oriented spaces (Joseph, 2006a, 2006b). EBD can be applied to other building types as well, though the current focus on healthcare facility design is likely due to two realities: (a) data is easier to collect in the healthcare arena, and (b) the financial stakes linked to occupant health are especially high (Rybkowski and Ballard, 2008).

In summary, different approaches have been used in the construction industry to consider customers' requirements better. These approaches range from advocating the active engagement of customers in early development phase (whether supported by tools or not) and the use of collected data on previous building performance to inform design decision making. Although the approaches are different, they are non-exclusive and project teams tend to use a combination of both to bring the customer voice into perspective.

6.3.2 From customers' requirements to product specification

As part of moving the industry towards a more customer-centred approach, different studies have focused on the translation of customers' requirements into product specifications by exploring the applicability of 'Performance-Based Design' in combination with Lean methods to support such translation (i.e. Koskela, Huovila, and Leinonem, 2002). The performance approach can be characterised as a practice of thinking and working in terms of ends rather than means. It is about describing what the building is expected to do, and not prescribing how it is to be realised (CIB, 1982). One method for implementing such an approach is to identify the users and the activities taking place in the building. Based on their requirements and the surrounding conditions (climate, existing buildings, etc.), quantitative performance requirements can be set. The technical solutions proposed during the design phase can then be verified against the set performance requirements. If they match, then the proposed technical solution is approved.

Early studies focused on testing the effectiveness of some Lean methods to support a performance-based approach to design, particularly the Quality Function Deployment (QFD) (Huovila and Seren, 1998; Leinonen and Huovila, 2000; Koskela, Huovila, and Leinonem, 2002; Lima, Formoso, and Echeeveste, 2008; Sampaio and Barros Neto, 2010; Björnfot and Bakken, 2013). QFD has been successfully used in new product development by Japanese manufacturing companies since the 1960s with the purpose of translating the customer's voice into specifications for a product (Akao, 1990).

QFD application to construction has led to some observed benefits, such as to support the active engagement of customers on defining priorities in the early stages of design (Leinonen and Huovila, 2000) and to structure the analysis of information from Post Occupancy Evaluations (Leinonen and Huovila, 2000; Lima, Formoso, and Echeeveste, 2008). However, limitations have also been identified: (a) its use can be laborious (Huovila and Seren, 1998); (b) its success is highly dependent on how well qualitative data about requirements is translated into measurable attributes (Lima, Formoso, and Echeeveste, 2008); and (c) it has an excessive focus on physical/functional solutions and fails to adequately consider human well-being (Björnfot and Bakken, 2013). Recent studies are breaking away from the rigid structure of QFD and looking for more organic alternatives to bridge the gap between customer requirements, product specifications and quality criteria by connecting upstream and downstream customers/stakeholders in identifying and discussing expectations, assuring that the design intent is kept and successfully realised during construction (Spencley et al., 2018).

There is also a movement towards the consideration of whole building standards and measures to guide the design of high-performance buildings – as opposed to only functional aspects (Sands, 2010; Sands and Abdelhamid, 2012). That includes focusing on requirements such as energy conservation, environment, safety, security, durability, cost–benefit, occupant productivity, functionality and operational considerations. For requirements that are easier to quantify, i.e. energy consumption and seismic requirements, technology adoption is playing an important role in simulating performance and allowing designers to make better-informed decisions (Arayici et al., 2018).

Kiviniemi (2005) suggested that technology can be beneficial to translate the customers' perspective into product specifications. Such support can be provided in different forms: (i) by creating a formal structure for modelling requirements; (ii) by enabling the creation of requirement templates, which may contain a large amount of information, being possible to define subsets of requirements for different types of projects; and (iii) by storing data that can be compared not only with design solutions but also with maintenance information throughout the building's life cycle. In this sense, more recent studies are exploring potential benefits of technology and especially automation for customer requirements management (i.e. Kim et al., 2015, 2013). A similar trend can be seen within the Lean community (i.e. Jylhä and Junnila, 2012; Baldauf, Miron, and Formoso, 2013; Beck et al., 2016; Soliman Junior et al., 2018).

The translation of requirements into product specifications has always been a challenge in the construction industry. Subject to a multitude of requirements, project teams are confronted with managing multiple customer requirements while still having to meet technical performance criteria. The future of requirements management research will likely focus on technology adoption to support the effort. There is a consensus in the literature about the need to achieve performance-based design via integrated BIM use in construction. However, current problems with interoperability of BIM and lack of clear guidance on how existing standards can be used are still posing challenges to effectively adopt technology in support of requirements management (Arayici et al., 2018).

6.3.3 Working within the boundaries of economically feasible solutions

Koskela and Ballard (2006) advocate that the role of project teams in construction is to help customers to clarify their needs and purposes and fulfil them within project constraints, e.g. time, costs. Target Value Delivery (TVD) is a Lean approach that involves a collaborative effort between suppliers and customers to make better decisions about their needs, understanding constraints and the (economic) consequences of their requests (Ballard, 2008, see also Chapter 8).

The origins of TVD can be traced back to Target Costing (TC), a practice used in new product development and popular in the car manufacturing industry in Japan. Originally introduced in Japan under the name of *Genka Kikaku*, TC is an approach to reduce the overall cost of a product over its entire life cycle, with the help of a multidisciplinary team, formed by the different departments of a company and the active contribution from the supply chain (Kato, 1993). TC has been defined as a 'system of profit planning and cost management that is price led, customer focused, design centred, and cross-functional' (Ansari, Bell, and Okano, 2007). In TC, the market price for a new product is not determined by adding a profit margin to cost (e.g. cost-plus approach). Instead, an allowable cost is determined by a target price less an appropriate profit margin (Kato, 1993).

One fundamental technique supporting that approach is Value Engineering (VE). VE provides the means for a customer to collaborate with the supply chain in defining what value is attached to different aspects of the performance specification, and then offers a structured way for pooling information on the cost and functional impact of design options, so that collective decisions can be made (Nicolini et al., 2000). Based on product performance criteria and expected functionalities, project teams can then analyse alternatives that will achieve the same results but present a lower cost of production, thus having a lower final cost.

TVD has been widely used in the United States (US), with the first successful application in the construction industry reported in 2004 (Ballard and Reiser, 2004). Since then, significant improvement in delivering built facilities for an agreed scope and within expected cost has been reported (Conwell, 2012). Also, several construction projects have reportedly used such approach to achieve final costs that were lower than market price (Zimina, Ballard, and Pasquire, 2012). The lessons learned from several applications in different projects have led to the publication and periodic revision of a benchmark to support its adoption in construction (e.g. Ballard, 2011).

The term 'Target Value Design' (later named Target Value Delivery) was coined by Macomber, Howell and Barberio in 2007. The intent was to replace the name 'Target Costing' as a reminder that project teams should not be focused on achieving cost predictability but also should be concerned about delivering value (or benefits) to customers. Recent documentation of practical application of TVD shows that along with typical elements of TVD (i.e. continuous cost estimating and cost transparency), some project teams use different strategies to define and steer design towards agreed goals beyond cost targets (Tillmann, 2018). These include design goals, programmatic requirements and priorities, as well as performance targets, i.e. seismic performance, energy consumption, durability and constructability requirements.

One of the key benefits of TVD is allowing designers to work within the boundaries of feasible solutions. Recent research is exploring the applicability of cost modelling software to support the engagement of customers in the process (i.e. Pennanen, Ballard, and Haahtela, 2011; Alwisy, Bouferguene, and Al-Hussein, 2018a, 2018b). These studies not only suggest improved methods for continuous cost estimating, but also provide better ways to clarify design priorities and assess the consequences of design decisions.

According to Silveira and Alves (2018), even though the practice of TVD is not widespread, a number of TVD-inspired practices are currently being used in several projects. Thus, efforts have been made so that processes, methods and benefits associated with TVD are better understood and effectively adopted by customers and project stakeholders (Alves, Lichtig, and Rybkowski, 2017). While TVD has been commonly adopted in Integrated Project Delivery (IPD) projects (see Chapter 9 by Alves, Lichtig for further details), studies suggest that the use of principles, such as shared incentives, better contractor selection, co-location, and training of key project personnel, can support achieving the level of collaboration required for successful TVD implementation even when projects are not signatories of a multi-party agreement (Melo et al., 2015). Despite the observed variability in its implementation, Nanda et al. (2017) observed that the use of TVD with IPD practices is preferred in the industry over more traditional delivery and project management methods.

6.3.4 The role of project integration on managing value

According to Matthews and Howell (2005), the major challenge to focus on value generation in construction lies in the way projects are traditionally set up. It is difficult to focus on generating value from projects when companies are bound by contracts that inhibit coordination, stifle cooperation and innovation, and reward individual contractors for both reserving good ideas and optimising their performance at the expense of others.

Recently, the industry has been moving towards more integrated delivery methods, i.e. Project Alliancing and Integrated Project Delivery (IPD). The impact of such integration on supporting value generation is the subject of study within the Lean community (i.e. Tillmann et al., 2012b). In addition, research analysing the impact of these integrated practices to project performance has demonstrated the benefits these practices bring for the delivery of higher-quality facilities at no significant cost premium (Asmar, Hanna, and Loh, 2013).

Thomsen et al. (2009) explain, for instance, that IPD brings key elements that help a team achieve greater alignment in goals and expectations. Some key elements are:

- Key customers, partners and team members are involved upfront, in the early phases of project development;
- Key partners (at least owner, general contractor and architect) make decisions by consensus;
- Teams are built with emphasis on trust, collaboration and learning – goals are mutually agreed and performance measured and improved throughout the project;
- Methods and techniques to support teamwork are agreed upon, i.e. Lean Construction tools and techniques;
- Risks are pooled with each party contributing contingency and sharing potential costs overruns; and
- Project savings and any unused contingency is shared at the end.

Past research indicates that, in projects with these characteristics, projects teams have greater support on understanding the project's value proposition, since there is an active pursuit of alignment among customers from early phases of the project up to completion (Tillmann, 2018; Tillmann et al., 2012b). One particular aspect that supports value generation is the governance structure that is set out in those types of projects, which allows owners to be active members of the team, and allows decisions to be made in a collaborative environment, involving – at least – owner, architects and contractors.

Such structure, coupled with the early engagement of project team members, allows key parties to be highly and actively involved in understanding and defining the project's value proposition as well as understanding what each party's key contribution to generating value is. Nonetheless, Melo et al. (2015) observe that the use of some IPD principles such as shared incentives, better contractor selection, co-location, and training of key project personnel, can facilitate the achievement of the level of collaboration required for the adoption of supporting techniques to value generation (i.e. TVD). Similarly, Kagioglou and Tzortzopoulos (2016) suggest that some value management techniques (i.e. Benefits Realisation) are more effective when organisational integration exists.

Structuring work in a more integrated manner is also allowing project teams to use technology more efficiently in support of value generation. With all major building specialties being involved from the early stages of design, together with the close relationship with the customers, building systems are not only designed with better consideration of customers' requirements, but are also better coordinated, thus avoiding potential gaps between design intent and the built facility.

The use of BIM has also played an important role in allowing for such a collaborative work environment by providing functionalities that help teams in dealing with some of the challenges of managing value. Some of these functionalities are (Sacks et al., 2009): (a) the possibility to visualise form (for aesthetic and functional evaluation); (b) rapid generation and evaluation of multiple construction plan alternatives; (c) maintenance of information and design model integrity; (d) online/electronic object-based communication; (e) automated generation of drawings and documents; and (f) collaboration in design and construction. Although some scholars have been dedicated to understanding the benefits of technology for supporting a more integrated form of organising work, opportunities for future research in this area still abound.

6.3.5 Considerations for managing value in construction projects

Although existing approaches follow different strategies, they generally have a common concern to generate knowledge about customers' requirements, which can be used to create solutions and products. A particular concern during the development of these methods has been to support the engagement or consideration of multiple customers (stakeholder) groups while also taking into account technical performance requirements. However, they differ in strategy in regards to advocating an active or passive participation of those customers in the design process. In addition, while some approaches are geared towards generating explicit design goals (exploring both stated explicit requirements and attempting to understand implicit ones), others are focused on helping designers to define and look for solutions within project limitations and constraints.

The different value management approaches reviewed in this chapter reflect the multifaceted characteristic of the concept of value and the need to consider and effectively manage customers' requirements within the limitations set by the circumstances of projects. It was observed that these approaches are rarely used in isolation. Rather, they have generally been used in a complementary fashion. Table 6.1 summarises the approaches, methods and techniques reviewed in this session and their main contribution to the concept of value.

A trend in the literature was observed, from a focus on design quality to include upstream activities, tackling intent, and investment priorities. Also, increasing interest has been shown on the consequences of integrating the supply chain to value generation and value management, as well as how technology advancement can support the effort.

Table 6.1 Summary of value management techniques and their support for operationalising the value concept

Approaches, Methods and Techniques	*Characteristics of value concept it* primarily *focuses on:*
Value management workshops, Design Quality Indicator, Benefits Realisation, Set-based Design and Choosing by Advantages	Intersubjective aspect resulting from a negotiation among different customers/ stakeholders on their perceived value of a design
Post Occupancy Evaluations, Evidence-Based Design, Hierarchical Value Maps	Customer satisfaction, Perceived value (uni-dimensional and multi-dimensional construct), Customer values
Quality Function Deployment, Performance-based design, Requirements management automation through BIM	Requirements management – translation of customers' voice into product specifications
Target Value Design, Cost modelling	Benefits/Sacrifices equation
Benefits Realisation, Through Life Management	Value as purpose fulfilment, tied to an ultimate intent

6.4 Concluding remarks and recommendations for future research

Since Koskela's early publications on value generation, over 25 years have passed and many contributions have been made to advance the understanding of value generation and value management in the Lean Construction literature. The aim of this chapter was to review and discuss these key contributions and their implication for project management in construction. In this session, the main findings are summarised as well as opportunities for future research.

a) Understanding for whom value is generated in construction

Several studies dedicated to understanding the customer in the construction industry were found, along with attempts to classify what has been called a 'complex customer'. Major contributions include the understanding of 'customers' as: internal vs. external, the complexity of their requirements, the role of the paying client, implications to affected stakeholders, i.e. users, wider society and observed changes in customer vs. supplier relationship brought about by a more integrated organisation of work.

Opportunities for research on this topic are diverse. One aspect that needs further clarification is the understanding of the building 'product' as a whole – a combination of physical space with human-enabled activities, and which establishes a relationship with the community as a whole. Those aspects create a complex chain of customers with distinct interests and/or requirements that need to be managed. Understanding who is relevant in the value equation and whose perspective should be considered remains an unanswered question.

b) Reflections about the concepts of value and perceived value

Contributions to understanding value and perceived value in construction are diverse, borrowing from different knowledge areas, i.e. manufacturing, marketing, economics and sociology. The reliance on several different bodies of knowledge may have contributed to the perception of ambiguity, with scholars yet not agreeing on a definition of value generation that would support its applicability to project management.

For this chapter, we offered the following definition: *generating value is the ability to deliver a built environment that is fit for purpose, i.e. fulfils intent and considers the needs of multiple customers (stakeholders) involved, within the limitations of resources (i.e. time and money) imposed by the circumstances of the project.*

Despite the perception of ambiguity brought by the review of contributions from other areas, the richness of concepts developed elsewhere has provided fertile grounds for developing applicable techniques to better understand the intricacies of customer-perceived value in the construction industry.

One of the major contributions is a distinction between personal 'values' and the 'value' perceived from a construction project. These have distinct roles in supporting value generation. While the first will influence how individuals perceive value, the latter has an implicit purpose of achieving certain goals, which are constructed through social interaction (representing the so-called intersubjective phenomenon) and supports the evaluation of trade-offs, negotiation and alignment over defining and setting the action to achieve the purpose of a project.

Major opportunities for future research are perhaps in the clarification of these concepts. Although helpful techniques have been developed based on the myriad of concepts related to customers' value, conceptual clarification can be supportive to further advance the implications of value generation and value management in construction projects.

c) Managing value in construction

Methods and techniques to bring the customer perspective into construction projects have been developed and tested for a long time. These efforts not only attempt to bring the customers' voice into play, but also to provide effective support for managing multiple and sometimes conflicting requests during the development process. The ability to translate the customers' voice into product specifications is also key in generating value and much has advanced to support that transition, especially with advances in technology and the possibility of automating part of the process. However, many challenges are yet to be overcome for that to become a standard practice in the construction industry.

Delivering value in construction projects will also depend on the ability to understand and successfully consider the limitations set by the project's circumstances (i.e. time and money). TVD is playing a significant role in supporting design development within the constraints of the project. Opportunities for further research in this topic also relate to technology adoption to facilitate updating relevant information to the design team. Efforts in this sense are still in their infancy stage.

Another topic is the impact of project delivery integration on value generation. This topic in itself is a fertile ground for further research. The changing relationship between suppliers and customers and how that can potentially affect value generation and value management needs further clarification. Also, the applicability of value management tools in a context in which design and construction are carried out in a more integrated manner is a topic worth exploring, as well as how technology will enable integrated teams to consider value generation in a more holistic perspective.

As a summary, throughout 25 years of Lean Construction, there has been a continuous effort to investigate the peculiarities of the construction industry and how to better adapt the concepts of value generation and value management in this context. While Lean thinking highlighted the importance of the customer perspective to project management practices, the exploration of multiple concepts from other disciplines has contributed to a better understanding of the implications that this focus poses in the construction industry.

Within the Lean Construction community, the term value has been used in different contexts and it is still subject to multiple interpretations. Further conceptual clarification is necessary to support customer-related efforts moving forward. Despite this challenge, a comprehensive list of contributions to further understand the perspective of external customers was reviewed and presented in this chapter, along with techniques and methods to support a greater focus on customers' perspective. A working definition of value generation in construction projects was offered, which distils some of the concepts reviewed in this chapter.

Many tools and techniques have contributed to the practical application of these ideas into practice. Perhaps, in the near future, much advancement will be seen about how technology adoption can further support these efforts. Opportunities for future research in this area are abundant.

Notes

1 Different authors use different terms to refer to customers (i.e. clients) or may choose to focus on a specific group of stakeholders, i.e. owner, user, paying client, investor. We adopt the term customers to represent all key stakeholders affected by the built environment.
2 Customer requirements refer to needs (objectives, wishes and expectations), which are translated into functions, attributes and other characteristics (Kamara, Anumba, and Evbuomwan, 2000, 2001).
3 A similar study was done by Salvatierra-Garrido, Pasquire, and Miron (2012). Those authors found that the term 'value' appears in publications from at least 12 different knowledge areas. They concluded that the concept still remains ill-defined and ambiguous. One explanation for that is perhaps the different uses of the term and its applicability to different contexts, in addition to a still ongoing pursuit to answer some questions related to its meaning to construction.

References

Akao, Y. (1990) *Quality Function Deployment QFD, Integrating Customer Requirements into Product Design*. Cambridge, MA, USA, Productivity Press.

Alves, T.C.L., Lichtig, W. and Rybkowski, Z.C. (2017) 'Implementing target value design: Tools and techniques to manage the process', *HERD: Health Environments Research and Design Journal*, 10(3), pp. 18–29.

Alwisy, A., Bouferguene, A. and Al-Hussein, M. (2018a) 'Factor-based target cost modelling for construction projects', *Canadian Journal of Civil Engineering*, 45(5), pp. 393–406. doi: 10.1139/cjce-2017-0289

Alwisy, A., Bouferguene, A. and Al-Hussein, M. (2018b) 'Framework for target cost modelling in construction projects', *International Journal of Construction Management*, pp. 1–16. doi:10.1080/15623599.2018.1462446

Ansari, S., Bell, J. and Okano, H. (2007) 'Target costing: Uncharted research territory', in *Handbook of Management Accounting Research*, Chapman, C.S., Hopwood, A.G. and Shields, M.D. (eds.), Oxford, UK: Elsevier, 2, pp. 507–530.

Arayici, Y., Fernando, T., Munoz, V. and Bassino, M. (2018) 'Interoperability specification development for integrated BIM use in performance based design', *Automation in Construction*, 85, pp. 167–181.

Arroyo, P. and Long, D. (2018) 'Collaborative Design Decisions', *Proc. 26th Annual Conference of the International Group for Lean Construction*, Chennai, India. Available at: www.iglc.net.

Arroyo, P., Tommelein, I.D. and Ballard, G. (2012) 'Deciding a Sustainable Alternative by Choosing by Advantages' in the AEC industry', *Proc. 20th Annual Conference of the International Group for Lean Construction*, San Diego, CA, USA. Available at: www.iglc.net.

Arroyo, P., Tommelein, I.D. and Ballard, G. (2013) 'Using 'Choosing by Advantages' to Select Tile From a Global Sustainable Perspective', *Proc. 21st Annual Conference of the International Group for Lean Construction*, Fortaleza, Brazil. Available at: www.iglc.net.

Arroyo, P., Tommelein, I.D., Ballard, G. and Rumsey, P. (2016) 'Choosing by advantages: A case study for selecting an HVAC system for a net zero energy museum', *Energy and Buildings*, *111*, pp. 26–36.

Asmar, M.E., Hanna, A.S. and Loh, W. (2013) 'Quantifying performance for the integrated project delivery system as compared to established delivery Systems', *Journal of Construction Engineering and Management*, 139(11), pp. 1–14.

Baldauf, J.P., Miron, L.I.G. and Formoso, C.T. (2013) 'Using BIM for Modeling Client Requirements for Low-Income Housing', *Proc. 21st Annual Conference of the International Group for Lean Construction*, Fortaleza, Brazil. Available at: www.iglc.net.

Ballard, G. (2008) 'The Lean Project Delivery System: An Update', *Lean Construction Journal*, pp.1–19. Available at: www.Leanconstruction.Org (Accessed: 28th Jul 2010).

Ballard, G. (2011) 'Target value design: Current benchmark', *Lean Construction Journal*, 6(1), pp. 79–84.

Ballard, G. and Reiser, P. (2004) 'The St. Olaf College Fieldhouse Project: A Case Study In Designing To Target Cost', *Proc. 12th Annual Conference of the International Group for Lean Construction*, Elsinore, Denmark. Available at: www.iglc.net.

Bascoul, A.M., Tommelein, I.D., Tillmann, P. and Muxen, S. (2018) 'Towards Facility Management Participation in Design: A UCSF Case Study', *Proc. 26th Annual Conference of the International Group for Lean Construction*, Chennai, India. Available at: www.iglc.net.

Beck, S., Schmalz, S., Heyl, J.V. and Binder, F. (2016) 'Optimizing the Value Stream – Application of Bim in Fm. Status Quo in Germany', *Proc. 24th Annual Conference of the International Group for Lean Construction*, Boston, MA, USA. Available at: www.iglc.net.

Bertelsen, S. and Emmitt, S. (2005) 'The Client as a Complex System', *Proc. 13th Annual Conference of the International Group for Lean Construction*. Sydney, Australia. Available at: www.iglc.net.

Björnfot, A. and Bakken, E.N. (2013) 'Quality Function Deployment (OFD) with a Human Touch', *Proc. 21st Annual Conference of the International Group for Lean Construction*. Fortaleza, Brazil. Available at: www.iglc.net.

Bonatto, F.S., Miron, L.I. and Formoso, C.T. (2011) 'Evaluation of Social Housing Projects Based on User Perceived Value Hierarchy', *Proc. 19th Annual Conference of the International Group For Lean Construction*. Lima, Peru. Available at: www.iglc.net.

Brito, J.N.D.S. and Formoso, C.T. (2014) 'Using the Means-End Approach to Understand Perceived Value by Users of Social Housing Projects', *Proc. 22nd Annual Conference of the International Group for Lean Construction*. Oslo, Norway. Available at: www.iglc.net.

Brown, T. (2008) 'Design thinking', *Harvard Business Review*, 86(6), pp. 1–10.

CIB Report. (1982) *Working with the Performance Approach in Building*. Rotterdam, The Netherlands, CIB Publication. Report number: 64.

Conwell, D. (2012) 'Sutter Health's Lean/Integrated Project Delivery Model'. A P2SL Workshop: *Owner Strategies for Project/Program Delivery*, Berkeley, CA, USA.

Tzortzopoulos, P., Cooper, R., Chan, P. and Kagioglou, M. (2006) 'Clients' activities at the design front-end', *Design Studies*, 27(6), pp. 657–683.

Dorst, K. (2011) 'The core of 'Design Thinking' and its application', *Design Studies*, 32(6), pp. 521–532. doi: 10.1016/j.destud.2011.07.006

Drevland, F. and Klakegg, O.J. (2017) 'A Taxonomy of Client Product Value in Construction Projects', *Proc. 25th Annual Conference of the International Group for Lean Construction*, Heraklion, Greece. Available at: www.iglc.net.

Drevland, F. and Lohne, J. (2015) 'Nine Tenets on the Nature of Value', *Proc. 23rd Annual Conference of the International Group For Lean Construction*, Perth, Australia. Available at: www.iglc.net.

Drevland, F., Lohne, J. and Klakegg, O.J. ((2018)) 'Defining an ill-defined concept - nine tenets on the nature of value', *Lean Construction Journal*, 2018. pp. 31–46.

Drevland, F. and Tillmann, P.A. (2018) 'Value for Whom?', *Proc. 26th Annual Conference of the International Group for Lean Construction*, Chennai, India. Available at: www.iglc.net.

Emmitt, S., Sander, D. and Christoffersen, A.K. (2004) 'Implementing Value Through Lean Design Management', *Proc. 12th Annual Conference of the International Group For Lean Construction*, Helsingør, Denmark. Available at: www.iglc.net.

Emmitt, S., Sander, D. and Christoffersen, A.K. (2005) 'The Value Universe: Defining a Value Based Approach to Lean Construction', *Proc. 13th Annual Conference of the International Group for Lean Construction*, Sydney, Australia. Available at: www.iglc.net.

Farbey, B., Land, F. and Targett, D. (1999) 'The moving staircase – problems of appraisal and evaluation in a turbulent environment', *Information Technology and People Journal*, 12(3), pp. 238–252.

Formoso, C.T. and Miron, L.I.G. (2017) 'Understanding value generation in complex urban regeneration projects', in *Future Challenges in Evaluating and Managing Sustainable Development in the Built Environment*, Brandon, Peter S., Lombardi, Patrizia; Shen, Geoffrey Q. (eds.), UK: Wiley Blackwell, pp. 231–251.

Gann, D.M., Salter, A.J. and Whyte, J.K. (2003) 'The design quality indicator as a tool for thinking', *Building Research and Information*, 31(5), pp. 318–333.

Giese, J.L. and Cote J.A. (2000) 'Defining consumer satisfaction', *Academy of Marketing Science Review*, 1(1), pp. 1–22. Vancouver.

Green, S.D. (1994) 'Beyond value engineering: Smart value management for building projects', *International Journal of Project Management*, 12(1), pp. 49–56.

Griffin, A. and Hauser, J.R. (1996) 'Integrating R&D and marketing: A review and analysis of the literature', *Journal of Production Innovation Management*, 13(3), pp. 191–215.

Gutman, J.A. (1982) 'Means-end chain model based on consumer categorization processes', *Journal of Marketing*, 46(2), pp. 60–72.

Hamilton, D.K. (2006) 'Evidence-based design supports evidence-based medicine in the ICU', *ICU Management Journal*, Belgium. 6(3), p. 31.

Hatoum, M.B., Mustapha, R.E., Nassar, C., Zaheraldeen, H. and Hamzeh, F. (2018) 'Lean Methods to Improve End User Satisfaction in Higher Education Buildings', *Proc. 26th Annual Conference of the International Group for Lean Construction*, Chennai, India. Available at: www.iglc.net.

Hentschke, C.S., Formoso, C.T., Rocha, C.G. and Echeveste, M.E.S. (2014) 'A method for proposing valued-adding attributes in customized housing', *Sustainability*, 6(12), pp. 9244–9267. doi: 10.3390/su6129244

Holbrook, M.B. (2006) 'Consumption experience, customer value, and subjective personal introspection: An illustrative photographic essay', *Journal of Business Research*, 59(6), pp. 714–725.

Huovila, P., and Seren, K.J. (1998) 'Customer-oriented design methods for construction projects', *Journal of Engineering Design*, 9(3), pp. 225–238.

ISO 13407. (1999) *Human-Centred Design Processes for Interactive Systems*. Geneva, Switzerland, Technical Specification International Organisation for Standardisation.

ISO 9241-210. (2010) *Ergonomics of Human-System Interaction - Part 210: Human-Centred Design for Interactive Systems*. Geneva, Technical Specification International Organisation for Standardisation.

Joseph, A. (2006a) *The Impact of the Environment on Infections in Healthcare Facilities*. Concord, MA, Center for Health Design.

Joseph, A. (2006b) *The Role of the Physical and Social Environment in Promoting Health, Safety, and Effectiveness in the Health Care Workplace*. Concord, MA, Center for Health Design.

Jylhä, T. and Junnila, S. (2012) 'Using the Kano Model to Identify Customer Value', *Proc. 20th Annual Conference of the International Group for Lean Construction*, San Diego, CA, USA. Available at: www.iglc.net.

Kagioglou, M. and Tzortzopoulos, P. (2016) 'Benefits Realisation: An Investigation of Structure and Agency', *Proc. 24th Annual Conference of the International Group For Lean Construction*, Boston, MA, USA. Available at: www.iglc.net.

Kamara, J.M., Anumba, J.M. and Evbuomwan, N.F.O. (2000) 'Establishing and processing client requirements: A key aspect of concurrent engineering in construction', *Engineering, Construction and Architectural Management*, 7(1), pp. 15–28.

Kamara, J.M., Anumba, J.M. and Evbuomwan, N.F.O. (2001) 'Assessing the suitability of current briefing practices in construction within a concurrent engineering framework', *International Journal of Project Management*, 19(6), pp. 337–351.

Kato, Y. (1993) 'Target costing support systems: Lessons from leading japanese companies', *Management Accounting Research*, 4(1), pp. 33–47.

Kelly, J. and Male, S. (1993) *Value Management in Design and Construction: The Economic Management of Projects*. London and New York, E & FN Spon.

Kim, T.W., Kim, Y., Cha, S.H. and Fischer, M. (2015) 'Automated updating of space design requirements connecting user activities and space types', *Automation in Construction*, 50, pp. 102–110.

Kim, T.W., Rajagopal, R., Fischer, M. and Kam, C. (2013) 'A knowledge-based framework for automated space-use analysis', *Automation in Construction*, 32, pp. 165–176.

Kiviniemi, A. (2005) *Requirements Management Interface to Building Product Model*. PhD Thesis. Espoo, Finland, Stanford University.

Koskela, L. (1992) *Application of the New Production Philosophy to Construction*. Espoo, Finland, Center for Integrated Facility Engineering, Department of Civil Engineering, Stanford University. Technical Report number: 72.

Koskela, L. (2000) *An Exploration Towards a Production Theory and its Application to Construction*. PhD thesis. Espoo, Finland, Technical Research Centre of Finland, VTT.

Koskela, L., Huovila, P. and Leinonem. (2002) 'Design management in building construction: From theory to practice', *Journal of Construction Research*, 3(1), pp. 1–16.

Koskela, L., Siriwardena, M. and Rooke, J. (2008) 'Through-Life Management of Built Facilities: Towards a Framework for Analysis', *Proc. 16th Annual Conference of the International Group For Lean Construction*, Manchester, UK. Available at: www.iglc.net.

Koskela, L.J. and Ballard, G. (2006) 'Should project management be based on theories of economic or production?', *Building Research & Information*, 32(2), pp. 154–163.

Kpamma, Z.E., Adjei-Kumi, T., Ayarkwa, J. and Adinyira, E. (2014a) 'Enhancing User-Involvement through A Multi-Criteria Decision Aid: A Lean Design Research Agenda', *Proc. 22nd Annual Conference of the International Group For Lean Construction*, Oslo, Norway. Available at: www.iglc.net.

Kpamma, Z.E., Adjei-Kumi, T., Ayarkwa, J. and Adinyira, E. (2014b) 'Creating, Sustaining and Optimising the Collaborative Realm for Participatory Design', *Proc. 22nd Annual Conference of the International Group for Lean Construction*, Oslo, Norway. Available at: www.iglc.net.

Kpamma, Z.E., Adjei-Kumi, T., Ayarkwa, J. and Adinyira, E. (2017) 'Strategic Considerations for Applying Choosing by Advantages in Design Process', *Proc. 25th Annual Conference of the International Group for Lean Construction*, Heraklion, Greece. Available at: www.iglc.net.

Leinonen, J. and Huovila, P. (2000) 'The House of the Rising Value', *Proc. 8th Annual Conference of the International Group For Lean Construction*, Brigthon, UK. Available at: www.iglc.net.

Levitt, T. (1960) 'Marketing Myopia', *Harvard Business Review*. pp. 45–56.

Li, P, Froeseb, T.M. and Bragerc, G. (2018) 'Post-occupancy evaluation: State-of-the-art analysis and state-of-the-practice review', *Building and Environment*, 133, pp. 187–202.

Lima, L.P., Formoso, C.T. and Echeeveste, M.E. (2008) 'Client Requirements Processing in Low-Income House-Building Using Visual Displays and the House of Quality', *Proc. 16th Annual Conference of the International Group for Lean Construction*, Manchester, UK. Available at: www.iglc.net.

Lima, L.P., Miron, L.I., Leite, F. and Formoso, C.T. (2009) 'Perceived Value In Social Housing Projects', *Proc. 17th Annual Conference of the International Group For Lean Construction*, Taipei, Taiwan. Available at: www.iglc.net.

Macomber, H., Howell, G. and Barberio, J. (2007) 'Target-Value Design: Nine Foundational Practices for Delivering Surprising Client Value', *The American Institute of Architects (AIA)*, 2007. Available at: www.Aia.Org/Nwsltr_Print.Cfm?Pagename=Pm_A_112007_Targetvaluedesign (Accessed: 26th Aug 2010).

Male, S., Kelly, S., Gronqvist, M. and Graham, D. (2007) 'Managing value as a management style for projects', *International Journal of Project Management*, 25(2), pp. 107–114.

Marshall, A. (1879) *The Pure Theory of* (Domestic) *Values*. UK, London School of Economics.

Matthews, O. and Howell, G.A. (2005) 'Integrated project delivery – An example of relational contracting', *Lean Construction Journal*, 2(1), pp. 46–61.

Melo, R.S., Do, D., Tillmann, P., Ballard, G. and Granja A.D. (2015) 'Target value design in the public sector: Evidence from a hospital project in San Francisco, CA', *Architectural Engineering and Design Management*, 12(2), pp. 125–137. doi: 10.1080/17452007.2015.1106398

Miron, L.I. and Formoso, C.T. (2003) 'Client Requirement Management In Building Projects', *Proc. 11th Annual Conference of the International Group For Lean Construction*, Virginia, USA. Available at: www.iglc.net.

Miron, L.I. and Formoso, C.T. (2010) 'Value Generation in Social Housing Projects: A Case Study on the City Entrance Integrated Program in Porto Alegre, Brazil', *Proc. 18th Annual Conference of the International Group for Lean Construction*, Haifa, Israel. Available at: www.iglc.net.

Monroe, K.B. (1990) *Pricing: Making Profitable Decisions*. New York, USA, Mcgraw-Hill.

Morais, M.V., Carneiro, T.M. and Neto, J.D.P.B. (2014) 'Improving Technique For Capturing Value in Low-Income Housing', *Proc. 22nd Annual Conference of the International Group for Lean Construction*, Oslo, Norway. Available at: www.iglc.net.

Nanda, U., Rybkowski, Z.K., Pati, S. and Nejati, A. (2017) 'A value analysis of lean processes in target value design and integrated project delivery: Stakeholder perception', *Health Environments Research & Design Journal*, 10(3), pp. 99–115. doi: 10.1177/1937586716670148

Nascimento da Silva, M. and Miron, L.I.G. (2017) 'Hierarchy of Value Perceived by Groups of Users About Their Neighbourhood', *Proc. 25th Annual Conference of the International Group For Lean Construction*, Heraklion, Greece. Available at: www.iglc.net.

Nicolini, D., Tompkins, C., Holti, R., Oldman, A. and Smalley, M. (2000) 'Can target costing and whole life costing be applied in the construction industry?: Evidence from two case studies', *British Journal of Management*, 11(4), pp. 303–324.

Norman, D. (1988) *The Design of Everyday Things*. New York, USA, Basic Books.

Oliver, R.L., Balakrishnan, P.V. and Barry, B. (1994) 'Outcome satisfaction in negotiation: A test of expectancy disconfirmation', *Organizational Behavior and Human Decision Processes*, 60(2), pp. 252–275.

Parrish, K., Wong, J., Tommelein, I.D. and Stojadinovic, B. (2008) 'Set-Based Design: Case Study on Innovative Hospital Design', *Proc. 16th Annual Conference of the International Group For Lean Construction*, Manchester, UK. Available at: www.iglc.net.

Pearce, D. (2003) *The Social and Economic Value of Construction: The Construction Industry's Contribution to Sustainable Development*. London, UK, nCRISP.

Pennanen, A., Ballard, G. and Haahtela, Y. (2011) 'Target costing and designing to targets in construction', *Journal of Financial Management of Property and Construction*, 16(1), pp. 52–63. doi: 10.1108/13664381111116089

Ravald, A. and Gronroos, C. (1996) 'The value concept and relationship marketing', *European Journal of Marketing*, 30(2), pp. 19–32.

Rokeach, M. (1973) *Beliefs, Attitudes and Values*. New York, USA, The Free Press.

Rokeach, M. (1981) *Understanding Human Values: Individual and Societal*. New York, USA, The Free Press.

Rooke, J.A., Sapountzis, S., Koskela, L.J., Codinhoto, R. and Kagioglou, M. (2010) 'Lean Knowledge Management: The Problem of Value', *Proc. 18th Annual Conference of the International Group for Lean Construction*, Haifa, Israel. Available at: www.iglc.net.

Rybkowski, Z.K. and Ballard, G. (2008) 'Using the 'Five Whys' As A Decision-Making Framework For Evidence-Based Design', *Proc. 16th Annual Conference of the International Group for Lean Construction*, Manchester, UK. Available at: www.iglc.net.

Sacks, R., Dave, B.A., Koskela, L. and Owen, R. (2009) 'Analysis Framework for the Interaction Between Lean Construction and Building Information Modelling', *Proc. 17th Annual Conference of the International Group for Lean Construction*, Taipei, Taiwan. Available at: www.iglc.net.

Salvatierra-Garrido, J., Pasquire, C. And Miron, L. (2012) 'Exploring Value Concept Through the IGLC Community: Nineteen Years of Experience', *Proc. 20th Annual Conference of the International Group for Lean Construction*, San Diego, CA, USA. Available at: www.iglc.net.

Sampaio, J.C.S. and Barros Neto, J.D.P.B. (2010) 'Value Generation and Its Relation With the Design Process', *Proc. 18th Annual Conference of the International Group for Lean Construction*, Haifa, Israel. Available at: www.iglc.net.

Sánchez-Fernández, R. and Iniesta-Bonillo, M.A. (2007) 'The concept of perceived value: A systematic review of the research', *Marketing Theory*, 7(4), pp. 427–451. doi: 10.1177/1470593107083165

Sands, M. and Abdelhamid, T.S. (2012) 'Whole-Building Measurement and Computing Science BIM for Lean Programming and Performance', *Proc. 20th Annual Conference of the International Group for Lean Construction*, San Diego, CA, USA. Available at: www.leanconstruction.org/learning/publications/lean-construction-journal/lcj-back-issues/2010-issue/

Sands, M.S. (2010) 'Standards and measures – whole-building metrics driving innovation and high performance', *Lean Construction Journal*, pp. 1–16.

Sanni-Anibire, M.O., Hassanain, M.A., and Al-Hammad, A.-M. (2016) 'Post-occupancy evaluation of housing facilities: Overview and summary of methods', *Journal of Performance of Constructed Facilities*, 30(5), doi: 10.1061/(ASCE)CF.1943-5509.0000868

Santos, A.D., Kristmann, V.B. and Fischer, S. (2004) 'Assessing Total Value Requirements on Low Income Houses in Brazil', *Proc. 12th Annual Conference of the International Group for Lean Construction*, Helsingør, Denmark. Available at: www.iglc.net.

Sapountzis, S., Harris, K. and Kagioglou, M. (2008) *Benefits Management and Benefits Realisation, a Literature Report. HaCIRIC.* Salford, UK, University of Salford.

Sapountzis, S., Yates, K., Lima J.B. and Kagioglou, M. (2010) 'Benefits realisation: Planning and evaluating healthcare infrastructures and services', in *Improving Healthcare Through Built Environment Infrastructure*, Tzortzopoulos, P.; Kagioglou, M. (eds.), Oxford, UK: Blackwell Publishing, pp. 166–195.

Saxon, R. (2005) *Be Valuable: A Guide to Value in the Built Environment.* London, UK, Constructing Excellence.

Schöttle, A., Arroyo, P. and Christensen, R. (2018) 'Demonstrating the Value of an Effective Collaborative Decision-Making Process in the Design Phase', *Proc. 26th Annual Conference of the International Group for Lean Construction*, Chennai, India. Available at: www.iglc.net.

Schöttle, A. and Tillmann, P.A. (2018) 'Explaining the Benefits of Team Goals to Support Collaboration', *Proc. 26th Annual Conference of the International Group for Lean Construction*, Chennai, India. Available at: www.iglc.net.

Sfandyarifard, E. and Tzortzopoulos, P. (2011) 'Supporting Value Generation in Children's Hospital Design through Participatory Approaches', *Proc. 19th Annual Conference of the International Group for Lean Construction*, Lima, Peru. Available at: www.iglc.net.

Shewhart, W.A. (1931) *Economic Control of Quality of Manufactured Product.* New York, USA, Van Nostrand.

Silveira, S., and Alves, T. (2018) 'Target value design inspired practices to deliver sustainable buildings', *Buildings*, 8(9), p. 116.

Soliman Junior, J., Baldauf, J.P., Formoso, C.T. and Tzortzopoulos, P. (2018) 'Using BIM and Lean for Modelling Requirements in the Design of Healthcare Projects', *Proc. 26th Annual Conference of the International Group for Lean Construction*, Chennai, India. Available at: www.iglc.net.

Spencley, R., Pfeffer, G., Gordon, E., Hain, F., Reed, D. and Marosszeky, M. (2018) 'Behavior-Based Quality, Case Study of Closing the Knowing-Doing Gap', *Proc. 26th Annual Conference of the International Group for Lean Construction*, Chennai, India. Available at: www.iglc.net.

Thomsen, C., Darrington, J., Dunne, D. and Lichtig, W. (2009) *Managing Integrated Project Delivery.* USA, CMAA College of Fellows.

Thomson, Derek S., Austin, Simon A., Devine-Wright, Hannah and Mills, Grant R. (2003) 'Managing value and quality in design', *Building Research & Information*, 31(5), pp. 334–345.

Thorp, J. (1998) *The Information Paradox: Realising the Business Benefits of Information Technology.* Toronto, Canada, McGraw-Hill.

Thyssen, M.H., Emmitt, S., Bonke, S. and Kirk-Christoffersen, A. (2010) 'Facilitating client value creation in the conceptual design phase of construction projects: A workshop approach', *Architectural Engineering and Design Management*, 6(1), pp. 18–30.

Tillmann, P. (2018) 'UCSF real estate lean project delivery guide: A guide for major capital projects'. Available at: https://escholarship.org/uc/item/7hk3m448.

Tillmann, P., Ballard, G., Tzortzopoulos, P. and Formoso, C.T. (2012b) 'How Integrated Governance Affects Value Generation Insights From An IPD Case Study', *Proc. 20th Annual Conference of the International Group for Lean Construction*, San Diego, CA, USA. Available at: www.iglc.net.

Tillmann, P., Tzortzopolous, P., Sapountzis, S., Formoso, C. And Kagioglou, M. (2012a) 'A Case Study on Benefits Realisation and Its Contributions for Achieving Project Outcomes', *Proc. 20th Annual Conference of the International Group for Lean Construction*, San Diego, CA, USA. Available at: www.iglc.net.

Tillmann, P.A., Tzortzopoulos, P. and Formoso, C.T. (2010) 'Analysing Benefits Realisation From a Theoretical Perspective and Its Contribution to Value Generation', *Proc. 18th Annual Conference of the International Group for Lean Construction*, Haifa, Israel. Available at: www.iglc.net.

Tillmann, P.A., Tzortzopoulos, P., Miron, L.I. and Formoso, C.T. (2011) 'The Challenges of Managing Stakeholder Requirements in a Urban Regeneration Project', *Proc. 19th Annual Conference of the International Group for Lean Construction*, Lima, Peru. Available at: www.iglc.net.

Topalov, Christian. (1979) *La Urbanización Capitalista (The Capitalist Urbanization)*. México City, Edicol. In Spanish.

Walch, J.M., Rabin, B.S., Day, R., Williams, J., Choi, K. and Kang, J.D. (2005) 'The effect of sunlight on postoperative analgesis medication use: A prospective study of patients undergoing spinal surgery', *Psychosomatic Medicine*, 67(1), pp. 156–163.

Ward, J. and Daniel, E. (2006) *Benefits Management – Delivering Value from IS & IT Investments*. West Sussex, UK, Wiley.

Winter, M., Smith, C., Morris, P. and Cicmil, S. (2006) 'Directions for future research in project management: The main findings of UK government-funded research network', *International Journal of Project Management*, 24(8), pp. 638–649.

Winter M. and Szczepanek,T. (2008) 'Projects and programmes as value creation processes: A new perspective and some practical implications', *International Journal of Project Management*, 26(1), pp. 95–103.

Woodruff, R.B. (1997) 'Customer value: The next source of competitive advantage', *Journal of the Academy of Marketing Science*, 25(2), pp. 139–153.

Woodruff, R.B., Schumann, D.W. and Gardial, S.F. (1993) 'Understanding value and satisfaction from the customer' s point of view', *Survey of Business-University of Tennessee*, 29, pp. 33–40.

Zhang, Y., Tzortzopoulos, P. and Kagioglou, M. (2016) 'Evidence-Based Design in Healthcare: A Lean Perspective With an Emphasis on Value Generation', *Proc. 24th Annual Conference of the International Group for Lean Construction*, Boston, MA, USA. Available at: www.iglc.net.

Zhang, Y., Tzortzopoulos, P. and Kagioglou, M. (2017) 'How Evidence-Based Design Supports Better Value Generation For End-Users', *Proc. 25th Annual Conference of the International Group For Lean Construction*, Heraklion, Greece. Available at: www.iglc.net.

Zimina, D., Ballard, G., and Pasquire, C. (2012) 'Target value design: Using collaboration and a lean approach to reduce construction cost', *Construction Management and Economics*, 30(5), pp. 383–398. doi: 10.1080/01446193.2012.676658.

7 Understanding waste in construction

Carlos Torres Formoso, Trond Bølviken and Daniela Dietz Viana

7.1 Introduction

A very high level of waste is assumed to exist in the construction industry. Although it is difficult to systematically measure all types of waste, studies from various countries have confirmed that waste represents a relatively large percentage of production costs in construction. A wide range of measures have been used to monitor waste, such as excess consumption of materials (Formoso et al., 2002), rework (Hwang et al., 2009; Love and Edwards, 2004; Zhao et al., 2010), building defects (Josephson and Hammarlund, 1999), unproductive time (Horman and Kenley, 2005), indirect work (Kalsaas, Gundersen, and Berge, 2014); and work-in-progress (Yu et al., 2009).

The elimination of waste has been largely used as a driver for improvement in companies that have adopted Lean. Waste measurement usually allows areas of potential improvements to be highlighted and the main causes of inefficiency to be identified (Ohno, 1988). Moreover, compared to traditional financial measures, some waste metrics (e.g. material waste, non-value-adding time) are more effective to support process management, as they allow operational problems to be quickly identified, and generate information that is more meaningful to the workforce, creating thus conditions to implement decentralised control.

Distinct types of waste have been measured in research studies, indicating that waste has been understood in different ways by the construction management academic community. Some studies have focused on material losses, often concerned with the environmental impact caused by construction and demolition material waste (Bossink and Brouwers, 1996; Gavilan and Bernold, 1994; Poon et al., 2004). In those studies, the conceptualisation of waste is rarely discussed: waste is often understood as debris that need to be removed from construction sites. For instance, Al-Hajj and Hamani (2011) adopted the definition of material waste as the by-products generated and removed from construction sites in different types of projects, such as new building, renovation, or demolition projects.

A similar problem is found in previous studies on the measurement of rework (e.g. Burati, Farrington, and Ledbetter, 1992; Hwang et al., 2009). Although the importance of this type of waste is widely recognised, it is not clear in the literature whether rework is simply a consequence of quality deviations or if it is also a result of change orders or uncompleted tasks. A major consequence of the lack of a clear definition in this type of waste is that it is not always possible to compare results from different studies (e.g. Burati, Farrington, and Ledbetter, 1992; Love, 2002).

By contrast, there are several publications that discuss the conceptualisation of waste in Lean, some of them published in the annual conferences of the IGLC (International Group for Lean Construction) (e.g. Bølviken and Koskela, 2016; Bølviken, Rooke, and

Koskela, 2014; Fireman, Formoso, and Isatto, 2013; Formoso et al., 2015, 2017; Koskela, 2004; Koskela, Bølviken, and Rooke, 2013; Koskela, Sacks, and Rooke, 2012; Viana, Formoso, and Kalsaas, 2012). Those studies relate waste to the occurrence of non-value-adding activities, i.e. activities that take time, resources, or space but do not add value from the perspective of the final customer (Koskela, 1992).

This chapter presents an overview on previous studies that have contributed to the understanding of waste in Lean. The main concepts and metrics adopted in the literature are presented, based on a systematic literature review considering papers published in the last 25 years. From a theoretical perspective, the idea of waste propagation is discussed, with the aim of identifying the types of waste that have the highest impact in the performance of production systems.

This study also has a practical ambition, which is to contribute to increasing the performance of the construction industry by providing a consistent and clear understanding of construction waste, useful for construction professionals.

Most of the contents of this chapter were produced in an IGLC research initiative, named Understanding Waste in Construction,[1] which has explored the conceptualisation and measurement of waste in construction, based on core concepts and principles of production management.

7.2 Brief history of the concept of waste

Koskela, Sacks, and Rooke (2012) presented an overview on the historical development and diffusion of the concept of waste, pointing out that it was discussed by political economists, scientists, and engineers as early as in the 19th century. In the initial developments, waste was understood from two perspectives: (i) from an instrumental perspective, resources (e.g. energy, machines or labour-force) are wasted when these are used but do not produce saleable products; and (ii) from a moral perspective, not using resources is a waste of time (e.g. energy available, labour, equipment) for all of humanity and for all of time (Koskela, Sacks, and Rooke, 2012). However, in neither perspective, value to the final client was considered, nor its relationship to price.

In the period 1880–1930, the concept of waste flourished as one of the key elements of the Scientific Management movement, which established the foundation of the Mass Production paradigm. At that time, the main focus was on labour inefficiency: waste in production was considered to be the difference between the optimal production that might have been achieved using the 'one best way' and the current level of efficiency (Koskela, Sacks, and Rooke, 2012). In fact, Taylor (1913) suggested that the economic loss caused by material waste is smaller than the one related to the inefficiency of human work. Ford (1926) also believed that human work should be the focus of waste elimination, based on the assumption that the value of materials depends, to a great extent, on the work that has been spent on them.

However, the concept of waste started to disappear from scholarly and professional literature in the third quarter of the 20th century. Two main reasons can be pointed out for this change: (i) the conceptualisation of production as transformation, originated in the field of Economics, in which waste is not visible at all, has been widely adopted in construction management (Koskela, Sacks, and Rooke, (2012); and (ii) as the Platonic epistemology (i.e. the search for universal truths in the ideal world) has dominated the fields of Engineering and Management, there is much less interest in waste, as this is a phenomenon that belongs to the natural and imperfect material world (Koskela et al., 2019).

The concept of waste started to re-emerge in the 1980s, due to the dissemination of the Lean Production Philosophy in the car industry and, later on, in other sectors. In fact, the elimination of waste plays a key role in the two key strategies adopted in that philosophy for improving efficiency: just-in-time (reduction of inventories) and autonomation (elimination of defects) (Ohno, 1988).

Since the foundation of IGLC in 1992, many successful initiatives for the implementation of Lean production concepts and principles in the construction industry have been reported. However, in the same period, few construction management and project management textbooks have emphasised the importance of waste elimination. Bølviken and Koskela (2016) discuss why waste reduction has not got off the ground in construction, pointing out that the explanation is not simply lack of talent or interest in the industry, but it is deeply rooted in how production and management are conceptualised (i.e. how we see and understand the world) and in which kind of production construction is (i.e. how the world is). In other words, understanding waste in construction requires a theoretical framework that is different from the foundations of traditional construction management, and also an in-depth understanding of the different characteristics of construction projects in relation to other types of production systems.

7.3 Previous academic studies on construction waste

This section presents the main results of a systematic literature review on construction waste. An initial version of this literature review was presented by Viana, Formoso, and Kalsaas (2012), and since then it was complemented by additional papers published between 2012 and 2017. The main sources of papers were IGLC conferences, the *Lean Construction Journal* and seven mainstream construction management journals.[2] After the initial selection of papers, a database was created in a citation manager, in order to check duplicates and to apply some quality criteria in the selection of papers. Some additional references cited in the selected papers were also included in the database. After several refinements in the search, 61 papers were selected. This set of papers was then analysed in detail, considering: (i) the concept of waste adopted and whether it was explicit or not; (ii) the kind of waste analysed; and (iii) the main contribution of the paper to the topic of construction waste, including new conceptualisations, systematic analyses of the causes of waste, definition of metrics, proposal of actions to reduce waste, and development of methods for waste control.

Many papers do not present a clear definition of waste. Only 41% of the papers explicitly present a conceptualisation of waste in a broad sense, and 16% defined only a specific kind of waste that was addressed, such as rework, making-do, or defects. In most papers produced by the IGLC community, waste is defined as the use of resources that do not add value from the perspective of the customer. It is based on the concept of process as a flow (Koskela, 1992), from which two alternatives for improving efficiency can be inferred: (i) to improve the efficiency of individual activities (for instance, by working harder or faster); or (ii) to reduce the share of non-value adding activities. The latter alternative is often pointed out as smarter, because it represents opportunities for more dramatic improvements than the first one, considering that the potential gains are often greater and the necessary effort for generating value can be reduced.

Bølviken, Rooke, and Koskela (2014) have extended the concept of waste beyond non-value adding activities, by considering not only the use of more resources than needed but also unwanted outputs from production. This is the case, for instance, of the negative environmental impact of construction projects, including the production of debris, sound pollution, and particulate matter emissions.

The main contributions provided by the set of papers analysed can be classified into: (i) identification of the causes of waste, and (ii) production of waste metrics. However, nearly half of the papers that investigated the causes of waste were mainly based on the perception of construction professionals obtained in surveys. Due to the lack of a clear conceptualisation of waste, in many of those survey-based studies there is a lack of in-depth analysis of the root causes of waste.

Regarding the measurement of waste, performing comparative studies between data from different studies is difficult for two reasons: (a) in several studies, the data collection procedures are not fully described; and (b) different conceptualisations are used to address the same type of waste (e.g. rework). In fact, a wide range of indicators has been used for measuring waste. Some are expressed as physical quantities, such as the volume of debris taken from the site (Gavilan and Bernold, 1994; Poon et al., 2004), while others are measured as costs, such as the cost of repairing defective products (Burati, Farrington, and Ledbetter, 1992; Ledbetter, 1994) and rework (Hwang et al., 2009). Time has also been often used as an important waste measure, especially when the aim is to identify the share of non-value adding activities (Forsberg and Saukkoriipi, 2007; Horman and Kenley, 2005; Kalsaas, 2010; Yu et al., 2009).

The contributions regarding actions for reducing or eliminating waste are also diverse. Some papers describe attempts to change practices in industry by implementing Lean techniques (Nahmens and Ikuma, 2011), while others use simulation models to support decision making by testing changes that could contribute to reducing the share of non-value adding activities (e.g. Park et al., 2011; Porwal and Hewage, 2011; Sacks, Esquenazi, and Goldin, 2007; Tommelein, Riley, and Howell, 1999).

Regarding the development of methods to identify and measure waste, the number of papers is relatively small and most of them focus on two types of methods: measurement of material losses, including direct and indirect waste (e.g. Formoso et al., 2002; Skoyles, 1976), and value stream mapping (VSM) to assess the share of non-value adding activities of a process and designing a future state of it (e.g. Choi et al., 2002; Yu et al., 2009).

Overall, the number of papers on waste identified in the systematic literature review is relatively small, considering the relevance of this topic for the construction industry. Most existing studies do not discuss the conceptualisation of waste at an abstract level, but simply adopt an operational definition of waste, in order to guide data collection.

7.4 Different taxonomies of waste

One way to understand the nature of waste is to devise a taxonomy of waste categories. The most cited waste taxonomy in the literature is the Seven Wastes proposed by Ohno (1988): (i) overproduction, (ii) time on hand (waiting), (iii) transportation, (iv) processing itself, (v) stock on hand (inventory), (vi) movement, and (vii) making defective products. That taxonomy was devised for the Toyota Production System to improve manufacturing. Ohno (1988) also points out that overproduction must be considered as the most important type of waste, because it is a major cause of several problems that exist in production systems, including other types of waste. However, Ohno (1988) does not offer an explicit or exact definition of waste and does not offer any theoretical or empirical arguments for the seven categories of waste (Bølviken and Koskela, 2016). Therefore, it seems that Ohno's categories of waste are not limited to the definition of resources that do not add value to customers, but extend the concept of waste to key events that result in different types of non-value adding activities.

Ohno's taxonomy has been adopted beyond manufacturing, demanding efforts of adapting the proposed waste categories to different contexts. Focusing on healthcare, Bush (2007) considers both staff and patient (customer) waiting under 'time on hand', although the two are substantially different in the original taxonomy, due to the fact that in a service industry, such as healthcare, the customer is actually processed by the system (Koskela, Bølviken, and Rooke, 2013). Bush (2007) considers transportation waste not only as the unnecessary movement of patients but also of files, supplies, and equipment. Bicheno and Holweg (2009) proposed a generic waste list for service industries that omits the key category of overproduction, presumably on the grounds that a service is consumed as soon as it is produced, making overproduction apparently impossible. In addition, waiting and movement are both now taken to refer to the customer experience, meaning that operational wastes are effectively hidden in this version (Koskela, Bølviken, and Rooke, 2013).

There are also other categories of waste that have been proposed in the literature, such as design of products that do not meet users' needs (Womack and Jones, 2004); unnecessary capital investment (Monden, 1983); theft and vandalism (Bossink and Brouwers, 1996); making-do (Koskela, 2004); and not listening and not speaking (Macomber and Howell, 2004). However, there is no evidence that Ohno's taxonomy or those additional categories of waste have been widely used in the construction industry to support improvement initiatives.

Bølviken, Rooke, and Koskela, (2014) proposed a taxonomy of waste for the construction industry based on the Transformation-Flow-Value (TFV) theory (Koskela, 2000). The main contribution of that taxonomy is that it includes also the value generation perspective, extending the concept of waste beyond the flow perspective, which seems to be the focus of Ohno's taxonomy.

In summary, it seems that the main role of existing classifications of waste is to call the attention of people to important problems in specific contexts, considering that the identification of waste is not always straightforward (Shingo, 1988). Some of the confusion about the development of waste taxonomies for different sectors seem to be related to the fact that these are not founded on core concepts and principles, such as the difference between the process and operation perspectives (Shingo, 1988); the nature of non-value adding work; the effect of work-in-progress, among others. By contrast, it seems that the phenomenon that we call waste is not something abstract, but rather specific and context-related. Therefore, waste must be understood and dealt with partly through general terms and theory, but partly by considering the specific context of a production system.

7.5 Propagation of waste

Based on the assumption that a production process can be consider as a chain of interdependent events (flow perspective), it is reasonable to expect not only to find singular waste events, but also chains of waste, i.e. chains of causes and effects in which one waste leads to another (Koskela, Bølviken, and Rooke, 2013). Therefore, the main disbenefit of a specific category of waste might not be the waste itself, but the entire chain of waste that is created. Ohno (1988) refers to such a chain as a vicious circle of waste generating waste and distinguishes between what he calls primary and secondary waste. For instance, both inventory and overproduction are considered by Ohno (1988) as primary waste in manufacturing, because it can cause other types of waste.

Similar to Ohno's definition of primary and secondary waste, Koskela, Bølviken, and Rooke (2013) proposed the terms core and lead waste. A core waste is a phenomenon that is both a waste in itself and the cause of other types of waste. By attacking a core waste, one can also eliminate the waste resulting from it. A lead waste is a dominant core waste, i.e. a core waste with substantial negative impact in the production system (Koskela, Bølviken, and Rooke, 2013). Based on this terminology, waste can either be caused by a core waste or by other causes that are not themselves waste.

There is also a strong connection between the occurrence of waste and variability. According to Hopp and Spearman (2000), there are two types of variability in production systems: (i) process-time variability refers to the time required to process a task at one workstation; and (ii) flow variability means the variability of the arrival of jobs to a workstation. Some categories of construction waste, such as quality deviation, making-do and accidents, can also be regarded as a major cause, although perhaps not the only one, for those two types of variability (Formoso et al., 2015). By contrast, by using the queuing theory, it is possible to demonstrate that variability is also a major cause of waste, such as longer lead times, higher level of work-in-progress, wasted capacity, and lost output (Hopp and Spearman, 2000). Therefore, considering the combined effect of variability and interdependence in construction projects, construction waste can be regarded as a complex parade of singular evanescent events, which are hard to be fully understood (Bølviken and Koskela, 2016).

7.6 Modelling waste networks

The phenomena of waste propagation can be broadly explained by considering a network of waste categories. If each category of waste results in other types of waste, it is possible to represent a network of waste, in which the different types of waste (nodes) are connected by cause–effect relationships. Some of the causal connections between nodes are unidirectional, but there might also be reciprocal relationships, i.e., waste A can lead to waste B at the same time as B leads to A. Given such a complex network, much research is necessary to identify and understand all connections between waste categories. However, the effort of modelling a waste network can help to identify core and lead types of waste, which should be the focus of improvement initiatives.

Based on an analysis of data from two previous studies (Formoso et al., 2002, 2017) that have attempted to measure waste as well as to assess its main causes and consequences, two networks are proposed to represent causal relationships between different types of waste. Figure 7.1 presents the set of waste categories and the connections between them identified in the study carried out by Formoso et al. (2002) on material waste, while Figure 7.2 presents a similar network for a study carried out by Formoso et al. (2017) regarding making-do. Both networks are limited to production control. If the focus of the analysis is another stage of construction projects, such as design, procurement, supply chain management, other waste categories should be identified.

Those waste networks have been divided into three main zones:

(i) Terminal waste: these are formed mostly by traditional categories of waste that are strongly related to the consequences of wasteful production processes. Some of these categories have been the focus of several measurement studies, such as material losses, rework, and non-value adding time;
(ii) Core production waste: these are the most relevant categories for production control, as they cause several other types of waste. In this zone, it is important to highlight lead production waste, whose elimination plays a very important role in performance improvement. Some core waste categories are based on concepts that are not widely known in the industry, such as making-do, work-in-progress, and unfinished work,

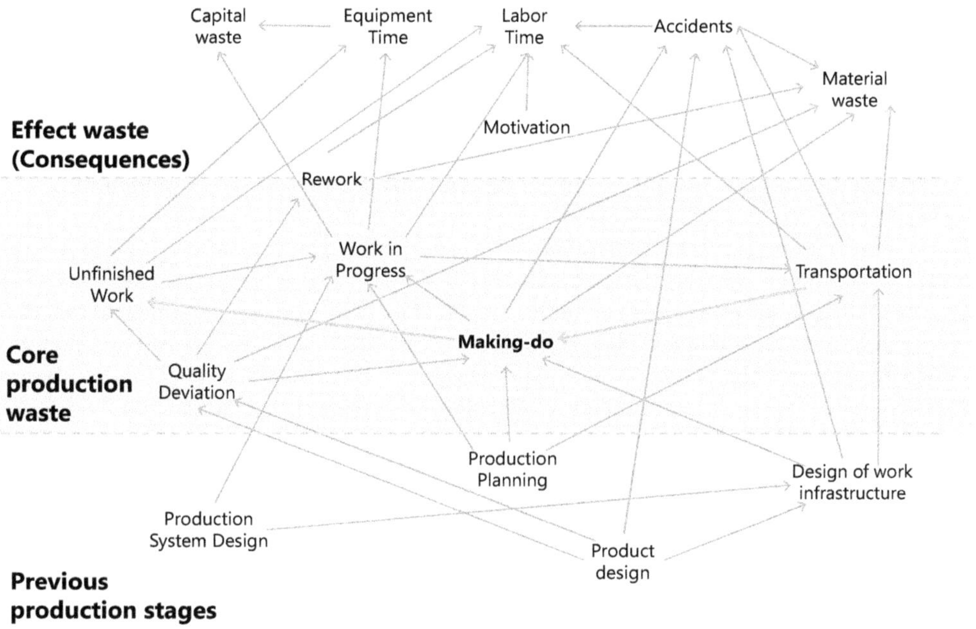

Figure 7.1 Causal network for material waste (based on Formoso et al., 2002)

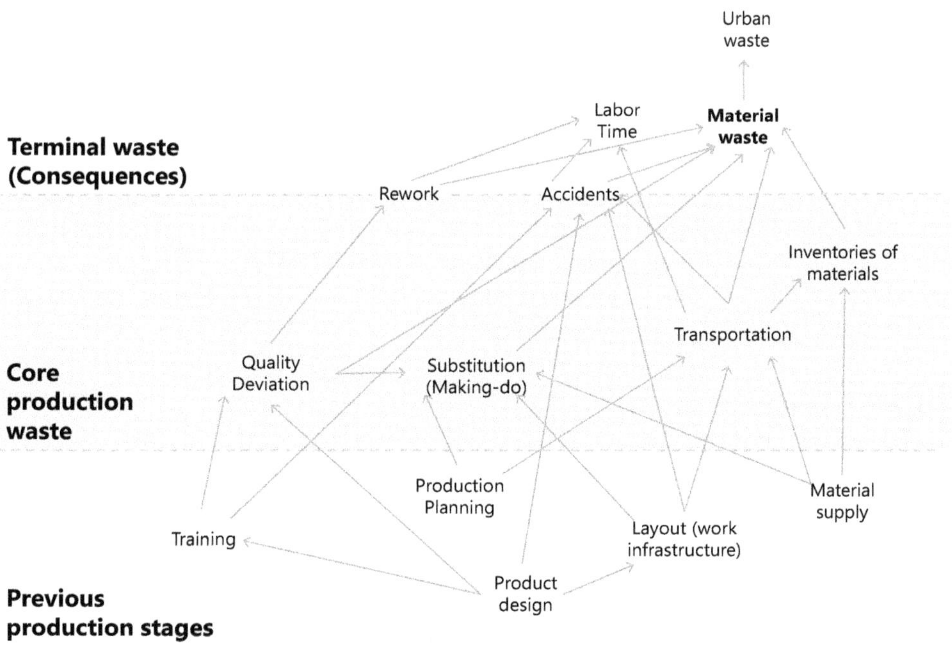

Figure 7.2 Causal network for making-do waste (based on Formoso et al., 2017)

but which are useful for showing non-obvious problems. Each lead production waste category has cause–effect relationships not only with terminal waste categories, but also with other core waste categories;

(iii) Previous production stages: in this zone, some of the previous production stages are represented, as failures in those stages may be root causes for different production waste categories. Understanding the relationships between previous production stages and waste categories is important for devising strategies for waste elimination.

These waste zones are fairly consistent with some conceptualisations adopted by other authors. Fernández-Solís and Rybkowski (2012) proposed three different waste concepts: discrete waste, synergistic, and systemic waste. The first one corresponds to terminal waste, and the second includes lead production waste and other categories that should be related to some of the previous production stages at the project level. Finally, systemic waste is related to the loose coupling of stakeholder organisations, which results in duplication of efforts, miscommunication, and misunderstanding at the transitions and transference of information. The concept of institutional waste, suggested by Sarhan, Pasquire, and King (2017), although based on another theoretical framework, is similar to synergistic waste, as it is a waste that also exists at the supply chain level.

Figure 7.3 represents an attempt to model the relationship between several waste categories that have been investigated in previous studies (Alarcon, 1997; Bølviken, Rooke, and Koskela, 2014; Fireman, Formoso, and Isatto, 2013; Formoso et al., 2002, 2017; Kalsaas, Gundersen, and Berge, 2014). It is not intended to be a well-established map of cause and effect relationships between different phenomena that lead to terminal waste. Instead, it can be regarded as an initial step towards a systematic and comprehensive analysis of waste in construction, which can be used as a reference for future research. Further work is necessary to refine this waste network, by providing further empirical evidence on existing cause–effect relationships and by improving the understanding of the relative importance of each waste category.

7.7 Which are the lead wastes in construction?

Based on the cause–effect relationships presented in Figure 7.3, five categories are proposed as the lead production waste for construction projects: (i) making do, (ii) work-in-progress; (iii) unfinished work; (iv) transportation; and (v) quality deviations. A brief discussion of each is presented as follows. Some traditional waste categories, such as rework, waiting, and material waste, have not been included because these are considered to be terminal waste.

7.7.1 Making-do

Making-do is a waste category proposed by Koskela (2004). It is defined as a decrease in performance when a task is started or continued even if a complete set of necessary inputs is not available. This concept was partly inspired by the idea of the complete kit proposed by Ronen (1992): the set of components, drawings, documents, and information needed to complete a given assembly, subassembly, or a process. Making-do is related to the concept of improvisation, as people may use whatever resources they have at hand to reach their goals in difficult and uncertain situations, or even redefine their objectives in line with the resources available (Cunha, 2004).

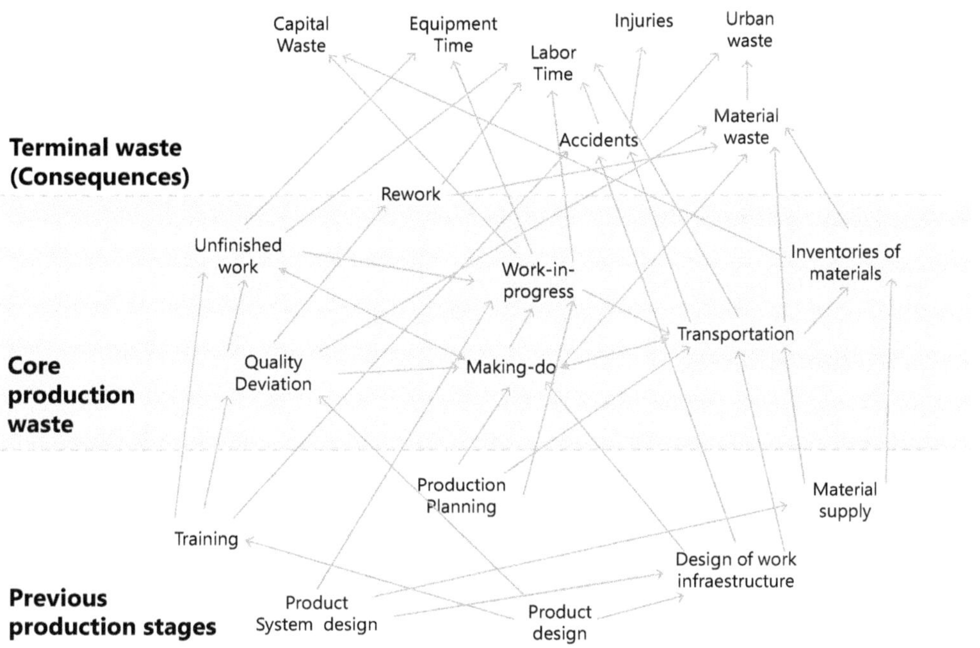

Figure 7.3 Cause–effect relationships between different waste categories – overview

If a task starts without the necessary inputs, there are two possible consequences: (i) to stop the execution of the task; or (ii) to start or continue the task even though one or more preconditions are missing. These two options will result in different types of waste and may have different short- and long-term effects. If there is a stoppage, the immediate consequence is waiting. The cost of stopping can be partly reduced by having a backlog of tasks that can be executed while waiting for the missing precondition to come into place. However, this results in an increase in work-in-progress. By contrast, stoppages could be regarded as improvement opportunities, by using the idea of 'stop the line', adopted in the Toyota Production System (Ohno, 1988). The disadvantage of this option is that additional costs that result from waste tend to be high in the short term. However, stopping the line avoids defective products being produced, creates high attention, and encourages the understanding and removal of root causes (Adler, Goldoftas, and Levine, 1999). It is therefore likely to be advantageous in the long term.

In the second option, there are two possible consequences, which may happen simultaneously. The first one is to informally reduce the scope of the task being performed. This is called task diminishment by Patton (2013) and implies that parts of the task are not done or done with inferior quality. For instance, if a carpenter chooses to build external walls with less or no insulation due to the lack of materials, waste in this situation comes in the form of value loss, as the final product will have defects or inferior quality.

The second possible consequence of making-do is a reduction in the performance of the production system, including: more work-in-progress, longer lead time, increase in the share of non-value adding activities, need for complex control systems, decline in overall productivity, decline in workers' motivation, poor quality, and decline in safety (Koskela, 2004; Ronen, 1992). Moreover, Formoso et al. (2017) provided evidence that making-do can be a major cause of other types of waste, such as material waste and accidents.

Based on data from two exploratory studies, Formoso et al. (2017) state that the incidence of making-do events is very high in some specific contexts. The most frequent types of making-do waste were related to the access and availability of working areas, and to the necessary infrastructure in terms of water and electricity supply, protection, and equipment and tools. In fact, the main causes of making-do were ineffectiveness in terms of providing adequate workspace infrastructure, planning and controlling the use of space, and lack of adequate information related to task procedures.

7.7.2 Work-in-progress

Hopp and Spearman, (2000) define work-in-progress (WIP) as the inventory between the start and end points of a product routing, but it does not include the stock at those points. WIP is, therefore, an inventory of unfinished or intermediate products, and has the same negative impacts as any kind of inventory: hides quality problems, generates non-value adding activities, such as transportation, moving, etc. (Ohno, 1988). Moreover, it is the WIP which turns the production process into a sequence of production islands hard to be managed (Hopp and Spearman, (2000): when production processes are separated from each other, there is a break in the flow of materials, often encouraging people to improve individual operations without much concern with the entire process.

According to Spearman, Woodruff, and Hopp (1990) some key benefits of pull systems can be achieved simply by limiting the level of work-in-progress:

(i) The chances for early detection of quality problems are improved, since when WIP levels are lower, so are flow times. Therefore, if a process produces defective items, it soon reaches a subsequent operation where the defect can be noticed;
(ii) It becomes easier to manage a shop floor. When WIP levels are low, operators waste less time searching through WIP for the next job to be processed; and
(iii) Reduced WIP makes it harder to hide or tolerate machine failures, defects, yield losses, theft, and unnecessary idle time.

Several studies have indicated that the amount of work-in-progress in construction sites is very high, due to several causes, including the adoption of large batches (Koskela, 1992), making-do (Formoso et al., 2017), lack of synchronisation between crews (Bashford et al., 2003), and poor logistics management of prefabricated building systems (Viana, 2015). In fact, similarly to over-production, a category of waste proposed by Ohno (1988), work-in-progress is caused by the fact that most supervisors think that it is better to keep workers producing with the aim of avoiding waiting, in order to maximise the rate of utilisation of production capacity.

7.7.3 Unfinished work

Unfinished work refers to the situation in which some small finishing tasks are left behind when a crew leaves a workplace (Fireman, Formoso, and Isatto, 2013). This is often caused by the fact that work packages from short-term plans are not properly inspected at the end of the control cycle (Leão, Isatto, and Formoso, 2016). In companies that use the Last Planner System (Ballard and Howell, 1998), this might cause a distortion in the

PPC (percent of plan completed) metric, as some packages are assumed to be completed, but additional work is still needed in the following week (Leão, Isatto, and Formoso, 2016). The tasks left for the following week are typically very small, being usually related to the need of rework or task completion, and contribute to the increase of work-in-progress (Fireman, Formoso, and Isatto, 2013).

Those small work packages are usually neglected in planning meetings and contribute to the fact that much of the work in construction sites is carried out informally, i.e. part of the work undertaken by crews is not formally included in operational plans, nor formally released by any other mechanism. Such small packages by nature have a large share of non-value adding activities, such as set up, moving, and cleaning (Fireman, Formoso, and Isatto, 2013; Leão, Isatto, and Formoso, 2016). Moreover, they tend to cause making-do events, and limit the implementation of systematic quality inspections.

Some previous studies on the implementation of the Last Planner System (see Chapter 3 by Ballard) provide additional evidence on the importance of informal work: (i) those small rework or finishing tasks are rarely included in short-term plans (Formoso and Moura, 2009); (ii) the total number of workers that are included in short-term plans does not correspond to the total number of workers on site (Fireman, Formoso, and Isatto, 2013; Formoso et al., 2017).

7.7.4 Transportation

Transportation waste is described by Ohno (1988) as materials handling activities that generate cost and do not add value. Previous studies have pointed out that a large percentage of construction work time is spent in transportation operations (Alarcón, 1997; Pérez and Costa, 2018). Formoso et al. (2002) pointed out that several types of resources may be wasted during transportation operations, such as materials, labour, and equipment. Therefore, this type of waste is concerned not only with unnecessary transportation operations, but also with other types of site logistical problems, such as set-ups, stoppages, and inefficient utilisation of equipment.

Pérez and Costa (2018) carried out a study with the aim of proposing a taxonomy for transportation waste, including the main causes and consequences for this type of waste. Based on two case studies, four main causes of transportation waste were pointed out: (i) obstacles in access routes, (ii) inadequate storage of materials, (iii) inappropriate equipment, and (iv) lack of labour to support operations. In another study, Bataglin et al., (2019) suggested that transportation waste is strongly related to inadequate site layout as well as to the lack of systematic planning of logistical operations on construction sites.

Regarding the consequences of transportation waste, previous studies suggested that it is a major cause of material waste (Bossink and Brouwers, 1996; Formoso et al., 2002), and also of labour related problems, such as unsafe working conditions and ergonomic problems (Pérez and Costa, 2018).

Therefore, transportation waste is related to several types of terminal waste, including material waste, waiting, inventories, and accidents. Moreover, there is also a strong connection between transportation and making-do waste. As pointed out by Formoso et al. (2017), making-do is often related to the access and availability of working areas. In fact, the main causes of making-do were the ineffectiveness of upstream processes in terms of providing the necessary equipment on time (e.g. scaffolds), planning and controlling the use of space, and making available information related to operations design and work-space infrastructure (Formoso et al., 2017).

7.7.5 Quality deviations

Quality is a multifaceted concept and cannot be assessed by a simple numeric measure (Hopp and Spearman, 2000). There are several definitions of quality, and most of them are heavily product oriented (e.g. performance, product reliability, conformance, and durability), based on the assumption that quality must be ultimately customer driven (Garvin, 1987). However, product quality is determined by several processes, such as design, procurement, production control, and customer service. Therefore, in order to make a production system effective in terms of delivering quality, customer concerns must be translated into measures and controls to support process management (Hopp and Spearman, 2000).

In construction, quality deviations continue to be pointed out as a major type of waste on many sites (Oakland and Marosszeky, 2017). Despite much effort in the industry for the implementation of quality management systems, often based on the Total Quality Management philosophy, results are still limited for different reasons: (i) implementation methods tend to be dogmatic and adopt a narrow perspective (Koskela, 1992; Leão, Isatto, and Formoso, 2016); (ii) quality management usually has the status of a secondary management track, being not properly integrated to production control (Fireman, Formoso, and Isatto, 2013), and (iii) quality management is often motivated by the need to disseminate a public image of company and to get contracts, by using ISO certification or national quality awards, instead of the urge for internal improvement (Koskela, 1992; Leão, Isatto, and Formoso, 2016). In fact, quality-related wastes have been detected even in companies that are considered to be well advanced in the implementation of Lean concepts and tools (e.g. Last Planner System) (Fireman, Formoso, and Isatto, 2013).

Processes with quality problems are characterised by excessive variability, poor deviation detection (i.e. long cycle time from detection to correction), and insufficient consideration of customer requirements (Koskela, 1992). Quality deviations represent a specific kind of waste, which might result in rework, unfinished work (Fireman, Formoso, and Isatto, 2013), or customer complaints, but also cause interruptions in physical flows (Koskela, 1992). Similarly, poorly defined requirements in internal customer–supplier relationships tend to increase the duration and cost of conversion activities, slowing down production flows (Koskela, 1992), causing other types of waste, such as transportation, work-in-progress, material waste, and waiting. Moreover, Formoso et al. (2017) pointed out that making-do has been identified as a major cause of quality deviations in construction sites, due to the effect of task diminishment (Patton, 2013).

7.8 Discussion and conclusions

Based on a comprehensive literature review on construction waste, it has been pointed out that the amount of waste tends to be fairly large in the construction industry, independently of the category of waste being investigated. However, despite the relevance of this topic for construction management, the number of papers and books on this theme is relatively small.

Regarding the conceptualisation of waste, previous studies have scarcely discussed the concept of waste at an abstract level, but have simply adopted operational definitions of waste, which are useful for guiding data collection. By contrast, some studies from the Lean Construction community have pointed out the need to use a more abstract conceptualisation of waste, based on the principle of reducing the share of activities that do not add value from the perspective of the client.

Several waste categories have been investigated in construction, most defined as terminal waste. In the manufacturing industry, the most used waste categories are the ones from the taxonomy proposed by Ohno (1988). It seems that the main role of existing waste classifications is to call attention to the most likely problems in a specific context, i.e. to play the role of drivers for improvement. Therefore, it seems that waste is a specific and context-related phenomenon: although general concepts and theories are helpful to describe and understand waste, the relative importance of each type of waste can only be assessed by looking at the specific context of a production system.

This chapter also discusses the idea of waste propagation, which can be represented by a network of waste categories, in which the different types of waste are connected by cause–effect relationships. The effort of modelling a waste network can help to identify lead waste, i.e., waste categories that have substantial negative impacts on the performance of a production system. These should be identified and be the main focus of improvement initiatives.

This study also proposed five categories of waste can be pointed out as lead waste in construction, briefly defined in Table 7.1: making-do; work-in-progress; unfinished work; transportation; and quality deviations. This proposition needs to be further tested in

Table 7.1 Definition of five lead waste categories

Waste category	*Brief definition*	*Main references*
Making-do	Decrease in performance when a task is started or continued even if a complete set of necessary inputs is not available. It may cause the stoppage of a task; or the execution of a task even though some of the necessary preconditions are missing.	Koskela (2004) Formoso et al. (2017)
Work-in-progress	It is an inventory between the start and end points of a product routing. It has the same negative impacts of any kind of inventory (e.g. hides quality problems, and generates non-value adding activities), and also turns the production process into a sequence of production islands hard to be managed.	Hopp and Spearman (2000) Viana (2015)
Unfinished work	It refers to the situation in which some small finishing tasks are left behind when a crew leaves a workplace. Those tasks are typically very small, being usually related to the need of rework or task completion, and may contribute to the increase of work-in-progress and of the share of non-value-adding activities.	Fireman, Formoso, and Isatto (2013) Leão, Isatto, and Formoso (2016)
Transportation	It is concerned with ineffective planning of transportation operations, which may result in waste of different types of resources during, such as materials, labour, and equipment (capital). It includes not only unnecessary transportation operations, but also other types of non-value adding activities, such as setting-up equipment, and inefficient use of equipment.	Perez and Costa (2018) Bataglin et al. (2019)
Quality deviations	Different types of problems can be associated with quality deviations: excessive variability in quality, poor deviation detection (i.e. long cycle time from detection to correction), and insufficient consideration of customer requirements.	Koskela (1992) Leão, Isatto, and Formoso, (2016)

empirical studies. Some traditional waste categories, such as rework, waiting, and, material waste, have not been included as these are considered to be terminal waste.

Based on the overview provided on previous work on construction waste, several suggestions for further research can be made:

(i) Refine the conceptualisation of some waste categories (e.g. rework, delays), based on core operations management concepts, making it easier to compare data from different construction projects, as well as to devise effective strategies for waste elimination;
(ii) Devise further models of waste networks by understanding cause–effect relationships between different waste categories, based on empirical data;
(iii) Develop studies to measure the magnitude of the impact of core waste categories downstream on terminal waste (i.e. the degree of waste propagation); and
(iv) Investigate how lead waste categories could be systematically monitored in production control systems, by introducing new operational measures or possibly by refining and exploring existing data in construction companies that have implemented Lean production concepts and principles.

Notes

1 This initiative has involved IGLC members from the following organizations: University of Huddersfield (UK), Technion (Israel), University of Agder (Norway), Veidekke (Norway) and Universidade Federal do Rio Grande do Sul (Brazil).
2 *Architectural Engineering and Design Management*; *Building Research & Information*; *Construction Management and Economics*; *Engineering, Construction and Architectural Management*; *Journal of Architectural Engineering*; *Journal of Construction Engineering and Management*; *Journal of Management in Engineering*.

References

Adler, P.S., Goldoftas, B. and Levine, D.I. (1999) 'Flexibility versus efficiency? A case study of model changeovers in the Toyota production system', *Organization Science*, 10(1), pp. 43–68.

Alarcón, L.F. (1997) 'Modelling waste and performance in construction', in Alarcon, L.F. (ed.) *Lean Construction*, Balkema, Rotterdam, pp. 51–66.

Al-Hajj, A. and Hamani, K. (2011) 'Material waste in the UAE construction industry: Main causes and minimization practices', *Architectural Engineering and Design Management*, 7(4), pp. 221–235.

Ballard, G. and Howell, G. (1998) 'Shielding production: Essential step in production control', *Journal of Construction Engineering and Management*, 124(1), pp. 11–17.

Bashford, H.H., Sawhney, A., Walsh, K.D. and Kot, K. (2003) 'Implications of even ow production methodology for US housing industry', *Journal of Construction Engineering and Management*, 129(3), pp. 330–337.

Bataglin, F.S., Viana, D.D., Formoso, C.T., Bulhões, I.R. (2019) 'Model for planning and controlling the delivery and assembly of engineer-to-order prefabricated building systems: Exploring synergies between Lean and BIM', *Canadian Journal of Civil Engineering*, V. 47, e-First Article: pp. 1–13.

Bicheno, J. and Holweg, M. (2009) *The Lean Toolbox: The Essential Guide to Lean Transformation*. 4th edition, Buckingham, UK, PICSIE Book.

Bølviken, T. and Koskela, L.J. (2016) 'Why Hasn't Waste Reduction Conquered Construction?', *Proc. 24th Annual Conference of the International Group for Lean Construction*, Boston, MA, USA. Available at: www.iglc.net.

Bølviken, T., Rooke, J. and Koskela, L. (2014) 'The Wastes of Production in Construction – A TFV Based Taxonomy', *Proc. 22nd Annual Conference of the International Group of Lean Construction*, Oslo, Norway. Available at: www.iglc.net.

Bossink, B.A.G. and Brouwers, H.J.H. (1996) 'Construction waste: Quantification and source evaluation', *Journal of Construction Engineering and Management*, ASCE, 122(1), pp. 55–60.

Burati J.L.J.; Farrington J.J. and Ledbetter W.B. (1992) 'Causes of quality deviations in design and construction', *Journal of Construction Engineering and Management*, 118(1), pp. 34–49.

Bush, R.W. (2007) 'Reducing waste in US health care systems', *Journal of the American Medical Association*, 297(8), pp. 871–874.

Choi, S., Ku, T.H., Yeo, D.H. and Han, S.H. (2002) 'Waste Elimination of Mucking Process of a Petroleum Storage Tunnel through the Value Stream Analysis', *Proc. 10th Annual Conference of the International Group of Lean Construction*, Manchester, UK, pp. 155–167. Available at: www.iglc.net.

Cunha, M.P. (2004) *Bricolage in Organizations*. Lisboa, Portugal, Instituto Nova Fórum, Universidade Nova de Lisboa.

Faniran, O.O. and Caban, G. (1993) 'Minimizing waste on construction project sites', *Engineering, Construction and Architectural Management*, 5(2), pp. 182–188.

Fernández-Solís, J.L. and Rybkowski Z.K. (2012) 'A theory of waste and value', *International Journal of Construction Project Management*, 4, pp. 2.

Fireman, M.C., Formoso, C.T., Isatto, E.L. (2013) 'Integrating Production and Quality Control: Monitoring making-do and Unfinished Work', *Proc. 21st Annual Conference of the International Group for Lean Construction*, Fortaleza, Brazil, 2, pp. 515–525. Available at: www.iglc.net.

Ford, H. (1926) *Today and Tomorrow*. Garden City, NY, USA, Doubleday.

Formoso, C.T., Bølviken, T., Rooke, J.A., Koskela, L.J. (2015) 'A Conceptual Framework for the Prescriptive Causal Analysis of Construction Waste', *Proc. 23rd Annual Conference of the International Group for Lean Construction*, Perth, Australia, pp. 454–461. Available at: www.iglc.net.

Formoso, C.T.; Moura, C.B. (2009) 'Evaluation of the Impact of the Last Planner System on the Performance of Construction Project', *Proc. 17th Annual Conference of the International Group for Lean Construction*, Taipei, Taiwan, pp. 153–164. Available at: www.iglc.net.

Formoso, C.T., Soibelman. L., De Cesare, C. and Isatto, E.L. (2002) 'Material Waste in Building Industry: Main Causes and Prevention', *Journal of Construction Engineering and Management*, 128(4), pp. 316–325.

Formoso, C.T., Sommer, L., Koskela, L., Isatto, E.L. (2017) 'The identification and analysis of making-do waste: Insights from two Brazilian construction sites', *Ambiente Construído*, 17(3), pp. 183–197.

Forsberg, A. and Saukkoriipi, L. (2007) 'Measurement of Waste and Productivity in Relation to Lean', *Proc. 15th Annual Conference of the International Group for Lean Construction*, Michigan, 46, pp. 67–76. Available at: www.iglc.net.

Gavilan, R.M. and Bernold, L.E. (1994) 'Source evaluation of solid waste in building construction', *Journal of Construction Engineering and Management*, 120(3), pp. 536–552.

Garvin, David A. (1987) "Competing on the Eight Dimensions of Quality."*Harvard Business Review* 65, no. 6 (November–December 1987).

Hopp, W. & Spearman, M. (2000) *Factory Physics: Foundations of Manufacturing Management*. Second, Boston, MA, USA, Irwin McGraw-Hill.

Horman, M.J. and Kenley, R. (2005) 'Quantifying levels of wasted time in construction with meta-analysis', *Journal of Construction Engineering and Management*, 131(1), pp. 52–61.

Hwang, B.-G., Thomas, S.R., Haas, C.T. and Caldas, C.H. (2009) 'Measuring the impact of rework on construction cost performance', *Journal of Construction Engineering and Management*, 135(3), pp. 187–198.

Josephson, P.-E., Hammarlund, Y. (1999) 'The causes and costs of defects in construction: A study of seven building projects', *Automation in Construction*, 8(6), pp. 681–687.

Kalsaas, B.T. (2010) 'Work-Time Waste in Construction', *Proc. 18th Annual Conference of the International Group for Lean Construction*, Haifa, Israel, pp. 507–517. Available at: www.iglc.net.

Kalsaas, B.T., Gundersen, M. and Berge, T.O. (2014) 'To measure Workflow and Waste. A Concept for Continuous Improvement', *Proc. 22nd Annual Conference of the International Group for Lean Construction*, Oslo, Norway, pp. 835–846. Available at: www.iglc.net.

Koskela, L. (1992) *Application of the New Production Philosophy to Construction*, Center for Integrated Facility Engineering, Department of Civil Engineering, Stanford University. Technical Report number: 72.

Koskela, L. (2000) *An Exploration towards a Production Theory and Its Application to Construction*, PhD thesis, Technical Research Centre of Finland, VTT.

Koskela, L. (2004) 'Making do-the Eighth Category of Waste', *Proc. 12th Annual Conference of the International Group of Lean Construction*, Helsingør, Denmark, pp. 1–10. Available at: www.iglc.net.

Koskela, L., Bølviken, T. and Rooke, J. (2013) 'Which are the Wastes of Construction?', *Proc. 21th Annual Conference of the International Group of Lean Construction*, Fortaleza, Brazil, pp. 3–12. Available at: www.iglc.net.

Koskela, L., Ferrantelli, A., Niiranen, J., Pikas, E. and Dave, B. (2019) 'Epistemological explanation of lean construction', *Journal of Construction Engineering and Management*, 145(2), pp. 18–29.

Koskela, L., Sacks, R. and Rooke, J. (2012) 'A Brief History of the Concept of Waste in Production', *Proc. 20th Annual Conference of the International Group of Lean Construction*, San Diego, CA, USA, pp. 321–330. Available at: www.iglc.net.

Leão, C.B., Isatto, E.L., Formoso, C.T. (2016) 'Proposta de modelo para controle integrado da produção e da qualidade com apoio da computação móvel', *Ambiente Construído*, 16(4), pp. 109–124.

Ledbetter, W.B. (1994) 'Quality Performance on Successful Project', *Journal of Construction Engineering and Management*, 120(1), pp. 34–46.

Love, P. (2002) 'Influence of project type and procurement method on rework costs in building construction projects', *Journal of Construction Engineering and Management*, 128(1), pp. 18–29.

Love, P. and Edwards, D.J. (2004) 'Determinants of rework in building construction projects', *Engineering, Construction and Architectural Management*, 11(4), pp. 259–274.

Macomber, H. and Howell, G. (2004) 'The Two Great Wastes', *Proc. 12th Annual Conference of the International Group for Lean Construction*, Helsingør, Denmark. Available at: www.iglc.net.

Monden, Y. (1983) *Toyota Production System: Practical Approach to Production Management.* 1st edition, Industrial Engineering Management, Norcross, Georgia, USA.

Nahmens, I. and Ikuma, L.H. (2011) 'Effects of lean construction on sustainability of modular homebuilding', Journal of Architecture Engineering, 18(2), pp. 155–163. doi: 10.1061/(ASCE)AE.1943-5568.0000054.

Oakland, J.S., Marosszeky, M. (2017) *Total Construction Management – Lean Quality in Construction Project Delivery*. Oxford, UK, Routledge.

Ohno, T. (1988) *Toyota Production System.* Cambridge, MA, USA, Productivity Pr.

Park, M., Yang, Y., Lee, H.-S., Han, S. and Ji, S.H. (2011) 'A floor-level construction material layout planning model considering actual travel path', *Journal of Construction Engineering and Management*, 138(7). doi: 10.1061/(ASCE)CO.1943-7862.0000493

Patton, J.R. (2013) *Task Diminishment: Construction value loss due to sub-optimal task execution.* PhD Thesis, Indiana State University.

Pérez, C.T. and Costa, D.B. (2018) 'Developing a taxonomy of transportation waste in construction production processes', *Built Environment Project and Asset Management.* doi: 10.1108/BEPAM-04-2018-0062

Poon, C.S., Yu, A.T.W., Wong, S.W. and Cheung, E. (2004) 'Management of construction waste in public housing projects in Hong Kong', *Construction Management & Economics*, 22(7), pp. 675–689.

Porwal, A. and Hewage, K.N. (2011) 'Building information modelling based analysis to minimize waste rate of structural reinforcement', *Journal of Construction Engineering and Management*, 138(8). Available at: 10.1061/(ASCE)CO.1943-7862.0000508

Ronen, B. (1992) 'The complete kit concept', *International Journal of Production*, 30(10), pp. 2457–2466.

Sacks, R., Esquenazi, A. and Goldin, M. (2007) 'LEAPCON: Simulation of Lean Construction of High-Rise Apartment Buildings', *Journal of Construction Engineering and Management*, 133(7), pp. 374–384. doi: 10.1061/(ASCE)0733-9364(2007)133:7(529).

Sarhan, S., Pasquire, C., King, A. (2017) 'The concept of institutional waste within the construction industry: A potential theoretical framework', *Lean Construction Journal*, 2017 issue, pp. 12–24.

Shingo, S. (1988) *Non-stock Production: The Shingo System for Continuous Improvement.* Cambridge, MA, USA, Productivity Press.

Skoyles, E.F. (1976) *Material Wastage: A Misuse of Resources.* Batiment International, Building Research and Practice, Taylor & Francis, London, pp. 232–243.

Spearman, M., Woodruff, D. and Hopp, W. (1990) 'CONWIP: A pull alternative to kanban', *International Journal of Production Research*, 28(5), pp. 879–894.

Taylor, F.W. (1913) *The Principles of Scientific Management.* New York, Harper and Brothers.

Tommelein, I.D., Riley, D.R. and Howell, G. (1999) 'Parade game: Impact of work flow variability on trade performance', *Journal of Construction Engineering and Management*, 125(5), pp. 304–310.

Viana, D.D. (2015) *Integrated Production Planning and Control Model for Engineer-to-order Prefabricated Building Systems.* Porto Alegre, Brazil, Universidade Federal do Rio Grande do Sul. Doctorate Thesis.

Viana, D.D., Formoso, C.T. and Kalsaas, B.T. (2012) 'Waste in Construction: A Systematic Literature Review on Empirical Studies', *Proc. 20th Annual Conference of the International Group for Lean Construction*, California. Available at: www.iglc.net.

Womack, J.P. and Jones, D.T. (2004) *Lean Thinking: Banish Waste and Create Wealth in Your Corporation.* Free Press, New York.

Yu, H., Tweed, T., Al-Hussein, M. and Nasseri, R. (2009) 'Development of lean model for house construction using value stream mapping', *Journal of Construction Engineering and Management*, 135(8), pp. 782–790.

Zhao, Z.Y., Lv, Q.L., Zuo, J. and Zillante, G. (2010) 'Prediction system for change management in construction project', *Journal of Construction Engineering and Management*, 136(6), pp. 659–669.

Part 2

Lean Construction approaches

8 Target Value Delivery

Glenn Ballard

8.1 Introduction

Target Value Delivery (TVD) is offered here as a fundamental delivery process for built environment projects. It pursues the most complete objective, namely providing acceptable net benefits throughout the life of the asset to be constructed. The term 'target value delivery' is used instead of 'target costing' because the cost target is but a means to value delivery. The name 'target value design' is also rejected because it suggests that steering to targets only occurs in design.

8.2 Target Value Delivery: background

TVD was adapted from product development's *target costing*, which is well known for the formula Revenue: Profit = Cost. When Toyota develops a new product, the allowable cost is determined by subtracting an acceptable profit from expected revenues over the life of the product (note that some products may increase the profitability of the entire product portfolio by filling a market niche that encourages buyer loyalty to the brand). What remains is the allowable cost, the most that can be spent over the life of the product and still earn an acceptable profit. This proactive approach can be applied in construction, with adjustments for the different relationships between buyers and users.

TVD is appropriate for all buyers of constructed assets: developers, producers, buyers of objects of art, and buyers of commodities and necessities. Determining an allowable cost is appropriate for each, but the criteria for allowable cost differ. For some, the criteria are both willingness and ability to pay. For others, it is only the latter. Each of these buyer types are described and arguments for TVD's appropriateness are provided in the following paragraphs.

Product development's target costing approach is used by real estate *developers* and others investing in the production of buildings, highways and other products of the built environment in order to either sell or lease them. Either way, it is advisable for them to consider the whole life benefits and costs, just as is done by Toyota and other new product developers, in order to make their products more attractive in the market.

Producers are those who buy constructed objects, which are not for sale or lease to others, but for their own use. Some of these producers are manufacturers that need factories to produce their own products. Others are law firms that need space within which to provide their services, or schools that need space within which to educate their students, or families that need space within which to live. TVD is also appropriate for them. Businesses buying means for their own production should carry out target costing because they must have profits in order to survive, and profit is what remains after subtracting costs from revenues. Further, there are specific requirements for production facilities, which, if falling short of those requirements, may be incapable of supporting the production of goods or services.

Families don't fit easily within that picture. Although some purchases of homes are made with an eye to price appreciation, it seems safe to assume that most are intended for own use, and that the latter are limited by ability to pay. The definition of allowable cost, which is the most a buyer is willing and able to pay in order to accomplish their purposes, covers families and others buying constructed assets for their own use, and consequently TVD is appropriate for them as well.

There are at least two other types of buyers of constructed assets to be considered: buyers of objects of art and buyers of commodities. We consider these in turn, starting with objects of art, which are typically design-driven rather than cost-driven at first, when funding is from donations influenced by design. Ultimately, of course, cost becomes a constraint on design, but that happens more or less quickly. The Sydney Opera House seems to have remained design-driven almost to its completion.

Commodities are products (goods or services) offered in the market that differ from competing offers only in price, and hence ability to pay is the sole criterion for allowable cost. That is not to be taken as if any price is acceptable. The proviso that products and services are truly commodities is critical. This is likely rarer than is assumed, and hence there is a risk that the buyer will pay more than the stipulated price through compromised quality or timeliness of delivery. Ability to pay is also the sole criterion for products or services that are needed in order to continue doing business.

8.2.1 Value engineering

One of the foundational concepts of target cost/TVD comes from value engineering (Miles, 1961[1]), namely that worth and cost are not correlated across the entire range of possible combinations, and hence steering design to cost targets is possible. This concept is also the foundation for the 'design to cost' initiative of the United States' Department of Defense from the 1970s (Michaels and Wood, 1989), which may have its roots in the early 1950s (Burroughs and Chenhall, 2012, p. 136).[2] It is well understood that value engineering was developed in the United States and subsequently adopted in Japan. Until Burroughs and Chenhall's article, it was widely assumed that target costing developed independently in Japan and was subsequently adopted in the West. In fact, target costing was developed interdependently. One of the primary reasons for not understanding this fact appears to be that the Japanese were more receptive to descriptive papers than were 'theory-based English-language publications' (Burroughs and Chenhall, 2012, p. 139). The term 'theory-based' is a code word for the disregard for relevance revealed in Koskela (2011).

Value engineering methods are used extensively in product development in search of functionality at less cost. TVD can also benefit from such methods in the design phase of construction projects, but goes beyond value engineering in setting and steering projects to targets for the entire project.

8.2.2 How target costing developed in construction

In Figure 8.1, we see the two factors in net benefits, namely benefits (the 'Healthcare outcomes' in the largest circle to the far right) and costs throughout the life of the constructed asset – in this case, a hospital. The circles to the left represent the costs incurred in acquiring the benefits. If cost of constructing the hospital is set at 1.0, design costs will be approximately

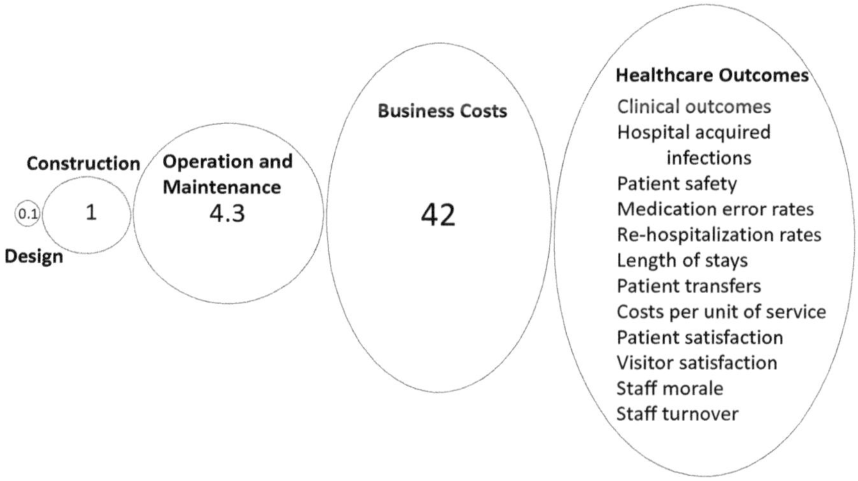

Figure 8.1 Whole Life Costs & Benefits (developed from Evans et al., 1998)

0.1, cost to operate and maintain the building over 20 years will be approximately 4.3, and cost to deliver healthcare in the hospital will be approximately 43. The rightmost circle must be larger than the sum of the 'cost' circles in order for the project to be an appropriate investment. Striving to provide ever more value with ever less waste is the Lean ideal. That is done in built environment projects by increasing benefits and reducing cost, respectively.

The application of target costing to construction was initiated by Boldt Construction in 2002 at St Olaf's College (Ballard and Reiser, 2004), followed within the next two years by Shawano Clinic for Thedacare (Toussaint, 2006), and Sutter Health's Fairfield Medical Office Building (Zimina et al., 2012). Sutter Health also applied the target costing approach on its Acute Rehabilitation Center project during this period (Ballard, 2006). The success of these early projects persuaded Sutter Health to include target costing as a 'Lean' management method for all its projects. The University of California's Project Production Systems Laboratory published a Current Process Benchmark for TVD to capture what had been learnt about target costing in construction up to that point (Ballard, 2006, and updated in 2011).

Sutter Health's Fairfield Medical Office Building project introduced a new development phase of target costing, in which it began to be applied along with Integrated Project Delivery (IPD). IPD is a contract in which owners, designers and builders share risks and rewards and are organisationally integrated, similar to the project alliancing model, but extended to the design and construction firms previously subcontracted to architects or general contractors, respectively (see also Chapter 9 for a detailed description of IPD). TVD is a production concept included in the Lean management methods specified in the first IPD contract (Lichtig, 2006). For that reason, IPD and its foundational concept, relational contracting, are introduced in the next section.

8.2.3 Target Value Delivery and relational contracting

TVD was categorised above as a production as opposed to a contracting concept. Nonetheless, relational contracting is relevant to TVD. Buying a constructed asset such as a building, refinery or highway can be understood as (1) a transaction, or an exchange of money for asset, and (2) entering into a relationship with others to help solve a problem or exploit an opportunity. The more uncertain and complex the project, the more needed is a relational contract, and the more needed is TVD as the project delivery process (Ballard et al., 2011). Projects are socio-technical production systems, and relational contracting speaks to the social dimension.

The classical theory of contracts sees them as transactions. The construct *relational contract* was created by Ian MacNeil (1973), whose research on historical forms of contract found that human agreements range over a spectrum that lies between the fully transactional and the fully relational, neither of which exists in nature. Even the marriage vow, used as illustration of a relational contract by MacNeil, has transactional elements, and the simplest transaction is made possible by a complex context of laws and customs. Every contract lies somewhere along the continuum, but historically the construction industry has been dominated by the classical theory.

MacNeil's continuum of contracts is echoed in the domain of construction projects, where it has been noted that all projects can be understood to lie along a continuum between the simple and certain on the one hand, and the uncertain and complex on the other hand (Ballard and Howell, 2005). The transactional view of contracts is promoted by two beliefs: 1) that design and construction services are commodities, distinguishable only in price, and 2) that competitive bidding yields the lowest price. These beliefs are at odds with the widely held view that projects are becoming ever more complex and uncertain,[3] posing challenges that require innovation, and thus requiring more sophisticated designers and constructors, working together more collaboratively. Projects at the simple/certain end of the spectrum approximate commodities because the ability to produce them is more widely available in the market – innovation in product design and improvement of previous processes may well be possible, but are less needed to achieve scope, cost and schedule certainty.

The expression 'Integrated Project Delivery' was coined by the owners of a company by that name formed in Orlando, Florida in the late 1990s (Matthews et al., 2003). Their stated motivation was to overcome what they saw as the primary obstacle to innovation and improvement, namely the inability to move resources, including money, across organisational boundaries in search of the best project-level investments. The companies that banded together to form IPD specialised in architecture, structural engineering, mechanical engineering, electrical engineering, general contracting and specialty construction companies with structural, mechanical and electrical fabrication and installation capabilities. They agreed to share rewards and risks, and to execute contracts as a team, regardless of who signed the contract with the client. The owners of these companies had worked together for 20 or more years on various projects in the Orlando area. Many were close friends and there was even some intermarriage between families. Even so, they all agreed that the problem 'Who pays? Who gains?' had prevented them from acting to improve project outcomes because they were paid for executing their scopes of work, not for project outcomes, and could only accommodate rather modest requests that reduced their own project profits. The company Integrated Project Delivery completed its first project six months early, 10% under budget, and won a quality award (Ballard et al., 2007; Appendix C, Case Study 5). Its contract structure was design-build, with the design-build team organisationally integrated and commercially tied together through shared risk and reward.

Although the name emerged in the late 1990s, the practice of IPD seems to have originated earlier in the North Sea with BP's Project Andrew (Knott, 1996). The extraction technology at the time was making the profitability of new wells very challenging. BP formed a team of companies, very similar to Orlando's Integrated Project Delivery, but in this case with the client sharing risk and reward with the engineers, fabricators and constructors. The result was innovation in every part of the project and a very successful outcome for all parties. This inspired other North Sea oil and gas companies to adopt the approach, which is said (in a private communication from Statoil managers) to have been equally successful until Statoil's Project Askar, where the budget was set at 50% of the conceptual estimate. Despite best efforts, the project finished well over budget, and the literature does not reveal any subsequent IPD projects in the oil and gas sector.

However, before Project Askar, the next chapter of this story unfolded in Australia, where Brown & Root, the lead contractor on Project Andrew, brought the alliancing model, which initially was applied to offshore platforms, the first of which was the Wandoo Offshore Oil Platform (The Warren Centre for Advanced Engineering: www.engineeringicons.org.au/engineering-icons/australian/wandoo-offshore-oil-platform/index.html). The alliancing model migrated onshore in response to the urgent need to clean up the waters in Sydney harbour prior to the 2000 Olympics (Pitsis et al., 2003). The Northside Tunnel Project was completed in 3.5 years, in half the time required for traditional delivery, in time for the Games to begin. From that success, project alliancing became the dominant form for delivery of public sector infrastructure projects in Australia (Walker and Hampson, 2008).

The Lean Construction Institute (LCI), formed in 1997 in the United States by Gregory Howell and the author of this chapter, had focused its first years on learning how to apply Lean concepts and methods to the construction industry and on increasing awareness about Lean in the industry. In late 1999, a call from Owen Matthews changed LCI's trajectory. Owen called from Orlando, Florida to tell us about a new company, namely Integrated Project Delivery. That prompted us to begin exploring forms of contract and organisational structures and cultures that might impede or promote the Lean philosophy and practice. In April 2003, LCI sponsored an International Symposium on Relational Contracting (Ballard and Howell, 2005) that brought together people and ideas from around the world. At the end of that symposium, Will Lichtig, then with a construction law firm in California, agreed to draft a form of relational contract for the United States. The Integrated Form of Agreement was published in 2004 and first used by Sutter Health in its Lean project delivery programme (Ballard et al., 2007). Subsequently, ConsensusDocs published their series 300 collaborative contract forms based on IFOA – Integrated Form of Agreement (www.consensusdocs.org/Catalog. Accessed October 22, 2017) and the American Institute of Architects initiated a series of 'integrated project delivery' contracts (see www.nspe.org/sites/default/files/resources/pdfs/Licensure/Resources/MFLResearch FellowshipIPDReport.pdf Accessed October 22, 2017).

8.3 Target Value Delivery: how it works

All projects involve commercial terms, an organisational structure, management methods and technology. As shown in Figure 8.2 below, Lean integrated project delivery specifies commercial terms that align interests, organisational integration (also called 'early contractor involvement', but extended to bringing upstream players into downstream processes), the use of Lean management methods, and the use of technologies best fit for delivering value and eliminating waste.

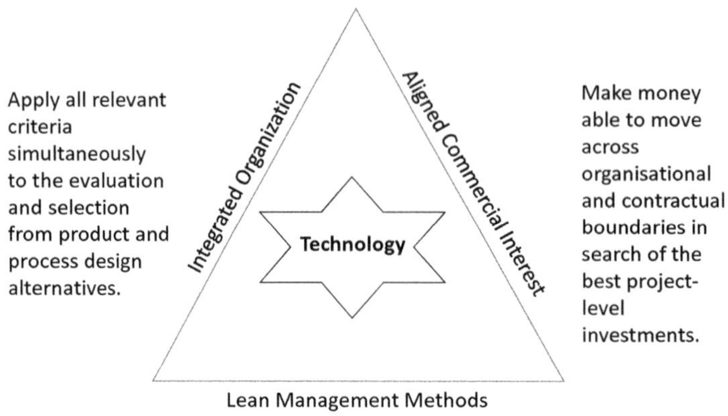

Figure 8.2 Lean Construction Triangle (Thomsen et al., 2011)

The reason for aligning commercial interests through shared risk and reward is to enable moving money and other resources where they are most needed in the project. The reason for organisational integration is to enable the consideration of all relevant criteria when developing and evaluating alternative designs for products or processes, i.e. what is to be built and used, and how to build and use it.

Aligned commercial interests, integrated organisations and Lean management methods are needed all the more as projects become increasingly complex and uncertain (Ballard et al., 2011). In such challenging projects, innovation is needed for successful project delivery, and innovation requires intense collaboration, which is discouraged when designers, builders and suppliers do not share risks and rewards. Multiple and competing criteria are normal in complex, uncertain situations, hence organisational integration is the more critical. Compared to traditional project management methods, Lean management methods are more suitable for projects where pathways to objectives, and even objectives themselves, are likely to change.

There are two primary types of contract structures – the design-build model used by the Orlando Florida team of companies (Darrington, 2011; Matthews and Howell, 2005), and the multi-party agreement model specified in the IFOA (Lichtig, 2006) and derived from project alliancing. In the first, the owner signs a contract with a single firm to design and construct (and, perhaps, also to maintain and operate), but that single firm is organised so that it involves all players in all phases of project delivery, key players share risk and reward, and the project is managed with the Lean philosophy and methods. In the IFOA model, similar to project alliancing, the owner signs a contract with the key project players,[4] the owner bears the risk of the project costs exceeding budget, while the key project players bear the risk of making less or no profit.

IPD's commercial terms (shared risk and reward) and organisational integration (downstream players in upstream processes and vice versa) incentivise the pursuit of targets and facilitate joint problem solving and team learning. TVD is the method used to deliver Lean IPD projects, but can be used, with some limitations, in other contract structures. Contract structures that allow for organisational integration and shared risk and reward, such as the Collaborative Design-Build model used by the Orlando team, are least limited, and in some cases may be referred to multi-party agreements[5]. This chapter describes target Value Delivery as it is used in IPD projects that end with the completion of construction.

The TVD process consists of:

1. Developing a project business case;
2. Validating the project business case/setting project targets;
3. Steering design to targets;
4. Steering construction to targets;

In the following, for the sake of simplicity, it is assumed that an owner is considering investing in an asset for their own use, e.g. a manufacturing facility for Proctor & Gamble, or a commercial office building to house a law firm.

8.3.1 TVD: project definition

In the project definition phase, two questions are answered: (1) What would we build if we could do so successfully? and (2) Can we build it successfully? Answering the first question involves deciding what's wanted and the corresponding allowable cost, then comparing the expected and allowable cost and deciding if it is feasible to close negative gaps. The second question is answered by assessing expected risks and the feasibility of managing those risks. A decision is then made either to revise the project or to fund it. In an IPD context, assessment and decision involve the team of designers and constructors who will risk some, or all, of their profits should the project go forward. For further information on this process, see Pennanen and Ballard (2008), Pennanen and Ballard (2011) and Ballard and Pennanen (2013).

8.3.2 TVD: steering to targets in design

Project targets typically extend beyond the scope/cost pair to include time, safety, environmental and social impact, and more. In preparation for steering design to cost targets, the project target cost (budget) is allocated to systems, subsystems and components, for each of which an allowable cost is set. This is done through some type of cost model that enables relating what is required in and from the asset-to-be-constructed with the cost to acquire and use it. In an IPD context, the profitability of the key companies on the project team is also included. Figure 8.3 below shows a graphic report on the progress of the design toward targets. At this point in the project, the cost reduction required to reach the profit goal is $3,753,005.

Also shown in Figure 8.3, in the lower-right quadrant, is the financial impact expected from risks and opportunities identified by the team, which attempts to generate opportunities that will exceed the cost of risks to deliver the project within budget and, hence, enable maximum profit for service providers.

Haahtela is a project management and cost consultancy located in Helsinki, Finland. It has developed and used a software, named TaKu, to help owners align what they want with their allowable cost. In Haahtela's process, the cost model is produced prior to design and considered 'King of the Hill' in the design phase until replaced. As such, it functions to keep a focus on designing within constraints, and also on enabling the target functionalities and capacities embedded in the cost model. As design decisions are made, they are incorporated in TaKu, replacing the original information. Once design is

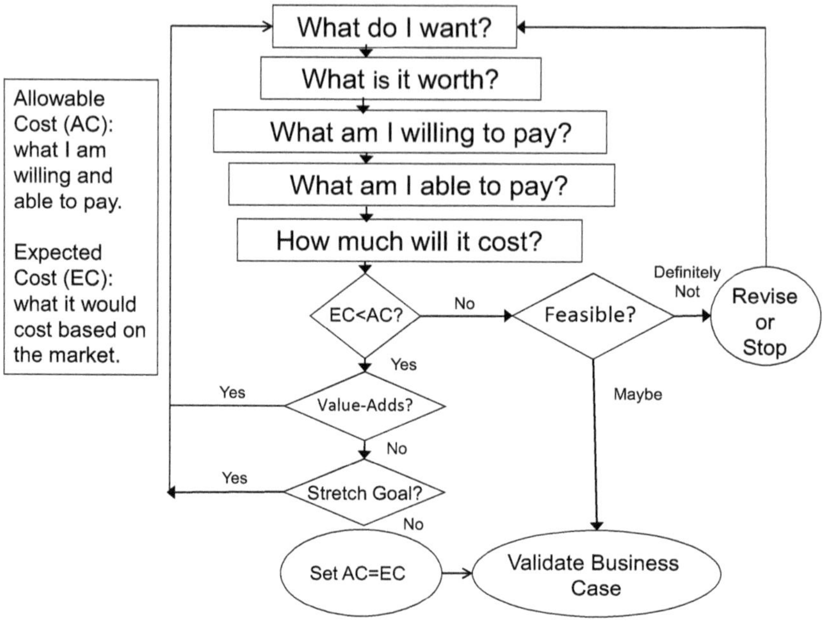

Figure 8.3 TVD Weekly Report (courtesy Sutter Health)

completed and fully captured, TaKu is used to translate functional components into procurement packages.

The total allowable cost for the project is allocated to teams representing the various systems that compose the asset. To take a building example, the TVD teams are formed for substructure, superstructure, envelope, mechanical, plumbing, electrical, controls, etc. TVD teams are cross-functional. Taking structural as an example, the team would consist of a structural engineer, a structural fabricator and a structural erector, plus a representative of the architect and the construction manager. Further, each team is to have estimating capability, so they can periodically report their cost estimate for their system. Teams can invite others to their meetings as needed to review and discuss design alternatives with cross-discipline impact.

These cross-functional teams try to produce the design for which they are responsible within their allocated portion of the total project cost. One method for reducing the cost of delivering specific functionalities is value engineering. More generally, this is where design creativity and ingenuity is at a premium. Even so, the allocation of target cost to systems and components is inexact, and consequently it will often be impossible, within the project limits of money and time, to design every element within its target cost. In that case, the cardinal rule of target costing applies, namely that only the client can change the project target cost. Dollars that are not needed for one element are used for an element that costs more than allocated.

This movement of money across contractual and organisational boundaries is obviously more or less difficult, depending on the commercial terms. It is easiest when the different cross-functional teams are reimbursed for cost of work. It is hardest when compensation is linked to performance of a fixed scope of work.

8.4 Lean management methods used in steering to targets in design

Steering to targets in the design phase involves additional Lean methods, including set-based design, Choosing by Advantages (see Chapter 10 by Arroyo for details), and A3 reports.

As previously explained, TVD facilitates the setting of targets for the project, including a target cost. The job of the design phase is to produce a design that will provide an asset with the buyer's desired functionalities and capacities, within their constraints. In so doing, the companies with profits at risk receive their agreed profits. In pursuing that objective, the TVD teams follow a set-based design methodology through the generation and evaluation of multiple alternative designs, choosing from them at the last responsible moment – that point in time, after which some alternatives are no longer viable (see Ward et al., 1995 for product development's 'set based engineering'; Parrish et al., 2008 for its adaptation to construction projects). To avoid extending project duration, an acceptable – even if not optimum – alternative is agreed early in the process, to be used if needed.

Value engineering methods are used to seek designs that deliver functionality at less cost. The most common value engineering method in use is function analysis (Miles, 1961). Multiple and competing criteria are normal in complex, uncertain situations, hence organisational integration is the more critical. The same is true for Choosing by Advantages, the method used to evaluate alternative designs. Its key characteristics are (1) criteria are not weighted, (2) subjective evaluation is deferred until there is agreement among stakeholders about the attributes of alternatives relative to criteria, and (3) cost is considered only after the total importance of advantages of each alternative has been agreed (see Suhr, 1999 for a description of the method by its inventor; Arroyo, 2014 for its original application in the design of construction projects).

A3 reports (Shook, 2008) are used to deliver proposals to decision makers and to report progress in solving problems. They are named after the paper size to which they are limited. The objective is to capture essential information that is easy to understand. Consequently, graphics are essential components, and the benefit-to-cost results of Choosing by Advantages evaluation are typically included as graphs.

8.4.1 Steering to targets in construction

Design is a potentiality that is made real in the construction phase of projects. Steering to targets in construction, as in design, involves frequent and public comparison of the current state against targets. To the asset's functionalities and capacities needed by the buyer are now added safety, quality, environmental impact, and profitability. These were targets for design, but their realisation is monitored in construction. Targets can be anything of value to project stakeholders, and any condition that must be satisfied in order to acquire that value.

Figure 8.4 is an example of posted information concerning progress toward targets. The report is issued monthly. One of the key metrics in Safety is the Lost Time Incident Rate, shown as currently on target. One of the People metrics is Local Resident Employee, currently just above target. The Quality target for Inspection Pass Rate of 100% is very nearly met. The Production target for Field Labor Savings is currently 19%, well above target. And finally, the amount in Contingency is just over $50 million versus the $61 million target – suggesting that the companies in the risk pool may not

Metric	Goal	Value	Value	Trending	Scale
SAFETY					
LOST TIME INCIDENT RATE	< 0.3 (GOAL)	0.2 (PROJECT TO DATE)	0 (NOV. INCIDENT COUNT)	— (TRENDING)	2.0% / 0.0%
TOTAL INCIDENT RATE	< 1.5 (GOAL)	1.5 (PROJECT TO DATE)	1 (NOV. INCIDENT COUNT)	↑ (TRENDING)	5.0% / 0.0%
RECORDABLE INCIDENT RATE	< 1.0 (GOAL)	1.4 (PROJECT TO DATE)	1 (NOV. INCIDENT COUNT)	↑ (TRENDING)	5.0% / 0.0%
CSI PARTICIPATION: FOREMAN	100% (GOAL)	83% (PROJECT AVERAGE)	97% (CURRENT MONTH)	↑ (TRENDING)	100% / 0%
PREDICTIVE SOLUTIONS PARTICIPATION: SUPTs, APMs, PEs	100% (GOAL)	87% (PROJECT AVERAGE)	95% (CURRENT MONTH)	↑ (TRENDING)	100% / 0%
PREDICTIVE SOLUTIONS PARTICIPATION: DIRECTORS, PMs, Sr. PMs	100% (GOAL)	90% (PROJECT AVERAGE)	89% (CURRENT MONTH)	↓ (TRENDING)	100% / 0%
PREDICTIVE SOLUTIONS PARTICIPATION: SAFETY MGR, SAFETY ENGR	100% (GOAL)	100% (PROJECT AVERAGE)	100% (CURRENT MONTH)	— (TRENDING)	100% / 0%
PEOPLE					
LOCAL BUSINESS ENTERPRISE ON PROJECT	> 14 % (GOAL)	14.31 (PROJECT AVERAGE)	15.0% (CURRENT MONTH'S AVG)	↓ (TRENDING)	50% / 0.0%
LOCAL RESIDENT EMPLOYEE (JOURNEYMEN & APPRENTICES)	> 30% (GOAL)	28.9% (PROJECT AVERAGE)	23.1% (CURRENT MONTH'S AVG)	↓ (TRENDING)	50% / 0%
PROJECT TEAM USE OF ALT. TRANSPORTATION	50% (GOAL)	50% (PROJECT AVERAGE)	65% (CURRENT MONTH)	↑ (TRENDING)	100% / 0%
TEAM SURVEY: PULSE REPORT SCORE	≥ 5.0 (GOAL)	5.6 (PROJECT AVERAGE)	5.7% (CURRENT MONTH)	↑ (TRENDING)	6 / 5.5 / 5
PULSE REPORT PARTICIPATION RATE	80% (GOAL)	45% (PROJECT AVERAGE)	71% (CURRENT MONTH)	↑ (TRENDING)	100% / 0%
QUALITY					
INSPECTION PASS RATE	100% (GOAL)	99.7% (PROJECT AVERAGE)	99.7% (CURRENT MONTH)	↓ (TRENDING)	100.00% / 50.00%
RE-INSPECTIONS OUTSTANDING FOR > 3 DAYS	0 (GOAL)	32 (PROJECT TOTAL)	6 (CURRENT MONTH)	↑ (TRENDING)	10 / 0
DEFECTS FOUND (DELIVERED & INSTALLED)	0 (GOAL)	7 (PROJECT TO DATE)	0 (CURRENT MONTH)	— (TRENDING)	10 / 0
PRODUCTION / SCHEDULE					
INTERMEDIATE MILESTONE PERFORMANCE (ON TIME OR FASTER)	15/15 (GOAL)	10/11 (PROJECT TO DATE)	0/0 (CURRENT MONTH)	— (TRENDING)	100% / 0%
TOTAL SCHEDULE SAVINGS	> 3% (GOAL)	0% (PROJECT TO DATE)	0% (CURRENT MONTH)	— (TRENDING)	50% / 0%
FIELD LABOR SAVINGS	> 10% (GOAL)	19.0% (LAST MONTH'S AVG)	20.0% (CURRENT MONTH'S AVG)	↑ (TRENDING)	50% / 0%
COST					
WORK IN PLACE VARIANCE	−5%≤WIP≤5% (GOAL)	−5% (PROJECT AVG VARIANCE)	−14% (CURRENT MONTH)	↓ (TRENDING)	100% / 0% / −100%
CONTINGENCY BALANCE	≥ $61,319,931 (GOAL)	$ 50,131,526 (LAST MONTH)	$48,234,833 (CURRENT MONTH)	↓ (TRENDING)	Millions: 100 / 0

Figure 8.4 VNGC Monthly Goal Progress Report – November 2016, Courtesy of Sutter Health

receive the maximum possible profit on this project – but note that the targets were set very aggressively. The inclusion of People goals is an indication that the Lean philosophy is fit for purpose to drive toward all three limbs of sustainability, social and environmental, as well as economic.

Location-based work structures, including takt structures (Frandson et al., 2013), are appropriate for and highly effective in construction. When implemented properly, they go a long way toward solving the problem of coordination between trades, and enable a focus on design of construction operations and coordination within those trades that perform multiple types of operations.

8.5 Benefits of Target Value Delivery

The Fairfield Medical Office Building project was Sutter Health's first Integrated Project Delivery project. Its success led Sutter Health to extend Lean integrated project delivery to the design and construction of its major hospital projects, culminating in the Van Ness and Geary Campus project in San Francisco, which promises to break all performance records as it nears scheduled completion in 2018.

There have been other reports on TVD, both with and without IPD. Zimina et al. (2012) provide case studies of two Sutter Health projects. UC Berkeley's Project Production Systems Laboratory, working with DPR Construction and representative companies from five of their IPD projects, carried out a five-year study dedicated to improving how TVD (then called 'target value design') is performed. Publications from that study include Ballard et al. (2015), Denerolle (2013), Do et al. (2014), Do et al. (2015a, 2015b). A more recent study of ten additional IPD projects is found in Cheng (2016). All but one of the projects included in these published reports were completed on time, within budget, and delivered buildings the owners considered fit for purpose. The only exception, reported in Ballard et al. (2015), hit all those targets except cost, which exceeded the budget by 6%. The primary reason for the breakdown was failure to follow the principles of IPD and TVD.

8.6 Conclusion

TVD has been described here as fundamental to the success of any project. The primary reason for that success is that cost is not dictated by what's wanted; hence, setting and pursuing cost targets is not solely a matter of making predictions, but rather of shaping one's own future. While shared risk and reward reduces the barrier to innovation caused by having seemingly independent financial interests, all projects can benefit from setting challenging but achievable targets for delivery of benefits to clients in the use of the constructed assets through their whole life. Because the companies comprising project teams are interdependent in designing and making (aka in production), their individual financial outcomes are likewise interdependent.

Creating product and process designs that are fit for purpose at less cost is what produces successful TVD projects. Benefits are increased by providing more valued functionalities and by increasing capacities to perform those functions. Costs are reduced through innovations in product design that make the building a better fit for performing its functions, and by innovations in process design that enable the product design to be transformed into a building with less waste.

Notes

1 Value engineering was developed and used by GE in WWII production, but first published in 1961 (Miles, 1966).
2 See quote from Shillinglaw (1967) referring to a defence contractor using a target costing methodology in the early 1950s.
3 Since 1996, the author and his colleague Gregory Howell have asked in every one of their many meetings with industry practitioners about the uncertainty and complexity of their projects. In every case, industry practitioners have said that their projects are on average toward the more dynamic (complex and uncertain) end of the continuum, and that they are getting more dynamic over time.
4 Only the owner, lead designer and main contractor sign the agreement in the Project Alliancing model. In IPD, the key engineering consultants and specialty contractors are also included in the multi-party agreement, and share the risk and reward.
5 Some public agencies are prohibited by law from signing multi-party agreements. Collaborative Design-Build provides an alternative. Owners or service providers (designers and constructors) sceptical of the other's ability to play their role in a multi-party agreement can use Collaborative Design-Build for testing and development. Service providers that want to take the initiative and not wait on owners can also use Collaborative Design-Build.

References

Arroyo, P. (2014). *Exploring decision-making methods for sustainable design in commercial buildings.* Doctoral dissertation, University of California, Berkeley, CA, USA.

Ballard, G. (2006). *Current Process Benchmark for Target Costing in Construction.* Berkeley, CA: Project Production Systems Laboratory, University of California Berkeley.

Ballard, G. (2011). Target value design: Current benchmark (1.0), *Lean Construction Journal*, 79–84.

Ballard, G., Dilsworth, B., Do, D., Low, W., Mosley, J., Phillips, P., Reed, D., Sargent, Z., Tillmann, P., Wood and Wood, N. (2015). *Proceeding of the 23rd annual conference of the International Group for Lean Construction.* Available at: www.iglc.net.

Ballard, G. and Howell, G.A., eds. (2005). Special issue on relational contracting and lean construction, *Lean Construction Journal*, 2, 1.

Ballard, G., Kim, Y.W., Azari, R. and Cho, S. (2011). *Starting from Scratch: A New Project Delivery Paradigm.* Construction Industry Institute, Research Summary 271, Austin, Texas.

Ballard, G., Kim, Y.W., Jang, J.W. and Liu, M. (2007). Roadmap for lean implementation at the project level, *The Construction Industry Institute.*

Ballard, G. and Pennanen, A. (2013). Conceptual estimating and target costing. *Proceedings of the 21st Annual Conference of the International Group for Lean Construction (IGLC)*, Research Report 234. Fortaleza, 31 July–2 August (pp. 217–226).

Ballard, G. and Reiser, P. (2004). The St. Olaf College Fieldhouse project: A case study in designing to target cost. *Proceedings of the 12th Annual Conference of the International Group for Lean Construction*, (pp. 234–249).

Burroughs, G. and Chenhall, R.H. (2012). Target costing: First and second comings, *Accounting History Review*, 22:2, 127–142. doi: 10.1080/21552851.2012.681124

Cheng, R., ed. (2016). Motivation and means: How and why IPD and lean lead to success. Available at: http://arch.design.umn.edu/directory/chengr/

Darrington, J. (2011). Using a design-build contract for lean integrated project delivery, *Lean Construction Journal*, 7, 85–91.

Denerolle, S. (2013). *The Application of Target Value Design to the Design Phase of 3 Hospital Projects.* Berkeley, CA: Project Production Systems Laboratory, University of California Berkeley. Available at: www.p2sl.berkeley.edu

Do, D., Ballard, G. and Tillmann, P. (2015a). The application of target value design in the design and construction of the UHS Temecula Valley Hospital. *Project Production Systems Laboratory*, University of California Berkeley. Available at: www.p2sl.berkeley.edu.

Do, D., Ballard, G. and Tommelein, I. (2015b). An analysis of potential misalignments of commercial incentives in integrated project delivery and target value design. *Proceedings of the 23rd annual conference of the International Group for Lean Construction*. Available at: www.iglc.net.

Do, D., Chen, C., Ballard, G., and Tommelein, I. (2014). Target value design as a method for controlling project cost overruns. *Proceeding of the 24th Annual Conference of the International Group for Lean Construction*. Available at: www.iglc.net.

Evans, R., Haryott, R., Haste, N., and Jones, A. (1998). *The Long Term Costs of Owning and Using Buildings*. London: *Royal Academy of Engineering*.

Ferry, D.J., Brandon, P.S. and Ferry, J.D. (1999). *Cost Planning of Buildings*. Hoboken, NJ: Wiley-Blackwell.

Frandson, A., Berghede, K. and Tommelein, I.D. (2013). Takt time planning for construction of exterior cladding. *Proceeding of the 21st Annual Conference of the Int'l Group for Lean Construction*.

Knott, T. (1996). *No Business as Usual: An Extraordinary North Sea Result*. London: British Petroleum Co.

Koskela, L.J. (2011). Fifty years of irrelevance: The Wild Goose chase of management science. *Proceeding of the IGLC-19 (International Group for Lean Construction)*, pp.85–96.

Lichtig, W.A. (2006). The Integrated agreement for Lean Project Delivery, *Construction Law*, 26, 25.

MacNeil, I.R. (1973). The many futures of contracts, *Southern California Law Review*, 47, 691.

Matthews, O. And Howell, G.A. (2005). Integrated project delivery: An example of relational contracting, *Lean Construction Journal*, 2:1, 46–61.

Matthews, O., Howell, G.A. and Mitropoulos, P. (2003), Aligning the Lean organization: A contractual approach. *11th Annual Conference on Lean Construction, Virginia Polytechnic Institute and State University*, Blacksburg, VA, USA. July.

Michaels, J.V. and Wood, W.P. (1989). *Design to Cost*. Vol. 3. Hoboken, NJ: John Wiley & Sons.

Miles, L. (1961). *Techniques of Value Analysis and Engineering*. New York, USA: McGraw-Hill.

Miles, L.D. (1966). Biography. Lawrence D. Milesvalue engineering reference center collection, Kurt F. Wendt Library, University of Wisconsin, 19 July 1966.

Parrish, K., Wong, J.M., Tommelein, I.D. and Stojadinovic, B. (2008). Set-based design: Case study on innovative hospital design. *16th Annual Conference of the International Group for Lean Construction (IGLC 16)*, Manchester, UK.

Pennanen, A. and Ballard G. (2008). Determining expected cost in the target costing process. *Proceedings of the 16th Annual conference of the International Group for Lean Construction*. Available at: www.iglc.net.

Pennanen, A. and Ballard, G. (2011). Target costing and designing to targets in construction, *Journal of Financial Management of Property and Construction*, 16:1, 52–63.

Pitsis, T.S., Clegg, S.R., Marosszeky, M. and Rura-Polley, T. (2003). Constructing the Olympic dream: A future perfect strategy of project management, *Organization Science*, 14:5, 574–590.

Shillinglaw, G. (1967). *Cost Accounting: Analysis and Control*. 2nd ed. Homewood, IL, USA: Irwin.

Shook, J. (2008). *Managing to Learn: Using the A3 Management Process to Solve Problems, Gain Agreement, Mentor and Lead*. Boston, MA: Lean Enterprise Institute.

Suhr, J. (1999). *The Choosing by Advantages Decisionmaking System*. Westport, CT: Greenwood Publishing Group.

Thomsen, C., Darrington, J., Dunne, D. and Lichtig, W. (2009). Managing integrated project delivery. *Construction Management Association of America (CMAA)*, McLean, VA, USA, 105.

Toussaint, S. (2006). *Thedacare Shawano Ambulatory Care Center: Lean Continuous Improvement Implementation*. Appleton, WI: Boldt Construction Company.

Walker, D. and Hampson, K., eds. (2008). *Procurement Strategies: A Relationship-Based Approach*. Hoboken, NJ: John Wiley & Sons.

Ward, A., Liker, J.K. and Cristiano, J.J., Sobek Ll, D.K. (1995). The second Toyota paradox: How delaying decisions can make better cars faster, *Sloan Management Review, Spring*, 36:3, 43–61.

Zimina, D., Ballard, G. and Pasquire, C. (2012). Target value design: Using collaboration and a lean approach to reduce construction cost, *Construction Management and Economics*, 30:5, 383–398.

9 Integrated Project Delivery

Basic tenets and recommendations

Thais da C. L. Alves and William Lichtig

9.1 Introduction

A recent report of the McKinsey Global Institute (2017) indicated that collaboration and contracting have the potential to boost construction productivity by an estimated 8 to 9% and result in 6 to 7% in cost savings in the industry. Considering that collaboration and contracting directly impact other areas related to design and engineering, procurement and Supply Chain Management, and on-site execution, the potential impact of improved contracting practices is much broader than these productivity improvements. For instance, the same report indicates that improvements in procurement and the construction supply chain could potentially result in a 7–8% global productivity improvement and 3–5% cost savings, and that long-term relationships fostered by Integrated Project Delivery (IPD) should become the norm in the industry.

IPD has been defined as

> a project delivery approach that integrates people, systems, business structures and practices into a process that collaboratively harnesses the talents and insights of all participants to optimize project results, increase value to the owner, reduce waste, and maximize efficiency through all phases of design, fabrication, and construction.
>
> (AIA, 2007)

The commercial elements of IPD agreements can be traced to previous experiences with partnering contracts (e.g. Project Partnering Contract (PPC) and New Engineering Contract (NEC)) and alliancing (e.g. Australia/New South Wales (NSW)) implemented in different parts of the globe (Gerrard, 2005; Young, Hosseini and Laedre, 2016). The goals of these underlying models were to reduce construction disputes and align project participants towards a common goal. The origin of project alliancing is credited to British Petroleum in a project in the North Sea in the early 1990s, which introduced the mechanism of 'pain-gain sharing' as an incentive for the team to search for solutions and established a joint organisation and a shared decision-making process to develop and implement solutions impacting the team (Lahdenperä, 2012). In the United States, these concepts were initially embraced by a group of contractors in Florida that formed a company named 'Integrated Project Delivery' to create a more fluid environment where stakeholders could cross specialisation lines to get the project done, share risks and profits, and innovate with the goal of improving the project (Matthews and Howell, 2005).

Building on these early experiences, Lichtig's (2005) work in developing the 'Integrated form of Agreement for Lean Project Delivery' (IFOA) recognised the potential of connecting the earlier focus of commercial terms and collaborative decision making with Lean design and construction methods. The IFOA was first developed to support Sutter Health's Lean Project Delivery initiative. After orienting the Sutter Health supply chain to the fundamentals of Lean Project Delivery and launching several pilot projects, Sutter Health routinely convened meetings referred to as its 'Vendor Forum', where members of the supply chain could share what they were learning, as well as what they saw as the challenges or obstacles to be overcome. During one of these sessions, a number of architects, engineers, and contractors raised a concern that the behaviours and methods that they were being asked to undertake were somewhat inconsistent with the contract terms of Sutter's standard contracts. After some discussion, it was agreed that Sutter would review its forms of agreement and look to align them with the basic aspects of Lean Project Delivery.

This effort, as later described by Ballard (2008), called the Lean Construction community's attention to the importance that contracts have in improving the delivery of capital projects. Up to that point, studies about delivery methods had quantified the benefits of more collaborative methods such as design-build (DB) when compared to the traditional design-bid-build approach (e.g. Konchar and Sanvido, 1998) and Lean researchers had focused their attention on improving the design, procurement, and construction processes; however, the link between the two was not very clear. By clearly drawing the link between contracts and the use of Lean as a philosophy grounded in solid concepts, and translated into functional principles and tools, Lichtig highlighted an entire new line of work for the Lean Construction community. Moreover, the fact that a prominent owner decided to use a relational type of contract with strong links to the Lean philosophy provided additional support and validation to the practicality and importance of Lean Construction as a viable means to transform the industry.

The basic form of the IFOA was first produced in 2004 and used by Sutter, HGA Architects, and The Boldt Company for the design and construction of a medical office building in Fairfield, California (AIA, 2012). Since that time, the document has evolved to meet the growing body of knowledge and reflect the experiences of the supply chain. It was also adopted as the template for industry standard ConsensusDocs® 300, which was most recently updated in 2015 (ConsesusDocs, 2015, 2018). This most recent iteration of the industry-standard IFOA embraces the Lean methods described throughout this book and established a commercial framework that supports and encourages the collaborative behaviours described throughout this chapter.

Considering this background, and recognising that Lean now has a more prominent and growing role in construction contracts, a review of the current literature on IPD is presented and a discussion of important tenets of IPD that can be replicated in projects with less collaborative contracting methods is discussed.

9.2 Literature review: IPD, Lean, and collaboration

The term IPD is sometimes loosely used in the literature to indicate a process where designers and builders work collaboratively, early in the life of a project, to better capture and address clients' needs. In this chapter, and given the current state of adoption, IPD is considered as a continuum – from a basic form of collaboration in a contract that suggests collaboration without necessarily putting in place specific mechanisms to make it happen (e.g. Franz and Leicht, 2016), or some of them (e.g. Franz and Leicht, 2012), all the way to fully executed IPD agreements embracing most or all the tenets outlined in the next sections of this chapter (e.g. Akron Children's Hospital, 2014; University of

Minnesota et al., 2016). While benefits can be realised anywhere along this continuum, the ultimate potential of IPD comes from its application as a system – where each of its elements or domains operate to support and reinforce the others. To better understand how such a mutually reinforcing system can be embodied in the contract, it is relevant to review the basic contractual domains addressed in IPD agreements. Additionally, the discussion is illustrated with Lean concepts, principles, and tools that help operationalise the IPD environment, and how collaborative governance can be used to set up and promote the behaviours expected in IPD contracts.

9.2.1 Lean concepts, principles, and tools applied to IPD

In fact, the discussion about collaborative construction contracts, whether IPD or not, exposed a broader audience to Lean tools that were being tried and tested under the close watch of practitioners, consultants, and researchers. Lean-related concepts, principles, and tools/processes/systems such as A3s and Choosing By Advantages (CBA), the Last Planner System (LPS®), and Target Value Design (TVD) began to be named (Tommelein and Ballard, 2016a) and requested in contracts.

This change was no accident, as IFOA contracts clearly outline organisational, commercial, and operational terms (Darrington, Dunne and Lichtig, 2009). The first two domains had always been addressed in construction contracts which outlined the organisation of the project and its lines of communication, as well as risk allocation and the financial terms of the agreement. The newly introduced Lean 'operating system' called attention to the definition of the production system that would deliver the project from design through construction. The contract outlined in broad strokes how that should happen using Lean as a foundation. Moreover, the operating system compelled the hyper-collaborative environment of IPD agreements where the client, designers, and constructors are viewed as equal participants in the governance of the project, with communication lines being viewed as a network, and risks being managed collectively. The notion of an intense collaborative environment is so strong in IPD projects that this topic is discussed in more detail in the next section. Using the work of Ansell and Gash (2007) about collaborative governance, we explore what can be learnt from their work to support collaboration in construction projects. Their work helps explain what happens in IPD projects and also provides some basic guidelines regarding ideal conditions for IPD agreements to start and flourish.

Figure 9.1 provides a visual representation of how the many concepts related to the design and operationalisation of IPD agreements relate to each other from a conceptual stage to an operational one. The terms addressed in Figure 9.1 are not exhaustive, but highlight two important ideas: the relationship between concepts, principles, and tools/systems/processes; and the identification of how IPD elements relate to one another in different levels from more abstract to more concrete ones. The elements highlighted in the figure are explained in detail in Koskela's (1992) seminal report on Lean Construction and its related concepts, principles, and tools, as well as more recent references such as Lichtig (2005) and Darrington, Dunne and Lichtig (2009) for IPD-specific terms, and Tommelein and Ballard (2016a) for a broad list of definitions comprising a Lean Construction glossary.

Concepts and principles in Figure 9.1 are identified as the soul or culture of the IPD environment. They are more abstract in nature, with the concepts representing an abstract, broad, tacit understanding of some of its elements at the base of the pyramid.

Concepts include the idea of value and waste elimination, as well as transparency, continuous improvement, and collaborative governance. Principles take a more specific shape as they usually have a verb attached to a concept, and indicate a direction in terms of how the concepts can be put into practice. These can be exemplified by 'increase transparency', 'increase output value by considering client's requirements', 'increase relatedness', and 'design to targets'. In fact, the principles just mentioned provide the foundation upon which IPD agreements are built. An increase in transparency and a strong drive to understand, and even challenge, clients' needs are essential in IPD projects. Participants need to work closely together using an open-book approach to cost, open/collocated working environments, and exercises to elicit clients' requirements (e.g. by building virtual and physical prototypes, developing pull plans) to be able to efficiently and proactively design to targets.

Finally, the tools are observable, concrete, operational, or specific materialisations of concepts and principles at the lower levels of the pyramid. The tools can be used to operationalise the goals underlying Lean concepts and principles and help practitioners solve specific problems related to production or design using LPS®, learning about how products and services flow through the system and identifying waste through value stream mapping, designing to meet a client's needs and wants with TVD, using A3s for problem-solving and documenting innovation, or using set-based design to explore multiple alternatives before committing to a final solution.

The distinction between the three levels outlined in Figure 9.1 underscores the importance of understanding what underlying concepts and principles are at play when tools, systems, and processes are used. Failure to understand these relationships might render the implementation of IPD and Lean-related tools ineffective because only the material/visible part – the body of the concept – is visualised and paid attention to. The need to understand the foundations of Lean concepts and their implementation has been documented in the literature as a crucial issue for the success of Lean implementation (Alves, Milberg and Walsh, 2012).

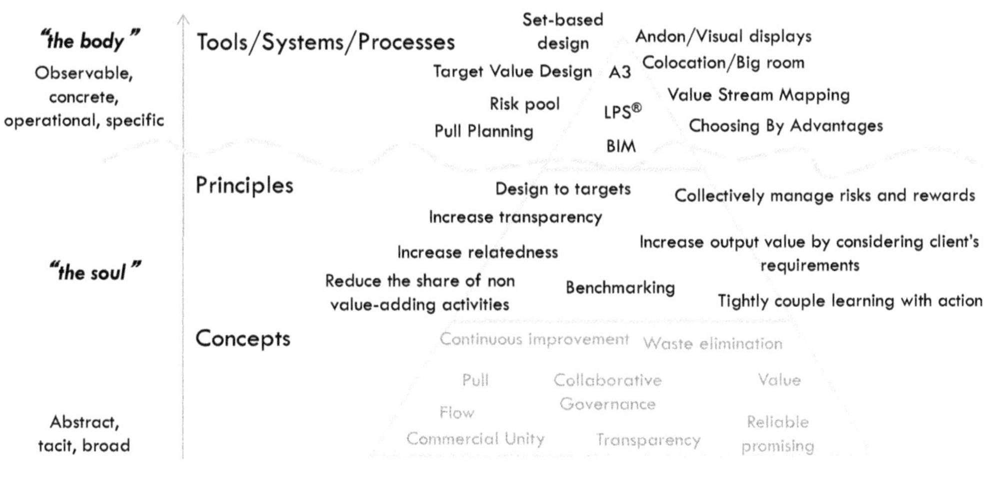

Figure 9.1 Relationship between Lean Construction concepts, principles, and tools/systems/processes supporting IPD (Adapted from Alves and Shah, 2018)

The tools included as part of the 'body' of the Lean Construction in Figure 9.1 support a system that builds and flourishes on transparency and the ability to properly capture and address the needs of internal and external clients of a project, throughout its life cycle. Most of these tools either render invisible attributes of a project or system visible, or contribute to clarifying and better communicating project decisions. The 'increase in transparency' provided by these tools directly supports other principles (or goals) outlined by Koskela (1992), for instance:

- 'Reduce the share of non-value adding activities', 'increase output value through systematic consideration of customer requirements', and 'simplify by minimising the number of steps, parts, and linkages' are observed by value stream mapping the system and making it visible regarding who needs to work on different hand-offs and when. Similarly, pull planning sessions (part of the LPS®) promote conversations and alignment among project participants and help teams structure project plans, phases, and deliverables according to major milestones to be achieved during design and construction;
- 'Reduce cycle times' is related to the use of CBA to evaluate and decide on solutions by carefully considering advantages associated with an alternative and vetting them before proceeding with the design. Along these lines, A3 reports are used to communicate alternatives and solutions in a clear and concise way, helping to ensure that participants have a good grasp of basic assumptions when alternatives are proposed. Moreover, the use of CBA and A3s supports the principle of 'tightly couple learning with action' by using these tools to document innovations and how solutions evolve over time, also supporting 'benchmark' efforts for future projects.

9.2.2 IPD contractual domains

IPD contracts address three basic project delivery domains, supported by mechanisms designed to reinforce the collaborative nature of these agreements. A detailed discussion about these domains, and the mechanisms associated with each, can be found in Darrington, Dunne and Lichtig (2009), whose discussion is presented here in a summarised format.

Commercial. IPD commercial terms are designed to promote the project stakeholders operating as a collective enterprise, aligning risk and reward structure to reinforce a focus on project-wide goals. Central to this structure is the risk-reward system that distributes gain and pain across the signatories of the IPD agreement as a proportion of their expected profit, which is put at risk, as well as through the shared management of project risks and contingencies. By rewarding parties as a team ('all for one and one for all') and letting them manage contingency in a transparent way, the agreement encourages parties to collectively manage risks and address them instead of just shifting responsibilities across a number of contractual lines commonly found in traditional delivery methods (Lichtig, 2005; Matthews and Howell, 2005). This process starts when the owner discloses the allowable cost for the project (the maximum amount it can afford to spend on the asset), followed by the team working towards better understanding (or even challenging) the client's wants and needs (Tillmann et al., 2012), and validating these to establish a market cost ('expected cost') during the conceptual design phase. The process continues with the team working towards a target cost, set below expected cost to spark innovation. The target cost is updated throughout the life of the project as the design starts being built and, ultimately, the resulting profit for the team is the difference between the allowable cost and the actual cost achieved (Alves, Lichtig and Rybkowski, 2017).

Organisational. The organisation is flat, networked, and cross-functional. It is governed by a 'Core Group' that includes, at a minimum, the owner, lead designer, and lead constructor. Often, major trade contractors and engineering consultants may be represented on the Core Group. The Core Group functions as a 'board of directors', making decisions by consensus. Instead of a traditional, siloed hierarchy, the remaining team members are typically organised into system-based, cross-functional clusters (e.g. structural, mechanical-electrical-plumbing, interior design, production) with the goal of addressing problems from a multidisciplinary perspective. Silos are broken and intense communication routines become part of the project's design and construction team's standard work (Lostuvali, Alves and Modrich, 2014). The organisational system leverages the commercial and operating system mechanisms to deliver the project (University of Minnesota, University of Washington, University of British Columbia and Scan Consulting, 2016).

Operating System. IPD replaces the traditional design and construction systems with a Lean operating system. The contract provides the direction on how the project should be managed during design and construction, and how it should serve clients' needs during operation, outlining Lean tools and systems that support a Lean Project Delivery System (Ballard, 2008). In this system, information generation, management, and dissemination follow visual management tenets to increase transparency and promote the investigation of alternatives that best suit the project, e.g. A3s, value stream maps, building information models (Tommelein and Ballard, 2016a; Alves, Lichtig and Rybkowski, 2017). A study focusing on the use of IPD and its impact on organisational culture found that the use of Lean tools and the knowledge gained through co-location and working with multidisciplinary teams had the most impact in the organisations involved (Suttie, 2013). By encouraging project teams to use the LPS® (Ballard, 2000), IPD supports the mindset that projects are networks of commitments and that promises made by team members collectively sustain project success throughout its life cycle. Furthermore, the LPS® promotes a number of Lean behaviours that are essential to the success of IPD projects, including: collaboration, open participation/communication, transparency, value identification, trust building and reliable promising, learning and continuous improvement (Fauchier and Alves, 2013). The team moves from a monitoring state, where tasks are observed as they progress, to a controlling state, where teams collaboratively and actively take charge of the project from the planning and scheduling phases throughout the execution of the plans. Finally, the Lean operating system dynamically addresses failures to meet targets and helps project participants learn from their mistakes, and prevent new ones and continuously improve their processes (Lichtig, 2005; Darrington, Dunne and Lichtig, 2009; Lostuvali, Alves and Modrich, 2014; Alves, Lichtig and Rybkowski, 2017).

To put IPD contracts into perspective regarding these three domains, Table 9.1 compares IPD to traditional delivery methods used in the United States and abroad, highlighting some of their main differences. Table 9.1 arranges the delivery methods according to the amount of pre-construction collaboration that is structurally integrated into the process. While each system can be implemented differently from what is presented and modified to promote collaboration, the chart seeks to provide an overall framework for considering the opportunity to collaborate and the use of the Lean tenets illustrated in Figure 9.1.

9.2.3 Collaborative governance

The literature backing the multiple facets of IPD include the discussion of transactional/relational contracts (Lahdenperä, 2012), motivation (Darrington and Howell, 2011), trust (Smith and Rybkowski, 2012; Pishdad-Bozorgi and Beliveau, 2016), and social network analysis (Hickethier, Tommelein and Lostuvali, 2013) to name a few. This chapter focuses on collaboration and how IPD fosters collaboration and improved performance using the definition of collaborative governance by Ansell and Gash (2007). They investigated collaborative governance in 137 cases related to policy sectors, and before starting their study they defined collaborative governance as (Ansell and Gash, 2007, p. 544): 'A governing arrangement where one or more public agencies directly engage non-state stakeholders in a collective decision-making process that is formal, consensus-oriented, and deliberative and that aims to make or implement public policy or manage public programs or assets.' Note that this definition addresses collaborative governance in the public sector environment, but can be translated to the project environment where IPD is implemented, substituting a few words to address the broad range of construction projects:

> A [project] governing arrangement where one or more [IPD signatories] directly engage [non-signatories] stakeholders in a collective decision-making process that is formal, consensus-oriented, and deliberative and that aims to make or implement [project-related] policy or manage [its related] programs or assets.

Ansell and Gash's definition underscores the importance of who starts the process, who is engaged along the way in the decision-making process and bears responsibility for its outcomes, and how the process unfolds to achieve consensus and ultimately implement what works best for the project. Moreover, they indicate that, along the way, project participants might behave like adversaries but will work together to shift their relationships to more collaborative ones (Ansell and Gash, 2007). This emphasises that stakeholders bring their individual interests to the discussion, but ultimately cooperate to address problems and explore solutions that address both individual and collective interests. This is also supported in the construction literature by the work of Tillmann et al. (2012), who documented the improved outcomes of 'integrated governance' in value generation. Along these lines, IPD agreements address four major conditions identified by Ansell and Gash's research (2007, p. 550) that need to be in place in order for collaborative governance to happen, namely: starting conditions, institutional design, facilitative leadership, and collaborative process.

- **Starting conditions.** At the outset, address the fact that people have incentives for and constraints on participation, which are influenced by power-resource-knowledge asymmetries and the prehistory of cooperation or conflict (Ansell and Gash, 2007). In the IPD environment, this suggests that: small players and newcomers should be invited and empowered by major stakeholders to participate and contribute to finding solutions, work on trust building to remediate any issues related to previous iterations (or lack thereof), and also be made more related to one another to promote interdependency (i.e. 'increase relatedness' big idea discussed below). It is also worth noting that IPD agreements are not executed immediately at the beginning of the project, but rather take some time to be developed as participants work with and get

Table 9.1 Comparison between delivery methods

	Design-Bid-Build (DBB)	*Construction Manager at Risk (CMAR)*	*Design-Build (DB)*	*Integrated Project Delivery (IPD)*
Commercial	• Lump Sum/Fixed Price. • O hires the A/E first, then through hard bid hires the GC. • GC hires STs and bears risks for cost & schedule and transfers to STs. • O bears risk for design. • Sequential design then building.	• Guaranteed Maximum Price (GMP) typically defined after design. • O hires the A/E first, then hires the GC through best value to start during the design phase. • GC hard bids trades and bears risks for cost & schedule and transfers to STs. • Some involvement of the GC and STs during design.	• Guaranteed Maximum Price (GMP), Target Value defined as design progresses. • O hires the DB entity through best value. • DB entity hires the STs, who might or might not participate as design builders. • DB entity designs and builds the project and assumes the combined risks.	• Target Value Design (TVD) process used to define owner value, then design and achieve the below-market target cost defined by the team. • Multiple parties (0, A, E, GC, STs) sign a single contract, work open-book based on actual cost, and assume shared risks and rewards. • GC is at same level as the A/E and other major STs.
Organisational	• Top-down, decision-making hierarchy. • Parties work in isolation over time and contractually. • Siloed approach in design and construction. Each party looks at their own interests.	• Top-down, decision-making hierarchy. • CMAR entity works during preconstruction with design team to manage the process and assist on design. • Some cross-pollination between the GC, A/E and STs during design.	• Top-down, decision-making hierarchy. • DB entity leads the work of the team, which might be collocated. • Multi-disciplinary approach to resolving design & construction issues.	• Collaborative, Core Group consensus-based decision-making. • Flat organisation. • Cross-functional teams, clusters defined to address design and construction in an integrated fashion. Collocated teams.
Operational	• Silent on Lean and other collaborative tools and processes.	• Owner might or might not require use of Lean tools.	• Use of Design Charrettes. • Lean tools and processes recently incorporated in the contracts.	• Lean Project Delivery System fully integrated.

Less — Collaboration → More

Legend: O = Owner; A/E = Architect/Engineer; GC = General Contractor; ST = Specialty Trades to know each other (University of Minnesota, University of Washington, University of British Columbia and Scan Consulting, 2016).

- **Institutional design.** This factor relates to participatory inclusiveness, forum exclusiveness, clear ground rules, and process transparency (Ansell and Gash, 2007). Ansell and Gash (2007, p. 555) indicate that this variable refers to 'the basic protocols and ground rules for collaboration, which are critical for the procedural legitimacy of the collaborative process'. It establishes the ground rules for relationships that will support collaborative governance through active participation and engagement of all stakeholders in the decision-making process. In the IPD environment, a number of methods support an institutional environment that promotes collaboration: co-location, use of visual management to convey information on site, and the use of Lean tools that promote transparency (e.g. value stream mapping, A3, diagrams, building information modelling, simulations (Alves, Lichtig and Rybkowski, 2017)). Additionally, the definition of clear goals, rules and responsibilities of participants, including the incentives and penalties to be used, are essential features of the agreement that need to be discussed at the onset of the project to avoid misalignments (Do, Ballard and Tommelein, 2015).
- **Facilitative leadership.** This variable also includes empowerment, the ability to establish clear common rules and goals, facilitate the trust-building process, and the exploration of mutual gains achieved with the success of the collaborative effort (Ansell and Gash, 2007). This variable finds broad support in the Lean literature in many different forms: a master/chief engineer to guide the product development process (Morgan and Liker, 2006); the need for a leader to guide the efforts to implement Lean and align the team during preconstruction in an IPD environment (Lostuvali, Alves and Modrich, 2014); the general contractor as the leader of the project, but also as a co-participant as changes are made in an IPD environment (Akron Children's Hospital, 2014); client appointing a subcontractor to lead the IPD process (University of Minnesota, University of Washington, University of British Columbia and Scan Consulting, 2016). These leaders need to be trusted and accepted by the project stakeholders, and can be selected based on previous experience with the client or emerge as the IPD process unfolds (Seed, 2014).
- **Collaborative process.** Finally, the collaborative process is influenced by the former three variables and comprises a cycle of sub-variables that influence one another including: *trust-building*, *commitment to the process* (including mutual recognition of interdependence, shared ownership of process, and openness to exploring mutual gains), *shared understanding* (including clear mission, common problem definition, and identification of common values), *intermediate outcomes* (including small wins, strategic plans, and joint fact-finding), *face-to-face dialogue* (good faith negotiation), closing the loop with *trust building* (Ansell and Gash, 2007). Smith and Rybkowski (2012) identified a number of characteristics that are relevant for trust building in construction, and questioned whether or not IPD is capable of supporting higher levels of trust when compared to traditional delivery methods. Pishdad-Bozorgi and Beliveau (2016) investigated in more detail how IPD agreements promote trust and concluded that there is a symbiotic relationship between IPD and trust, which builds upon practices such as colocation, transparency, peer performance evaluation, and liability, amongst others. The development of shared understanding during the early days of the project, usually through face-to-face conversations related to the TVD process, support the commitment to the process to develop what is best for the client, and

not to individual companies (Alves, Lichtig and Rybkowski, 2017). Cross-functional conversations between owners, contractors, suppliers, and end users also help team members identify, understand, and collaborate to address each other's constraints. Consequently, teams develop designs and plans that reflect each team's intent and foster higher levels of commitment, as discussed in the literature about the Last Planner System (Ballard, 2000). Alongside the rules defined as part of the institutional design of IPD projects, and related tools to increase transparency and promote the elements outlined above, spelling out the 'Five Big Ideas' (Lichtig, 2005) sets the groundwork for a collaborative environment that reflects what is espoused by Ansell and Gash: *collaborate really collaborate, network of commitments, tightly couple learning with action, optimise the whole, increase relatedness.*

The commercial environment established in IPD agreements removes barriers between stakeholders by creating a pool of shared risk and reward to incentivise team members to work together and take risks alongside one another to improve project outcomes (Matthews and Howell, 2005). The use of the LPS® and TVD as part of the operating system supports the view that projects are networks of commitments, promoting interdependency/relatedness during design and construction, and shared ownership of results achieved (Ballard, 2000; Tommelein and Ballard, 2016b; Alves, Lichtig and Rybkowski, 2017). The use of LPS® promotes an environment where planning is done collaboratively and controlling is done collectively. Stakeholders are accountable for their promises in an open and transparent environment, with an emphasis on learning about their failures or their inability to meet the targets set in an effort to improve future performance. Adding support to the importance of these items, when acting in concert in a project, Ansell and Gash (2007, p. 543) revealed that they 'found that a virtuous cycle of collaboration tends to develop when collaborative forums focus on "small wins" that deepen trust, commitment, and shared understanding.' By tightly coupling learning with action and sharing what the team accomplishes throughout the project, these small wins achieved by the team have the potential to make the bond among stakeholders stronger over time.

After understanding the domains, underlying collaborative principles, and Lean concepts, principles, and tools associated with IPD, the next section discusses the results of recent studies, which have quantified the benefits of IPD agreements.

9.3 Benefits of IPD

The benefits of IPD have been quantified in different studies, most notably by El Asmar, Hanna and Loh (2013), who compared the results of 35 projects with different delivery methods and concluded that IPD projects performed significantly better in a set of 14 metrics in six categories: quality, schedule, project changes, communication among stakeholders, environmental, and financial performance. The projects analysed in their study were delivered in a better quality without additional cost or time involved. Moreover, the same authors developed the Project Quarterback Indicator (PQR) based on seven performance areas, which were elicited in a survey to evaluate the performance of IPD projects (El Asmar, Hanna and Loh, 2016).

The following seven factors and their respective weight in the PQR results represent a combined indicator to assess the performance of projects using different delivery methods: (1) customer relations [0.23], (2) safety [0.17], (3) schedule [0.16], (4) cost

[0.13], (5) quality [0.12], (6) profit [0.11], and (7) communication [0.09] (El Asmar, Hanna and Loh, 2016). Interestingly, about a third of the PQR is attributed to customer relations and communication, which are items not usually measured as performance indicators in projects, but are reported by survey respondents as important attributes to be considered when successfully delivering a project. These two attributes might serve as good predictors of project success according to the PQR; however, additional research is needed to confirm that. Given the importance of these attributes in the definition of the PQR, it is no surprise that PQR results show that more collaborative delivery methods displayed better performance when assessed using this performance metric. Along these lines, IPD performed better than DB – which, in turn, performed better than Construction Management at Risk (CMAR), while Design-Bid-Build was the worst performer of the four delivery methods.

Another study illustrating how collaboration pays off in delivering construction projects, a comparison between traditional delivery methods versus more collaborative delivery methods, revealed that cost performance (difference between budget and actual cost divided by budgeted cost) improved, and that change order costs were reduced (Kulkarni, Rybkowski and Smith, 2012). Along the same lines, Tillmann et al.'s (2012) study revealed that the integrated governance promoted by IPD agreements contributes to a co-creation process in which value is added collaboratively to the project. Their work concluded, after the investigation of an IPD project, that client and suppliers worked together to define value, challenge each other's assumptions, and implement the project strategic goals through operational decisions made by the project team. The study conducted by the University of Minnesota et al. (2016) also supports the idea that IPD projects are more effective in delivering what clients want, or even exceeding baseline expectations by reinvesting contingency and budget savings in 'added value items'. The study also found that IPD projects fostered more collaborative cultures, with aligned teams delivering more profitable projects for the parties involved. Teams also indicated that: projects were fun and exciting; parties would enter an IPD agreement again; different stakeholders could relate to each other's work and appreciate the challenges they faced collectively. Another study conducted by Dodge Analytics found that 'high Lean intensity projects were three times more likely to complete ahead of schedule and two times more likely to complete under budget' (Lean Construction Institute, 2017), which adds support to the symbiotic relationship between Lean and IPD, and how Lean tools help IPD teams achieve their goals.

However, in spite of the documented benefits achieved with the use of IPD agreements, Ballard et al. (2015) documented a case in which the shared risk and reward structure of an IPD project did not prevent a negative outcome and the team learnt from it. Among the study's recommendations, the team concluded that, in order to better support the shared governance and risk model, only companies whose work represents a small risk or can be decoupled from the rest of the project delivery should be excluded from the risk pool. This is to better align the commercial terms with the actual project risk posed by a trade contractor's work. Ensuring that trade contractors who have a significant impact on the collective outcome are included in the risk pool should increase that company's commitment to the collective outcome – like other risk-reward pool signatories. These signatories should also be empowered to terminate the contracts of other members of the risk-reward pool who are not performing as defined by the agreement. In addition, lessons learnt from unintended misalignments between IPD and TVD can also be used to improve IPD agreements in order to provide the following benefits for those involved (Do, Ballard and Tommelein, 2015): allow savings in different cluster teams to be considered as part of the shared risk-reward pool; disburse profits in a timely fashion to reward those who delivered a successful project; and require that the signatories of the risk-reward pool disclose and explain the rationale for their overhead rates and potentially cap the number of extra hours to be charged to the pool.

The importance of proper management of the financial risk involved in IPD agreements cannot be overemphasised, and the same can be said for other risks related to professional liability. As team members waive their rights to file claims against each other, except for negligence (Lichtig, 2005), and also negotiate with insurers new forms of insurance to cover the project (Akron's Children's Hospital, 2014), frivolous lawsuits and costly claims are removed from the process. This represents a significant improvement when IPD is compared to traditional delivery methods. The 'all for one and one for all' motto, which is also operationalised in the risk-reward pool, indicates that organisations should be looking after each other and helping to keep the project's costs and goals on track.

In order to address the question of whether IPD is the proper delivery method for specific organisations, a recent guide (CIDCI, IPDA and Charles Pankow Foundation, 2018) discusses in detail specific phases of IPD agreements and basic structures that need to be in place for these to succeed. The guide emphasises that owners must be engaged throughout the project and have their own organisations aligned in terms of what they need and what is best for the project. Additionally, it highlights the need to properly assemble the teams and have them aligned via workshops and define the organisational structures and contractual mechanisms to support their work.

9.4 Implementing IPD: basic concepts, principles, and tools

At its core, IPD agreements and Lean methods promote simplicity, focusing on enhancing communication, collaboration, and transparency. In essence, design is a series of decisions, documented in the Building Information Model (BIM) or on paper, and later built in the physical world. To the extent that projects can improve the speed, quality, and durability of decisions, outcomes are improved. The Lean infrastructure created to support IPD projects enables improved decision making.

Project participants in collocated offices can communicate more frequently, plan and design together, and quickly resolve issues to avoid long paper trails. The combination of Lean and IPD is not about technology; simple tools such as planning boards and A3s, and open communication, are used to communicate while being simple and reducing the time to identify, solve, and learn from issues. Honest conversations are held frequently and information is freely shared to build commitment among owners, users, builders, designers, and suppliers. The result is an open flow of information and increased speed in resolving issues.

Owners have adopted different approaches in their move from more traditional project delivery models to IPD. Some owners have chosen to quickly interrupt the way they do business and opt for embarking on full-fledged IPD contracts by setting their teams up for a fast-track learning and intense collaborative environment. Alternatively, other owners have elected to move to an IPD delivery model by incrementally improving their contractual processes and team structure, developing 'IPD-lite contracts'.

Different levels of IPD are suggested in the literature, giving practitioners flexibility to decide where they want to start. In part, the point of entry needs to be calibrated to the owner's culture and willingness to embrace change. The following levels defined by NASFA et al. (2010, iii) address this issue:

- Level 1: Typical (collaboration not contractually required), IPD as a philosophy. More traditional delivery methods like CMAR might use some sort of partnering guidelines to define a charter to promote collaboration. The discussions presented in Alves and Britt (2011) and Britt et al. (2014) also exemplify a project using CMAR, which adopted collaborative ideas related to IPD projects and defined a core group to address issues in the project and align team members, engaged trades in the decision-making process, and used communication boards and other Lean tools to promote transparency;
- Level 2: Enhanced (some contractual collaboration requirements), IPD as a philosophy. Franz and Leicht (2012) describe the use of a 'collaboration addendum' to implement IPD concepts on the design and construction of campus facilities. The California State University system has recently developed a collaborative DB document to promote some IPD tenets from the procurement phase through design and construction (The California State University, 2017). The Saxum project in Israel is another example, which used some IPD tenets including a modified agreement for risk-reward sharing (gains shared amongst general contractor, architect, and project manager), no big room, architect in a separate contract, and partial use of BIM (Korb et al., 2016). Finally, the AIA contract A295 represents a transitional IPD agreement for those who are interested in starting but who might not be ready to commit to a full IPD contract (O'Connor Jr., 2009);
- Level 3: Required (collaboration required by a multi-party contract), IPD as a delivery method. This level finds examples in the IFOA described by Lichtig (2005) for Sutter Health, which includes a number of projects described in the literature (Lostuvali, Alves and Modrich, 2014; University of Minnesota, University of Washington, University of British Columbia and Scan Consulting, 2016), the ConsesusDoc 300 (2015), and also for the Akron's Children's Hospital (2014).

Once organisations have defined where they will start their journey, a number of enablers, in addition to those already discussed, can be put in place to facilitate the move from traditional project delivery systems along the IPD continuum and toward the IPD ideal state. What follows is a list of approaches that can be used to move towards IPD environments. The list is organised according to when the item might be first used in the process, and it is meant to draw attention to some of the less well-known items associated with IPD projects. The LCI publication, 'Transforming Design and Construction: A Framework for Change', provides a guide to many of the key learnings that will accelerate a team's ability to embrace and implement IPD. Although a comprehensive implementation guide does not yet exist, LCI is currently developing such a guide, with a view to publication during the winter of 2018.

9.4.1 Selecting the team

There are different ways that team selection takes place in collaborative projects using IPD tenets. A few of them are indicated below:

- Potential team members might be invited via a request for proposal (RFP) and after submitting the required documents for an initial evaluation by the owner. A few are short-listed, given a small stipend to develop solutions for the project, and later invited to present their initial thoughts to the owner. During this stage, the owner might also decide to engage with the team in an exercise that might resemble a design charrette to understand how the team works and how it might work with the owner as a partner throughout the project. The client might use another round of presentations and interactions with their team to visualise the solutions that resulted from the previous phases, to see how the team handled the owner's inputs, and finally make a selection based on what was presented throughout the RFP process;
- The owner trusts a selected contractor and relies on this contractor to bring the other members of the team;
- The owner works to develop partners in the community by selecting national design and construction partners and pairing them with local companies to develop the local industry (University of Minnesota, University of Washington, University of British Columbia and Scan Consulting, 2016; Fischer et al., 2017);
- The owner uses CBA to select partners (University of Minnesota, University of Washington, University of British Columbia and Scan Consulting, 2016);
- The owner defines a 'Lean Project Delivery Guide' outlining the entire process from selection, through to design and construction in order to attract and educate partners about their processes (Tillmann et al., 2012; Universal Health Services, 2017).

9.4.2 Workshops and conditions of satisfaction (CoS)

Use workshops to elicit client's wants and needs by asking lots of questions, clarifying what 'success' looks like in terms that participants can understand, and adding the goals of the project delivery team. The CoS keep the team focused on achieving what the client needs and values (Seed, 2014). These workshops also help the team build alignment toward the project's goal (University of Minnesota, University of Washington, University of British Columbia and Scan Consulting, 2016). Examples of CoS and how they are defined can be found in more detail at the LCI's website, which shares the Universal Health Services (2017) Lean Project Delivery Guide, and in Fischer et al. (2017), as they discuss how projects measure and control how well they are delivering value to the client through the definition of targets based on clients' needs translated into design solutions.

9.4.3 Signatories and timing to execute the IPD agreement

Depending on the adopted contracting strategy, not all parties who are part of the risk and reward pool need to sign the multi-party agreement. The findings outlined in the University of Minnesota et al.'s (2016) study reveal that the multi-party agreement was usually signed by more than three parties; however, the risk and reward pool had a much broader membership. In some cases, all the signatories of the IPD contract were part of the risk and reward pool, while in others the pool had as many as three times the number of additional participants (e.g. in the Akron's Children's Hospital, the IPD contract was signed by five companies, whereas the risk and reward pool was signed by 24 parties). In the same study, the authors found that teams with more clearly defined points of contact (POCs) with the owner might enjoy more consistent communication and trust amongst each other.

Moreover, IPD contracts are often developed and signed after parties have begun working together. Coupling the contract discussions, opening the team to candid conversations about risk and business objectives, helps to promote the development of the hyper collaborative IPD environment. Conversely, if project work is proceeding simultaneously, it often helps illuminate the benefits achieved by collaboration. On some projects described in the literature, teams have waited six months or more to execute the agreement, when design is well under way, and in some cases after construction has started (University of Minnesota, University of Washington, University of British Columbia and Scan Consulting, 2016).

9.4.4 Incentives

A relevant discussion on the relationship between contracts and production is presented in Koskela, Howell and Lichtig's (2006) paper, which discusses the need to revisit the way risks are defined and managed in contracts. Koskela, Howell and Lichtig emphasise that contracts should not be used simply to avoid risks and/or shift them, but fundamentally change the way parties collaborate to: enable good things to happen, prevent bad things from happening, and adjust the system when necessary to address issues as they happen. Along these lines, Darrington and Howell (2011) highlight the importance of defining the compensation structures and motivation schemes that promote desired behaviours in IPD projects. They suggest, for instance, that: agents should define how they will develop their tasks; goals should be set with participation of the agents involved in doing the work; there should be regular communication to share feedback about the performance of those developing tasks; and intrinsic motivation should be recognised. These studies are very much aligned to what is proposed by Ansell and Gash (2007) to support the collaborative governance, and also what is required for IPD projects to work.

Studies have revealed that parties with small stakes in IPD projects tend to be less committed to the project (Ballard et al., 2015; Do, Ballard and Tommelein, 2015), or might find motivation in other areas (University of Minnesota, University of Washington, University of British Columbia and Scan Consulting, 2016). That said, membership in the risk-reward pool, previously addressed, as well as specific metrics to evaluate the project, whether related to cost, schedule, sustainability, life-cycle savings, or other expressions of 'value,' need to be specified to create a framework for ongoing evaluation and continuous improvement. Participants in IPD projects are not just driven by financial incentives, even though they might play a major role in how they behave (University of Minnesota, University of Washington, University of British Columbia and Scan Consulting, 2016). Addressing fairness and reciprocity in the way that an incentive programme is developed is very telling about how the parties consider individual autonomy and intrinsic motivation to develop their tasks. It might help build trust, as the owner signals that they trust the team to do an outstanding job to deliver their project (Darrington and Howell, 2011).

The fact that these projects are usually challenging and often provide an extraordinary learning experience, which sharply contrasts with the isolation, risk shifting, and claim-laden mentality of traditional delivery methods (Ashcraft, 2014), might be responsible for a shift in mindset and commitment. In fact, Pink (2009) highlights that individuals performing complex tasks are mostly intrinsically motivated by concepts of purpose, autonomy, and mastery. This adds support to the fact that IPD projects might benefit from the work of individuals and companies who see beyond the monetary compensation to perform well and find motivation in the work done in multidisciplinary teams and learning about Lean tools (Suttie, 2013).

9.4.5 Operational terms and the use of Lean from design to construction

The use of Lean methods to support construction planning and execution in the industry at large appears to be more advanced and more frequently used by practitioners when compared to their use in design. This is not much different in IPD projects. Recently, the study conducted by the University of Minnesota, University of Washington, University of British Columbia and Scan Consulting (2016) revealed that Lean was mostly used to improve construction, but not used as often to improve design. Their study also showed that some industry practitioners perceive IPD and Lean as synonyms, which also indicates a lack of understanding regarding the distinction between the levels shown in Figure 9.1 and what they can accomplish when viewed together.

In fact, the use of Lean methods throughout project delivery helps participants to focus on operational goals to be achieved by the team and to establish an operational system that supports a collaborative environment. The extensive use of Lean practices during preconstruction of a healthcare project, and how they were used to manage design, is discussed in Lostuvali, Alves and Modrich (2014). Furthermore, the use of TVD is another staple of more recent collaborative delivery methods, and not exclusively in IPD contracts. The TVD process also benefits from a range of Lean tools that promote communication and provide more transparency to projects by diligently eliciting clients' requirements and following through to ensure that they are being considered during the design and construction phases (Akron Children's Hospital, 2014; Alves, Lichtig and Rybkowski, 2017). Along these lines, the University of Minnesota, University of Washington, University of British Columbia and Scan Consulting (2016) also found that teams with high Lean-implementation experience (Akron, Mosaic, T. Rowe, Wekiva) also had strong collaborative team cultures, good communication, and positive outcomes. Interestingly, these teams tended to rate their project's complexity lower than typical, which may have been the result of lower complexity or that the Lean processes helped to clarify and make the project appear less complex.

It is worth noting that, more recently, contracts designed to promote collaboration, whether IPD or not as discussed earlier in this chapter, have explicitly called for the use of Lean tools as part of the means and methods to manage the project from design to construction. These contracts set the stage for deep collaboration among team members by explicitly using words and methods that allude to collaborative practices, in contrast to traditional contracts which are silent about collaboration (Alves and Shah, 2018).

Moreover, building the team's capability in Lean tools that enhance the team's ability to plan and communicate has been credited as an enabler of the IPD culture (University of Minnesota, University of Washington, University of British Columbia and Scan Consulting, 2016). Not all team members arrive with the same level of understanding of what it takes to perform well in an IPD environment, thus enhancing participants' ability to quickly learn about the project, communicate effectively with other partners, and quickly engage with the team, might help build the culture that supports IPD projects and also speed up the learning curve for newcomers (Napolitano and Cerveró-Romero, 2012). In this sense, the LPS, which provides the backbone for collaborative planning systems, serves as a catalyst to promote Lean behaviours expected from project participants (Fauchier and Alves, 2013). In the University of Minnesota, University of Washington, University of British Columbia and Scan Consulting (2016) study, pull planning, one of the elements of LPS, was reportedly used and found valuable in all ten IPD projects investigated.

The choice to use IPD contracts is aimed at reducing conflicts and changing the culture of how projects are designed and built (University of Minnesota, University of Washington, University of British Columbia and Scan Consulting, 2016). In this environment, Knapp, Long and Howell (2014) underscore the role owners play in IPD projects by saying that the change to IPD projects is one of the greatest changes in the construction industry since the use of the Critical Path Method (CPM) and Computer Aided Design (CAD). The owner's representative and their role has to be clearly defined in the agreement, and the person taking this charge should be committed to IPD and how s/he can support the team to collaboratively deliver the project. However, it should not be taken for granted that every project should work under a multi-party agreement for different reasons.

Changing the culture of a project team might start with the goal of executing an IPD contract, but it might also start with a contract that strongly mentions and supports collaboration using multiple IPD tenets, e.g. having a core group to address strategic decisions and using colocation. The terms of an IPD contract set the stage for collaboration (e.g. use of TVD, big room, collaborative planning using the LPS) and incentivise participants to stay on track due to their commitment to the risk and reward pool (e.g. shift from individual to project's profit/success). Tillmann et al. (2012) reported that, due to the growing demand of owners expecting a change in how projects are delivered, some have taken the charge to define how teams should implement Lean. Tillmann et al. reported that

> Sutter Health developed a document entitled "Sutter Health's Lean Project Delivery System", which is a guideline for Sutter's project managers. Thus, it is also a role of the Core Group to make sure the adequate lean techniques and methods are used in practice to achieve an expected Lean model.

9.4.6 Developing IPD contracts and their teams through education

The Form-Storm-Norm-Perform developmental stages in small teams (Tuckman, 1965) provide a framework to understand what happens in IPD projects from the time teams are invited or selected to participate in the project to the time they execute the contract and fully engage to design and build the project. In the early stages, the team is forming and storming in order for its members to get to know each other; later on, the members start settling down to build a culture that promotes symbiotic relationships to support the project. In this environment, formal training is credited with establishing a culture that supports the IPD environment (University of Minnesota, University of Washington, University of British Columbia and Scan Consulting, 2016). Teams might use the time they take to make sense of the IPD contract to learn about Lean and how it benefits projects. In fact, education of practitioners and the industry at large about Lean tenets has been identified as a critical factor in bridging the gap between Lean Construction education, research, and practice (Alves, Milberg and Walsh, 2012), as well as the implementation of alternative project delivery methods (Lines, Sullivan and Smithwick, 2014; Forero et al., 2015; Fischer et al., 2017).

Considering that trade contractors and team membership invariably change, training has to be a constant during the life of a project. Workshops are a great resource early on when the team is eliciting clients' needs, exploring alternatives, and exploring how to work together (Alves, Lichtig and Rybkowski, 2017). The goal is to keep participants current in terms of the Lean knowledge available to the team, to bring newcomers up to speed as they join the project, and to constantly reinforce expected behaviours to avoid falling back to the old way of doing business (University of Minnesota, University of Washington, University of British Columbia and Scan Consulting, 2016).

In an example provided in Universal Health Services' (UHS) CoS (UHS, 2017) for 'project team participation and satisfaction', the document indicates different ways in which learning can take place: 'Two visitors (owners/industry colleagues/additional team firm employees) in the Corona Big Room per month'; 'two or more educational presentations in the Corona Big Room per month'; 'every team member an active participant in at least one lean organization'. Holding book club meetings to read about the Lean philosophy and how its tools are applied also represents an alternative method to educate team members (Lostuvali, Alves and Modrich, 2014).

Another important component in the education of IPD teams is the figure of a mentor, champion, coach, or 'sensei', as mentioned in the literature related to Lean implementation in various phases of a project (Morgan and Liker, 2006; Lostuvali, Alves and Modrich, 2014). In the IPD environment, the general contractor has to learn not to act as the sole leader of the team, who tells people what to do, but instead see their role as a co-participant of the process and one of the contract signatories (University of Minnesota, University of Washington, University of British Columbia and Scan Consulting, 2016). Trade contractors have to step up and be engaged in not just designing and installing parts of the project, but also planning how things will happen.

While working to learn about the client's needs, and how users will use and operate the building, team members make suggestions which are shared broadly with those involved, receive feedback, and continue to learn from the process. With that approach, teams embrace the mantra 'fail early, often, and learn from it'. The teams promote open discussions with trade contractors early in the project to help them see the whole picture of what they are building, as they are brought in during collaboration meetings and invited to discuss the project. Participants will be heard and feel empowered as they realise how their work fits with the rest of the project and can identify potential issues with the work that they will do. Additionally, team learning is promoted by the use of A3s or other documents to document innovations, failures, and lessons learnt (Alves, Lichtig and Rybkowski, 2017).

9.5 Final remarks

This chapter discussed the basic domains of IPD contracts and how they enjoy a symbiotic relationship with Lean concepts, principles, and tools. IPD has gained considerable attention from the industry in the United States and beyond in the past 12 years; however, much has to be done to change the way projects are delivered in the construction industry. The implementation of IPD and IPD-lite contracts has advanced the conversation in the industry towards better value delivered and less waste billed to clients, and the use of Lean to advance these causes. The explicit indication of the Lean Project

Delivery System as part of IPD contracts, and their use as a requirement, has been pushing the industry to learn more about Lean and to implement it in various projects.

The three basic domains of IPD – commercial, organisational, and operational – provide a baseline to understand these contracts and illustrate how they have dramatically changed the contractual landscape in construction projects. The commercial terms include the definition of a risk-reward pool aligned with the TVD process, which is not seen in other delivery methods. The organisational structure is more dynamic and flat, and the general contractor is viewed as a participant and integrator of the team rather than the hierarchical leader of the project. Traditional specialty silos, often shown in a work breakdown structure, are also reintegrated in the IPD organisation.

Moreover, the operational system discussed as part of IPD contracts brought attention to how production systems are designed and how tools beyond CPM could be specified in construction contracts. The indication of Lean tools to deliver projects in this highly collaborative environment is a differentiator of IPD agreements. Lean tools help promote transparency and alignment, which are essential to the collaborative governance found in IPD projects. These contracts radically embrace Lean tenets, which have been found essential to successful IPD projects according to recent research (University of Minnesota, University of Washington, University of British Columbia and Scan Consulting, 2016). The pursuit of more transparent and collaborative project environments is supported by the use of 'design to target' principles, open book commercial structures, collocated teams, and extensive use of prototypes. Given the results achieved in IPD projects, transitional IPD agreements have slowly started to embody some of these features as they become more 'true and tested' in different settings in the industry.

While some might argue that IPD should not be required or even recommended for every project, the simple fact that this discussion is taking place draws the industry's attention to the need to change delivery methods towards more collaborative ones and has made owners more aware of better ways to get what they need.

Additional issues remain regarding the use of IPD and will likely be addressed by the Lean Construction community and the industry at large. Some of these are discussed as follows.

Defining membership and terms of the risk-reward pool – recent research reveals that there is room for improvement concerning how risk-reward pool membership is defined and who, or what type of work, should be part of it (Ballard et al., 2015; Do, Ballard and Tommelein, 2015). Accountability by its members remains an important topic in order to avoid participants taking advantage of the risk-reward pool for the benefit of their own organisations, or substantially falling behind in terms of cost and schedule and jeopardising the profits of the other members.

Further explaining, using theoretical constructs – how the concepts and principles that form the soul of IPD, as illustrated in Figure 9.1, support and benefit from collaborative governance's basics elements, i.e. starting conditions, institutional design, facilitative leadership, and collaborative processes. The guide recently published by CIDCI, IPDA and Charles Pankow Foundation (2018) addresses some of these elements as they relate to alignment and team/organisational structure, from a practical standpoint like other publications (e.g. AIA, 2007, 2012; NASFA et al., 2010; Akron Children's Hospital, 2014; University of Minnesota, University of Washington, University of British Columbia and Scan Consulting, 2016; Universal Health Universal Health Services, 2017). The idea of collaborative governance, followed closely by commercial transparency and shared risk and reward, are arguably the most important elements of IPD agreements, which are not clearly found in other delivery methods indicated in Table 9.1.

Designing insurance policies to promote IPD and transitional IPD projects – large owners have leverage to work with insurers to develop policies that support collaborative environments and which are adequate to their projects (e.g. Akron Children's Hospital, 2014). However, the industry at large still lags in this respect and might benefit from additional discussion on this topic.

Extending IPD to the supply chain – Promoting supply chain integration including different phases and goals from designing, to building, and operating facilities. The design of simulations to teach and study how structures mimic different levels of engagement, and how risk and reward are shared throughout the supply chain, might result in additional practical and theoretical insights regarding extending IPD to the supply chain. From a practical standpoint, this would address the investigation of multiple Lean principles associated with IPD, e.g. 'tightly coupling learning with action', 'collectively manage risks and rewards', 'increase relatedness', 'reduce the share of non-value adding activities', and 'increase value by considering client's requirements' to name a few.

Use IPD in the public sector – the use of multi-party agreements, implementing pain-gain sharing mechanisms, and internal resistance, are still viewed as major obstacles to using IPD in the public sector. Some IPD tenets conflict with current procurement methods and rules used by public owners (Ashcraft, 2014; Kim et al., 2016) and need to be addressed through changes in regulations (McKinsey Global Institute, 2017). Additionally, a better understanding of the IPD propositions by owners (Center for Innovation in the Design and Construction Industry [CIDCI], Integrated Project Delivery Alliance [IPDA] and Charles Pankow Foundation, 2018) represents another area of research that might have been underestimated as it relates to the public sector. Research on the intersection of public policies and how they address capital projects and their clients' needs might hold the key to designing IPD agreements for the public sector.

Educating the industry at large about the benefits of IPD and how Lean can help – this topic has been advocated in many ways by the Lean community, represented largely by the LCI (industry) and the IGLC (academia), as well as research groups such as the Project Production Systems Laboratory (P2SL) at UC Berkeley. Collaborations across the globe have been developed between research institutions, owners, and private companies, and this should continue as more innovative ways to teach and implement Lean are developed.

Global implementation of IPD – previous studies have either surveyed or tested IPD agreements in Finland (Liikennevirasto, 2013), the Middle East (Korb et al., 2016; Rached et al., 2014), Colombia (Forero et al., 2015), and China (Li and Ma, 2017) to name a few. As these studies are developed, additional traits of IPD contracts might surface as they are used in different countries and types of projects and will go through adaptations to match the characteristics of a global industry. These studies will likely relate to more abstract issues addressed at the bottom of Figure 9.1, as they are tied to idiosyncrasies of the construction industry in different countries. Successfully implementing IPD across the globe will require adaptation in order to address different cultures and how they shape human and organisational behaviours. This is a good starting point for the definition of a theory outlining and explaining global versus local principles for IPD implementation.

References

AIA. (2007) *Integrated Project Delivery: A Guide.* Washington, DC, AIA National, AIA California Council, The American Institute of Architects, p. 57.

AIA. (2012) *IPD Case Studies.* Washington, DC, AIA, AIA Minnesota, School of Architecture, University of Minnesota.

Akron Children's Hospital. (2014) *L3: Building a New Hospital Facility.* Akron, OH, USA: Akron Children's Hospital.

Alves, T.C.L. and Britt, K. (2011) 'Working to Improve the Lookahead Process', *Proc. 19th Annual Conference of the International Group for Lean Construction*, Lima, Peru. Available at: www.iglc.net.

Alves, T.C.L., Lichtig, W. and Rybkowski, Z.C. (2017) 'Implementing Target Value Design: Tools and Techniques to Manage the Process', *HERD: Health Environments Research and Design Journal*, 10(3), p. 18–29.

Alves, T.C.L., Milberg, C. and Walsh, K. (2012) 'Exploring Lean Construction Practice, Research, and Education', *Engineering, Construction and Architectural Management, 19(5)*, pp. 512–525.

Alves. T.C.L. and Shah, N. (2018) 'Analysis of Construction Contracts: Searching for Collaborative Practices', in ASCE, *2018 Construction Research Congress*, New Orleans, LA, USA, 148–157.

Ansell, C. and Gash, A. (2007) 'Collaborative Governance in Theory and Practice', *Journal of Public Administration Research and Theory*, 18(4), pp. 543–571.

Ashcraft, H.W. (2014) 'The Transformation of Project Delivery', *The Construction Lawyer*, 34, pp. 35–41 and pp. 57–58.

Ballard, G. (2008) 'The Lean Project Delivery System: An Update', *Lean Construction Journal*, pp. 1–19.

Ballard G., Dilsworth B., Do D., Low W., Mobley J., Phillips P., Reed D., Sargent Z., Tillmann P. and Wood N. (2015) 'How to Make Shared Risk & Reward Sustainable', *Proc. 23rd Annual Conference of the International Group for Lean Construction*, Perth, Australia. Available at: www.iglc.net.

Ballard, G.H. (2000) *The Last Planner System of Production Control.* PhD Thesis, The University of Birmingham, Birmingham, UK. Available at: http://leanconstruction.org.uk/media/docs/ballard2000-dissertation.pdf.

Britt, K., Alves, T.C.L., Reed, D. and Gracz, B. (2014) 'Lessons Learned from the Make Ready Process in a Hospital Project', *Proc. 22nd Annual Conference of the International Group for Lean Construction*, Oslo, Norway. Available at: www.iglc.net.

The California State University (2017) 'Contract General Conditions for Collaborative Design-Build Major Projects', *Office of the Chancellor, Capital Planning, Design, and Construction*, Long Beach, CA. Available at: www.calstate.edu/cpdc/cm/contract_docs.shtml (Accessed: 11th Aug 2017).

Center for Innovation in the Design and Construction Industry [CIDCI], Integrated Project Delivery Alliance [IPDA] and Charles Pankow Foundation (2018) Integrated Project Delivery: An Action Guide for Leaders. Available at: www.pankowfoundation.org/integrated-project-delivery-an-action-guide-for-leaders-now-available/ (Accessed: 5th Nov 2018).

ConsensusDocs (2015) *Standard Multi-Party Integrated Project Delivery (IPD) Agreement.* Available at: www.consensusdocs.org/Resource_/FileManager/300.pdf (Accessed: 30th April 2018).

ConsensusDocs (2018) Contract Series: 300 Collaborative. Available at: www.consensusdocs.org/Catalog/collaborative (Accessed: 30th April 2018).

Darrington, J., Dunne, D. and Lichtig, W. (2009) 'Organization, Operating System, and Commercial Terms', in Chuck Thomsen (ed.), *Managing Integrated Project Delivery*, CMAA, McLean, VA, USA, pp. 10–47.

Darrington, J.W. and Howell, G.A. (2011) 'Motivation and Incentives in Relational Contracts', *Journal of Financial Management of Property and Construction*, 16(1), pp. 42–51.

Do, D., Ballard, G. and Tommelein, I.D. (2015) 'An Analysis of Potential Misalignments of Commercial Incentives in Integrated Project Delivery and Target Value Design', *Proc. 23rd Annual Conference of the International Group for Lean Construction*, Perth, Australia. Available at: www.iglc.net.

El Asmar, M., Hanna, A.S. and Loh, W. (2013) 'Quantifying Performance for the Integrated Project Delivery System as compared to Established Delivery Systems', *Journal of Construction Engineering and Management*, 139, p. 11.

El Asmar, M., Hanna, A.S. and Loh, W. (2016) 'Evaluating Integrated Project Delivery Using the Project Quarterback Rating', *J. of Construction Engineering and Management*, 142, pp. 1.

Fauchier, D. and Alves, T.C.L. (2013) 'Last Planer® System is the Gateway to Lean Behaviors', *Proc. 21st Annual Conference of the International Group for Lean Construction*, Fortaleza, Brazil. Available at: www.iglc.net.

Fischer, M., Ashcraft, H.W., Reed, D. and Khanzode, A. (2017) *Integrating Project Delivery*. New Jersey, USA, Wiley.

Forero, S., Cardenas, S., Vargas, H. and Garcia, C. (2015) 'A Deeper Look Into the Perception and Disposition to Integrated Project Delivery (IPD) in Colombia', *Proc. 23rd Annual Conference of the International Group for Lean Construction*, Perth, Australia. Available at: www.iglc.net.

Franz, B. and Leicht, R.M. (2012) 'Initiating IPD Concepts on Campus Facilities with a 'Collaboration Addendum', *ASCE 2012 Construction Research Congress*, pp. 61–70.

Franz, B.W. and Leicht, R.M. (2016) 'An Alternative Classification of Project Delivery Methods Used in the United Stated Building Construction Industry', *Construction Management and Economics*, 34(3), pp. 160–173.

Gerrard, R. (2005) 'Relational Contracts: NEC in Perspective', *Lean Construction Journal*, 2(1), pp. 80–86.

Hickethier, G., Tommelein, I.D. and Lostuvali, B. (2013) 'Social Network Analysis of Information Flow in an IPD-Project', *Proc. 21st Annual Conference of the International Group for Lean Construction*, Fortaleza, Brazil. Available at: www.iglc.net.

Kim, Y., Rezqallah, K., Lee, H.W. and Angeley, J. (2016) 'Integrated Project Delivery in Public Projects: Limitations and Opportunity', *Proc. 24th Annual Conference of the International Group for Lean Construction*, Boston, MA, USA, 7. Available at: www.iglc.net.

Knapp, S., Long, D. and Howell, G. (2014) 'The Role of the Owner's Representative on IPD Projects', *Proc. 22nd Annual Conference of the International Group for Lean Construction*, Oslo, Norway. Available at: www.iglc.net.

Konchar, M. and Sanvido, V. (1998) 'Comparison of U.S. Project Delivery Systems', *Journal of Construction Engineering and Management*, ASCE, 124(6), pp. 435–444.

Korb, S., Haronian, E., Sacks, R., Judez, P. and Shaked, O. (2016) 'Overcoming 'But We're Different': an IPD Implementation in the Middle East', *Proc. 24th Annual Conference of the International Group for Lean Construction*, Boston, MA, USA, 7. Available at: www.iglc.net.

Koskela, L. (1992) *Application of the New Production Philosophy to Construction*. Center for Integrated Facility Engineering, (CIFE), Department of Civil Engineering, Stanford University. Technical Report number: 72.

Koskela, L., Howell, G.A. and Lichtig, W. (2006) Contracts and Production, *CIB W92 Symposium on Sustainability and Value through Construction Procurement*, Salford, UK, pp. 332–339.

Kulkarni, A., Rybkowski, Z.K. and Smith, J. (2012) 'Cost Comparison of Collaborative and IPD-Like Project Delivery Methods Versus Competitive Non-collaborative Project Delivery Methods', *Proc. 20th Annual Conference of the International Group for Lean Construction*, San Diego, CA, USA. Available at: www.iglc.net.

Lahdenperä, P. (2012) 'Making Sense of the Multi-party Contractual Arrangements of Project Partnering, Project Alliancing and Integrated Project Delivery', *Construction Management and Economics*, 30(1), pp. 57–79.

Lean Construction Institute (2017) Lean projects are three times more likely to complete ahead of schedule according to Dodge Data and Analytics' Research. Available at: www.leanconstruction.org/press-release/lean-projects-three-times-likely-complete-ahead-schedule-according-dodge-data-analytics-research/ (Accessed: 27th Nov 2017).

Li, S. and Ma, Q. (2017) 'Barriers and Challenges to Implement Integrated Project Delivery in China', *Proc. 25th Annual Conference of the International Group for Lean Construction*. Heraklion, Greece. Available at: www.iglc.net.

Lichtig, W.A. (2005) 'Sutter Health: Developing a Contracting Model to Support Lean Project Delivery', *Lean Construction Journal*, 2(1), pp. 105–112.

Liikennevirasto (2013) 'Welcome to the Tunnel!'. Available at: www2.liikennevirasto.fi/rantatunneli/print/rantatunneli_print_en.pdf (Accessed: 21st Nov 2017).

Lines, B., Sullivan, K. and Smithwick, J. (2014) 'An Action Research Approach to Implementation of Alternative Project Delivery Methods within Architectural, Engineering, and Construction Owner Organizations: Overcoming Resistance through Education', *2014 ASCE Construction Research Congress*, pp. 2084–2095.

Lostuvali, B., Alves, T.C.L. and Modrich, R. (2014) 'Learning from the Cathedral Hill Hospital Project during the Design and Preconstruction Phases', *International Journal of Construction Education and Research*, 10(3), pp. 160–180.

Matthews, O and Howell G.A. (2005) 'Integrated Project Delivery an Example of Relational Contracting', *Lean Construction Journal*, 2(1), pp. 46–61.

McKinsey Global Institute. (2017) *Reinventing Construction: A Route to Higher Productivity*. McKinsey & Company, p. 155.

Morgan, J.M. and Liker, J.F. (2006) *The Toyota Product Development System*. New York, USA: Productivity Press.

Napolitano, P.T.S. and Cerveró-Romero F. (2012) 'Meta-organization: The Future for the Lean Organization', *Proc. 20th Annual Conference of the International Group for Lean Construction*, San Diego, CA, USA. Available at: www.iglc.net.

NASFA, COAA, APPA, AGC and AIA (2010) Integrated Project Delivery for Public and Private Owners. Available at www.agc.org/projectdelivery (Accessed: 12th Oct 2010).

O'Connor Jr, P.J. (2009) Integrated Project Delivery: Collaboration through New Contract Forms. Available at: www.faegrebd.com/webfiles/AGC-IPD%20Paper.pdf.

Pink, D. (2009) *Drive: The Surprising Truth about What Motivates Us*. New York: Riverhead Books.

Pishdad-Bozorgi, P. and Beliveau, Y.J. (2016) 'Symbiotic Relationships between Integrated Project Delivery (IPD) and Trust', *International Journal of Construction Education and Research*, 12(3), pp. 179–192.

Rached, F., Hraoui, Y., Karam, A. and Hamzeh, F. (2014) 'Implementation of IPD in the Middle East and its Challenges', *Proc. 22nd Annual Conference of the International Group for Lean Construction*, Oslo, Norway. Available at: www.iglc.net.

Seed, W.R. (2014) 'Integrated Project Delivery Requires a New Project Manager', *Proc. 22nd Annual Conference of the International Group for Lean Construction*, Oslo, Norway. Available at: www.iglc.net.

Smith, J.P. and Rybkowski, Z. (2012) 'Literature Review on Trust and Current Construction Industry Trends', *Proc. 20th Annual Conference of the International Group for Lean Construction*, San Diego, CA, USA. Available at: www.iglc.net.

Suttie, J.B. (2013) 'The Impacts and Effects of Integrated Project Delivery on Participating Organisations With a Focus on Organisational Culture', *Proc. 21st Annual Conference of the International Group for Lean Construction*, Fortaleza, Brazil. Available at: www.iglc.net.

Tillmann, P., Ballard, G., Tzortzopolous, P. and Formoso, C. (2012) 'How Integrated Governance Contributes to Value Generation – Insights From an IPD Case Study', *Proc. 20th Annual Conference of the International Group for Lean Construction*, San Diego, CA, USA. Available at: www.iglc.net.

Tommelein, I.D. and Ballard, G. (2016a) *Lean Construction Glossary*. Berkeley, CA: Project Production Systems Laboratory, University of California, Berkeley.

Tommelein, I.D. and Ballard, G. (2016b) *Target Value Design: Introduction, Framework, and Current Benchmark (Report)*. Berkeley, CA: Project Production Systems Laboratory, University of California, Berkeley.

Tuckman, B.W. (1965) 'Developmental Sequence in Small Groups', *Psychological Bulletin*, 63(6), pp. 384–399. (reprinted in *Group Facilitation: A Research and Applications Journal*, number 3, Spring 2001).

Universal Health Services (2017) Lean Project Delivery Guide. Available at: www.leanconstruction.org/learning/uhs-lean-project-delivery-guide/ (Accessed: 1st Nov 2017).

University of Minnesota, University of Washington, University of British Columbia and Scan Consulting (2016) Motivation and Means: How and Why IPD and Lean Lead to Success. Available at: www.leanconstruction.org/wp-content/uploads/2016/02/MotivationMeans_IPDA_LCI_Report.pdf (Accessed: 13th April 2017).

Young, B., Hosseini, A. and Laedre, O. (2016) 'The characteristics of Australian infrastructure alliance projects', *Energy Procedia*, 96, pp. 833–844.

10 Choosing by advantages and collaborative decision making

Paz Arroyo

10.1 Introduction

Poor performance on construction projects can be partially explained by the use of unsound decision-making processes based on short-term cost analysis (Flyvbjerg, Bruzelius, and Rothengatter, 2003). These processes often lack collaboration and fail to account for multiple stakeholders views, an inefficient dynamic which is amplified in the context of a highly fragmented construction industry characterised by poor communications. Choosing By Advantages (CBA) is a multiple-criteria decision-making method originally developed by Suhr (1999), which counters these detrimental trends by supporting context-specific, collaborative decisions and helps decision-makers choose alternatives that provide the best value for the available budget (Correa et al., 2017; Nguyen, Lostuvali, and Tommelein, 2009; Parrish and Tommelein, 2009).

Suhr (1999) states that decision-making methods produce decisions, those decisions trigger actions, and, finally, those actions cause outcomes. The outcome, therefore, relies as much on the decision-making method as on the decision itself (see Figure 10.1).

Not all methods are the same and, as such, it is important to identify characteristics that support sound, transparent, and collaborative decision-making. CBA presents several characteristics that make it appropriate for collaboration. CBA helps decision-makers organise information in a transparent fashion, guides conversations to reveal concerns from stakeholders, encourages decision-makers to anchor their assessments to the context of the decision, prevents double-counting information, facilitates consensus, provides structure to articulate the rationale behind a decision, and is easily documented.

CBA has been tested and used successfully in construction projects (Arroyo and Long, 2018; Arroyo, Tommelein, and Ballard, 2013; Kpamma et al., 2014; Lee, Tommelein, and Ballard, 2010; Nguyen, Lostuvali, and Tommelein, 2009; Parrish and Tommelein, 2009), generating significant benefits in terms of cost and time reductions while increasing owner satisfaction. However, although its use has been increasing, CBA is not widely utilised. The potential for expansion and application to improve the construction industry

Figure 10.1 Cause–effect model of decision making (adapted from Suhr, 1999)

is vast. CBA can be implemented in a range of decision-making situations including procurement decisions, design decisions, and production decisions, among others, and can be applied to both simple and complex decisions involving any number of alternatives (Suhr, 1999). In fact, Arroyo et al. (2018) developed an algorithm to use CBA for exploring a thousand design alternatives for the core and shell of a hotel building. The algorithm reduces the number of trade-offs decisions required by designers, while maintaining decisions based on the importance of advantages.

The CBA system consists of definitions, models, principles, and a suite of methods designed to support decisions of every complexity (Suhr, 1999). Rather than defaulting to the cheapest option, CBA evaluates value and cost separately by focusing on the advantages of each alternative. By comparing advantages and identifying factors that will reveal 'significant differences between alternatives' rather than identifying factors based on 'preconceptions', this approach empowers decision-makers to identify the option which provides the highest value, and thus to proceed through the project with confidence (Correa et al., 2017).

Decisions are often made in silos without consulting all parties affected or considering the full scope of impact. Choices made in this context are assumptive and risky: they may be later found infeasible as assumptions are proved wrong, overlooked information comes to light, and misunderstandings occur between team members (Ballard, 2000b). Project participants often state that they were excluded from a decision or were unsure why the project took a certain direction. This lack of communication creates waste and frustration at different levels of the organisation. Without shared understanding, teams can neither make sound decisions nor commit to proceed with corresponding action. Through the use of CBA, teams are guided to identify the advantages of each alternative and to use that information as the foundation for determining which choice offers the most value, given a budget constraint. Additionally, contractual methods such as Integrated Project Delivery (IPD) or Integrated Form of Agreement (IFOA) that are based on Lean thinking require teams to collaborate and engage in a transparent decision-making process (see Chapter 9 IPD by Alvez and Lichtig for more information). Decision-making often requires the consideration of concerns from diverse stakeholders, and CBA is especially suited for including such a range of perspectives.

Lean philosophy offers several techniques to manage a successful decision-making process, such as *Nemawashi*, characterised by a slow, deliberate decision-making process followed by rapid implementation, or Set Based Design (SBD), which recommends exploring multiple alternatives until the last responsible moment (Ward et al., 1995). Both methods account for the importance of timing when making decisions on a construction project; understanding that decisions must be made in a timely manner while taking relevant information into account is key for project success. The qualities of a Lean decision-making process can be summarised as providing transparency, building consensus, and allowing for continuous learning (see Figure 10.2). The Lean Construction community adopted CBA as a decision-making method because it connects well with Lean principles and ideas and CBA can provide all the qualities of a Lean decision-making process. Decision-making is not a branch of mathematics (Suhr, 1999) and, despite available data and technologies, the responsibility of choice remains with the people involved.

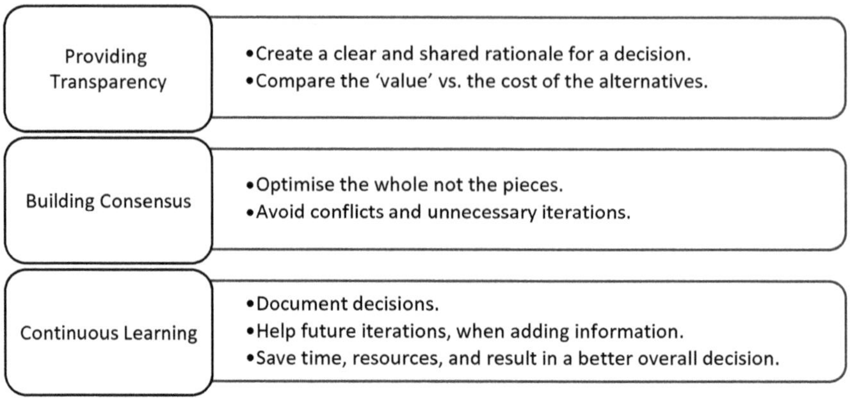

Figure 10.2 Qualities of a Lean decision-making process (Arroyo, 2014)

This chapter provides an overview of how a CBA system works in practice, its connection with Lean, research outcomes related to Lean and CBA, the benefits of using CBA, a demonstrative case study, recommendations for implementation, and future research opportunities.

10.2 How does CBA work?

CBA is a system for making decisions based on the importance of beneficial differences (i.e. advantages) between alternatives. It has definitions, models, principles and a set of methods (Suhr, 1999). This section presents definitions, basic principles of the method, an overview of the decision-making phases, and a description of the CBA Tabular Method.

In order to provide a clear and consistent language to explain and implement CBA, the definitions presented in Table 10.1 are used.

Table 10.1 CBA definitions (modified from Suhr, 1999)

Term	*Definition*
Alternatives	Two or more persons, things, or plans (e.g. construction methods, materials, building designs, or construction systems) from which one must be chosen.
Factor	An element, part, or component of a decision. When assessing sustainability, factors should represent social, environmental, and economic aspects. However, in CBA cost is treated separately from the rest of the factors.
Criterion	A decision rule or a guideline. A 'must' criterion represents conditions each alternative must satisfy. A 'want' criterion represents preferences of one or multiple decision-makers.
Attribute	A characteristic, quality, or consequence of one alternative.
Advantage	A benefit, gain, improvement, or betterment. Specifically, an advantage is a beneficial difference between attributes of two alternatives.

10.2.1 Basic CBA principles

CBA is composed of several principles (Suhr, 1999):

- The pivotal cornerstone principle: Decision makers must learn and skilfully use sound methods;
- The fundamental rule: Decisions must be based on importance of advantages;
- The principle of anchoring: Decisions must be based on the relevant facts;
- The methods principle: Different types of decisions call for different types of methods.

10.2.2 Decision-making phases

The CBA system divides the decision-making process into five phases (Suhr, 1999): (I) the stage-setting phase, (II) the innovation phase, (III) the decision-making phase, (IV) the reconsideration phase, and (V) the implementation phase (Figure 10.3).

During (I) the stage-setting phase, decision-makers determine the purpose, scope, and circumstances of the decision; discover the concerns of the stakeholders; identify the needs and preferences of the stakeholders; begin establishing 'must' and 'want' criteria; and identify every participant who will be involved in the decision. In addition, during this stage, leaders should ensure that the team possesses the skills necessary to successfully carry out the process of CBA.

During (II) the innovation phase, decision-makers ensure that CBA training has been adequately provided; formulate alternatives (we cannot choose the best alternative if we do not formulate it); and determine and display attributes related to each alternative (it may take days, weeks, or years).

During (III) the decision-making phase, decision-makers summarise the attributes, determine the advantages, and decide the importance of each advantage. This is an assessment of value, which is typically subjective as each person interprets the decision differently. Decision-makers also evaluate costs during this phase.

During (IV) the reconsideration phase, decision-makers reconsider the decision by answering questions such as 'Can the decision be improved?' 'Are there other alternatives to consider?' 'Are there other relevant factors to consider?' and by reviewing information from the previous phases (I, II, III).

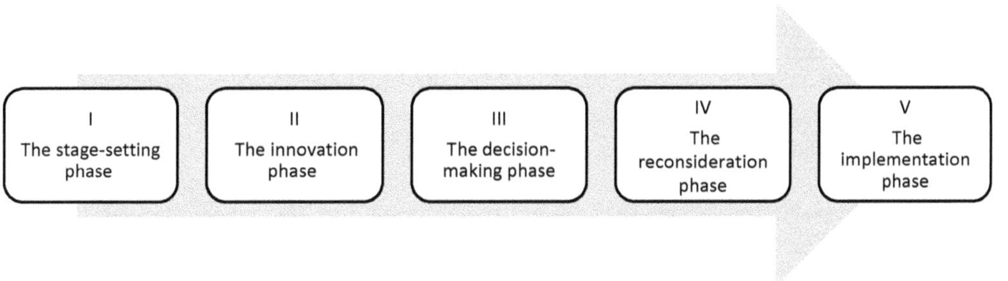

Figure 10.3 Decision-making phases (adapted from Suhr, 1999)

Finally, during (V) the implementation phase, decision-makers establish and carry out an implementation plan based on their decision. Decision-makers commit to move forward together and do their part to have a successful implementation.

10.2.3 The choosing by advantages methods

CBA includes a variety of methods to be applied depending on the specifics of the situation that requires evaluation. All these methods fit into the decision-making phase of the five-part methodology outlined above (Figure 10.3). According to Suhr (1999), the CBA methods are:

For simple decisions:

1. Recognition response CBA;
2. Instant CBA;
3. Simplified two-list method;

For complex decisions:

4. Two-list method;
5. Tabular method (two or more alternatives).

This chapter explains the Tabular Method for complex decisions because it is most often used and is the most detailed step-by-step guideline of a CBA system; the other CBA methods are simplifications that can be easily understood after the introduction of the Tabular Method. To further learn about other CBA methods, refer to Suhr (1999).

Figure 10.4 shows the seven steps required to implement the CBA Tabular Method.

The steps involved in the application of the CBA Tabular Method are detailed below (Suhr, 1999):

1) Identify alternatives to be assessed in the decision process;
2) Define factors that differentiate the alternatives. Note that cost is not included as a factor, but rather is treated as a constraint to the decision, and is incorporated in the final step of this method;
3) Define the criterion for each factor by which the alternatives are judged. The criteria define preferences, such as 'more is better';
4) Summarise the attributes, or characteristics, of the alternatives. These can be qualitative or quantitative, but they are not judgemental statements. Underline the least preferred attribute in each factor according to the criterion set in step 3;
5) Decide the advantages of each alternative. In this step, the least-preferred alternative within each factor category is used as a baseline against which every other alternative is compared and ranked within that same category. Ultimately, each alternative will be ranked in terms of its potential advantage in each factor;
6) Decide the importance of the advantages (IofA). The IofA corresponds to a value that is given to the alternatives, given the relevance of the advantages assessed by the team. The sum of the IofAs for each alternative represents the total importance of advantages for that alternative. The following process is used to weight the advantages:

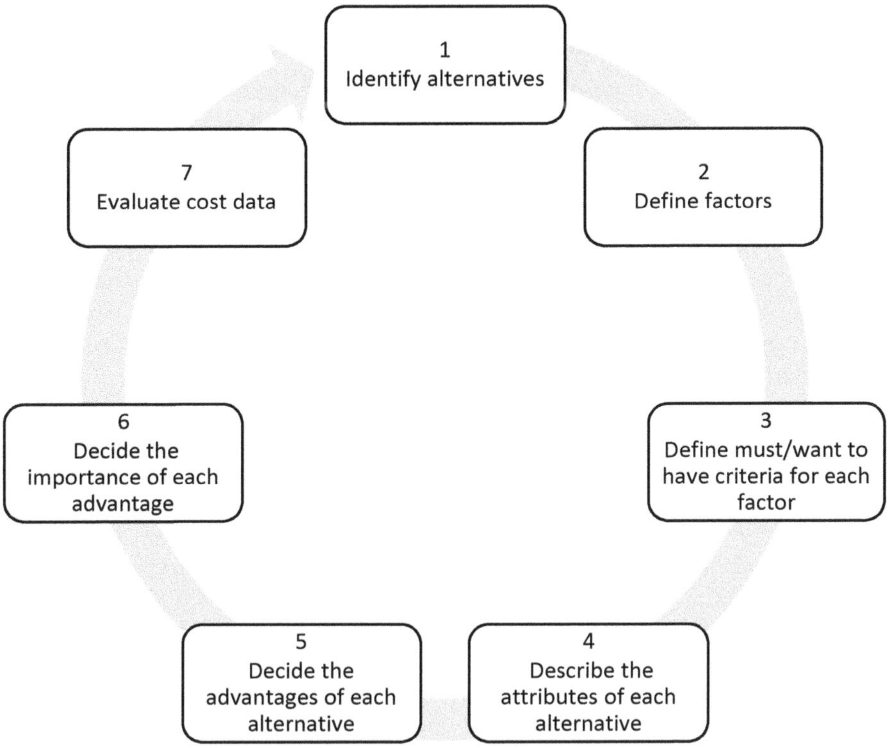

Figure 10.4 Steps to implement CBA Tabular Method (modified from Suhr, 1999)

i. Circle (or highlight) the most important advantage per each factor;
ii. Select the 'paramount advantage', that advantage with greatest importance across all factors, and assign the highest IofA weight to it;
iii. Weigh the most important advantage of each factor, comparing it to the paramount advantage;
iv. Weigh the importance of the remaining advantages.

7) Evaluate cost data vs. IofAs. In the case that the alternatives have the same cost, the alternative with the higher IofAs is clearly preferable. If alternatives have different costs, a graph is drawn relating the IofAs and costs of each alternative. The reconsideration phase should also be considered formally in the decision-making process.

10.3 How is CBA connected with Lean?

According to Liker and Meier (2006), *Nemawashi*, one of The Toyota Way principles, requires teams to 'Make decisions slowly by consensus, thoroughly considering all options; implement decisions rapidly'. Furthermore, Liker and Meier (2006) recommend that a team should not commit to a single direction without thoroughly considering every alternative – a strategy which aligns with Set-Based Design (SBD). However, once the decision is made, teams should commit to proceed quickly and continuously down the path of action. This, of course, does not mean that decisions should not be questioned or changed; if new, relevant information is available, decisions can be reconsidered by the team.

Nemawashi specifically refers to the process of discussing problems and potential solutions with all stakeholders, meaning all affected by the decision, to collect ideas and achieve consensus from which to move forward. Liker and Meier (2006) state that, although this process is time consuming, once the decision is made teams are prepared to implement the action plan without delay. This has been proven in practice on construction decisions (Arroyo and Long, 2018).

Lean design practices are highly collaborative and designed to enhance the function of the project as a whole rather than isolate individual parts. Lean approaches and tools such as Target Value Design (TVD) (Ballard, 2009; see Chapter 8 by Ballard), Set-Based Design (Ward et al., 1995), A3-Thinking (Sobek II and Smalley, 2011), and CBA help the design team increase value for the customer and minimise waste. Lean approaches and tools can be more effective with the use of relational contracts such as those used in Integrated Project Delivery (IPD – see Chapter 9 by Alvez and Lichtig) where the commercial interests of the parties are aligned, establishing an environment for effective collaboration.

Figure 10.5 shows a simplification of a theoretical TVD strategy and the relationship between synergic Lean methods to be used in design. TVD principles should be used to identify target values and target costs.

For every opportunity identified to improve the existing baseline design, the team should use a systematic decision-making process, which includes a SBD method that encourages the exploration of alternative approaches for design solutions by carrying them along in parallel paths, gradually eliminating each until one preferred approach is arrived at (Ward et al., 1995). SBD prevents premature selection of an alternative by developing alternative design sets until the last responsible moment, and thus avoiding re-work or suboptimal value to the client (Ward et al., 1995).

Ballard (2000b) expands on this idea, defining positive and negative iterations for the design team. According to Ballard (2000b), iteration is essential for generating value in design processes. However, not all iteration generates value. Iteration that can be

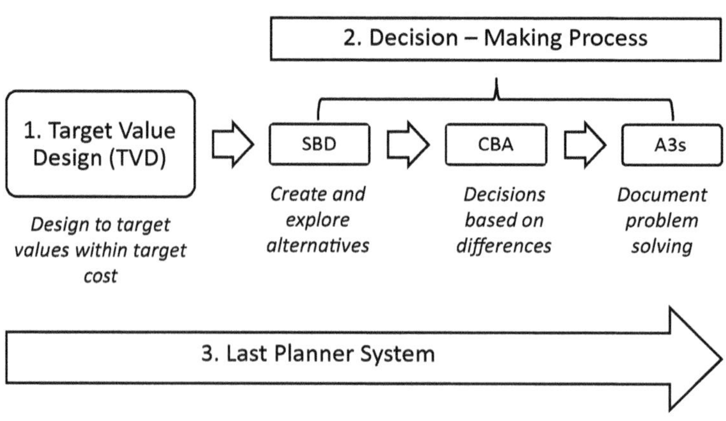

Figure 10.5 Lean decision-making process (Arroyo and Long, 2018)

eliminated without value loss is waste (negative iteration). Late changes in the design process due to lack of agreement by the stakeholders or failure to incorporate all relevant perspectives is an example of negative iteration. Often, it may be preferable to delay a decision rather than to make one prematurely or without the right people involved.

The sets of alternatives are collaboratively determined by the design team based on their plan. The entire process is in accordance with The Last Planner® System (LPS), as proposed by Ballard (2000a). This process informs when decisions need to be made by pull planning decisions, identifying constraints to make them, and promoting clear commitments from the design team,

In addition, CBA is used as the decision-making method and included in A3 reports to manage and report the decisions, recommendations, or documentation. A3-thinking is a problem-solving method based on a 'plan-do-check-act' (P-D-C-A) process, which is used extensively in the Toyota Production System (Sobek II and Smalley, 2011). This structured approach to problem-solving clearly defines the problem, desired outcomes, hypothesised solution, and steps to implement the solution. Using A3 reports, the team is prompted to summarise the entire issue on an A3-sized piece of paper, which allows for feedback among stakeholders. An A3 report refers to an 11 x 17 inch piece of paper divided into several structured sections. The exact structure depends upon the type of A3 and the needs of the situation. A general example follows the pattern: (1) Background; (2) Current Situation and Problem; (3) Goal; (4) Root Cause Analysis; (5) Action Items/Implementation Plan; (6) Check of Results; and (7) Follow Up (Sobek II and Smalley, 2011).

10.4 CBA and Lean Construction research

This section presents a brief history of Lean and the thinking behind CBA based on the author's experience in the Lean Construction industry. Figure 10.6 shows the recent history of papers published by the Lean Construction community (above timeline) as well as papers that have influenced the author's view of the decision-making process in general (below timeline).

A summary of the research on Lean Construction and CBA follows:

CBA was originally developed by Jim Suhr in the 1980s based on the cornerstone principle that *decisions must be based on the importance of advantages* (1981). Suhr invented the method while working as an engineer at the US Forest Service, seeking better ways to collaborate with multiple stakeholders involved in decisions on US National Parks. In addition, his work was inspired by the theory of general semantics developed by Korzybski in the 1920s. Suhr's book *Choosing by Advantages Decision Making System* was published in 1999.

CBA earned the attention of Lean Construction practitioners and researchers in the late 2000s when several research papers were published demonstrating the use of CBA in construction projects. Specifically, the Cathedral Hospital Project in Northern California provided the context for research such as Parrish and Tommelein (2009), who used CBA and Set Based-Design (SBD) for rebar design, Nguyen, Lostuvali, and Tommelein (2009) who used CBA for choosing a viscous damping wall system, and Grant (2007) who used the CBA method for green roof design decisions. On a separate project, Lee, Tommelein, and Ballard (2010) used CBA when choosing a superstructure design for a guideway infrastructure project in the public sector, highlighting the synergy between CBA, TVD, and SBD. Additionally, Arroyo, Tommelein and Ballard (2012) used CBA for choosing a wall system; Thanopoulos (2012) used CBA for choosing building systems (Telecommunication,

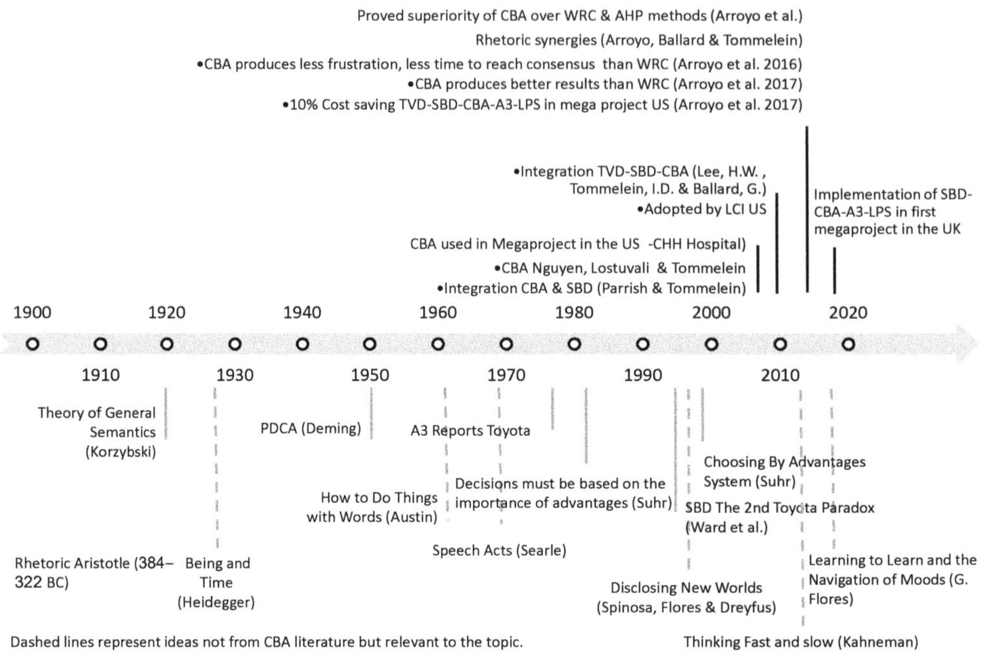

Figure 10.6 CBA connection with Lean and other influences

Lighting, and HVAC systems); and Arroyo, Tommelein, and Ballard (2013) used CBA for selecting tiles from a global and sustainable perspective.

Several research publications have compared CBA with traditional decision-making methods used in design and construction. Arroyo, Tommelein, and Ballard (2012; 2014) explain the advantage of the CBA method compared to the traditional Weighting Rating and Calculating (WRC) methods, also known as the weighted sum method. CBA helps decision-makers to focus on important differences between the alternatives, while WRC focuses on what factors are more important, which is an abstract judgement basing decisions on preconceptions rather than actual differences between the alternatives. WRC asks conflicting questions such as is safety more important than productivity? Whereas, CBA asks stakeholders to understand the factual differences between the alternatives in terms of safety and productivity before discussing trade-offs.

In 2014, Arroyo published her PhD dissertation comparing Multiple Criteria Decision Making methods (MCDM) for sustainable design on construction projects (Arroyo, 2014). Compared with WRC, the Analytical Hierarchy Process (AHP), Outranking and Optimisation methods, CBA is recommended for reaching consensus, providing transparency, and supporting continuous learning. In summary, CBA is a new way of thought based on the principle that decisions should be based on the positive differences (advantages) between the alternatives, as opposed to traditional methods which focus on weighting factors relevant to the decision-makers without considering actual differences offered by the alternatives. The methodological difference is comparable to that between Platonic and Aristotelian schools of thought: AHP and WRC processes are based on the notion that there is a natural preference order regardless of the alternatives, while CBA implies that preferences are based in a context given by the realities of available alternatives.

The synergy between CBA and rhetorical theory was explored by Arroyo, Ballard, and Tommelein in 2014, with particular focus on the application of rhetoric tools (logic, ethos, and pathos) during the implementation of CBA. Also, in 2014 more applications of CBA in design decisions were published, such as Kpamma et al. (2014) on the use of CBA for collaborative design decisions in construction, while Arroyo, Tommelein, and Ballard (2014) used CBA for choosing structural systems.

In a 2015 publication, Arroyo, Tommelein, and Ballard compared CBA and AHP, a structured technique for organising and analysing complex decisions that was developed in the 1970s by Thomas L. Saaty, based on principles of mathematics and psychology (Saaty, 1980). The research demonstrated that, despite the widespread use of AHP in construction, CBA presented significant advantages in terms of transparency of the decision trade-offs when making detailed design decisions. Arroyo, Tommelein, and Ballard (2015) demonstrate flaws of AHP, by presenting a case where the final decision changes (rank order reversal) when eliminating information that does not differentiate the alternatives. This does not happen in CBA because the method is focused on the importance of the advantages.

Schöttle, Arroyo, and Bade (2015) explore the use of CBA in the tendering procedure, and how it may be applied for pubic owners. Schöttle, Arroyo, and Haas Georgiev (2017) continue to explore this topic in a case study for a hospital project in California. Haapasalo et al. (2015) applied CBA to select the optimal contract type for road maintenance.

Karakhan, Gambatese, and Rajendran (2016) applied the CBA system to select fall-protection measures. This paper begins to explore the use of CBA for safety purposes in construction. In the same year, Arroyo et al. (2016b) demonstrate the use of CBA for choosing an HVAC system for a net zero energy (NZE) museum. In addition, controlled experiments by Arroyo et al. (2016a) demonstrated that CBA requires less time to achieve consensus and produces less frustration than WRC. This study presents the first evidence of CBA benefits compared to WRC. In this study, practitioners were asked to make several decisions with WRC and later they were introduced to CBA. The study measures statistically significant improvements using the multiple baseline testing method.

Martinez, Tommelein, and Alvear (2016) demonstrate the use of CBA for formwork system selection. This application was in the context of social housing in Ecuador. Interviews with project participants indicated that the discussion sessions of CBA generated a social process in which debate, argumentation, and rhetoric played an important role in the final resolution.

Schöttle and Arroyo (2016, 2017) used CBA for public bidding procedures based on an algorithm to pre-anchor trade-offs such that the weights of advantages could be published before receiving proposals in the bidding process.

Correa et al. (2017) demonstrated that CBA leads to decisions that deliver higher value at a lower cost than those made with WRC methods, based on an experimental study in collaboration with the Catholic University of Chile and Stanford University. In addition, under the same collaborative effort, Arroyo et al. (2018) developed a new method to use CBA with the capacity to consider thousands of alternatives based on an algorithm to generalise preference curves without assuming overall linearity of trade-offs, as most traditional methods do.

Murguia and Brioso (2017) used CBA and 4D models to select the best construction-flow option in a residential building in Peru. The result showed improvement in collaboration and the transparency of the decision-making process.

Strategic considerations for applying CBA during the design process have been explored by Kpamma et al. (2017) and Schöttle, Arroyo, and Christensen (2018). CBA has been used to make safety decisions in construction, such as Nnaji et al.'s (2018) use of CBA for selecting safety technologies for highway construction, and Zuluaga, Albert and Arroyo's (2018) use of CBA for fall protection devices for the North Carolina Department of Transport (DOT).

In 2018, Koskela, Arroyo, and Ballard showed a favourable comparison between the principles of legal proceedings and Lean practices such as TVD, SBD, CBA, and A3s as they are applied during the design process.

The benefits of the implementation of CBA specifically in the context of construction projects have also been studied. Arroyo and Long (2018) presented a construction project demonstrating an 11% savings in overall cost and significant cost reductions achieved when using CBA and A3 systematically, compared to a trial and error approach previously used by the designers. The original process was highly inefficient, confrontational, and frustrating; CBA helped the design team make better decisions in a more efficient manner while reducing conflicts and frustrations. Arroyo and Molinos-Senante (2018) used CBA for choosing wastewater treatment plants, comparing this process to previous results using AHP.

Several opportunities for CBA implementation exist in the engineering world. Despite the Lean community's adoption of CBA in 2007 and the growing number of publications about its effects (Figure 10.7), the rest of the engineering sector has been practically blind to it. The lack of a standardised decision-making process is problematic for construction management teams throughout the industry, often resulting in wasteful iterations and suboptimal decisions. However, knowledge of the measurable benefits of implementing CBA is relatively recent and, unfortunately, largely confined to a few universities throughout the world. The construction industry may use CBA more as more education is available and as more benefits are documented in practice.

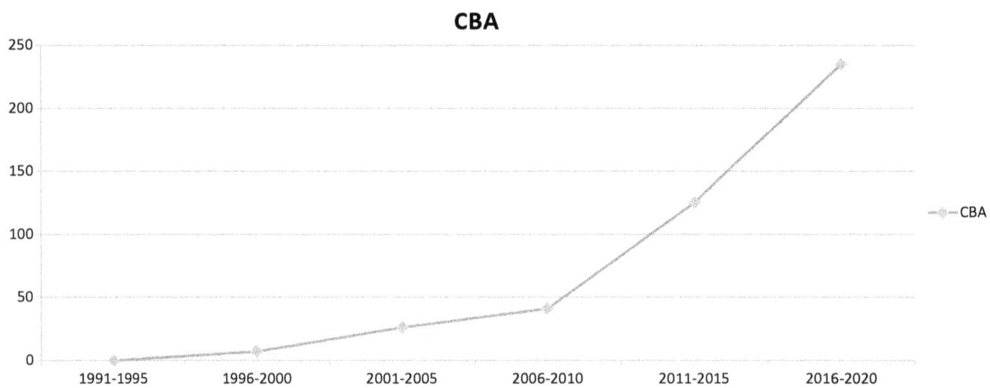

Figure 10.7 CBA Publications (Google scholar publications by January 2018)

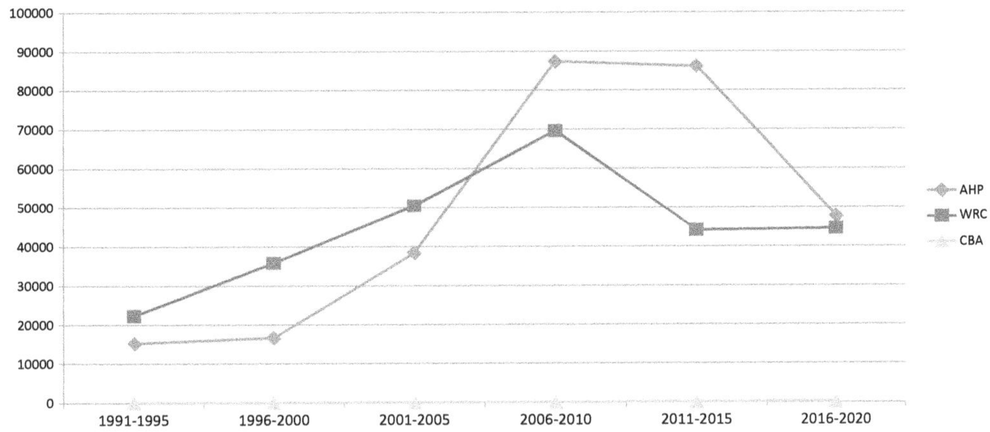

Figure 10.8 AHP, WRC, and CBA Publications (Google scholar publications by January 2018)

Figure 10.8 shows the predominance of traditional decision-making methods in engineering practice. According to data collected from Google Scholar, AHP and WRC are significantly more documented in academic research.

10.5 Benefits of choosing by advantages

Practitioners make several decisions throughout the design and construction of a project. The impacts of these decisions are relevant for the client, for society, and for the environment. A transparent and collaborative process that accounts for the involvement of several stakeholders is therefore crucial for the success of any project. CBA allows for this integration of different perspectives in the decision process and has several more demonstrated benefits such as:

Creating transparency: The practitioners' difficulty to explain the rationale of a decision made while they were involved on a project is a common indicator of poor decision-making processes. Additionally, many decisions are based solely on cost, disregarding important differences between the alternatives. CBA is a transparent method for choosing alternatives that allows everyone involved to understand the rationale for a decision, clarifies the value vs. the cost of the alternatives, and optimises decisions across the project (Arroyo, 2014; Parrish and Tommelein, 2009).

Building consensus: When making decisions in big projects, practitioners appear to use 'decide, present, and defend' approaches which are seldom collaborative and are narrowly targeted rather than seeking to optimise the entire project design. Building consensus is desirable for avoiding conflicts and unnecessary iterations in the design phase. CBA helps in building consensus among stakeholders (Arroyo, 2014; Martinez, Tommelein, and Alvear, 2016). This is critical in the case of sustainable design, which involves multiple stakeholders with conflicting interests.

Supporting continuous learning: Many decisions are not documented, or are documented in a way that is not transparent, which does not allow for continuous learning. If the design team is able to clearly identify the rationale for a decision, future iterations, especially those adding information (e.g. new alternatives or new factors) to the design process, will be better understood and guided. CBA saves time and resources, and results in a better overall decision than made otherwise (Arroyo and Long, 2018).

Supporting other Lean tools and methods, allowing for increasing value and reducing cost: When CBA is used with other Lean tools and methods in a collaborative contract, the project can obtain more benefits, thus increasing value for all stakeholders and reducing costs for the client. This is achieved by grounding the design in a target cost and value, involving all relevant stakeholders in decisions, exploring alternatives, and delaying decisions until the design team understands the consequences of each alternative (Kpamma et al., 2017; Schöttle, Arroyo, and Christensen, 2018).

Allowing for innovation in design: By evaluating multiple alternatives, practitioners understand the advantages of each, and may be able to identify or create new, innovative alternatives based on the CBA analysis (Arroyo and Long, 2018; Correa et al., 2017).

10.6 CBA case study

This case study applies CBA for choosing an HVAC system for a museum in San Francisco, California. The objective is to demonstrate the applicability of CBA in a real design project. This case study is based on a research paper developed by Arroyo et al. (2016b) using a case study method. The researcher gathered public information on the project, and conducted a series of interviews with the mechanical engineer, owner, and building operations engineer to analyse the potential use of CBA. Results showed how CBA would have helped organise the information towards more transparent and collaborative decisions. Feedback from project stakeholders was incorporated into the research results.

The project consisted of a new museum campus, the size of which is 330,000 ft^2 including indoor and outdoor exhibit space, and 1.5 acres of freely accessible public space. The project has several sustainable goals that were incorporated into the design. The design team targeted to minimise water usage, and to educate visitors about energy-efficient design.

10.6.1 Step-by-step CBA application to choose an HVAC system

The following sections present the application of CBA to this problem.

10.6.1.1 Step 1: Identify alternatives

Choosing an HVAC system is comprised of many decisions (Lechner, 2008). The design team needed to choose the primary energy source for heating systems (e.g. gas or electricity), the distribution system (e.g. radiant slab or fan coil unit), and the heat sink for pumping heat from the building (e.g. the outdoor air or a body of water).

This case study analyses three different but interdependent decisions: (1) the energy source for the heating system, (2) the energy source for the cooling system, (3) and the sink to pump heat from the building (bay water or air using a cooling tower). Table 10.2 shows the alternatives that were considered by the design team.

All options include 3 inches roof insulation, energy efficient lighting and basic daylighting controls. In addition, the mechanical engineer recommended the use of a radiant floor system as a distribution system for all alternatives.

The following presents the description of the three alternatives considered in this test case.

Table 10.2 Design alternatives for energy sources and heat sink

	Alternative 1	*Alternative 2*	*Alternative 3*
Heating Source	Natural gas boilers	Bay water-coupled electric heat pumps sourced by utility company	Bay water-coupled electric heat pumps with PV panels
Cooling Source	Electric chiller source by utility company	Bay water-coupled electric heat pumps sourced by utility company	Bay water-coupled electric heat pumps with PV panels
Heat Rejection	Cooling tower	Bay water	Bay water

Alternative 1: Natural gas boiler and chiller system with a cooling tower. This alternative is commonly used in commercial buildings and is considered the standard way of designing HVAC systems. In this case, the heating source is a natural gas boiler, and the cooling source is an electric chiller. Heat rejection is carried out through a cooling tower, which evaporates water.

Alternative 2: Electric heat pump sourced by utility company with bay water. This alternative takes advantage of the position of the building to use bay water for cooling and heating. The idea is to exchange heat between the cooler bay water and the hot water coming from the building, after the bay water is released. In this case, the heating and cooling source will be bay water coupled with electric heat pumps connected with a local utility company. Heat rejection is accomplished though bay water recirculation.

Alternative 3: Electric heat pump sourced by PV system with bay water. In this case, the heating and cooling source will be bay water coupled with electric heat pumps (same as alternative 2) connected with a PV system as energy source instead of the local utility company. Heat rejection is accomplished though bay water recirculation (same as alternative 2).

Finally, Table 10.3 presents the initial and life cycle costs of these three alternatives.

Table 10.3 Initial and life cycle cost of the three alternatives (Integral Group, 2007)

	Alternative 1: Natural gas boiler and chiller system with cooling tower	*Alternative 2: Electric heat pump sourced by utility company with bay water.*	*Alternative 3: Electric heat pump sourced by PV panel system with bay water.*
Initial Cost	$1,154,250	$1,654,250	$6,173,399
Annual Operation Cost of Energy	$269,755	$245,146	$0
Annual Maintenance Costs	$1,500	$6,000	$6,000
Replacement Costs HVAC (every 40 years)	$912,500	$596,250	$596,250
Life Cycle Cost (50 years)	$15,857,614	$14,956,841	$7,218,712

The PV cost calculations include available incentives from PG&E ($0.37/kWh for 5 years which equals $3.25 million for alternative 3). This amount is approximate and the exact incentive depends on the timing of the project, and the type and layout of PV installed.

10.6.1.2 Step 2: Define factors

In CBA, the design team needs to identify factors that will help differentiate between alternatives. The process of identifying factors may be iterative, since the design team may find new information when it develops a more detailed design of the alternatives. Based on the available data, seven factors help to differentiate between the alternatives. The factors represent the views of the owner, the architect, the mechanical engineer, the operation manager and users (Table 10.4).

The factors and the rationale for selecting them are as follows:

(1) *Experience in using HVAC systems:* The alternatives have important differences in how often they are utilised. The majority of commercial buildings use cooling towers, and water-based systems are infrequently utilised. This information is important for the purpose of having a proven and reliable system.
(2) *Space requirements:* The alternatives have important differences in space requirements, depending on the use of a cooling tower. Space is scarce and could be used either for the museum displays or for a cooling tower.
(3) *Contribution to goal of NZE:* The alternatives differ in how they contribute to the goal of NZE, e.g. using a PV panel system allows for on-site generation, compared to natural gas which would need to be sourced elsewhere.
(4) *Water usage:* The alternatives differ in their requirements for municipal fresh water. The cooling tower recirculates fresh water and loses significant quantities in evaporation as opposed to the bay-water system, which does not require fresh water from public sources.
(5) *Maintainability:* The alternatives are different in terms of maintenance requirements, e.g. using a bay-water system poses special challenges for maintenance vs. standard maintenance procedures for a cooling tower.
(6) *CO_2 emissions:* The alternatives have important differences in terms of CO_2 emissions during the operation and maintenance phase.
(7) *Noise:* The alternatives differ in terms of noise generated due to airflow movement; a cooling tower emits more noise than a bay-water system, which does not require cooling tower.

10.6.1.3 Step 3: Define the 'must' and 'want to have' criteria for each factor

For each factor, the design team needs to agree on criteria to judge the alternatives. Some attributes have a standard evaluation, in which case it is easy to establish a criterion (e.g. for CO_2 emissions, the criterion may be 'the lower the CO_2 emissions the better'). In other cases, the design team needs to describe what it wants (e.g. for experience using the HVAC systems, the criterion may be 'the more reliable the system is, the better'). The criteria do not represent a trade-off, and CBA does not assume that every increment of performance is equally valuable. Table 10.4 presents the factors and criteria considered in this decision in the first column.

Table 10.4 CBA Tabular Method

Factor (Criterion)	*Alternative 1: Natural gas boiler and chiller system with a cooling tower*	*Alternative 2: Electric heat pump sourced by utility company with bay water*	*Alternative 3: Electric heat pump sourced by PV panel system with bay water*
1. Experience using this HVAC system	Att.: Typically used in commercial buildings. This is the standard way.	Att.: It is not that common to use bay water for HVAC systems – unanticipated problems may arise.	Att.: It is not that common to use bay water for HVAC systems – unanticipated problems may arise.
(The more reliable the system is, the better)	Adv.: *It is more reliable than alt. 2 and 3.* Imp.: 50	Imp.:	Adv.: Imp.:
2. Space requirements	Att.: Cooling tower uses a lot of space. 370 ft^2 approximately.	Att.: The condenser occupies the least amount of space	Att.: The condenser occupies the least amount of space
(The less space the HVAC system uses, the better)	Adv.: Imp.:	Adv.: *It saves around 370* ft^2 Imp.: 20	Adv.: *It saves around 370* ft^2 Imp.: 20
3. Contribution to goal of NZE	Att.: It requires external energy by using natural gas and electricity from the grid. It does not allow for NZE.	Att.: Lower power consumption than air-cooled systems, especially at peak load. It requires external energy by using natural electricity from the grid. It does not allow for NZE.	Att.: Lower power consumption than air-cooled systems, especially at peak load. It allows the building to produce the same energy that it consumes.
(The more the alternative contributes to achieve the NZE target, the better)	Adv.: Imp.:	Adv.: slightly better than alt. 1. Imp.: 20	Adv.: significantly better than alt. 1. Allowing for NZE Imp.: 100
4. Water usage	Att.: It requires the use of evaporative cooling towers, which uses an estimated 2 million gallons of fresh water per year	Att.: Used water is returned to the bay	Att.: Used water is returned to the bay
(The less water the system uses, the better)	Adv.: Imp.:	Adv.: *It saves 2 million gallons of fresh water per year.* Imp.: 35	Adv.: *It saves 2 million gallons of fresh water per year.* Imp.: 35
5. Maintainability	Att.: Easy standard maintenance	Att.: Hard maintenance. Bio fouling will be produced	Att.: Hard maintenance. Bio fouling will be produced
(The easier to maintain, the better)	Adv.: *maintenance is easier and less frequent than alt 2 and 3.* Imp.: 60	Adv.: Imp.:	Adv.: Imp.:
6. CO_2 emissions	Att.: 574,740 lb. CO_2 per yr.	Att.: 363,198 lb. CO_2 per yr.	Att.:

(*Continued*)

Table 10.4 (Cont.)

Factor (Criterion)	*Alternative 1: Natural gas boiler and chiller system with a cooling tower*		*Alternative 2: Electric heat pump sourced by utility company with bay water*		*Alternative 3: Electric heat pump sourced by PV panel system with bay water*	
(The lower the CO_2 emissions, the better)	Adv.:	Imp.:	Adv.: Avoids 211,542 lb. CO_2 per yr.	Imp.: 40	Adv.: *Avoids 574,740 lb. CO_2 per yr.*	Imp.: 80
7. Noise	Att.: <u>It may produce noise problems associated with high airflow required for air-cooled systems.</u>		Att.: it is a quiet system		Att.: it is a quiet system	
(The less noise, the better it is)	Adv.:	Imp.:	Adv.: *Less noisy than alt 1.*	Imp.: 10	Adv.: *Less noisy than alt 1.*	Imp.: 10
Total IofAs		110		125		245

10.6.1.4 Step 4: Summarise the attributes of each alternative

We used data available from Integral Group and from the interviews to determine the attributes of each alternative. The least preferred attributes are underlined and are to be used as comparison points to describe advantages.

10.6.1.5 Step 5: Decide the advantages of each alternative

Once the attributes are summarised, the design team needs to apply the criteria to identify the advantages. Table 10.4 presents the advantages (Adv.) of each alternative for each factor.

10.6.1.6 Step 6: Decide the importance of each advantage

This part of the process is collaborative, and decisions are reached through discussion within the design team. In this case, for the purpose of exemplifying the use of CBA, the subjective values of the Importance of Advantages (IofAs) were assigned by the researcher and not by the design team. Table 10.4 presents the IofAs (Imp.). The most important advantage for each factor is shown in italics.

Determining the IofA weights was based on the following rationale:

- Alternative 3 offers a significantly greater advantage than Alternative 1 in terms of contributing to the NZE goal (this can be identified as the 'most important advantage'). In fact, based on the data, Alternative 3 is the only option that could achieve an NZE building, which is a primary goal of the project, so it was assigned an IofA of 100. It is noted that Alternative 2 is preferred over Alternative 1 with regards to contributing to NZE because it does not require natural gas. However, that seemed less important when compared to the paramount advantage offered by Alternative 3, so the researcher assigned an IofA of 20;
- Compared to Alternative 1, Alternative 3 would avoid 574,740 lbs. of CO_2 per year and Alternative 2 would avoid 211,542 lbs. of CO2 per year. The researcher thus assigned Alternatives 3 and 2 with an IofA of 80 and 40, respectively;

- Alternative 1 has two advantages over Alternative 2 and 3. First, it is more reliable than Alternatives 2 and 3, which, although somewhat important, that advantage is not as highly prioritised as the paramount advantage. Accordingly, the researcher assigned an IofA of 50. Second, Alternative 1 needs less frequent easier maintenance than Alternatives 2 or 3, which is also important, so the researcher assigned an IofA of 60;
- Lastly, Alternatives 2 and 3 have three equal advantages over Alternative 1. First, by not requiring a cooling tower, they save approximately 370 ft^2 compared with Alternative 1. This advantage seems relatively unimportant when compared with the paramount advantage; therefore, the researcher assigned an IofA of 20. Second, Alternatives 2 and 3 both save 2 million gallons of freshwater annually compared to Alternative 1. This seems somewhat important when compared to the paramount advantage, so the researcher assigned an IofA of 35. Third, Alternatives 2 and 3 are both less noisy than Alternative 1, which does not seem very important when compared to the paramount advantage, so the researcher assigned an IofA of 10.

10.6.1.7 Step 7: Evaluate cost data

The design team can plot total IofAs against first cost (Figure 10.9) and total IofAs against life cycle cost (Figure 10.10) for the three alternatives. Figure 10.9 shows that the first cost follows an almost linear curve with increasing IofAs. The design team needs to decide whether they can and are willing to pay more for Alternative 3, which has the highest IofA.

Figure 10.10 shows that, in the long-term comparison of life cycle cost and IofA, Alternative 3 is better than Alternatives 1 and 2. Since Alternative 3 also has the advantages of achieving NZE goal and reducing CO_2 emissions to almost zero (for the operation of the HVAC system), it makes sense for the design team to select Alternative 3.

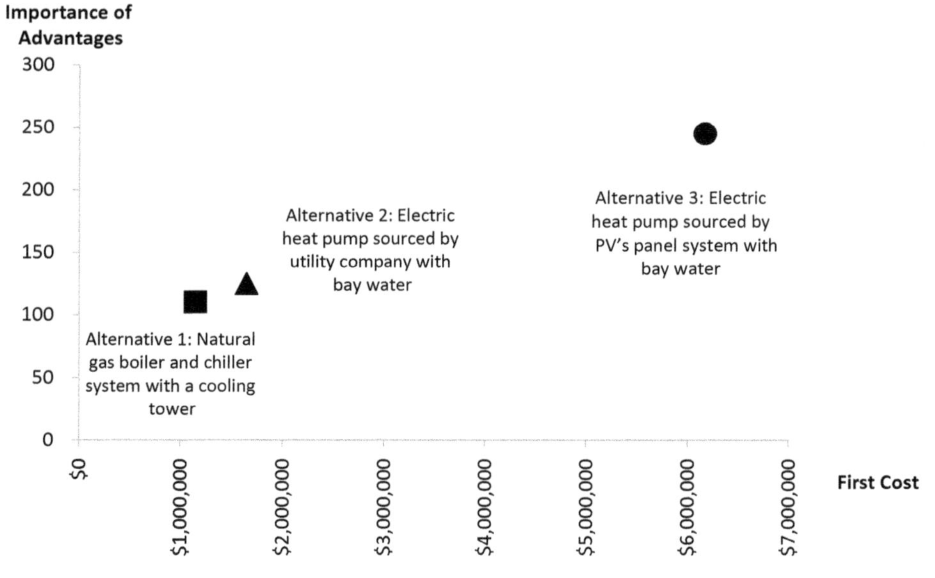

Figure 10.9 IofA vs. first cost

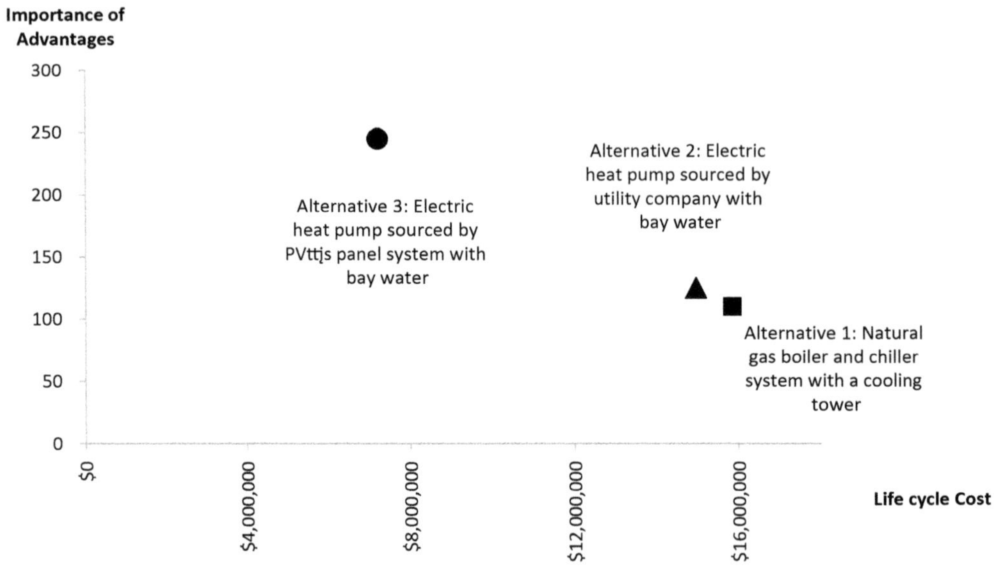

Figure 10.10 IofA vs. life cycle cost

10.6.1.8 Case study conclusion

This case study shows how CBA can be used to organise relevant information, making it easy for the design team to make subjective judgements and present trade-offs in terms of cost vs. 'value'. CBA helps document the rationale behind the decision-making process; aiding designers in learning from past experiences improves designs for future projects. Documenting decisions with CBA also helps retain knowledge of the project, which is crucial particularly when people leave the design team.

The CBA method creates a transparent framework for decision-making and builds consensus within a design team while allowing for multiple and varied perspectives (e.g. those of the owner, architects, mechanical engineer, operation manager, and users). Different perspectives are discussed in terms of factors, criteria, attributes, and a shared understanding of the advantages before estimating the IofAs. Building consensus is most challenging when determining the IofAs; CBA helps the design team through this process, specifically discouraging the tendency of making decisions based on assumptions or previous experiences. Instead, CBA requires the design team to understand the differences between the alternatives in the particular decision context, decide which alternative offers advantages for each factor, understand the importance of these advantages in comparison to one another, and, finally, decide on the importance of the advantages.

10.7 Recommendations to implement a collaborative decision-making process

Based on the literature, the following are recommendations to implement a collaborative decision-making process:

- Involve the right stakeholders, who are impacted by the decision and who have relevant technical knowledge on the impacts of the decision. Collaborative decision-making works when the right stakeholders are involved at the right time (Ballard, 2000b; Suhr, 1999);
- Defer decisions until the last responsible moment as described in SBD method (Ward et al., 1995);
- Make decisions from the perspective of long-term benefits and needs;
- Make decisions slowly by consensus and then implement them quickly; this concept is known as *Nemawashi* (Liker and Meier, 2006);
- Separate value from cost, and allow money to move across contractual and system boundaries. One of the most common mistakes in decision-making is to consider cost as a regular factor (Schöttle and Arroyo, 2016; Schöttle, Arroyo, and Haas Georgiev, 2017; Suhr, 1999);
- Be transparent. No method ensures the right decision, but an honest and transparent decision-making process is more likely to recommend a beneficial decision (Suhr, 1999);
- Give weight to advantages rather than factors or attributes (Suhr, 1999).

We embody incredible potential when we oversee our own decisions which give us the ability to learn from others and enact change. Learning CBA decision-making methods as a mechanical step-by-step procedure is not enough. Learning CBA requires training, coaching, and practice to develop a new skill.

Project teams need commitment to transform organisations through informed action, proven methods, and collaboration. This includes willingness to share information and change the construction industry history of hiding information and not sharing rationale. Moreover, the decision-making process is more than providing information; it is a collaborative process of discovery. If information is not shared, then the best alternatives may not be discovered by a team. The potential for the implementation of methods such as CBA is large, as the industry has yet to adopt these changes on a broad scale.

10.8 Future research

As discussed, CBA is a fairly new concept in the AEC industry compared to traditional methods, and therefore more research is needed before it can evolve into a widespread practice. Based on the current research, there is immense potential for CBA to achieve further success in improving decisions on construction projects by instituting a sound, transparent, and collaborative decision-making process. Suggestions for future research are presented below:

- Understand how language & moods affect decisions. According to Heidegger, human beings are always attuned to their moods and we can only perceive the world from the perspective in which we are at a particular time (Heidegger, 1996). Our possibilities are also influenced by our moods (Flores, 2016). Projects are not free of the influence of the mood of their stakeholders and, as research has proved, moods are contagious (Kohn et al., 2013). Therefore, the alternatives presented for a particular project team are also influenced by their moods. Understanding the impact of productive vs. unproductive moods in how project teams see possible alternatives may have a great impact on project outcomes. In addition, conversations during the CBA process can also modify the moods;

- Understand how neuroscience & cognitive bias affect collaborative decisions. Research on psychology and neuroscience in the last decades has shown how the way we present information affects the result of our decisions. In particular, research performed by Kahneman and Tversky shows how we are influenced by the order in which the information is presented, and how it is framed. Their work provides insights into how cognitive biases work (Kahneman, 2011). In construction management, the decision-making process does not consider the influences of biases, and CBA may or may not help with the avoidance of biases;
- Understand how technology, such as virtual reality, may help stakeholders better perceive future options. Several construction decisions are hard to evaluate because of stakeholders' lack of understanding of a future space. For many people without a degree in construction, it is hard to visualise dimensions from reading plans. On the other hand, designers may find it hard to understand interactions with building users, such as doctors' and patients' movements in a hospital. Virtual reality provides the opportunity to create experiences that are hard to describe as attributes. The ability to provide such experiences could facilitate decision-making during the design process and reduce negative iteration;
- Develop and evaluate CBA training efficiency and efficacy. Learning CBA requires an effort. Several approaches can be taken such as a 3-day workshop, as Jim Suhr did several times (Suhr, 1999). It also can be taught and coached on a project basis (Arroyo and Long, 2018), or it can be taught remotely. The impact of different training strategies has rarely been documented and analysed.

References

Arroyo, P. (2014) *Exploring Decision-Making Methods for Sustainable Design in Commercial Buildings*, PhD thesis, University of California, Berkeley, USA.

Arroyo, P., Ballard, G. and Tommelein, I.D. (2014) 'Choosing by advantages and rhetoric in building design: relationship and potential synergies', *Proc. 22nd Annual Conference of the International Group for Lean Construction*, Oslo, Norway, pp. 391–408.

Arroyo, P., Fuenzalida, C., Alberti, A. and Hallowell, M. (2016a) 'Collaborating in decision making of sustainable building design: An experimental study comparing CBA and WRC methods', *Energy and Building*, Elsevier, 128, pp. 132–142.

Arroyo, P. and Long, D. (2018) 'Collaborative design decisions', *Proc. 26th Annual Conference of the International Group for Lean Construction*, Chennai, India.

Arroyo, P. and Molinos-Senante, M. (2018) 'Selecting appropriate wastewater treatment technologies using a choosing-by-advantages approach', *Science of the Total Environment*, 625, pp. 819–827.

Arroyo, P., Mourgues, C., Flager, F. and Correa, M.G. (2018) 'A new method for applying choosing by advantages (CBA) multicriteria decision to a large number of design alternatives', *Energy and Buildings*, 167, pp. 30–37.

Arroyo, P., Tommelein, I., and Ballard, G. (2012) 'Deciding a sustainable alternative by "choosing by advantages" in the AEC industry', *Proc. 20th Annual Conference of the International Group for Lean Construction*, San Diego, CA, USA, pp. 41–50.

Arroyo, P., Tommelein, I. and Ballard, G. (2014) 'Comparing AHP and CBA as decision methods for choosing problem in detailed design', *ASCE, Journal of construction engineering and management*, 141(1), p. 04014063.

Arroyo, P., Tommelein, I. and Ballard, G. (2015) 'Selecting globally sustainable materials: A case study using choosing by advantages', *ASCE Journal of construction engineering and management*, 142(2), p. 05015015. Available at: 10.1061/(ASCE)CO.1943-7862.0001041.

Arroyo, P., Tommelein, I., Ballard, G. and Rumsey, P. (2016b) 'Choosing by advantages: A case study for selecting an HVAC system for a net zero energy museum', *Energy and Building*, ELSEVIER, 111, pp. 26–36.

Arroyo, P., Tommelein, I.D. and Ballard, G. (2013) 'Using "Choosing by advantages" to select tile from a global sustainable perspective', *Proc. 21st Annual Conference of the International Group for Lean Construction*, Fortaleza, pp 309–318.

Ballard, G. (2000b) 'Positive vs negative iteration in design', *Proc. 8th Annual Conference of the International Group for Lean Construction*, Brighton, UK, pp. 17–19.

Ballard, G. (2009) *An update on target value design*, Lean Design Forum-Lean Construction Institute & Project Production System Laboratory. Available at: www.leanconstruction.org.

Ballard, G.H. (2000a) *The Last Planner System of Production Control*, PhD Thesis, The University of Birmingham, Birmingham, UK.

Correa, M.G., Arroyo, P., Mourgues, C. and Flager, F. (2017) 'Comparing choosing by advantages and weighting, rating and calculating results in large design spaces', *Proc. 25th Annual Conference of the International Group for Lean Construction*, Heraklion, Greece, pp. 259–266.

Flores, G.P. (2016) *Learning to Learn and the Navigation of Moods: The Meta-Skill for the Acquisition of Skills*, Pluralistic Networks Publishing, San Bernardino, CA.

Flyvbjerg, B., Bruzelius, N. and Rothengatter, W. (2003) *Megaprojects and Risk: An Anatomy of Ambition*, Cambridge University Press, Cambridge, UK.

Grant, E. (2007) *A Decision-Making Framework for Vegetated Roofing System Selection*, PhD Thesis, Architecture and Design Research, Virginia Polytechnic Institute and State University, Blacksburg, VA, USA.

Haapasalo, H., Aapaoja, A., Björkman, S. and Matinheikki, M. (2015) 'Applying the choosing by advantages method to select the optimal contract type for road maintenance', *International Journal of Procurement Management*, 8(6), pp. 643–665.

Heidegger, M. (1996) *Being and Time: A Translation of Sein und Zeit*, SUNY Press, Albany, NY.

Kahneman, D. (2011) *Thinking, Fast and Slow*, Straus and Giroux, New York, USA.

Karakhan, A., Gambatese, J. and Rajendran, S. (2016) 'Application of choosing by advantages decision-making system to select fall-protection measures', *Proc. 24th Annual Conference of the International Group for Lean Construction*, Boston, USA.

Kohn, N., Falkenberg, I., Kellermann, T., Eickhoff, S.B., Gur, R.C. and Habel, U. (2013) 'Neural correlates of effective and ineffective mood induction', *Social Cognitive and Affective Neuroscience*, 9(6), pp. 864–872.

Koskela, L., Arroyo, P. and Ballard, G. (2018) 'Designing as a court of law', *Proc. 26th Annual Conference of the International Group for Lean Construction*, Chennai, India, pp. 614–624.

Kpamma, Z.E., Adjei-Kumi, T., Ayarkwa, J. and Adinyira, E. (2014) 'Creating, sustaining and optimising the collaborative realm for participatory design', *Proc. 22nd Annual Conference of the International Group for Lean Construction*, Oslo, Norway, pp 475–485.

Kpamma, Z.E., Adjei-Kumi, T., Ayarkwa, J. and Adinyira, E. (2017) 'Strategic considerations for applying choosing by advantages in design process', *Proc. 25th Annual Conference of the International Group for Lean Construction*, Heraklion, Greece, pp 507–514.

Lechner, N. (2008) *Heating Cooling, Lighting: Sustainable Design Methods for Architects*, 3rd, John Wiley & Sons, Hoboken, NJ, p. 68.

Lee, H.W., Tommelein, I.D. and Ballard, G. (2010) 'Lean design management in an infrastructure design-build project: A case study', *Proc. 18th Annual Conference of the International Group for Lean Construction*, Haifa, Israel, pp 113–122.

Liker, J.K. and Meier, D. (2006) *The Toyota Way Fieldbook*, The McGraw-Hill Companies, New York.

Martinez, E., Tommelein, I. and Alvear, A. (2016) 'Formwork system selection using choosing by advantages', *Construction Research Congress,2016*, pp. 1700–1709.

Murguia, D. and Brioso, X. (2017) 'Using "Choosing by advantages" and 4D models to select the best construction-flow option in a residential building', *Procedia Engineering*, 196, pp. 470–477.

Nguyen, H.V., Lostuvali, B. and Tommelein, I.D. (2009) 'Decision analysis using virtual first-run study of a viscous damping wall system', *Proc. 17th Annual Conference of the International Group for Lean Construction*, Taipei, Taiwan, pp 371–382.

Nnaji, C., Lee, H.W., Karakhan, A. and Gambatese, J. (2018) 'Developing a decision-making framework to select safety technologies for highway construction', *ASCE, Journal of Construction Engineering and Management*, 144(4), p. 04018016.

Parrish, K. and Tommelein, I.D. (2009) 'Making design decisions using choosing by advantages', *Proc. 17th Annual Conference of the International Group for Lean Construction*, Taipei, Taiwan, pp. 501–510.

Saaty, T.L. (1980) *The Analytic Hierarchy Process*, McGraw-Hill, New York, USA.

Schöttle, A. and Arroyo, P. (2016) 'The impact of the decision-making method in the tendering procedure to select the project team', *Proc. 24th Annual Conference of the International Group for Lean Construction*, Boston, MA, USA, pp. 23–32.

Schöttle, A. and Arroyo, P. (2017) 'Comparison of weighting-rating-calculating, best value, and choosing by advantages for bidder selection', *ASCE, Journal of Construction Engineering and Management*, 143(8) Available at: https://ascelibrary.org/doi/abs/10.1061/%28ASCE%29CO.1943-7862.0001342 (Accessed: Aug 2017).

Schöttle, A., Arroyo, P. and Bade, M. (2015) 'Comparing three methods in the tendering procedure to select the project team', *Proc. 23rd Annual Conference of the International Group for Lean Construction*, Perth, Australia, pp. 267–276.

Schöttle, A., Arroyo, P. and Christensen, R. (2018) 'Demonstrating the value of an effective collaborative decision-making process in the design phase', *Proc. 26th Annual Conference of the International. Group for Lean Construction*, Chennai, India.

Schöttle, A., Arroyo, P. and Haas Georgiev, C. (2017) 'Applying choosing by advantages in the public tendering procedure', *Proc. 25th Annual Conference of the International Group for Lean Construction*, Heraklion, Greece, pp. 45–52.

Sobek II, D.K. and Smalley, A. (2011) *Understanding A3 Thinking: A Critical Component of Toyota's PDCA Management System*, CRC Press, Boca Raton, FL.

Suhr, J. (1999) *The Choosing by Advantages Decision making System*, Greenwood Publishing Group, Westport, CT.

Thanopoulos, T. (2012) *Lean Decision making and Design Management for Sustainable Building Systems and Controls: Target Value Design, Set-Based Design, and Choosing by Advantages*, Master of Engineering Thesis, Engineering and Project Management, Dept. of Civil and Environmental Engineering, University. of California, Berkeley, CA, USA.

Ward, A., Liker, J.K., Cristiano, J.J. and Sobek, D.K. (1995) 'The second Toyota paradox: How delaying decisions can make better cars faster', *Sloan Management Review*, 36(3), p. 43.

Zuluaga, C.M., Albert, A. and Arroyo, P. (2018) 'Protecting bridge maintenance workers from falls: Evaluation and selection of compatible fall protection supplementary devices (FPSD)', *ASCE, Journal of Construction Engineering and Management*, 144(8), p. 04018073.

11 Lean Construction

A management model for interdependencies in detailed design

Bo Terje Kalsaas

11.1 Introduction

While the design process in construction makes up a relatively small share of construction costs (about 10%), it is extremely important from the perspective of the building's life cycle, including customer value, maintenance and operation costs (Evans et al., 1998; Gilbertson, 2006). Koskela, Bølviken and Rooke (2013, p. 9) view 'the design-production-use process as a chain where value is created as a potential in design, is embodied in production and is realised in the intended use by the client.'

The management of design processes is often designated as Lean Design Management (LDM) by the Lean Construction community, e.g. Koskela, Ballard and Tanhuanpää (1997) and Uusitalo et al. (2017). Uusitalo et al. (2017) provide a review of LDM processes, methods and techniques, arguing that, despite an increased application of LDM, many processes take place in isolation from one another. The authors therefore call for an integrated framework which specifies a structure or model for design execution. This chapter aims to provide an integrated method to support design management at detailed design. The concept of design in this chapter includes both architectural and engineering design.

The management of the design process is complex due to characteristics such as reciprocal interdependencies and the related iterations, learning, creativity and gradual maturity and the often-fragmented design process involving several consulting companies, the client and construction companies, as well as their subcontractors (Kalsaas and Moum, 2016; see also Chapter 2). An increasing level of interest about the subject has been indicated by the amount of publishing on both design management as well as understanding design and engineering e.g. Pikas, Koskela and Seppänen (2017); Pikas et al. (2018); Koskela et al. (2019).

There are several methods of design management, one of which is Virtual Design and Construction (VDC) (Kunz and Fisher, 2012; Fisher et al., 2017; Fosse, Ballard and Fischer, 2017). However, in the integrative approach to product organisation and process, there is no trace of systematic documentation on how the elements are linked, seen for example in the Last Planner System (LPS), which is made up of elements such as master plan, phase plan, lookahead schedule, production plan and measurement of PPC (production planned complete) (Ballard, 2000a, see also Chapter 3). When it comes to LPS, practical experience indicates that, although this method is fruitful for production, it is not comprehensive with regard to design management processes. See, for example, Hamzeh, Ballard and Tommelein (2009) and Kalsaas, Bonnier and Ose (2016). We can relate the limitations of LPS to the iterative nature of design in complex projects.

Design management in the chapter is studied in a 'design-build' contract setting, where the general design-build contractor is responsible for construction and the actual process of creating detailed design and procurement, which prepares the working basis for the subsequent construction process.

This chapter has been organised as follows. First, an account is given of the method utilised, which connects the chapter to Lean Construction. It is assumed that design management should be based on an understanding of design as a phenomenon, which is discussed as a background to the method. Next, a case study containing theory-related discussions is presented. The most important findings are extracted from the case study and, in combination with theory, support the constructed model for design management. The model's applicability and relevance are discussed as a prelude to the chapter's conclusion.

11.2 Method

Design Science Research (DSR) has been applied as the methodical framework, as it is aimed at developing a solution to a problem with practical and theoretical relevance (March and Smith, 1995). Koskela et al. (2019) relate DSR to Shewhart's ideas (1931) of using induction from empirical experimentation as the primary form of reasoning, which 'is centered around the complementarity of theoretical knowledge and empirical observations as sources of engineering/design knowledge, requiring interaction between induction and deduction'. Moreover, it is argued that all types of reasoning are applied in Lean Construction, including regression and abduction, and that contextual knowledge is the best source of knowledge. Furthermore, Hevner et al. (2004, p. 82) argue that the fundamental principle of DSR states 'that knowledge and understanding of a design problem and its solution are acquired in the building and application of an artefact', a process from which they derive seven guidelines: (1) design as artefact; (2) problem relevance; (3) design evaluation; (4) research contribution; (5) research rigor; (6) design as search process; (7) research communication.

This chapter is primarily based on the first two guidelines of DSR. An innovative, purposeful artefact (1) is constructed based on theory and a case study. Problem relevance (2) is based on findings from the case study and discussed. Evaluation (3) will be developed through future research. Moreover, the study takes up novelty in research contributions (4) in terms of how it might solve a familiar problem more effectively.

Data has been collected from the construction of a high school (2016–2019) with ambitious environmental goals (BREEAM-NOR Outstanding).[1] Data was gathered during detailed design, between December 2016 and February 2018, being supplemented in April 2019. The school is expected to be completed in June 2019. A total of 14 semistructured interviews were completed with participants in the value chain related to detailed design: production manager, design manager, building site manager, two architects, structural engineer, electrical engineer, HVAC engineer, three representatives from the client organisation (the head, building technology, technical systems), ventilation and automation subcontractor, piping subcontractor and electrical subcontractor.

The researcher participated in 18 design meetings, mostly as an observer. In addition to observing, the researcher conducted a large number of short, informal conversations with the participants, and the data includes information from BIM and BREEAM coordinators as well as a coordinator for the project's technical components.

11.3 The phenomenon of design

Throughout the evolution of design practice, there have been countless attempts to tackle the challenge of understanding, mastering and explaining the processes behind our constructed environment (Kalsaas and Moum, 2016). For example, in the 1960s the first generation of design methodologists stated that the design process was sequential and linear in nature; however, this belief has long been challenged by others (Lundequist, 1992). The understanding of the architectural design process as a complex universe of predictable and unpredictable interactions, interrelations and interdependencies between actors and their actions relates to observations of the architectural design practice made by researchers such as Cuff (1991), Schön (1991), Kalay (2004) and Lawson (2006).

According to Cuff (1991), the design process itself is a social construction where buildings are collectively conceived, which can be related to a highly complex universe where predictable and unpredictable interactions, interrelations and interdependencies between actors and processes create our physical environment (Moum, 2009). Kalay (2004) refers to design as a cyclical relationship between problem solving and puzzle making. Lawson (2006) describes the design process as 'a negotiation between the problem and solution through the three activities of analysis, synthesis and evaluation' and challenges the comprehension of the design process as a sequence of activities. He regards the design process as simultaneous learning about both the nature of the problem and range of possible solutions. In the beginning of the design process, neither the architect, engineer nor client know exactly what the building will look like, which problems will arise, or even what the requirements to be fulfilled are. Schön (1991) characterises design practice as a reflective dialogue between the designer and design situation, emphasising the crucial role played by tacit knowledge in this process. Further, while this intuitive knowledge shared by the practitioners involved in architectural design is crucial, it is difficult for others to grasp.

In addition, it is common in architecture and engineering to start with the client's specifications and then to develop both concept and practical solutions in several stages/phases. This is an example of backward reasoning starting from user requirements and then creating a design to fulfil them (Koskela et al., 2019). In design, one must often return to the problem by trying once more to solve it in a different way, i.e. a new iteration. Reciprocal interdependency (Thompson, 1967) drives iteration in design, as the conversation between interdependent specialists must go through an indeterminate number of cycles to achieve alignment. Koskela and Kagioglou (2006) refer to the concept of iteration arising as a new idea in the 1980s based on the observation that, when working, designers move between goals and means instead of following a linear path in a one-way process. A significant shift in the comprehension and management of iterations took place with the arrival of agile methods in software development (Schwaber and Beedle, 2002).

Ballard and Koskela (2013) link the works on rhetoric and design by Kaufer and Butler (1996) to the concept of wicked problems (Churchman, 1967; Rittel, 1988). Because of the complex interdependencies, the effort to solve one aspect of a wicked problem may either reveal or create other problems. The phrase was originally used in social planning and is contrasted with 'tame problems', which we can say are more linear in nature, where the concept of cause and effect is well known.

Wicked problems are related to the Cynefin framework for complexity (Snowden, 2000), including the denoted 'complex' and 'chaotic' domains where, in the beginning, cause and effect are unknown. Skaburskis (2008) refers to a conversation with Horst Rittel (e.g. Rittel, 1988) during which they discuss whether or not wicked problems can be solved. Rittel argued that they could be solved and that he did so every day. However, he also pointed out the fact that, 'We can't solve them by reference to the logic inherent in the problem.' This citation corresponds well with the complex domain referred to in Cynefin, where cause and effect is not known beforehand but emerges after actions have taken place.

Conklin (2006) generalised the concept of problem wickedness to areas other than social planning and policy: (1) The problem will not be understood until after the formulation of a solution; (2) Wicked problems have no stopping rule; (3) Solutions to wicked problems are neither right nor wrong; (4) Every wicked problem is essentially novel and unique; (5) Every solution to a wicked problem is a 'one shot operation'; (6) Wicked problems have no given alternative solutions. The relationship of wicked problems to design has been discussed in Stolterman (2008) and Rittel (1988), who argue that design is a wicked problem. In his book *Managing the design factory*, Reinertsen (1997) also indirectly refers to wicked problems, arguing that designers are primarily driven by the desire to design best-in-class solutions, and that these tasks are therefore rarely delivered ahead of time because the solutions can always be improved on. Moreover, several of the characteristics of rhetoric (Kaufer and Butler (1996) in Ballard and Koskela, 2013) that relate to design are familiar with wicked problems, e.g. 'only better or worse "answers", not right or wrong ones. Test: fitness for purpose.'

11.3.1 Design management

Koskela and Howell (2002) present Last Planner (Ballard, 2000a) and Scrum (Schwaber and Beedle, 2002) as two practical responses to the failure of conventional project management methods. For instance, in the theoretical comparison of these two methods, the authors emphasise 'transformation' and 'flow' for Last Planner and 'flow' and 'value generation' for Scrum. The equivalent for conventional methods is 'transformation', as per Koskela's (2000) TFV theory.

The Scrum method originates from software projects that have similarities in design with respect to iterations and dynamics and which are conceptualised to have the potential to handle reciprocal interdependencies between disciplines. There are examples of utilising Scrum-inspired methods in design management. A German experience has been reported by Demir and Theis (2016), and a Norwegian example has been reported by a Statsbygg and KHiB team (2017). However, these examples have no theoretical basis; therefore, a more detailed description of Scrum features relevant to the model that is intended to be developed appears below.

Scrum is one of several agile methods addressing variability and differs from traditional project management in that it does not work with Work Breakdown Structure (WBS), but rather functionalities, and the decision-making function is decentralised (Schwaber, Sutherland and Beedle, 2016). Agile methods share a certain number of attributes, such as being incremental and focusing on iterative development cycles to complete projects (Abbas, Gravell and Wills, 2008). Moreover, agile methods assume that variability cannot be removed; this is why the focus is not on minimising or eliminating change, but rather managing it. The world is perceived as being uncertain.

Another significant attribute is that Scrum delivers functionality that can be rapidly tested. It starts by listing the software functionalities desired by the owner/customer. This list is denoted as a product backlog which can be presented in terms of *user stories*. Changes are frequently made to the product backlog, which is important for registering new ideas and output from learning processes. Two work cycles are defined: a daily scrum (stand-up meeting) and sprints (2–4 weeks). A sprint is a list of prioritised functionalities transferred from the product backlog via a sprint backlog (Schwaber, Sutherland and Beedle, 2016). Various roles and meetings are standardised in order to handle issues such as customer value, uncertainty, learning and cost control.

Ballard (2000a) argues that the LPS applies to both production and design management; however, more recent work indicates the need for having something more than the Last Planner for managing design (Hamzeh, Ballard and Tommelein, 2009; Kalsaas, Bonnier and Ose, 2016). Arguably, the main reason for this lies in the variability surrounding the design of complex projects, which have a high level of interdependencies leading to iterations and calls for changes. This assertion will be verified in the case study presented later in the chapter.

VDC (Fisher et al., 2017; Kunz and Fisher, 2012) has been paid a great deal of attention in Norwegian design management. The concept revolves around Building Information Modelling (BIM) and other related design processes. The developers have added 'integrated' to concurrent engineering, denoted ICE, which has a central part in VDC. Koskela et al. (2019) relate concurrent engineering (also known as simultaneous engineering) (Eastman, 2012; Jo, Parsaei and Sullivan, 1993) and collaborative engineering (Lu, Elmaraghy and Wilhelm, 2007) to the disciplinary scope in Lean Construction. VDC addresses the integration of product organisation and processes and applying interactive digital screens for ICE problem-solving sessions. Kunz and Fisher (2012) prescribe the critical path method for scheduling purposes in VDC, which is a method Koskela et al. (2019) related to the traditional construction management method. However, VDC is also combined in practice with Last Planner as the scheduling technique (see Fosse and Ballard, 2016).

There are different design management methods of coordinating reciprocal interdependencies (Thompson, 1967) in design (mutual adjustment), including: (1) conversing on the telephone, (2) participating in telephone and video meetings, (3) conducting design meetings, (4) working in a Big Room/I-room (with reference to VDC), and (5) participating in ICE meetings/gatherings. ICE meetings, as they are described by Fosse and Ballard (2016), are those that are run on a strict schedule and take place in an I-room/Big Room. Arguably, it is difficult to understand that these meetings comply with ICE because there is hardly more time for mutual adjustment and joint problem solving than is found in traditional design meetings utilising BIM technology and conducting group meetings for problem solving. ICE in this respect is, however, a more efficient way of organising meetings compared to traditional practice.

Clearly, it is beneficial that the designers can sit together and work, as this provides positive conditions for effective communication and joint problem solving. Nevertheless, it may be difficult from a cost perspective to achieve this situation in medium-sized projects, of which the case study is an example. The method proposed in this chapter for organising the design process is meant to fill this need; in addition, it is an alternative for large projects.

11.3.2 From theory to case study

The theory addressed above regarding architectural and engineering design in the AEC industry demonstrates that it is an iterative process that gradually causes the design to mature through often strong reciprocal interdependencies between various architectural and engineering disciplines. Through using the wicked problem concept and the complex domain in Cynefin, we can understand that there is no best practice or solution for solving design challenges; rather, there are several good ones. Many iterations often provide better solutions, especially if the design is complex in nature.

Coordination in the form of mutual adjustment is significant when coordinating design processes, playing a central role in the design management task. Consequently, in the case study itself, emphasis has been placed on both identifying dependencies and discovering how sequential and reciprocal interdependencies are coordinated through planning and mutual adjustment. Arguably, integrated processes for this type of work are regarded as crucial because they both effectively guide a project's iterative knowledge processes, as well as ensure its progression towards completion.

11.4 Case study: construction of a high school

The planned school will offer students training in practical skills as well as academic subjects that will prepare them for university studies. The school building includes approximately 200 technical systems, including 10,000 digital components, with several depending on each other to achieve optimal operating conditions. For instance, these conditions are needed for teaching a skill such as mechanics, where teachers and students need proper ventilation and methods for handling oil waste, painting, etc., in their workshop classroom.

11.4.1 Background and contracting

This is a publicly funded construction project in Norway with a budget of approximately 800 million NOK (100 million USD). The floor area is 18,000 square metres with more than 500 rooms. The project is being built in accordance with environmental classification BREEAM-NOR Outstanding, and the client regards the project as a contribution to the green shift. The client has entered into a contract with supplier groups following a competitive dialogue process that resulted in a fixed price design-build contract. The main justification for the selected procurement method is that the client wanted to use the market to find a way to reach the project's ambitious environmental target. In brief, the following contracting process took place:

1. Prequalification of company groups consisting of general contractor, architect, construction engineering and technical subcontractors. The owner selected six groups to participate in the competition;
2. Each group prepared an outline of how they would satisfy the owner's requirements. Costs were not announced during this phase. Three groups were selected by a jury for further participation in the competition;
3. Dialogues with the three winners were conducted by the client, and a winner was then nominated. The architect of the group that won believes that they received a high score because they demonstrated their willingness and ability to make changes to the location of different functions in the building according to the users' needs;
4. The winning company group developed a pre-design, including technical descriptions and costs;
5. In accordance with a political resolution by the owner, the winning group was chosen, and the budget and primary objectives were approved. Thereafter, a contract was signed for a fixed-price design-build contract. Allowances were made in the budget for a certain amount of interaction between the winning group and owner following the conclusion of the contract.

In relation to the design process, it is worth noting that it was a group of professionals that won the contract together, and there was reason to believe that this factor would be positive for the group's future ability to cooperate in the detailed design phase (Swärd, 2016; Wondimo et al., 2018). However, from an organisational viewpoint, the execution model was a traditional design-build contract so that the architect and building disciplines, including the environmental consultant, worked for the general contractor, whilst the technical disciplines, including (MEP) electrical, plumbing and ventilation/automation, each worked for their respective subcontractor. The general contractor outsourced all the design and engineering, which is common in Norway, and many of the design and engineering disciplines came from different consultancy firms. The design process was led by the general contractor's design manager, and design meetings were set to take place every 14 days. The frequency of the client-general contractor meetings was also 14 days.

The detail design process was organised in accordance with LDM principles (Munthe-Kaas et al., 2015; Fosse and Ballard, 2016; Uusitalo et al., 2017). Initially, reverse scheduling technique was applied by using post-it notes to identify the sequence of the design and engineering tasks. A dialogue matrix was also used in each design meeting to identify the short-term need for input from/to the various disciplines and general contractor (the dialogue matrix is explained later). Secondly, BIM was a key part of the process and was actively used in design meetings as a backdrop to discuss problem solving and problem areas. The usage of BIM included pointing, selecting and zooming in and out, at times being supplemented with drawing sketches on a board. It was also used to a certain extent for both an analysis and procurement basis. The design manager believes that he could not have managed the project without BIM.

A 'project hotel' was also utilised as communication technology. This was a web-based platform in which participants shared documents and communicated with one another. Design issues needing to be resolved were set up in the project hotel, which included deadlines for task completion and the individual responsible for completing them. The idea was also that participants were to be able to monitor any progress and close issues upon their completion. There was a system in place that emitted individual warnings in the event of any changes being entered, which is based on the idea that the participants needed this information for coordination purposes.

11.4.2 Complexity

The project was perceived as being 'enormously complex' by the leading architect, who pointed out that 'nothing was the same' in the 18,000 square metres that were to be built. The architect stated that it was a huge task to familiarise oneself with the floor plan. A 200-page report was created by the group from the pre-design process in order to record the design, and 'we learnt a lot about high schools'. Several informants pointed out that BREEAM complicated the design process because it was initially paid a great deal of attention at the expense of other issues, thereby creating a significant amount of additional work. The general contractor had hired a specialist to be the link between the project-based BREEAM design work and the UK-based BREEAM certification organisation. In a professionally open forum outside the project, using the project as a case example, there was discussion as to how to best divide the project into a manageable process. Reference was then made to the metaphorical question, 'How do you divide an elephant?' The design manager replied that it was difficult to divide anything when 'we don't even know what kind of animal we are talking about'. Arguably, this is a statement that confirms the abovementioned assertion that it is fruitful to regard design in complex projects as a wicked problem.

11.4.3 Social environment and commercial interests

Several actors from different companies mentioned that there was a good social environment in the design group. For example, one member remarked that 'You are allowed to say that you are unsure about a solution', indicating that members had a high level of trust in one another. During an interview, the architect pointed out that, in this particular project, we were allowed to 'do our job', while in certain other projects the general contractor tried to be the architect. 'The contractor was good at coordinating all of our groups' and 'The architect truly realised that we were all in the same boat' are other examples of statements showing the members' mutual respect and trust. However, there were also disputes. For example, the design of the bearing system was completed just in time before construction was to start, and the general contractor and structural engineer got into a disagreement about whose fault this was. Moreover, the contractor claimed the solutions had increased the costs through increasing steel usage and creating a more labour-intensive building method.

In relation to the design interaction among the architects, engineers and contractors, it should also be noted that the client only wanted minimum participation in the design meetings. There was consensus between the designers and engineers, as well as the general contractor, that they needed a venue for discussion where the client would not participate. Similarly, the client was reluctant to participate in decision making during design meetings out of fear of being made responsible for the contractor's risk. These are mechanisms that are a consequence of design-build contracts with fixed prices, which are transactional contracts. In order to expand on the effect of the current contractual format, consider a statement made by one of the engineers, who mentioned that, 'From time to time, I am asked to choose solutions that are not the best for the project.' He continued by saying, 'This could, for example, apply to the choice of a low-quality plastic component rather than a metal one.'

Moreover, data indicate that the parties involved sometimes attempted to shove costs on one another. For instance, air ducts (800 mm diameter) did not reach all the way through the energy centre because there was not enough room to do so; however, the general contractor refused to widen this centre. The solution was to remove one wall; however, if that happened, then there would no longer be any way to support the duct. In the end, a steel mount had to be used. This situation raised the question as to who should assume the extra costs for fixing this problem.

Limited involvement on the part of the client in the design and engineering can lead to additional iterations and delay. However, in one instance a representative of the client was observed participating in the discussion of a problem. A need had arisen to move a wall to make room for a shaft containing ventilation ducts; yet, if these were installed, the room would become smaller than specified. The client intervened in the lengthy discussion that took place, stating that, 'It doesn't matter to us. Just do it that way.' This is a minor example of the fact that the project contractual model has a significant impact on the design and engineering process when it comes to the number of iterations and to maintaining or increasing customer value.

A client representative responsible for receiving the delivery of a large number of technical systems to the school was not satisfied with the process, which he described in an interview a few months before the final delivery was made. He felt he was 'kept in the dark' about information required for following up the many stages of the process and doing what was expected of him. Regarding the final delivery, he commented that he could only 'hope for the best'.

Regarding customer value, maintenance and operations were not listed on the design meeting agenda and no information was gathered on the subject apart from at the start of the meeting, when the client was allowed to participate. However, maintenance and operation issues are included in the BREEAM requirements. Koskela et al. (2019) argue that, in Lean Construction, all life cycle stages should be addressed through concurrent engineering. Interestingly, the completed building project has received prizes for its contribution to the green energy shift.

11.4.4 Design and production demands

The general contractor's quality assurance system required that a design must be completed nine weeks prior to construction; however, this was not the case. Occasionally, the design was only completed a few days ahead of production, and not always by then, either. A subcontractor described that, 'The other day, we got a telephone call early in the morning from a manager at the construction site, who said that "We're casting the floor in room X this afternoon. Do you need to do anything there first?"' They did. The fat and oil separators were each to be assembled. The subcontractor continued, 'We managed to do it in the nick of time.' In another example, concrete had to be chiselled out due to delayed designs for stair shafts, where an external supplier of precast concrete units was responsible for the design. In the case of the stair shafts, there was no time to carry out an analysis in BIM, and the technical design engineers had to adjust to what was supplied by the concrete unit provider. Recesses that were previously designed for pipes and ducts needed to be removed from the BIM model when these spaces became occupied by the reinforcement and nib fixtures.

11.4.5 Interdependencies and contingency

Late changes to the construction method created challenges for design. The original design was to build one floor at a time in the traditional style throughout the entire building. This was modified to build section by section, finishing from bottom to top, based on the BREEAM requirement to avoid exposure to moisture during the construction period. Moreover, there was a late change from on-site built exterior walls to prefabricated ones. This meant that decisions regarding matters such as solar shading, U-value windows and access control on exterior doors now needed to be made quickly. The design was further accelerated by the fact that the general contractor would get a low procurement price if the vendor could produce the goods during a period of low demand. Moreover, the late decision to switch from on-site moulded stair shafts to prefabricated ones created many additional iterations for the engineering. Seen from the general contractor's point of view, the change in stair shafts was necessary because of the work schedule for production. Yet, making these big changes created conflicts with the Lean Construction principle of facilitating the removal of constraints surrounding the design work (Ballard, 2000a).

Another example of problematic coordination is the late design freeze of the bearing system. This created additional iterations and waste when, for example, the structural engineer increased the dimension of a pillar without warning, which forced the other technicians to move a sewage conduit and pipes/ducts. Design freeze is a term related to the level of maturity in design and its related learning processes. Maturity is related to learning, which showed up in statements such as 'The project is a maturing process for the owner, the users and all of us.' The interfaces are especially important, a fact confirmed by the BIM coordinator, who stated that, 'It makes little sense to make a report on clashing when I don't know the maturity of the components.'

One engineer highlighted the fact that 'The greatest challenge is to get a hold of information early enough in the process', pointing out that the general contractor created his procurement model too late. It was decided that cross-laminated timber structural floors were to be procured from Austria, steel from Poland and roofing from Germany, while both prefabricated exterior walls of half-timber and walls for stair shafts were to be delivered by Norwegian suppliers. All the abovementioned suppliers carried out their own engineering work and had important interfaces with both the structural group and other groups. The structural engineer could have easily become 'the third wheel on the wagon', according to the respondent. For example, for precast concrete units, the HVAC engineer communicated back and forth with the precast supplier's engineering staff about recesses for pipes and ducts. The structural engineer stood between these two groups and needed to draw the plans, and there were often many iterations with which they had to contend. The informant pointed out that 'The number of sub-consultants increases the complexity.'

The user process provided a constant challenge that extends into the detailed design phase. In this context, the architect makes several changes that rearrange functions, an example of which is the relocation of a dental clinic. This moving process resulted in pipes and ducts having to be rerouted and the bearing system penetrated. The structural engineer had to check that the bearing system was adequate, update the BIM and make new drawings. Another example the informant refers to is the change in location of heavy equipment in a workshop; here, because the locations of electricity units and drains in the ground were affected both directly and indirectly, the foundations were also affected. A third example was the relocation of the kitchen on the fifth floor ('first it's out, and then it's in'), which had consequences for the building's piping system. The informant indicated that, because of these changes, 'I held back awhile with drawing the details.'

One technical engineer felt that the architect often 'didn't understand what we were doing'. While changing a function looks acceptable in a drawing, it can turn engineering work upside down. The HVAC engineer needed to replace the generator when functions were moved. For example, the most energy-efficient generators are those that draw back in some of the exhaust fumes (stale indoor air inside that is replaced by fresh outdoor air), but they cannot be used if this exhaust contains cooking and paint fumes. Relocating the kitchen, therefore, necessitates using a different type of generator, which then affects the energy consumption. This may, in turn, generate a need for more solar panels (related to BREEAM); alternatively, that energy use can be reduced elsewhere.

There are strong dependencies between electrical energy and HVAC. For example, the HVAC engineer told the electrical engineers about everything that needed power, such as ventilation, piping, the HVAC-system and fire extinguishers. These needs were then modelled and drawn up by the electrical engineer. The design-build contract encompasses a wide scope of equipment (for instance, interactive monitors), and their location guides the electrical engineer's placement of points for power and telecommunications. Perhaps unsurprisingly, this is also an area of great interest to clients. However, their decisions about these matters come 'too late, time and again', according to the informant, who pointed to an ongoing situation where 'We need working drawings completed in 3.5 weeks, but we don't have the information we need to do the work.' He then added, 'We often have to start on the electrical design before having the basic data in place so that we can finish on time.'

One informant found it frustrating to start building before the design and engineering elements were fully completed, declaring that 'changes are death'. The reason for this dramatic statement is that there almost always arises the need for change on the upper floors after the basement concrete has been poured, covering piping systems, drains and technical installations. The statement was also made that it was 'hard to pick up changes in time', meaning that it is difficult to have an overview of every consequence that will occur because of changes. The question as to how far ahead engineering work should be in design was commented upon by an HVAC engineer, who asserted that, 'If we are way behind, we get our wings clipped in relation to changes.' There is a kind of competition when it comes to getting enough space for the different equipment, pipes and ducts (some ducts alone being 1000 mm in diameter), which can be a driver for negative iterations (Ballard, 2000b).

11.4.6 Problematic and deficient coordination

A Last Planner phase plan that was intended to serve as a project schedule for the design work was prepared in several rounds through a collaborative planning process. The design manager tried out several alternatives for making up the phase plan, which can be quite complicated to read when working with design, as the network of sequential interdependencies must be understood and interpreted. In addition, reciprocal interdependencies must be dealt with in one way or another during this process.

The observations provide reason to believe that the schedule worked badly as an aid for controlling the design work's progress. Over a long period of time, there was scarcely any link between this main progress schedule for design and the dialogue matrix for short-term planning (two weeks). Indeed, in many design meetings the phase schedule was never mentioned; it was, for all practical purposes, a dead schedule. Nonetheless, it should be mentioned that the preparation itself led to substantive discussions among the disciplines represented at these meetings, which we can relate to learning processes.

Arguably, the main reason that the phase schedule collapsed is that it is a schedule that manages sequential interdependencies and not reciprocal ones, and that planning is not a correct coordination method for reciprocal interdependencies, as referred to in this chapter's theoretical section. According to Thompson (1967), mutual adjustment is the coordinating mechanism for reciprocal interdependencies.

The dialogue matrix was more suitable for managing reciprocal interdependencies because each group reports what they need from the other groups up to the next design meeting, and what they themselves will complete. Several of the groups pointed out that, while they were satisfied with the dialogue matrix method, they lacked information as to what they should prioritise with regard to the site production, and that they had to make these priorities themselves. Some of the engineers missed the traditional memo reports from meetings containing action lists, which had previously been an important part of these reports.

The general design meeting was held every 14 days, after which group meetings were held to solve specific problems involving certain areas, which is an example of mutual adjustment occurring in order to coordinate reciprocal interdependencies. According to observations and interviews, this arrangement seemed to work appropriately.

Figure 11.1 shows the difference between closed and open issues in the web-based project hotel from week 44/2016 to week 20/2017.

Figure 11.1 shows how the gap between the number of issues to be solved and closed expands. A strongly increasing number of open cases indicates multitasking for the designers, and we know that multitasking is not favourable for efficient flow. See the critique put forward by Goldratt (1997) to the critical path method. On the contrary, Figure 11.1 indicates a situation where we move in a direction where there is little control, which looks similar to the chaotic domain in the Cynefin framework (Snowden, 2000) and where action must be taken in order to try and stabilise the situation.

This is precisely what the design manager has tried to do; new rules were created for the issues to be solved, which were then uploaded into the project hotel. Up to that point, both small and large issues had been entered with no organisation whatsoever. The main reason for the somewhat chaotic situation concerning management was, for instance, the weak definitions of issues that were entered. What exactly is being delivered? One example of a response to this question came from an engineer, who, when asked if a deliverable had been completed, replied only that 'This is something that goes on all the time', without attempting to further specify the issue. This practice is in conflict with 'the DNA of the Toyota production system', in which a central rule is to 'highly specify all work' with regard to content, sequence, timing and outcome

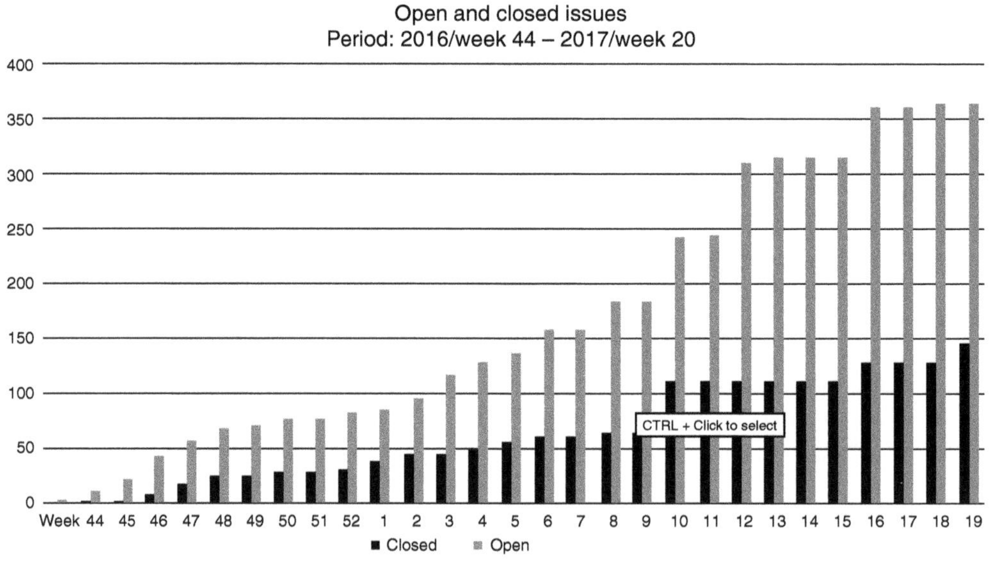

Figure 11.1 Time sequence of open and closed issues in the project hotel

(Spear and Bowen, 1999, p. 98). It appears that some of the designers are not used to planning, resulting in the progress being more characterised by 'organic growth' and 'muddling through', a point that will be further illustrated in the next section.

An engineer stated during his interview that, 'I work a bit here and there and everywhere.' Another engineer vented his frustration, declaring that, 'If any of the (groups) fail to deliver, then nothing happens. The deliverable is just moved up to the next meeting.' Furthermore, he adds that, 'We've talked about something and agreed on this and that, but then all this is often forgotten at the next meeting.' Observations support this point, and even the group meetings between the general contractor and a certain design group could, at times, come to nothing and end without any progress having been made.

After changing the use of the project hotel in order to cover larger issues, its ability to provide overview improved; however, the use of emails sent to individuals not involved with the hotel increased considerably. While the project hotel sent out emails containing updates to 'everybody', several informants pointed out that they needed to get a separate email stating that these updates applied to them. The interviews confirmed that informants felt that there was a problem getting important information distributed through several channels, and that it was 'difficult to find out where decisions were being made'. After having completed several interviews and observations, it was impossible to identify any logging of decisions before this action was introduced at the close of 2017, as a change meant improving the design management. The change was implemented during a period of data collection after asking the interviewees about how decisions had been logged up to that point. A client representative was also concerned with logging decisions made about the design work, as it related to their responsibilities concerning Health and Environmental Safety.

11.4.7 Lessons learnt from the case study

The case study exemplifies a long series of iterations, learning processes, changes, decisions and interdependencies, all of which were challenging to coordinate. Major findings related to coordination and efficient design flow are summarised in Table 11.1 The main challenge was that the applied method for planning and management did not meet the needs for coordinating reciprocal interdependencies.

However, the design team did manage to design the desired object, despite the challenges presented by how the processes were being managed. One element that might explain this situation is that the practice is institutionalised, meaning that there are established roles and ways of interacting. Established ways of interacting have grown over time that comprise a kind of pattern for coordination that prevents chaos (Lawrence, Suddaby and Leca, 2009). Dubois and Gadde (2002) relate the same phenomenon to the notion of Community of Practice (CoP), which develops a 'shared understanding of what is done and how it is done' (p. 626); as such, a strong CoP serves as an informal coordination mechanism. Designers are skilled people who know their tasks and what is expected of them. For example, when they were not given priorities by the general contractor, they made their own priorities based on their own assumptions. Yet, the challenge is that we easily end up with processes characterised by negative iterations, such as 'fire fighting' and 'muddling through', which can cause inadequate design and production challenges, limiting the creation of customer value.

The case study exemplifies that the type of contract has, as expected, a great deal of influence on the design process, even if the requirements for the design as such are independent of contract type. In this particular case, there was early involvement from the design team, being a positive factor for the overall process (Swärd, 2016; Wondimo et al., 2018). However, once the contract had been won, this became transformed into a traditional fixed price design-build contract to be completed.

In this case, the designers in the detailed design phase also worked with several projects simultaneously; they spent up to two hours commuting to and from the construction site. Hence, there were significant costs for the designers related to the design meetings. The experience from the case study is that having design meetings every second week appeared to be too infrequent in the period covered. This point is indicated by Figure 11.1, where peaks in the number of new issues to be solved every two weeks may be identified, while at the same time the number of issues to be closed increases significantly. Respondents confirmed that, although it would have been useful to have more frequent meetings, there would also have been a trade-off related to time and cost. It is quite natural that a higher level of complexity increases the demand for a high level of meeting frequency in order to be able to sort out reciprocal interdependencies.

In the next section, an outline based on case study experiences of a theoretically and empirically informed model (structure) used to organise planning and control in detailed design management is offered. This model has been developed to meet the challenges related to what has been summarised and discussed in connection to the case study.

11.5 The artefact: management model for interdependencies in detailed design

The model is based on one main idea with two complementary elements:

- Divide the overall design task into work packages (modules) that encapsulate the complicated elements. The modules that are characterised by strong interdependencies are designed, through concurrent engineering, by the self-organised teams. The work packages include one or more sprints (in the sense of Scrum). While some work packages/sprints are single-discipline based, most are multidisciplinary. This approach is expected to allow for creativity and joint problem-solving of complicated and complex design issues;
- The sprints are conducted in accordance with a rigid time schedule to ensure progress. The rigid schedule is prepared based on participants' involvement.

A work package is a concept taken from WBS; therefore, sprints are related to work package units, meaning they are derived from Scrum as applied in software development. What is similar, though, is the respective teams' autonomy to make decisions. The rigid time schedule is, however, inspired by the Scrum concept's delivery of functionalities, which is addressed below.

In addition to the Scrum-based concept, the five principles of Last Planner are applied (Ballard, Hammond and Nickerson, 2010). These include: (1) plan in greater detail the closer you come to project completion; (2) plan in cooperation with the individuals who are to perform the work; (3) identify and remove hindrances for planned tasks; (4) work out reliable duties so that the work will be performed as agreed; (5) learn from situations where problems with task completion arise. Furthermore, the model comprises the perspectives found in the TFV-model: transformation logic in design, focus on flow in the design process, in addition to developing customer value.

We can think of the complex element's encapsulation as industrialisation. For example, products are designed in modules/systems with defined content and interface attributes, and where the complexities and interdependencies between individual parts are safeguarded at the factory rather than, for example, at the construction site.

11.5.1 Rigid time schedules

Although a rigid time schedule is not part of the Scrum concept, the researcher finds it necessary to include it in order to address the issue of progress. The delivery of functionalities makes up the basis for milestones in Scrum, but the system has been criticised for being less functional when it comes to strategic time management (Schwaber, Sutherland and Beedle, 2016).

Kalsaas and Kristensen (2018) propose a method for creating integrated milestones based on a pull-oriented delivery logic. This approach regards requirements for design set by production considerations as central, which is closer to the ideas of milestones being based on functionalities delivery, as found in SCRUM.

Based on key building elements, e.g. facade, foundation, bearing system, ventilation, access control, etc., integrated milestones may be prepared for design based on production demands. Further, the idea includes planning the inclusion of sprints and work packages within the milestone structure itself, as is found in the method of Last Planner. Additionally, gradual maturity in design is important, e.g. the Level of Development approach is applied in combination with BIM (Grytting et al., 2017).

11.5.2 Self-organised teams

The project is divided into work packages consisting of sprints, each lasting 1–2 weeks. The sectioning in work packages is based on interdependency and maturity levels. Two types of sprints occur: multidisciplinary and single disciplinary. Self-organised teams manage the sprints, including internal decision-making processes. The discipline that delivers the design output is 'lead'. The teams are self-organised, but they deliver well-defined content within a rigid time schedule (concurrent engineering among the groups). Design backlog (ideas and sprints under preparation) is applied, and items that emerge during the process are entered here. From this perspective, the design manager serves the groups and has a unique responsibility with respect to taking care of customer value, obtaining relevant information from both the client and general contractor and, most importantly, 'wearing the client's hat' when interacting with the groups.

One important principle in the model is to divide the design into modules and work packages. This action must typically be taken in consultation with the design groups in order to provide the participants with the best possible working conditions. One way of proceeding is to have small group meetings with the individual designers before coming together in a plenary session. When experiences are harvested from project to project, it will presumably be easier to divide the 'elephant' according to the first principle of the LPS, 'plan in greater detail the closer you come to project completion' (Ballard, Hammond and Nickerson, 2010). A decomposition of the main elements should, however, be in place before breaking them down in detail the closer the execution date draws near. The product backlog is an important aid in setting the sprints on hold until they become sufficiently mature, thereby confirming the above-mentioned Last Planner principle addressing constraint analysis and providing assurances as to quality.

Reports are supplied by the teams in design meetings on either a weekly basis or every second week. A strict agenda is practised with regard to time in reference to ICE meetings, including the following points:

- Present sprints are presented for approval (decision);
- Update the reversed schedule milestone plan, including sprints/work packages if necessary (only minor adjustments for milestones);
- Update the backlog;
- Discuss interfaces between sprints;
- Measure Percent Plan Completed (PPC) and execution of root cause analysis;
- Set aside time for discussion of what has been learnt from the last round;
- Prioritise new sprints based on the reversed schedule milestone plan; constraint analysis (preparation) is part of this step;
- The teams complete the day by holding their own meetings.

The example in Figure 11.2 is illustrated as a scrum board. The participant in bold print is the sprint lead. The sprint backlog can be expanded to include more maturity levels. When the maturity level 'freeze' has been achieved, any subsequent changes are subject to separate change controls.

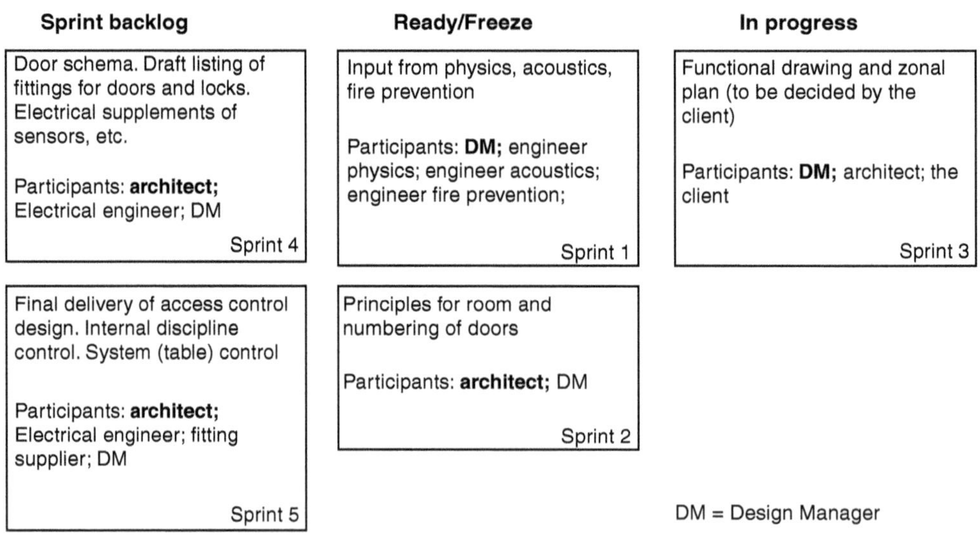

Figure 11.2 Example of a work package (access control doors) and associated sprints, Courtesy of Kai Haakon Kristensen (unpublished material)

Table 11.1 Comparison of identified challenges and attributes related to the constructed model

Identified problems in the case project applied to constructing the model	*Attributes in the constructed model used for solving problems in detailed design management*
Late changes to building method, which caused several iterations in design and increased complexity.	Building method freeze is initially identified as a milestone.
Late contracting of suppliers of building parts where the supplier is responsible for design, which creates uncertainty in related disciplines.	Milestones for latest possible contracting.
Examples of integration collapse between production and design, which led to 'putting out fires' and extremely short-time buffers between design and production.	Pull-based milestones based on the needs of drawing production.
Limited interaction between the design group and client in design meetings due to differing commercial interests in the fixed-price design-build contract.	The client's experts should participate in design meetings in order to immediately remove uncertainty when possible.
No systematic attempt to manage gradual design maturity.	Milestones are identified for maturity at various levels as well as degrees of maturity of sprints/work packages.
No clear strategy for change management.	Levels of maturity are included in the created model, as in the example in Figure 11.2. At the maturity level, 'freeze' is implemented in a strict change control, where no single discipline is unilaterally allowed to make changes to the design.
Failed attempt to use Last Planner's phase plan to coordinate reciprocal interdependencies. To find a balanced level of detail in the phase plan is important. A too detailed level can be difficult to handle is the experience from the case study.	Last Planner is used to develop milestones and insert work packages and sprints. Reciprocal interdependencies are managed in the model by self-organised teams when sprints are executed. This is concurrent engineering.
Design planning was, to a high degree, attempted to be conducted on the general contractor's terms and, to a lesser degree, on the MEP contractor's terms.	The basis of the proposed model is that planning is performed by participants who are equal. However, the type of contract may influence the ease or difficulty of this being achieved.
Work tasks with corresponding deliverables were often imprecisely specified.	It is essential to specify deliverables so that it is possible to control that they have been executed. See Figure 11.2.
Lack of consistent structure for communicating information in the design team generated contingency for the participants (web-based project hotel, email, etc.).	It is commonplace for all planning and control to have standardised communication channels.

The teams can continuously handle sprints in a suitable project hotel. The idea relating to the project hotel can, in turn, be related to Thompson's (1967) definition of reciprocal interdependency, where it is the status of the object that determines the next action to be taken. The hotel can host sprints defined as objects; agreed actions regarding an object may be seen by the group members and assist dispersed groups in making decisions about what can be done in the next step in order to develop the object for it to move toward the predefined status. It requires, however, that all communication outside the meeting takes place in the project hotel channel and is updated in accordance with the decisions made in meetings.

11.5.3 Problem relevance

Table 11.1 identifies challenges from the case study with comments regarding management presented in the constructed model for design planning and control in the detailed design phase.

The constructed model has been tested for problem relevance in discussion with the designers and design manager that have been interviewed, some of whom thought that they were already working in this manner through holding group meetings following plenary project design meetings in addition to more ad hoc meetings for mutual adjustment and problem solving. Interestingly, HVAC and electrical engineers, along with subcontractors for pipes, ventilation/automation and electrical, took the initiative to hold separate meetings for problem solving.

However, they were all keen to obtain improved management structure for design processes and, as such, were interested in learning more about the model. The background for this interest may be found in the identified challenges in the studied case, including lack of decision making and input regarding priorities held by the general contractor, inaccurately identifying design deliveries, design planning based more on the general contractor's need and less on the subcontractor's need, lack of consistency between short- and long-term design planning, and vague links between production and design, etc. (see Table 11.1 for more details). A full-scale evaluation of the proposed model will be a topic of future research.

11.6 Conclusion

This chapter aimed at contributing to an appropriate structure for the management and control of complex design and engineering processes. The context has been set to the detailed design phase in design-build contracts, where the general contractor owns the design process and has a back-to-back contract with the MEP subcontractors.

The model has been theoretically constructed and has recorded challenges encountered in a case study. The case study findings confirm the theoretical assumption that the main cause of failure in managing design processes is that design as a phenomenon often relates strongly to interdependencies, iterations and design as complex and wicked problems.

The proposed organisation is based on two complementary ideas: (1) encapsulation of the complex and reciprocal aspects of work packages and sprints, where each sprint is controlled by self-organised teams; (2) progress of the specified deliverables from the self-organised teams is delivered according to a rigid time schedule.

The objective is to simplify the management of the design process by improving its organisation. The self-organised groups are composed of individuals who are the most integrated in order to come up with solutions to reciprocal design problems. While the structure should ensure the handling of degrees of maturity and leave room for creativity, it should do so within an accepted timeframe.

Experiences show that, even in LDM, work situations quickly become rather chaotic with a low degree of control. Interestingly, the design process in this case may be described as being one characterised by 'muddling through' and 'organic growth', and management seems to lack an understanding of how important design is for value creation, whether it is for buildability or customer value.

The case analysis indicates that the applied design theory is beneficial for the study of the design process. Therefore, in order to operationalise the design theory of design management, the most central elements are to understand design as a wicked problem where iterations (Scrum) and decomposition are central to 'taming' the design. Moreover, the five principles of Last Planner have been applied to the proposed model for design management. However, Last Planner as a stand-alone concept for design management has been confirmed by the case study to be insufficient for managing iterations and reciprocal interdependencies.

Note

1 The case study has been conducted as part of a four-year research project entitled, *Integrated methods for design management* (2013–2017) and funded by the Norwegian Research Council.

References

Abbas, N., Gravell, A.M. and Wills, G.B. (2008) 'Historical roots of Agile methods: Where did 'Agile thinking' come from?' *International Conference on Agile Processes and Extreme Programming in Software Engineering*, Springer, Berlin, Heidelberg, pp. 94–103.

Ballard, G. (2000a) *The Last Planner System of Production Control*, PhD Thesis, Birmingham, UK: The University of Birmingham.

Ballard, G. (2000b) 'Positive vs negative iteration in design', *Proc. 8th Annual Conference of the International Group for Lean Construction*, Brighton, UK, pp. 17–19.

Ballard, G., Hammond, J. and Nickerson, R. (2010) 'Production control principles', *Proc. 17th Annual Conference of the International Group for Lean Construction*, Haifa, Israel, pp. 489–499.

Ballard, G. and Koskela, L. (2013) 'Rhetoric and design', *International Conference on Engineering Design*, ICED 13, Seul, Korea.

Churchman, C.W. (1967) 'Wicked problems', *Management Science*, 14(4), Guest Editorial, pp. B141–B142.

Conklin, J. (2006) *Dialogue Mapping: Building Shared Understanding of Wicked Problems*, Chichester, UK: Wiley Publishing.

Cuff, D. (1991) *Architecture: The Story of Practice*, Cambridge, MA: The MIT Press, pp. 1–16, 247–263.

Demir, S.T. and Theis, P. (2016) 'Agile design management – The application of scrum in the design phase of construction projects', *Proc. 24th Annual Conference of the International Group for Lean Construction*, Boston, MA, pp. 13–22.

Dubois, A. and Gadde, L-E. (2002) 'The construction industry as a loosely coupled system: Implications for productivity and innovation', *Construction Management and Economics*, 20(7), pp. 621–631.

Eastman, C.M. ed. (2012) *Design for X: Concurrent Engineering Imperatives*, New York, USA: Springer Science & Business Media.

Evans, R., Harryott, R., Haste, N. and Jones, A. (1998) *The Long-Term Costs of Owning and Using Buildings*, London, UK: Royal Academy of Engineering.

Fisher, M., Ashcraft, H., Reed, D. and Khanzode, A. (2017) *Integrating Project Delivery*, Hoboken, NJ: John Wiley & Sons.

Fosse, R. and Ballard, G. (2016) 'Lean design management in practice with the last planner system', *Proc. 24th Annual Conference of the International Group for Lean Construction*, Boston, MA, pp. 33–42.

Fosse, R., Ballard, G. and Fischer, M. (2017) 'Virtual design and construction: Aligning BIM and lean in practice', *Proc. 25th Annual Conference of the International Group for Lean Construction*, Heraklion, Greece, pp. 499–506.

Gilbertson, A.L. (2006) 'Briefing: Measuring the value of design', *Proceedings of the Institution of Civil Engineers - Municipal Engineer*, 159(3), pp. 125–128.

Goldratt, E.M. (1997) *Critical Chain*, Great Barrington, MA: North River Press.

Grytting, I., Svalestuen, F., Lohne, J., Sommerseth, H., Augdal, S. and Lædre, O. (2017) 'Use of LoD decision plan in BIM-projects', *Procedia Engineering*, 196, pp. 407–414.

Hamzeh, F.R., Ballard, G. and Tommelein, I.D. (2009) 'Is the last planner system applicable to design? A case study', *Proc. 17th Annual Conference of the International Group for Lean Construction*. Taipei, Taiwan, pp. 165–176.

Hevner, A.R., March, S.T., Park, J. and Ram, S. (2004) 'Design science in information systems research', *MIS Quartely*, Management Information Systems Research Center, University of Minnesota, 28(1), pp. 75–107.

Jo, H.H., Parsaei, H.R. and Sullivan, W.G. (1993) 'Principles of concurrent engineering', in Parsaei, H.R. and Sullivan, W.G. (eds), *Concurrent Engineering*, Boston, MA: Springer, pp. 3–23.

Kalay, Y.E. (2004) *Architecture's New Media: Principles, Theories, and Methods of Computer Aided Design*, Cambridge, UK: MIT Press.

Kalsaas, B.T., Bonnier, K.E. and Ose, A.O. (2016) 'Towards a model for planning and controlling ETO design projects', *Proc. 24th Annual Conference of the International Group for Lean Construction*, Boston, MA.

Kalsaas, B.T. and Kristensen, K.H. (2018) 'New approach to developing integrated milestones for planning and production control', *Proc. 26th Annual Conference of the International Group for Lean Construction*, Chennai, India, pp. 869–878.

Kalsaas, B.T. and Moum, A. (2016) 'Design and engineering understood as processes of learning', *Proc. CIB World Building Congress 2016*, Tampere University of Technology, pp. 210–221.

Kaufer, D.S. and Butler, B.S. (1996) *Rhetoric and the Arts of Design*, Mahwah, NJ: Lawrence Erlbaum.

Koskela, L. (2000) *An Exploration Towards a Production Theory and its Application to Construction*, Espoo, Finland: VTT Technical Research Centre of Finland. VTT Publication 408. Available at: www.vtt.fi/inf/pdf/publications/2000/P408.pdf.

Koskela, L., Ballard, G. and Tanhuanpää, V. (1997) 'Towards lean design management', *Proc. 5th Annual Conference of the International Group for Lean Construction*, Gold Coast, Australia, pp. 1–13.

Koskela, L., Bølviken, T. and Rooke, J. (2013) 'Which are the wastes of construction?' *Proc. 21st Annual Conference of the International Group for Lean Construction*, Fortaleza, Brazil, pp. 3–12.

Koskela, L. and Howell, G. (2002) 'The theory of project management: Explanation to novel methods', *Proc. 10th Annual Conference of the International Group for Lean Construction*, Gramado, Brazil, pp. 1–11.

Koskela, L. and Kagioglou, M. (2006) 'The proto-theory of design: The method of analysis of the ancient geometers', *Proc. 9th Design 2006 – International Design Conference*, Dubrovnik, Croatia, pp. 53–60.

Koskela, L.J., Ferrantelli, A., Niiranen, J., Pikas, E. and Dave, B. (2019) 'Epestemological explanation of lean construction', *Journal of Construction Engineering and Management*, 145(2), 04018131.

Kunz, J. and Fisher, M. (2012) *Virtual Design Cases & Studies*, Working Paper 097, Stanford, CA, USA: Center for Integrated Facility Engineering.

Lawrence, T.B., Suddaby, R. and Leca, B. (2009) *Institutional Work: Actors and Agency in Institutional Studies of Organizations*, New York: Cambridge University Press.

Lawson, B. (2006) *How Designers Think - The Design Process Demystified*, 4th ed., Oxford, UK: Architectural Press.

Lu, S.Y., Elmaraghy, S.G. and Wilhelm, R. (2007) 'A scientific foundation of collaborative engineering', *CIRP Annals*, *56*(2), pp. 605–634.

Lundequist, J. (1992) *Projekteringsmetodikens Teoretiska Bakgrund*, Stockholm, UK: KTH Reprocentral.

March, S.T. and Smith, G.F. (1995) 'Design and natural science research on information technology', *Decision Support Systems*, 15(4), pp. 251–266.

Moum A. (2009) 'The role of BIM in the architectural design process - learning from practitioners' stories', in Underwood, J. & Isikdag, U., eds. *Handbook of Research on Building Information Modelling and*

Construction Informatics: Concepts and Technologies, Hershey, PA: Information Science Reference, IGI Global, pp. 587–618.

Munthe-Kaas, T.S., Hjelmbrekke, H., Lohne, J. and Lædre, O. (2015) 'Lean design versus traditional design approach', *Proc. 23rd Annual Conference of the International Group for Lean Construction*, Perth, Australia, pp. 578–588.

Pikas, E., Koskela, L. and Seppänen, O. (2017) 'Design management in a design office: Development of the model for 'to-be'", *Proc. 25th Annual Conference of the International Group for Lean Construction*, Heraklion, Greece, pp. 555–562.

Pikas, E., Koskela, L., Tredal, N., Knotten, V. and Bølviken, T. (2018) 'The dual nature of design management', *Proc. 26th Annual Conference of the International Group for Lean Construction*, Chennai, India, pp. 647–657.

Reinertsen, D.G. (1997) *Managing the Design Factory: A Product Developer's Toolkit*, New York: Free Press.

Rittel, Horst W.J. (1988) *The Reasoning of Designers*, Working Paper A-88-4, Stuttgart, Germany: Institute fur Grundlagen der Planung.

Schön, D.A. (1991) *The Reflective Practitioner - How Professionals Think in Action*, Aldershot, UK: Ashgate.

Schwaber, K. and Beedle, M. (2002) *Agile Software Development with Scrum*, Upper Saddle River, NJ: Prentice Hall.

Schwaber, K., Sutherland, J. and Beedle, M. (2016) The scrum guide TM. The Definitive Guide to Scrum: The Rules of The Game.

Shewhart. (1931) *Economic Control of Quality of Manufactured Product*, New York: D. Van Nostrand Company.

Skaburskis, A. (2008) 'The origin of 'Wicked problems'", *Planning Theory & Practice*, 9(2), pp. 277–280.

Snowden, D.J. (2000) 'Cynefin: A sense of time and space, the social ecology of knowledge management', in Despres, C. and Chauvel, D., (eds.) *Knowledge Horizons: The Present and the Promise of Knowledge Management*, London and New York: Routledge, pp. 237–266.

Spear, S. and Bowen, H.K. (1999) 'Decoding the DNA of the toyota production system', *Harward Business Review*, 77(5), pp. 95–106.

Statsbygg and KHiB team (2017) Lean strategy a brief overview. KHiB - booklet no 2, Bergen Academy of Art & Design, Revised 05/ 12-2017, Oslo, Norway.

Stolterman, E. (2008) 'The nature of design practice and implications for interaction design research', *International Journal of Design*, 2(1), pp. 55–65.

Swärd, A. (2016) 'Trust, reciprocity, and actions: The development of trust in temporary inter-organizational relations', *Organization Studies*, 37(12), pp. 1841–1860.

Thompson, J.D. (1967) *Organizations in Action. Social Science Bases of Administrative Theory* 2003 ed., New York: McGraw-Hill.

Uusitalo, P., Olivieri, H., Seppänen, O., Pikas, E. and Peltokorpi, A. (2017) 'Review of lean design management: Processes, methods and technologies', *Proc. 25th Annual Conference of the International Group for Lean Construction*, Heraklion, Greece, pp. 571–578.

Wondimo, P.A., Hosseini, A., Lohne, J. and Laedre, O. (2018) 'Early contractor involvement approaches in public project procurement', *Journal of Public Procurement*, 18(4), pp. 355–378.

12 Lean as an appropriate approach for managing production in refurbishment projects

Sergio Kemmer and Lauri Koskela

12.1 Introduction

Refurbishment[1] is a key topic in the current construction research agenda in the UK. This is due to the vital role that refurbishments play in meeting the sustainable targets set by the government (DECC, 2008; Itard et al., 2008; Ravetz, 2008) and also due to the significant part, more than a third, that those type of projects represent in the total output of the construction industry (ONS, 2018).

Refurbishments present different features in comparison to new build projects (CII, 2009; CIRIA, 1994; Egbu, 1994; Krizek, Lo, and Hadavi, 1996; Rahmat, 1997; Sanvido and Riggs, 1991; Whiteman and Irwig, 1988). Such particular features are likely to produce management issues that are unique to the refurbishment project environment. The existing asset is the most distinguishable characteristic. Another distinct feature concerns the possibility of carrying out a project in a building that remains in operation (Construction Industry Research and Information Association, 1994).

Managerial challenges typical of refurbishments include coping with a large number of constraints such as space limitations for construction and material storage, noise and dust control, non-disruption to existing utilities, user and public safety, among others[2] (Singh, 2007). Additional levels of uncertainty and variability are generated by the lack of as-built drawings (Construction Industry Research and Information Association, 1994; Quah, 1992; Sanvido and Riggs, 1991) as well as the unforeseen site conditions (Howard et al., 2009; Quah, 1992; Singh, 2007). Furthermore, there is a need to deal with projects involving a high degree of interdependence between elements related to refurbishments such as the existing operations in the building, tenants, neighbours, maintenance and construction personnel (Construction Industry Research and Information Association, 1994; Howard et al., 2009; Krizek, Lo, and Hadavi, 1996; Sanvido and Riggs, 1991). Moreover, there are concerns related to health and safety due to dangerous work, especially when it involves demolition and structural instability (Anumba et al., 2004; Construction Industry Research and Information Association, 1994; Egbu, 1995) and hazardous material (e.g. asbestos) that might be encountered in the existing building (Construction Industry Research and Information Association, 1994; Egbu, 1997; Sanvido and Riggs, 1991).

A consequence of many of the discussed characteristics is that the complexity in refurbishments is often higher than what is found in new build projects. In this respect, Noori et al. (2016) list factors such as unforeseen site conditions, site access, lack of space, availability of material, design change, defective design and plan, and lack of information during the design phase. The increased complexity of refurbishment in comparison to new build projects produces further management challenges. The lack of understanding of the complex nature of refurbishment projects can lead to bad managerial decisions and ultimately to poor project performance.

There is a consensus in the literature that production control systems should match the context in which they are operating (Henrich, Tilley, and Koskela, 2005; Melles and Wamelink, 1993; Yamin and Harmelink, 2001). According to Sanvido and Riggs (1991), appropriate management procedures should be adopted to cope with the intrinsic features of refurbishments in order to enable the achievement of project objectives. CIRIA (1994) contends that such specific characteristics influence the selection and control of all kinds of resources, including managerial means. Egbu, Young, and Torrance (1996) stress the importance of adopting adequate managerial practices so that improved efficiency and effectiveness of refurbishment operations can be achieved. Following the same line of thought, Egbu, Young and Torrance (1998, p. 319) argue that 'refurbishment works demand an ability to deal with non-continuous and complex processes' and also emphasise that 'the process of planning should be adjusted to fit project characteristics'.

Despite the recommendations to match the managerial approach to the project context, several studies have indicated that construction organisations have predominantly used a traditional approach (e.g. Critical Path Method) for managing refurbishment projects (Construction Industry Research and Information Association, 1994; Egbu, Young, and Torrance, 1998; Henrich, 2009; Krizek, Lo, and Hadavi, 1996; McKim, Hegazy, and Attalla, 2000; Rahmat, 1997; Singh, 2007). This refers to production planning and control processes, as well as managerial practices and techniques usually used for managing refurbishment construction projects. The problem is that the use of such a management approach is often not appropriate for dealing with the dynamic nature of complex production settings (Ballard and Howell, 2004; Koskela et al., 2002; Laufer, Denker, and Shenhar, 1996). As a result, problems in managing refurbishments have been reported in several countries (Kemmer and Koskela, 2012).

Indeed, there is evidence in the literature showing that the performance of refurbishment projects is not satisfactory. According to Naaranoja and Uden (2007), Finnish renovation projects presented poor project performance in terms of cost, time, and also failures concerning the fulfilment of customer requirements. Krizek, Lo, and Hadavi (1996) conducted an investigation in a multiphase reconstruction project in the United States and pointed out difficulties in communication between project members, as well as problems to manage production leading to schedule overruns. Rahmat (1997) examined 67 refurbishment projects and found that 53.3% exceeded the target cost and 52.5% exceeded the target time. McKim, Hegazy, and Attalla (2000) carried out a survey in Canada and reported a poor performance of reconstruction projects in terms of cost, schedule, and quality, in comparison with new building projects. In all studies mentioned, except the first, tools such as bar charts and CPM were reported as the main techniques adopted. In conclusion, as noted by Henrich (2009, p. 98), 'the construction industry has been using outdated production methods or using them in a wrong context'.

The aim of the chapter is to show why Lean can be considered as an appropriate approach for managing production in refurbishment projects. To that end, the chapter is organised in two parts. First, light is shed on the construction management approach commonly used in refurbishment projects and on the reasons why it fails, by identifying the root causes of such failure. The second part introduces the theoretical background upon which the production management approach in refurbishment projects should be based. Several parts of this chapter were expanded from what was presented in Kemmer and Koskela (2012) and Kemmer, Koskela, and Nykänen (2013).

12.2 Typical approaches to construction management in refurbishments

Sanvido and Riggs (1991) identified critical factors for the success of refurbishment projects such as the need to have in place the right project team,[3] deemed by those authors as the most important factor, and also contract incentives, partnering arrangements, scope management, plant knowledge, communication, special procurement and preplanning strategies, and high level management support. In terms of effective management techniques used in refurbishments, they found limited information, but preplanning (in strategic and operational terms) was highlighted as an important technique.

Egbu (1994) acquired an overview of particular issues and characteristics related to the management of refurbishment works within the UK construction industry. For instance, variation and change orders to the works, keeping the site tidy, cost control, maintaining site safety and welfare standards, and programming and scheduling were identified as the most frequently occurring characteristics in managing refurbishment work. In turn, cost control, dust control, the influence of tenants on the regular progress on site, pricing of the works, and variation/change orders to the works are identified as the most challenging refurbishment characteristics faced by managers. Furthermore, Egbu (1995) investigated the degree of difficulty associated with managing refurbishment tasks. Forecast and planning, analysis of project risks and uncertainty, and competitive tendering, were perceived as the most difficult management tasks in refurbishment projects. Table 12.1 presents the top ten ranking.

Egbu, Young, and Torrance (1996, 1998) identified the planning and control techniques used by refurbishment organisations. Data was collected through a case study carried out in a hospital and hotel refurbishments in England (Table 12.2).

Table 12.1 Management tasks perceived as most difficult in managing refurbishment works (adapted from Egbu, 1995)

Rank	*Job dimensions (tasks)*
1	Forecast and planning
2	Analysis of project risk and uncertainty
3	Competitive tendering
4	Budgetary control
5	Managing time
6	Quality control and assurance
7	Health and safety
8	Site security
9	Use of computer technology
10	Managing conflict/crisis

Table 12.2 Planning and control techniques used in refurbishment projects (adapted from Egbu, Young, and Torrance, 1996, 1998)

Rank	*Hospital refurbishment*	*Hotel refurbishment*
1	Schedules	Bar chart
2	Critical Path Method (CPM)	Project cost-value reconciliation
3	Project cost–value reconciliation	Labour (actual and forecast) reconciliation
4	Bar chart	Material (actual versus forecast) reconciliation
5	Milestone date programming technique	Schedules
6	Labour (actual versus forecast) reconciliation	Plant (actual versus forecast) reconciliation
7	Plant (actual versus forecast) reconciliation	Critical Path Method (CPM)

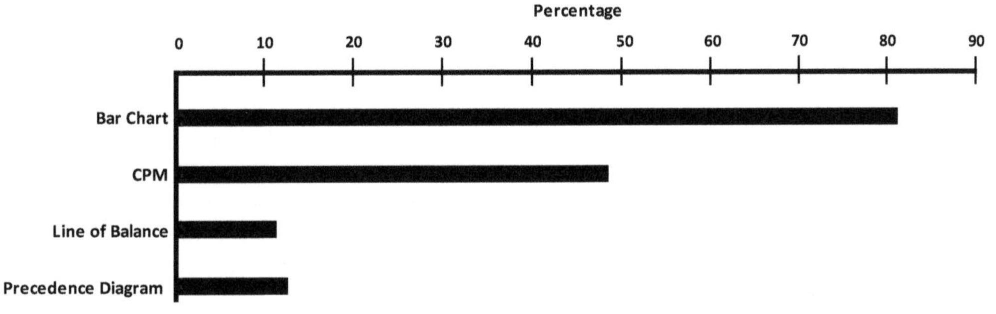

Figure 12.1 The planning techniques commonly used by refurbishment contractors (adapted from Rahmat, 1997)

Moreover, a survey conducted by Rahmat (1997) among 103 construction firms in the United Kingdom provided additional evidence on the type of planning techniques most used in refurbishments (Figure 12.1).

Table 12.2 and Figure 12.1 show clearly the predominance of traditional planning techniques such as bar charts and the Critical Path Method (CPM). The use of such techniques for managing construction in refurbishment projects is also mentioned in the guide to the management of building refurbishment produced by CIRIA (1994), as well as in the studies developed by Krizek, Lo, and Hadavi (1996) and McKim, Hegazy, and Attalla (2000).

Singh (2007) conducted interviews with contractors and subcontractors involved in renovation projects and learnt that the production planning process is carried out on an informal basis. No formal documentation or standardised process was identified for dealing with the constraints intrinsic to such projects (e.g. poor as-built drawings, unforeseen conditions, presence of hazardous materials, etc.), regardless of the influence that those project conditions might have on project performance (Krizek, Lo, and Hadavi, 1996; McKim, Hegazy, and Attalla, 2000; Singh, 2007).

Some could contend that the aforementioned managerial practices are no longer in use, implying that the management of refurbishment works has evolved. However, the study carried out by Henrich (2009) indicates that it has not. His findings from two case studies conducted on refurbishment sites confirm that construction companies are still using traditional planning practices for managing production; for example, planning is centralised and contains excessive details at early stages, there is a lack of involvement of stakeholders (managers, subcontractors, suppliers, labourers, etc.) in the creation of plans, and CPM-based software packages are used to define the critical path and estimate project duration. Moreover, managers use pre-estimated plans to push activities to the production regardless of the system status, and the communication between planning and production is performed in a one-way fashion. Furthermore, there was no evidence of a continuous improvement programme implemented on site.

As a result of the use of those inadequate managerial practices, Henrich (2009) identified several types of waste within those refurbishment projects, such as: waiting time, the use of wrong equipment, rework, unnecessary transport, double handling, space conflicts between materials, equipment and assembly crews, and disruptions during refurbishment works. It is argued that such wastes are likely to result in low productivity, project delays, cost overruns, and tenants' annoyance.

The findings from the study conducted by Henrich (2009) provide a comprehensive description of the traditional construction project management approach used in refurbishment projects. Clearly, there is an emphasis on managing tasks on refurbishment sites by applying scheduling, particularly by implementing CPM. However, the findings imply that the use of conventional management methods leads to wastes and ultimately poor project performance. This insight poses the following question: why does the traditional construction management approach fail? A close look at the literature on construction management reveals that the root cause of such a failure lies deeper than in an inappropriate selection of managerial practices.

12.2.1 Why does the traditional construction management approach fail?

The use of the CPM for production planning and control in construction projects has been criticised for a long time. Peer (1974) cites its limitations for site management by arguing that plans are quickly out of use, even prior to the start of works on site. He also contends that the lack of adherence to reality demands permanent and time-consuming adjustments of the plan to real situations. Birrell (1980) argues that the CPM does not produce satisfactory results because it is not compatible with the essence of the construction process. The following observation clarifies such incompatibility (Birrell, 1980, p. 403):

> When such an approach is used to plan a process (e.g. the construction process) that is dynamic and contains many individual resources which are controlled by a diverse group of human participants, the 'plan it and pass it on' attitude is insufficient for successful execution of the process.

Laufer and Tucker (1987) also point out also several weaknesses of the CPM in planning construction projects such as the inability to ensure the continuity for gangs on-site and to deal with non-sequential activities, which are typical in construction sites (i.e. installation type of work), the difficulty to apply the CPM models in practice, and also the failure to recognise the uncertainty and variability inherent to construction.

It seems that the traditional planning approach is not appropriate for managing complex projects such as refurbishments. However, despite the several shortcomings aforementioned, there is a fundamental issue that has not been referred to as yet, namely theory. Koskela (2000) argues that traditional construction project management fails because it is not based on a sound theoretical foundation. He contends that this approach is essentially based on a reductionist view that is incapable of coping with complex production settings such as the ones found in the construction sector. He concludes that construction project management needs a suitable theory of production capable of dealing with the intrinsic construction peculiarities.

Koskela and Howell (2002) also emphasise the importance of a sound theory for enabling a better project management. They address this topic in two parts: theory of project (which refers to production-oriented processes) and theory of management (which refers to project management processes). By analysing the principles of traditional project management and comparing them to the theories of operations management in general, they conclude that the theory of project rests on the transformation theory (or view) of production. According to this view, production is conceptualised as a transformation of inputs to outputs (Figure 12.2).

This conceptual model is underpinned by the following principles (Koskela, 2000): the total transformation can be decomposed into smaller transformations, the cost of production can be minimised by minimising the cost of each decomposed transformation, the need of physical or organisational buffering to shield production from the external environment, and the value of the output of a process is associated with the value (or costs) of inputs to that process.

While the transformation view is helpful for specific purposes (e.g. determining which tasks are needed in a project), it fails to recognise the existence of other activities (Figure 12.3) in production (non-value-adding activities such as moving, waiting, and inspection) as well as acknowledging that value is created in production by having the output to conform to the customer's requirements and not by just transforming input into outputs (Koskela, 2000).

Figure 12.2 Transformation model (adapted from Koskela, 2000)

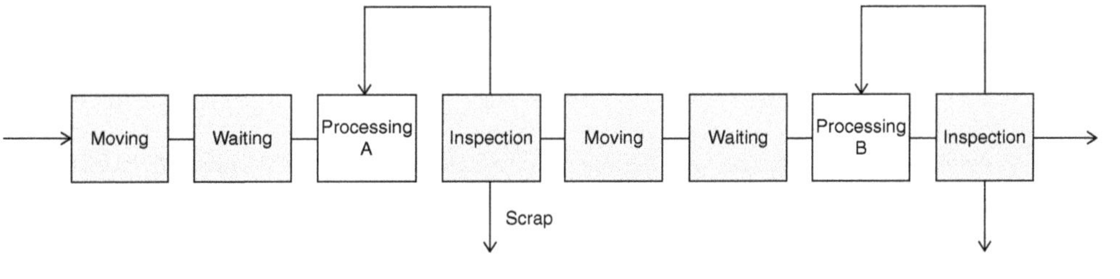

Figure 12.3 Production as a flow process (adapted from Koskela, 2000)

Regarding the theory of management, Koskela and Howell (2002) conducted an analysis of conventional practices based on the core project management processes, namely planning, execution, and controlling. In terms of planning, they found that the prevailing approach is based on a management style called management-as-planning (Johnston and Brennan, 1996). The underlying assumption is that projects can be successfully managed through the preparation and implementation of plans. While effective for managing simpler projects, this approach is not capable of dealing with projects subjected to high levels of uncertainty (Laufer, Denker, and Shenhar, 1996).

Regarding execution, Koskela and Howell (2002) conclude that traditional practices follow the concept of dispatching in manufacturing (Emerson, 1917). The dispatching model is basically concerned with assigning tasks or jobs to machines or work crews, which is done usually by a central authority. The main problem with this model is that it uses pre-made plans to push tasks to execution with no consideration to the production system status. In addition, there is only one-way communication between management and production.

In terms of control, Koskela and Howell (2002) contend that classical project controls correspond to the thermostat model, namely action is taken only when a variance against the standard performance is identified. They stressed that such a model does not focus on finding the reasons for deviations. As a result, the root causes of problems are not eliminated.

Ballard and Howell (2004) argue that the main problem with the conventional construction project management (i.e. detailed CPM schedules, thermostat model, competitive bidding) is that it 'assumes that variability in workflow is outside management control and so does not attempt to systematically reduce variability' (Ballard and Howell, 2004, p. 40).

In summary, the traditional construction management approach fails because it neglects the context in which the refurbishment projects take place and because it is not based on an appropriate theory. The importance of acknowledging refurbishments as complex projects, as well as having a sound theory of production as a reference basis for managing and improving construction management, is addressed in the following.

12.3 Towards an appropriate construction management approach in refurbishment projects

12.3.1 Refurbishments as complex projects

Williams (2002) contends that there are two dimensions of complexity in project management, namely structural complexity and uncertainty (Figure 12.4). The former concerns to the number of elements and the degree of interrelatedness between these elements, while the latter can be understood in terms of the goals and methods of the project, which can be uncertain. He argues that both concepts (i.e. structural complexity and uncertainty) together produce 'the difficultness and messiness of the overall project' (Williams, 1999, p. 271).

Interdependence and uncertainty have been recognised as key features in construction for a long time (Crichton, 1966). Gidado (1996) argues that the construction process is characterised by a large number of interconnected parts. He also considers uncertainty as a component of project complexity that can be related to the following aspects (Gidado, 1996, p. 216):

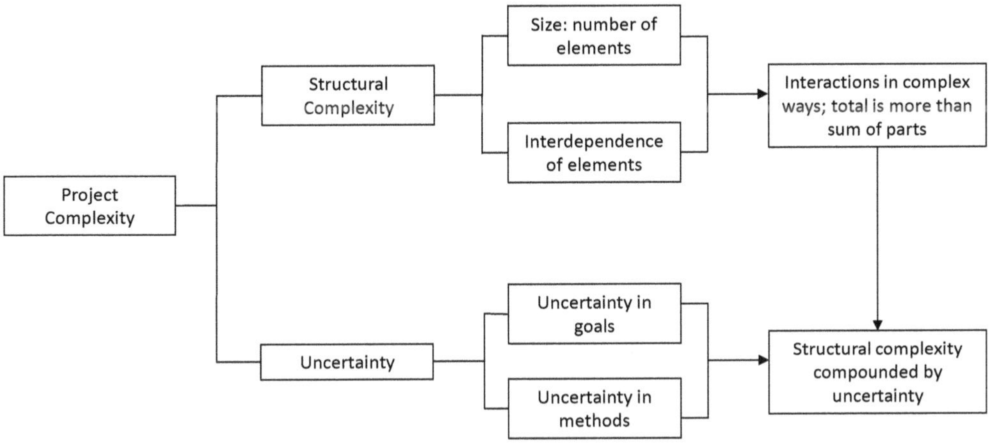

Figure 12.4 Dimensions of project complexity (adapted from Williams, 2002)

- Lack of complete specification for the activities to be executed;
- Unfamiliarity of the inputs and/or environment by management;
- Lack of uniformity of work;
- Unpredictability of the environment (e.g. refurbishment of very old buildings having no as-built drawings).

Interestingly, the definitions of complexity encountered in the literature are similar to the perceptions of people in industry. Gidado (1996) carried out interviews with construction practitioners and found that they view a complex project as follows:

- A project having a large number of different systems that need to be put together and/or with a significant number of interfaces between elements;
- A project involving works in a confined site with access difficulty and requiring many trades to work in close proximity and at the same time;
- A project with a considerable degree of intricacy which is difficult to specify clearly how to achieve a desired goal or how long it would take;
- A project that demands a large number of details on how it should be executed; and
- A project that requires effective coordination, control, and monitoring from start to finish.

According to Koskela (2000), the peculiarities of the construction sector such as one-of -a-kind nature of projects, site production, and temporary organisation, contribute to added complexity and uncertainty to production. Also, in complex projects, customer requirements are not clearly defined at the outset, therefore making use of techniques that push for early decisions and local optimisation is to misunderstand the project's nature (Koskela et al., 2002). Bertelsen (2003a, 2003b) and Bertelsen and Emmitt (2005) contend that complexity in construction projects stems from several sources such as the design process, the project organisational structure (i.e. several stakeholders with different objectives each one), the operations on site (i.e. parallelism), the client's often irrational behaviour, the weather, etc. Those characteristics show that construction projects are generally complex by nature.

In refurbishment projects, both dimensions of complexity cited by Williams (2002) are encountered. These can be better understood by examining the typical features of refurbishments (Table 12.3).

It is worth mentioning that the complexity found in refurbishment projects might vary depending on the project type. For example, occupied hospital refurbishments tend to be more complex to manage than a renovation of an unoccupied house as they involve a larger number of stakeholders (e.g. subcontractors, patients, hospital staff members, etc.), as well as particular characteristics that have to be considered for the development of planning and control systems such as patient safety, staff flow, adherence to regulatory agency guidelines, etc. (Worley and Hohler, 2008).

The acknowledgment of refurbishments as complex projects has important implications from a project management standpoint. Baccarini (1996, p. 201) emphasises this by outlining the following reasons:

Table 12.3 Refurbishment features that influence complexity in project management

Dimension of complexity (Williams, 2002)	*Refurbishment features*	*References*
Structural Complexity	Refurbishment projects usually involve a large number of subcontractors	Rahmat (1997)
	Refurbishments involve a large number of constraints such as space limitations (e.g. storage, new equipment, construction), restricted access, pollution control, etc.	Whiteman and Irwig (1988); CIRIA (1994), Rahmat (1997) Krizek, Lo, and Hadavi (1996), Singh (2007)
	Refurbishments might happen in an occupied building, hence temporary structures may be necessary to safeguard users and operations in the existing building	CIRIA (1994), Krizek, Lo, and Hadavi (1996), Rahmat (1997)
	Safety issues due to the interface of workers with existing operations in the building and involvement of building users who may include the public	CIRIA (1994); Howard et al. (2009)
Uncertainty	Unforeseen conditions more prominent. The amount of problems is not revealed until dismantling and stripping work have started	Quah (1992); Singh (2007); Howard et al. (2009)
	As-built drawings are often unavailable and when they are available normally they are inaccurate	Sanvido and Riggs (1991); Quah (1992); CIRIA (1994)
	Discovery of unforeseen conditions continue to happen during construction phase	CIRIA (1994), Krizek, Lo, and Hadavi (1996)
	Project scope can be unclear due to limited information about the existing building	Sanvido and Riggs (1991); CIRIA (1994)

- Project complexity has to be taken into account when determining planning, coordination, and control requirements;
- Project complexity hinders the clear identification of goals and objectives of major projects;
- Complexity is an important criterion in the selection of an appropriate project organisational form;
- Project complexity influences the selection of project inputs, e.g. the expertise and experience requirements of management personnel;
- Complexity is frequently used as a criterion in the selection of a suitable project procurement arrangement;
- Complexity affects the project objectives of time, cost, and quality. Broadly, the higher the project complexity, the greater the time and cost.

Bertelsen and Koskela (2003) recommend the use of a complexity approach for improving the understanding of construction processes. Bertelsen (2003a, 2003b) argues that acknowledging construction projects as complex settings leads to a new way of managing projects, which is quite different from the traditional project management approach. In conclusion, as stated by Bertelsen and Koskela (2005, p. 70), 'the complex and often turbulent world of project production should take the project complexity and dynamics as an outset and design and operate its management systems from that perspective'.

In summary, the views on complexity provide the basis for understanding refurbishments as complex projects as well as the practical implications of such interpretation. Attributes such as number of elements in a project, interdependence between those elements, and uncertainty should be taken into account when defining the management approach to be adopted in a project.

12.3.2 Managing complex and uncertain projects

As discussed in the previous section, understanding construction projects as a complex phenomenon leads to particular ways of management. Several authors have proposed a number of approaches to deal with the complexity and uncertainty inherent to construction.

Laufer (1991) proposed an approach to cope with project uncertainty on construction planning. It includes an early and systematic diagnostic phase, which involves an overall assessment of the project environment covering an analysis on technological and organisational aspects. Besides this, it comprises the following measures (Laufer, 1991, p. 58):

- Narrowing down uncertainty by gathering more information;
- Adjusting to uncertainty by deferring and splitting decisions as much as possible, thus gaining more complete or stable information;
- Absorbing uncertainty when postponement is no longer possible by making flexible, robust, and less sensitive decisions (this solution might at times be sought before deferring decisions).

Bertelsen and Koskela (2003) also argue that it is important to understand the project environment and propose an assessment based on the analysis of four characteristics, namely project complexity, project's internal and external setting, and project organisation. The underlying idea of this exercise is to increase the awareness of the participants regarding the nature of the project in order to enable discussions and actions to support the management of complexity. In a subsequent paper, Bertelsen and Koskela (2005) support the aforesaid idea by stating that improvements in productivity in complex and uncertain projects emerge through a better knowledge of the project situation, but they should also come from the development of methods suitable to handle uncertainty (e.g. Last Planner System). They also propose the following strategies for managing complexity:

- Build in buffers (slack, margin) for absorbing the impacts of complexity;
- Reduce the complexity or dynamics seen from an operational point of view;
- Codify the procedures to be used and train in performing these procedures under stress;
- Improve the system's own capability to act on the given situation without orders from the management.

Howell, Laufer, and Ballard (1993) recognise the uncertain nature related to the process of setting project objectives and provide recommendations on how to deal with this phenomenon. The main message put forward by those authors is to avoid the assumption that objectives must be completely fixed at the outset, because this notion does not accord with the environment in which projects normally take place. Thus, they suggest a more flexible approach, in which the objectives and their supporting premises are tested and refined gradually by taking into consideration the information available at the time (e.g. constraints related to the means to achieve objectives). In line with that, Laufer, Denker, and Shenhar (1996) also suggest that the timing of decisions and their degree of completeness, the planning horizon, and the degree of detail of the plan to the completeness and stability of information, should be adjusted to uncertainty.

Baccarini (1996) points out that integration is a method of dealing with this complex scenario. This refers to the prioritisation of aspects such as coordination, communication, and control. Following the same line of thought, Laufer, Denker, and Shenhar (1996) recommend a project management style based on elements such as integration, systemic management, and mainly simultaneous management. As a criticism of the traditional management-as-planning[4] style, Bertelsen (2003a) proposes the use of supplementary approaches for organising, planning, and controlling the construction process. This involves, for instance, the consideration of other elements to improve project management such as self-organisation and cooperation.

Saurin, Rooke, and Koskela (2013) and Saurin et al. (2013) have also devised guidelines for the management of complex social-technical systems based on a thorough literature review of studies related to complex systems theory. These include:

- Give visibility to processes and outcomes;
- Encourage diversity of perspectives when making decisions;
- Anticipate and monitor the impact of small changes;
- Design slack;[5]
- Monitor and understand the gap between prescription and practice;
- Create an environment that supports resilience.

In conclusion, as the complexity in projects often cannot be eliminated, it must be recognised and mitigated through the use of appropriate managerial practices. For instance, the planning approach should be designed to handle the uncertainty and variability inherent in construction, and it should be based on elements that are helpful in supporting the management of complexity such as cooperation, coordination, integration, self-organisation, and communication. The following prescriptions summarise the main recommendations for managing complex construction projects:

- Mitigate uncertainty by better understanding the project environment through collaboration between project participants;
- Adjust the decision-making process and planning levels to uncertainty, i.e. the greater the uncertainty, the lower the level of details;
- Engage project participants in the planning and control processes in order to ensure that different perspectives are considered;
- Foster collaboration between project participants through the implementation of collaborative and decentralised planning and control;
- Use buffers to shield production against the innate uncertainty in complex construction projects;
- Increase process transparency by employing visual management tools.

Next, the importance of a theory of production is clarified. Also, the theory selected by the authors to provide guidance on the management of refurbishment works is presented.

12.4 Theory of production in construction

It was contended earlier in this chapter that a sound theory of production is required for underpinning effective construction management. Thus, the indication of such theory is primary. This section starts by elucidating the importance of a theory of production. Then, a theoretical foundation is presented that is suggested to serve as a reference basis for managing and improving production in refurbishments. Also, managerial tools and techniques are proposed as candidate solutions for improving the management of refurbishments.

12.4.1 Why is a theory of production necessary?

Koskela (2000) argued that a theory, generally, provides an explanation of observed behaviour, hence contributing to the understanding and also prediction of future behaviour. Moreover, he contended that it serves to identify the sources of further progress and, once it has been made explicit, it can be tested to prove its validity. The specific functions of a theory of production management are described by other authors. Kochikar and Narendran (1994) argued that, on the basis of a theory, tools for analysing, designing, and controlling can be built. Heim and Compton (1992) pointed out that a theory, when shared, provides a common language or framework, through which the cooperation of people in collective undertakings is facilitated and enabled. Fenves (1996) stated that a theory can be seen as a condensed piece of knowledge, and thus it empowers novices to do things that formerly only experts could do. Lillrank (1995) contended that innovative practices can be transferred to other settings by first abstracting a theory from that practice and then applying it in target condition.

12.4.2 TFV theory of production

Koskela (2000) conducted an investigation on the existing theories of production and concluded that the three main approaches available in the literature diverged in terms of the conceptualisation of production as well as the principles for managing it. While one approach conceptualised production as transformation (T), the others deemed it as flow (F) or value (V) generation. In order to integrate these different views, Koskela (2000) proposed that production could be conceptualised by using these three perspectives simultaneously, thus giving rise to the TFV theory of production. The features of this integrated view on production are summarised in Table 12.4.

The conceptualisation of production as a flow process suggested by Koskela (2000) explicitly acknowledges the existence of two types of activities, i.e. transformation or value-adding activities, and non-transformation or non-value-adding activities (waste). This distinction is particularly important when the core of the flow concept of production is analysed. In the flow view, time is seen as one resource of production, thus reducing the consumption of such resource is a major goal. Since those two types of activities consume time, the elimination of time waste by cutting the share of non-value-adding activities is promoted as the main principle in the flow view. Thus, compressing lead time by eliminating waste from production processes is a basic improvement according to this view. Another interesting insight promoted by the introduction of time as a resource is the fact that time is impacted by the uncertainty and variability inherent in production, hence reducing the variability and uncertainty within flow processes must be seen as an essential goal (Koskela, 2000).

The authors of this chapter argue that the compression of lead time and the reduction of variability are the most powerful principles of production management for driving improvements in refurbishments. The latter is instrumental for addressing the uncertainties related to

Table 12.4 Integrated TFV view on production (adapted from Koskela, 2000)

	Transformation view	*Flow view*	*Value generation view*
Conceptualisation of production	As a transformation of inputs into outputs	As a flow of material, composed of transformation, inspection, moving, and waiting	As a process where value for the customer is created through fulfilment of his requirements
Main principles	Getting production realised efficiently	Elimination of waste (non-value-adding activities)	Elimination of value loss (achieved value in relation to best possible value)
Methods and practices (examples)	Work breakdown structure, MRP, organisational charts	Continuous flow, pull production control, continuous improvement	Methods for requirements capture, Quality Function Deployment
Practical contribution	Taking care of what has to be done	Taking care that what is unnecessary is done as little as possible	Taking care that customer requirements are met in the best possible manner
Suggested name for practical application of the view	Task management	Flow management	Value management

the existing asset, and for supporting the management of constraints related to the existing facility and production phase. The former, namely the compression of lead time, is appropriate to a project environment normally pressed for time to avoid causing problems to the users of the existing building.

The underlying idea of the TFV theory is that production has to be managed from these three viewpoints simultaneously (Koskela, 2000). In this respect, it is essential that managers have a balanced approach regarding each perspective of production. It is worth mentioning that the Lean production principles have served as a theoretical inspiration for the formulation of the TFV theory (Koskela et al., 2002). Therefore, from now on the term *Lean* will be used in the chapter as a reference to this integrated view of production.

12.4.3 Lean Construction approach to refurbishment projects

While the Lean theory has been well tested in new construction projects, in the refurbishment sector the same level of practical application has not been undertaken or reported (Kemmer and Koskela, 2012; Kemmer, Koskela, and Nykänen, 2013). Table 12.5 presents studies where the use of Lean concepts and principles in the management of refurbishment projects was reported.

The majority of cases presented in Table 12.5 concern the implementation of Lean principles in the construction phase of refurbishment projects (e.g. Ballard, 2000; Bryde and Schulmeister, 2012; Haarr and Drevland, 2016; Horman et al., 2003; Kemmer et al., 2016; Pereira and Cachadinha, 2011; Singh, 2007; Vrijhoef, 2016). Yet, there are also studies that focus on design management for refurbishments (e.g. Mitropoulos and Howell, 2002; Tuholski et al., 2009; Tuholski and Tommelein, 2008).

Findings from the studies mentioned in Table 12.5 indicate that the Lean theory can be applied to improve the management and performance of refurbishment projects. Reported benefits include, for example, reliable workflow on-site (Ballard, 2000), improved design process (Tuholski and Tommelein, 2008) and logistics (Pereira and Cachadinha, 2011), enhanced coordination and integration between project participants (Ladhad and Parrish, 2013), and increased productivity (Vrijhoef, 2016). In addition, the studies provide guidance on how to implement Lean in refurbishment projects. A summary of the main recommendations mentioned in Table 12.5 is presented as follows:

- Foster collaboration, engagement, and early involvement of key project members, including building users, in the design and production planning processes;
- Reduce uncertainty by accelerating the discovery of the existing conditions of the building and by identifying key project constraints in a collaborative manner;
- Compress lead time by mitigating uncertainty and variability in the design and construction processes and by reducing the share of non-value-adding activities;
- Ensure continuous flow on-site by integrating production planning with users and operational requirements and by safeguarding users and operations in the existing building;
- Use managerial methods and practices suitable for coping with variability in production;
- Maintain an effective and constant communication with the users of the building;
- Use visual aids to improve communication and decision-making.

Despite the effectiveness of Lean principles and tools for improving the management of refurbishment projects, some authors also indicated that adaptations might be necessary for

Table 12.5 Lean concepts and principles applied to the management of refurbishment projects (adapted from Kemmer and Koskela, 2012)

Authors	*Research developed*
Ballard (2000)	The study highlighted the importance of collaboration of all stakeholders in the planning process, especially in the production of detailed phase schedules. Other issues stressed in the study include the need for a more transparent lookahead and the need for more explicit learning from analysis and action on reason for failures.
Mitropoulos and Howell (2002)	In order to address design challenges that cause iteration and rework, the authors prescribed the following strategies: early involvement of key project members, prototyping, the identification of key constraints, and an accelerated discovery of existing conditions of the building.
Horman et al. (2003)	The planning method used in the project segmented it into small batches to improve production workflow. Buffers were recommended to help smooth the varying levels of work between project zones. Yet, in order to meet a key requirement for the office space (i.e. layout flexibility), a standardised design layout was employed for enabling adaptation and reconfiguration to customise to the specific customer's needs.
Singh (2007)	The main contributions of the study are: a method to quantify the impacts of constraints on production schedule, a list of potential constraints for renovation projects and the definition of critical activities in those projects. Formalisation of constraints before planning crew assignments is also recommended by the author.
Tuholski and Tommelein (2008)	The use of Design Structure Matrix (DSM) is prescribed for improving the design process. For supporting the identification of interactions at different levels and for improving communication across the design team, a colour-coded cross-functional diagram of the design process is also recommended.
Tuholski et al. (2009)	Cross-functional charts are recommended for improving project delivery through identification of functional parties with interrelated material or information handoffs. The need for work structuring early on in the delivery process, the importance of promoting collaboration between design and suppliers, and the need for integrating process with product development are also guidelines prescribed by the authors.
Ho (2009)	The main contribution of the study is an automated method to identify interactions between tenants and crews in renovations of an occupied building in order to reduce disruptions on-site. For enabling the identification of such interactions, requirements for the representation of location, organisation, and activity renovation planning information and for the reasoning methods to utilise this information are proposed.
Lahtinen et al. (2009)	Recommendations were prescribed for improving communication and cooperation between the project team and users of the building, such as the engagement of users in the renovation process, the formation of a multi-disciplinary team to bring different views to the project environment, and the use of several channels of communications between users and project team.
Pereira and Cachadinha (2011)	A model for a practical application of Lean in rehabilitation works was proposed. It comprises five steps: selection of production processes subject to improvements, creation of current value stream map (VSM), analysis of VSM and improvement plan, elaboration of future state map, and implementation and assessment of results.

(*Continued*)

Table 12.5 (Cont.)

Authors	*Research developed*
Bryde and Schulmeister (2012)	The main contribution of the study is an assessment on the utility of Lean principles to a refurbishment project. Findings indicate that not all Lean tenets are easily applied to the context of managing refurbishments. The importance of considering cultural, commercial, and people-related issues when developing and implementing Lean is also stressed in the study.
Ladhad and Parrish (2013)	For ensuring energy savings on retrofit projects, the authors proposed a number of guidelines, such as: learn from previous projects, develop a complete understanding of the existing facility, conduct value analysis, involve the project team early and often, use Lean principles as a foundation for management, etc.
Haarr and Drevland (2016)	Based on an analysis of failures to implement Lean methods during the construction phase in a rehabilitation project in Norway, the authors made recommendations, such as involve downstream players in upstream decisions, foster a proper understanding of Lean principles across the project team, and implement Lean tools gradually.
Kemmer et al. (2016)	The authors prescribed adaptations to the Last Planner System in order to suit the project context, namely retrofit of social housing. For reducing user's disruptions throughout the construction process, the use of 4D BIM models and line of balance are recommended.
Vrijhoef (2016)	In order to enable continuous workflow and a one-piece-flow approach, a number of strategies were put in place, such as: use of multi-skilled teams, transfer of activities between workers, use of autonomous self-controlled teams, and just-in-time delivery. Prefabrication is also suggested by the author of the study to improve productivity.

allowing an effective implementation of Lean tools and techniques. For instance, Bryde and Schulmeister (2012) reported the difficulty in applying Lean principles to the refurbishment context, thus suggesting that modifications to the implementation approach need to be considered to suit the project context. The study developed by Kemmer et al. (2016) confirmed that supposition. In order to implement the Last Planner System in a retrofit of small houses to be executed within a short period of time, some adaptations had to be made. These are summarised in Table 12.6.

As illustrated in Table 12.6, the master plan and phase planning were devised simultaneously, and the constraints were listed for the entire retrofit duration so that uncertainties could be anticipated and disruptions avoided. In order to bring stability to production and rapid learning, the short-term plan was carried out on a daily basis as well as the PPC measurements and root cause analysis.

It is noteworthy that there are also examples of Lean implementations in projects that have similar characteristics to refurbishment in terms of project complexity. For instance, the studies developed by Schramm, Rodrigues, and Formoso (2006) and Cuperus, Wamelink, and Resodihardjo (2010) do not refer directly to refurbishments, but they approach construction projects with complex characteristics, akin to the refurbishment context. Both studies indicate that significant benefits can be achieved through the use of managerial practices in line with the Lean theory such as Production System Design, Last Planner System, Visual Management, Line of Balance, etc. Clearly, lessons can be learnt from those experiences in order to improve the way production is managed in refurbishment projects.

Table 12.6 Last Planner's adaptations to suit the retrofit context (adapted from Kemmer et al., 2016)

LPS elements	*LPS's adaptations*
Long-term (master plan) Phase planning	Devised in a collaborative fashion by the delivery team through the use of post-it notes and a location-based chart
Medium-term (lookahead plan)	Constraints were listed for the entire project duration
Short-term (commitment plan)	Devised on a daily basis to register the assignment of tasks to crews on-site
Learning	Daily measurements of Percent of Plan Complete (PPC) along with root cause analysis

Based on the analysis of the literature, a set of managerial tools and techniques in consonance with Lean concepts is suggested as candidate solutions for initial testing. These are: Production System Design, the Last Planner System, Prototyping, Line of Balance, Multiskilling, Visual Management, Prefabrication, Modularisation, Building Information Modelling (BIM), Standardisation, and Mass Customisation. This list is not exhaustive; rather, it serves as a starting point for practical testing and validation. Table 12.7 connects the list of prescriptions described earlier in this section to the managerial solutions cited above in this paragraph.

Table 12.7 Link between prescriptions on how to implement Lean in refurbishment projects and Lean managerial solutions

Prescriptions	*Lean managerial solutions*
Foster collaboration, engagement, and early involvement of key project members, including building users, in the design and production planning processes	Production System Design, Last Planner System, BIM
Reduce uncertainty by accelerating the discovery of the existing conditions of the building and by identifying key project constraints in a collaborative manner	Prototyping, Production System Design, Last Planner System, BIM
Compress lead time by mitigating uncertainty and variability in the design and construction processes and by reducing the share of non-value adding activities	Prototyping, Last Planner System, Standardisation, Modularisation, BIM, Prefabrication, Mass Customisation
Ensure continuous flow on-site by integrating production planning with users and operational requirements and by safeguarding users and operations in the existing building	Production System Design, Last Planner System, BIM
Use managerial methods and practices suitable for coping with variability in production	Production System Design, Last Planner System, Line of Balance, Multiskilling
Maintain an effective and constant communication with the users of the building	Visual Management
Use visual aids to improving communication and decision-making	Cross-functional charts, Line of Balance, Prototyping, BIM

12.4.4 Discussion

Several conclusions concerning project complexity and theory of production were drawn based on the literature. Firstly, a thorough understanding of the project environment is essential for the development of a suitable production management system, given that complexity depends on the project context. This interpretation is particularly important as the refurbishment sector comprises different types of projects, such as houses, hospitals, offices, department stores, etc. Presumably, there are different managerial approaches to different projects.

Secondly, the theoretical underpinning of project management should be seen as the basis for reaching better practical results. As noted by Koskela and Howell (2002), a deficient theoretical foundation leads to a myriad of problems throughout the life cycle of a project, consequently resulting in poor project performance. In this regard, the transformation view of production, which underpins the traditional construction management approach commonly applied in refurbishments, is ineffective because it does not recognise the uncertainty and variability inherent in production; hence, it does not seek to reduce them systematically. Besides, it fails to focus on the creation and delivery of value for customers.

Thirdly, the TFV theory (Koskela, 2000) is the theoretical foundation that should serve as a reference basis for managing and improving production in refurbishment projects. It is a more powerful reference basis as it integrates the three major views of production, namely transformation, flow, and value. It is contended that this theory has a better approach regarding waste in comparison to traditional management. Indeed, in the flow view of production, the elimination of waste through the reduction of the share of non-value-adding activities is a major goal. Moreover, the compression of lead time and the reduction of variability, principles associated with the flow view of production, arguably are particularly convenient for the refurbishment context because they address typical features of those projects such as uncertainties related to the existing asset, high number of constraints, and short construction timeframes. Besides, there are several managerial tools and techniques that support the practical implementation of the TFV theory. Indeed, findings from the studies mentioned in Table 12.5 indicate that there is a potential for using Lean managerial solutions for improving performance of refurbishment projects. These are: Last Planner System, Production System Design, Prototyping, Visual Management, Multiskilling, Line of Balance, Prefabrication, Building Information Modelling, etc.

Lastly, an important aspect observed in those studies refers to the use of managerial methods and practices that consider principles of flow process design, control, and improvement. These include: Last Planner System, cross-functional charts, design structure matrix, multiskilling, identification of key project constraints, early involvement of key project members, standardisation, and prototyping. It is argued that these methods can be considered as a means of taking Lean principles into account, such as waste reduction, pull planning, variability reduction, continuous flow, process transparency, collaboration, learning, and improvement.

12.5 Conclusions

The use of the traditional construction management approach for managing production in refurbishment projects is no longer tenable because it is not based on the best theory available. As stressed by Koskela and Howell (2002, p. 300), 'it rests on a faulty understanding of the nature of work in projects'.

The management of construction in refurbishment projects must be based on the TFV theory (Koskela, 2000). It is argued that the integration of the three concepts of production (transformation, flow, and value) makes it a more powerful reference basis for modelling, designing, controlling, and improving complex production systems. It is, therefore, the reference basis upon which managerial methods and practices should be underpinned.

It is argued that the flow view of production is particularly valuable for improving production management in refurbishment projects. Empirical evidence suggests that flow-concept related principles are instrumental in enhancing performance of the construction phase. Thus, it is contended that the refurbishment context offers an interesting opportunity to apply principles such as waste reduction, lead time compression, and variability reduction.

A list of candidate solutions for managing production in refurbishment projects was suggested to be tested and validated by construction organisations with the purpose of ensuring that their use improves production performance.

Notes

1 The meaning of the term 'refurbishment' is rarely questioned. However, for clarity, it is appropriate to state that the authors follow the interpretation made by Egbu, Young and Torrance (1998, p. 316): 'refurbishment refers to such works as improvements, adaptation, upgrading, rehabilitation, restoration, modernization, conversion, retrofit, and repair which are carried out on existing buildings for a variety of reasons. This definition, however, excludes such works as cleaning, decorating, and emergency maintenance work.'
2 Singh (2007) classifies the constraints in refurbishment projects into nine categories: utility, physical, pollution, uncertainty, coordination, regulatory, traffic, schedule, and safety.
3 According to Sanvido and Riggs (1991), to have in place the right project team goes beyond just having a team with appropriate experience and skills. To those authors, this involves a team with the ability to work together, as well as having the attitude, flexibility, and responsiveness to cope with the typical changes inherent in refurbishment projects. Also, it refers to a team that is put together at the beginning of the project and stays together until the end of it.
4 In simple terms, the management-as-planning approach assumes that the project can be considered as an ordered phenomenon, thus susceptible to be represented and manipulated by plans. A thorough explanation of this subject can be found in the paper by Johnston and Brennan (1996).
5 According to Saurin, Rooke, and Koskela (2013) and Saurin et al. (2013), this prescription is mostly intended to reduce tight couplings in order to absorb the effects of variability. Slack might take on a number of forms, such as redundant equipment, cross-trained workers, underutilised space, excess of labour and machinery, and generous time margins for task completion. Those authors also stress that slack may have side effects, such as contributing to maintaining problems hidden and disguising small changes.

References

Anumba, C., Marino, B., Gottfried, A. and Egbu, C. (2004) *Health and safety in refurbishment involving demolition and structural instability*, Health & Safety Executive, Research Report number 204, Norwich, UK: HSE Books. Available at: www.hse.gov.uk/research/rrpdf/rr204.pdf (Accessed: 4th Nov 2012).

Baccarini, D. (1996) The concept of project complexity – a review, *International Journal of Project Management*, 14(4), pp. 201–204.

Ballard, G. (2000) *The last planner system of production control*, PhD Thesis, The University of Birmingham, Birmingham, UK. Available at: http://leanconstruction.org.uk/media/docs/ballard2000-dissertation.pdf (Accessed: 7th Jan 2011).

Ballard, G. and Howell, G. (2004) Competing construction management paradigms, *Lean Construction Journal*, 1(1). Available at: www.leanconstruction.org/about-us/publications/lean-construction-journal/lcj-back-issues/volume-1-issue-1-october-2004/(Accessed: 17th May 2012).

Bertelsen, S. (2003a) 'Complexity – Construction in a new perspective', *Proc. 11th Annual Conference of the International Group for Lean Construction*, Blacksburg, VA. Available at: http://iglc.net/Papers/Details/230 (Accessed: 17th May 2011).

Bertelsen, S. (2003b) 'Construction as a complex system', *Proc. 11th Annual Conference of the International Group for Lean Construction*, Blacksburg, VA. Available at: http://iglc.net/Papers/Details/231 (Accessed: 17th May 2011).

Bertelsen, S. and Emmitt, S. (2005) 'The client as a complex system', *Proc. 13th Annual Conference of the International Group for Lean, Construction*, Sydney, Australia. Available at: http://iglc.net/Papers/Details/351 (Accessed: 31st May 2011).

Bertelsen, S. and Koskela, L. (2003) 'Avoiding and managing chaos in projects', *Proc. 11th Annual Conference of the International Group for Lean Construction*, Blacksburg, VA. Available at: http://iglc.net/Papers/Details/229 (Accessed: 17th May 2011).

Bertelsen, S. and Koskela, L. (2005) 'Approaches to managing complexity in project production', *Proc. 13th Annual Conference of the International Group for Lean, Construction*, Sydney, Australia. Available at: http://iglc.net/Papers/Details/350 (Accessed: 17th May 2011).

Birrell, G.S. (1980) Construction planning-beyond the critical path, *Journal of the Construction Division*, 106(3), pp. 389–407.

Bryde, J.D. and Schulmeister, R. (2012) Applying Lean principles to a building refurbishment project: Experiences of key stakeholders, *Construction Management and Economics*, 30(9), pp. 777–794.

Construction Industry Institute (2009) *Front end planning of renovation and revamp projects - Implementation resource 242-2*, Austin, TX: Construction Industry Institute.

Construction Industry Research and Information Association (1994) *A guide to the management of building refurbishment*, Report number 133. London, UK: CIRIA.

Crichton, C. (1966) *Interdependence and uncertainty. A study of the building industry*, London, UK: Tavistock Publications Ltd.

Cuperus, Y., Wamelink, H. and Resodihardjo, G. (2010) 'Reducing fit-out time in a Netherlands housing project', *Proc. 18th Annual Conference of the International Group for Lean, Construction*, Haifa, Israel. Available at: http://iglc.net/Papers/Details/676 (Accessed: 31st May 2011).

Department of Energy and Climate Change. (2008) Climate change act. Available at: www.decc.gov.uk/en/content/cms/legislation/cc_act_08/cc_act_08.aspx (Accessed: 03rd June 2011).

Egbu, C.O. (1994) *Management education and training for refurbishment work within the construction industry*, PhD thesis, Salford, UK: The University of Salford.

Egbu, C.O. (1995) Perceived degree of difficulty of management tasks in construction refurbishment work, *Building Research and Information*, 23(6), pp. 340–344.

Egbu, C.O. (1997) Refurbishment management: Challenges and opportunities, *Building Research and Information*, 25(6), pp. 338–347.

Egbu, C.O., Young, B.A. and Torrance, V.B. (1996) Refurbishment management practices in the shipping and construction industries – lessons to be learned, *Building Research and Information*, 24(6), pp. 329–338.

Egbu, C.O., Young, B.A. and Torrance, V.B. (1998) Planning and control processes and techniques for refurbishment management, *Construction Management and Economics*, 16(3), pp. 315–325.

Emerson, H. (1917) *The twelve principles of efficiency*, New York: The Engineering Magazine.

Fenves, S.J. (1996) 'The penetration of information technologies into civil and structural engineering design: State-of-the-art and directions toward the future', in Kumar, B. & Retik, A. (Eds.) *Information representation and delivery in civil and structural engineering design*, Edinburgh, UK: Civil Comp Press, pp. 1–5.

Gidado, K.I. (1996) Project complexity: The focal point of construction production planning, *Construction Management and Economics*, 14(3), pp. 213–225.

Haarr, K.J. and Drevland, F. (2016) 'A mandated lean construction delivery system in a rehab project – a case study', *Proc. 24th Annual Conference of the International Group for Lean, Construction*, Boston, USA. Available at: http://iglc.net/Papers/Details/1246 (Accessed: 22nd May 2017).

Heim, J., and Compton, W.D. (1992) *Manufacturing systems: Foundations of world-class practice*, Washington, DC: National Academy Press.

Henrich, G. (2009) *Development of a tool for diagnosing production management efficiency on construction sites*, PhD thesis, Salford, UK: The University of Salford.

Henrich, G., Tilley, P. and Koskela, L. (2005) 'Context of production control in construction', *Proc. 13th Annual Conference of the International Group for Lean, Construction*, Sydney, Australia. Available at: http://iglc.net/Papers/Details/364 (Accessed: 12nd Jan 2012).

Ho, P. (2009) *An Automated Method to Identify Occupant Interactions in Renovations of Occupied Buildings*. Stanford, CA: Center for Integrated Facility Engineering, Stanford University.

Horman, M.J., Messner, J.I., Riley, D.R. and Pulaski, M.H. (2003) 'Using buffers to manage production: A case study of the Pentagon Renovation Project', *Proc. 11st Annual Conference of the International Group for Lean, Construction*, Virginia, USA. Available at: http://iglc.net/Papers/Details/240 (Accessed: 7th January 2011).

Howard, S.L., Gibson Jr., G.E., Whittington, D.A. and Cui, Q. (2009) *Analysis supporting front end planning for renovation and revamp projects, Part 1 - Research report 242-11*, Austin, TX: Construction Industry Institute.

Howell, G., Laufer, A. and Ballard, G. (1993) Uncertainty and project objectives, *Project Appraisal*, 8(1), pp. 37–43.

Itard, L., Meijer, F., Vrins, E. and Hoiting, H. (2008) *Building renovation and modernisation in Europe: State of the art review - Final report*, Delft, Netherlands: OTB Research Institute for Housing, Urban and Mobility Studies, 31.

Johnston, R.B. and Brennan, M. (1996) Planning or organizing: The implications of theories of activity for management of operations, *Omega, International Journal of Management Science*, 24(4), pp. 367–384.

Kemmer, S., Biotto, C., Koskela, L. and Fazenda, P.T. (2016) 'Implementing Last Planner in the context of social housing retrofit', *Proc. 24th Annual Conference of the International Group for Lean, Construction*, Boston, USA. Available at: http://iglc.net/Papers/Details/1302 (Accessed: 25th Nov 2016).

Kemmer, S. and Koskela, L. (2012) 'Developing a lean model for production management of refurbishment projects', *Proc. 20th Annual Conference of the International Group for Lean, Construction*, San Diego, USA. Available at: http://iglc.net/Papers/Details/788 (Accessed: 4th Nov 2012).

Kemmer, S., Koskela, L. and Nykänen, V. (2013) *Towards a lean model for production management of refurbishment projects*, Espoo, Finland: VTT Technical Research Centre of Finland. Available at: http://usir.salford.ac.uk/29292/1/T94_%282%29.pdf (Accessed: 17th May 2014).

Kochikar, V.P. and Narendran, T.T. (1994) On using abstract models for analysis of flexible manufacturing systems, *International Journal of Production Research*, 32(10), pp. 2303–2322.

Koskela, L. (2000) *An exploration towards a production theory and its application to construction*, PhD thesis, Espoo, Finland: Technical Research Centre of Finland, VTT.

Koskela, L. and Howell, G. (2002) 'The underlying theory of project management is obsolete', *Proc. 10th Annual Conference of the International Group for Lean, Construction*, Gramado, Brasil. Available at: www.pmi.org/learning/library/underlying-theory-project-management-obsolete-8971 (Accessed: 12nd Jan 2014).

Koskela, L., Howell, G., Ballard, G. and Tommelein, I. (2002) 'The foundations of lean construction', in Best, R. and de Valence, G. (Eds.) *Design and construction: Building in value*, Oxford, UK: Butterworth Heinemann, 211–226.

Krizek, R.J., Lo, W., and Hadavi, A. (1996) Lessons learned from multiphase reconstruction project, *Journal of Construction Engineering and Management*, 122(1), pp. 44–54.

Ladhad, A. and Parrish, K. (2013) 'The role of lean practices for zero net energy retrofits', *Proc. 21st Annual Conference of the International Group for Lean, Construction*, Fortaleza, Brasil. Available at: http://iglc.net/Papers/Details/921 (Accessed: 22nd May 2015).

Lahtinen, M., Salonen, H., Lappalainen, S., Huttunen, J., & Reijula, K. (2009) Renovation of a "Sick Building": The challenge of attaining the confidence of occupants. *American Journal of Industrial Medicine*, *52*(5), 438–445.

Laufer, A. (1991) Construction planning in uncertain environments, *International Journal of Project Management*, 9(1), pp. 53–60.

Laufer, A., Denker, G.R. and Shenhar, A.J. (1996) Simultaneous management: The key to excellence in capital projects, *International Journal of Project Management*, 14(4), pp. 189–199.

Laufer, A. and Tucker, R.L. (1987) Is construction project planning really doing its job? A critical examination of focus, role and process, *Construction Management and Economics*, 5(3), pp. 243–266.

Lillrank, P. (1995) The transfer of management innovations from Japan, *Organization Studies*, 16(6), pp. 971–989.

McKim, R., Hegazy, T. and Attalla, M. (2000) Project performance control in reconstruction projects, *Journal of Construction Engineering and Management*, 126(2), pp. 137–141.

Melles, B., and Wamelink, J.W.F. (1993) *Production control in construction: Different approaches to control, use of information and automated data processing*, Doctoral thesis, Delft University of Technology.

Mitropoulos, P. and Howell, G.A. (2002) Renovation projects: Design process problems and improvements mechanism, *Journal of Construction Engineering and Management*, 18(4), pp. 179–185.

Naaranoja, M. and Uden, L. (2007) Major problems in renovation projects in Finland, *Building and Environment*, 42(2), pp. 852–859.

Noori, A., Saruwono, M., Adnan, H. and Rahmat, I. (2016) 'Conflict, complexity, and uncertainty in building refurbishment projects', Yusoff M., Hamid N., Arshad M., Arshad A., Ridzuan A., Awang H. (Eds) *International Civil and Infrastructure Engineering Conference (InCIEC 2015)*, Springer, Singapore.

Office for National Statistics (2018) Construction Output in Great Britain: November 2018. Table 01 – Construction output main figures, Great Britain. Available at: www.ons.gov.uk/businessindustryandtrade/constructionindustry/bulletins/constructionoutputingreatbritain/november2018 (Accessed: 15th Jan 2019).

Peer, S. (1974) Network analysis and construction planning, *Journal of Construction Division*, 100(3), pp. 203–210.

Pereira, D. and Cachadinha, N. (2011) 'Lean construction in rehabilitation works – suitability analysis and contribution for the definition of an application model', *Proc. 19th Annual Conference of the International Group for Lean, Construction*, Lima, Peru. Available at: http://iglc.net/Papers/Details/1136 (Accessed: 7th Jan 2016).

Quah, L.K. (1992) Comparative variability in tender bids for refurbishment and new build work, *Construction Management and Economics*, 10(3), pp. 263–269.

Rahmat, I.B. (1997) *The planning and control process of refurbishment projects*, PhD thesis, London: University College London.

Ravetz, J. (2008) State of the stock – What do we know about existing buildings and their future prospects? *Energy Policy*, 36(12), pp. 4462–4470.

Sanvido, V.E., and Riggs, L.S. (1991) *Managing retrofit projects*, Technical Report Number 25, Austin, TX: Construction Industry Institute.

Saurin, T.A., Rooke, J. and Koskela, L. (2013) A complex systems theory perspective of lean production, *International Journal of Production Research*, 51(19), pp. 5824–5838.

Saurin, T.A., Rooke, J., Koskela, L. and Kemmer, S. (2013) 'Guidelines for the management of complex socio-technical systems: An exploratory study of a refurbishment project', *Proc. 21st Annual Conference of the International Group for Lean, Construction*, Fortaleza, Brazil. Available at: http://iglc.net/Papers/Details/945 (Accessed: 22nd May 2014).

Schramm, F.K., Rodrigues, A.A. and Formoso, C.T. (2006) 'The role of Production System Design in the management of complex projects', *Proc. 14th Annual Conference of the International Group for Lean, Construction*, Santiago, Chile. Available at: http://iglc.net/Papers/Details/447 (Accessed: 31st May 2011).

Singh, Y.P. (2007) *A framework for production management of renovation projects*, Master dissertation, East Lansing, MI: Michigan State University.

Tuholski, S.J., Gursel, A.P., Tommelein, I.D. and Bomba, G. (2009) Lean Comparison Using Process Charts of Complex Seismic Retrofit Projects, *Journal of Construction Engineering and Management*, 135(4), pp. 330–339.

Tuholski, S.J. and Tommelein, I.D. (2008) 'Design Structure Matrix (DSM) implementation on a seismic retrofit', *Proc. 16th Annual Conference of the International Group for Lean, Construction*, Manchester, UK. Available at: http://iglc.net/Papers/Details/615 (Accessed: 12nd Jan 2012).

Vrijhoef, R. (2016) 'Effects of Lean Work Organization and Industrialization on Workflow and Productive Time in Housing Renovation Projects', *Proc. 24th Annual Conference of the International Group for Lean, Construction*, Boston, USA. Available at: http://iglc.net/Papers/Details/1283 (Accessed: 17th May 2017).

Whiteman, W.E. and Irwig, H.G. (1988) Disturbance scheduling technique for managing renovation work, *Journal of Construction Engineering and Management*, 114(2), pp. 191–213.

Williams, T. (2002) *Modelling complex projects*, West Sussex, UK: John Wiley and Sons, Ltd.

Williams, T.M. (1999) The need for new paradigms for complex projects, *International Journal of Project Management*, 17(5), pp. 269–273.

Worley, D.J. and Hohler, S.E. (2008) OR Construction Project: From planning to execution, *AORN Journal*, 88(6), pp. 917–941.

Yamin, R.A. and Harmelink, D.J. (2001) Comparison of linear scheduling model (LSM) and Critical Path Method (CPM), *Journal of Construction Engineering and Management*, 127(5), pp. 374–381.

13 Extended roles of construction supply chain management for improved logistics and environmental performance

Ruben Vrijhoef

13.1 Introduction

For the past few decades, the construction industry has seen the introduction of supply chain management (SCM). SCM originates from the domains of Lean production, and from purchasing and logistics management (e.g. Cooper, Lambert and Pagh, 1997; Lamming, 1996). From both perspectives, and notably from a Lean view, SCM contributes to achieving flow, maximised value and minimised waste in the value stream that runs through the supply chain (Womack and Jones, 1996). However, the applications of SCM in construction have been rather limited to efficiency improvements on-site, while it also has a potential for improving efficiency of supply chains and logistics to site, as well as reducing environmental effects (e.g. Elfving, Ballard and Talvitie, 2010; Hamzeh et al., 2007; Vrijhoef, 2015).

Following the current global environmental needs to act upon climate change and improve air quality, particularly in cites, the aim of this chapter is to extend the existing roles and further investigate the possibilities that construction supply chain management (CSCM) offers to improve logistics processes and environmental performance as a result. In particular, construction logistics account for high percentages of heavy vehicle movements and emissions of greenhouse gasses (GHG) in cities (Transport and Travel, 2010). These kinds of symptoms typically result from causes earlier in the supply chain, which should be solved 'higher up' in the supply chain and in the 'control hierarchy' (Cox and Ireland, 2002). As a response, central and local authorities tend to strengthen regulations and restrictions for construction firms to take actions and mitigate environmental effects and take social responsibility into account (Sullivan, Barthorpe and Robbins, 2010; Transport for London, 2017).

Construction firms tend to struggle with the additional requirements and how to implement them in their supply chains. Also, suppliers encounter various problems in the way construction supply chains operate (Dainty, Millett and Briscoe, 2001). Suppliers are faced with unpredictability in the supply because of a lack of stability of orders and a lack of detail in planning. Because of last-minute planning, adjustments and buffers in production capabilities and stock levels are required in order to respond to demand of materials. The separated subcontracting and purchasing of materials delivered by various transport firms who are not managed in a coherent way lead to a low utilisation rate of trucks to site. This is mostly caused by last-minute or dedicated supplies to various construction sites without overall coordination of supply chains (Sullivan, Barthorpe and Robbins, 2010).

In this chapter, first a view is taken on CSCM and the identification of logistics as an aspect of it, based on literature. Logistics has been identified as a basic ingredient of SCM both in the original and in the contemporary understanding. Next, the existing four roles of CSCM are revisited. The existing roles are found to be limited in their aim to suggest improvements merely for production efficiency from an economic perspective. However, current trends in society also stress the need for industries to improve sustainability and mitigate environmental effects of economic activities, including construction and transport. Construction and transport together account for one third of global energy-related CO_2 emissions (UN, 2017). To this aim, the role of logistics in CSCM is underlined to have positive effects on process efficiency as well as environmental performance. The chapter identifies four factors of influence to improve both process efficiency and environmental performance. Evidence of the factors and their impact is presented in four practical cases that apply advanced logistics measures in construction projects. The comparative analysis of the case studies give rise to a wider understanding of CSCM, both for production efficiency on-site as well as supply chain processes and the environmental effects for the environment. As a final result, the chapter discusses and proposes the reformulation of extended roles of CSCM, merging its aims for improved performance of site production, logistics and environmental impact.

13.2 Characteristics and problems of construction supply chains

In construction supply chains, multiple flows run across functional and corporate boundaries, including materials and information foremost, but also finance, equipment, labour and other production resources (Saad, Jones and James, 2002). In this chapter, the focus is on the material flow in the supply chain, distinguishing three characteristic problems in construction supply chains (Vrijhoef and Koskela, 2000):

- It is a *converging* supply chain directing all materials to the construction site where the object is assembled from incoming materials. The 'construction factory' is set up around the single product, in contrast to manufacturing systems where multiple products pass through the factory and are distributed to many customers;
- It is, apart from rare exceptions, a *temporary* supply chain producing one-off construction projects through repeated reconfiguration of project organisations. As a result, the construction supply chain is typified by instability, fragmentation, and especially by the separation between the design and the construction of the built object;
- It is a typical *make-to-order* supply chain, with every project creating a new product or prototype. There is little repetition, again with minor exceptions. The process can be very similar, however, for projects of a particular kind.

As a result of the characteristics of the construction supply chain, the material flow is often hindered and much waste is caused, particularly on interfaces between firms along the supply chain (Vrijhoef and Koskela, 2000). In other words, improving the interfaces in the supply chain has a waste and cost reduction potential in which improved logistics is one of the solutions. When taking the whole supply chain into consideration, and all possible solutions, the amount of avoidable costs should be considerably higher (O'Brien, 1998). For those solutions to be beneficial, however, they must be part of the agreements made earlier among firms and implemented well before projects are initiated and operations start (Dainty, Millett and Briscoe, 2001).

However, purchasing is often organised separately from planning, logistics and site management and, therefore, the amounts, order and packaging of material delivered to site are frequently insufficient or hard to handle (Dainty, Millett and Briscoe, 2001). In a study on the implementation of Lean production in construction component manufacturing, Koskela and Leikas (1997) found that there is a tendency to place construction materials orders with missing information due to incomplete design data. In general, it has been found that several of the problems observed in the supply chain were due to earlier and external root causes within or beyond the scope of the supply chain itself (Figure 13.1). Design information is often inadequate and difficult issues are not detailed to the level needed for site installation. Typically, changes are caused by unavailable, late, wrong and incomplete information and are often not communicated (Chan, Scott and Chan, 2004). On the other hand, the suppliers sometimes cause problems for the contractor, for instance when materials are delivered in a different order than the order in which the materials have to be installed (Ala-Risku and Kärkkäinen, 2006).

In many cases, economic considerations are the root causes of operational problems. It was often found that the purchasing price is a dominating criterion for supplier selection, and that subcontractors are predominantly selected on the basis of price and subsequent room for improvement is restricted (e.g. London, 2008; Wegelius-Lehtonen, 2001). Also, decision-making on the improvement of logistics is often constrained to those solutions one has experience of. It is customary to use material inventories as buffers against disruption. Similarly, nearly all actors in the supply chain add a time buffer to their schedule, and thus unduly inflating the build-up of time and reducing time for improvement or proper coordination of logistics (Vrijhoef and Koskela, 2000).

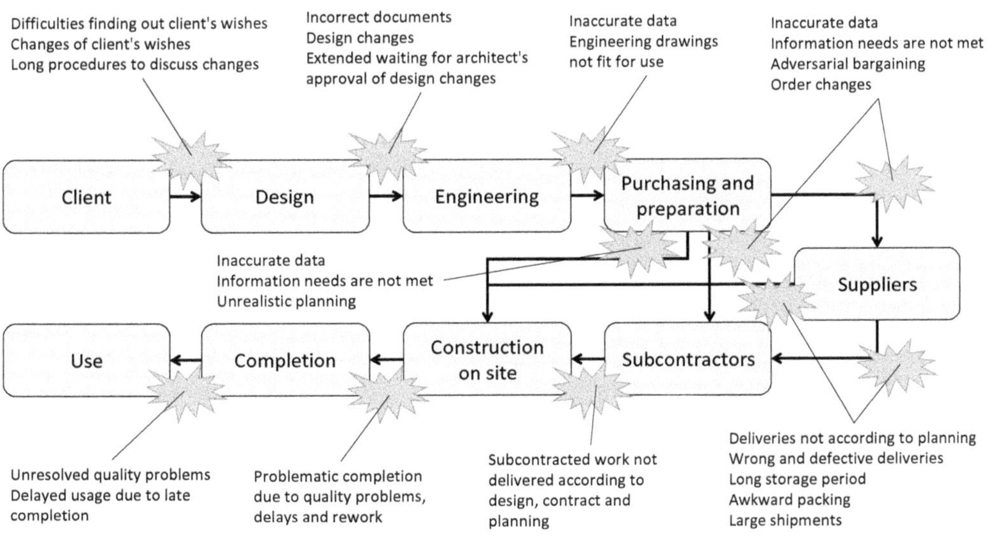

Figure 13.1 Typical problems in the construction supply chain (adapted from Vrijhoef, 2011)

Prior research justifies the conclusion that waste and problems in construction supply chains are extensively present and persistent, and therefore complicated to resolve. This is partly because of construction's ingrained culture and classic approach to the supply chain (Fellows, 2009). Due to the interdependence across corporate boundaries in the supply chain, waste and problems are often caused in other parts of the supply chain. In addition, the fragmented control of the construction supply chain hampers the effective resolution of waste and problems, and their causes (Vrijhoef and Koskela, 2000).

13.3 Reviewing supply chain management literature

13.3.1 Origin and relevance of supply chain management

The concept of SCM originated and flourished in the manufacturing industry. The first visible signs of SCM were in the just-in-time delivery in the Toyota Production System (Monden, 2011). This system aimed to regulate supplies to the Toyota motor factory just in the right – small – amount, just at the right time. The main goal was to drastically decrease inventories, and to effectively regulate the suppliers' interaction with the production line. Another stimulus for SCM originated in the field of quality control. In Japanese culture, it is suggested that total quality management and working with the supplier as a supply chain partner in a long-term relationship of loyalty and trust would improve the quality and decrease the costs of production (Dahlgaard, Khanji and Kristensen, 2008).

The basic approach in SCM is viewing the entire supply chain rather than just the next stage, and aims to increase transparency and alignment throughout the supply chain, regardless of functional or corporate boundaries (Mentzer et al., 2001). The basic idea of SCM is to recognise the interdependencies in the supply chain, and act upon them, so that processes and entities are controlled and improved in a connected way. The aim is to create flow through the supply chain, and value creation without barriers in the value stream from the supply base to client (Womack and Jones, 1996).

Since its emergence in the Japanese automotive industry, the conceptual evolution of SCM has resulted in an autonomous status of the concept in industrial management (Cooper, Lambert and Pagh, 1997). In addition to the original application in the automotive industry, SCM has found its way into other industrial and non-industrial sectors such as aerospace and retail (Harland, 2015). Along with this spread of SCM approaches, additional concepts have influenced the conceptual evolution of SCM leading to its present comprehensive understanding of SCM, bundling an array of sub-concepts including logistics.

13.3.2 Basic roles of construction supply chain management

Previously, four major roles of CSCM have been recognised, dependent on whether the focus is on the supply chain, the construction site, or both (Vrijhoef and Koskela, 2000). Figure 13.2 shows these four areas of focus. The roles as identified are not mutually exclusive, but rather viewed jointly:

1. Firstly, the focus may be on the impacts of the supply chain on-site activities. The goal is to reduce costs and duration of site activities. In this case, the primary consideration is to ensure dependable material and labour flows to the site to avoid disruption to the workflow. This may be achieved by simply focusing on the relationship

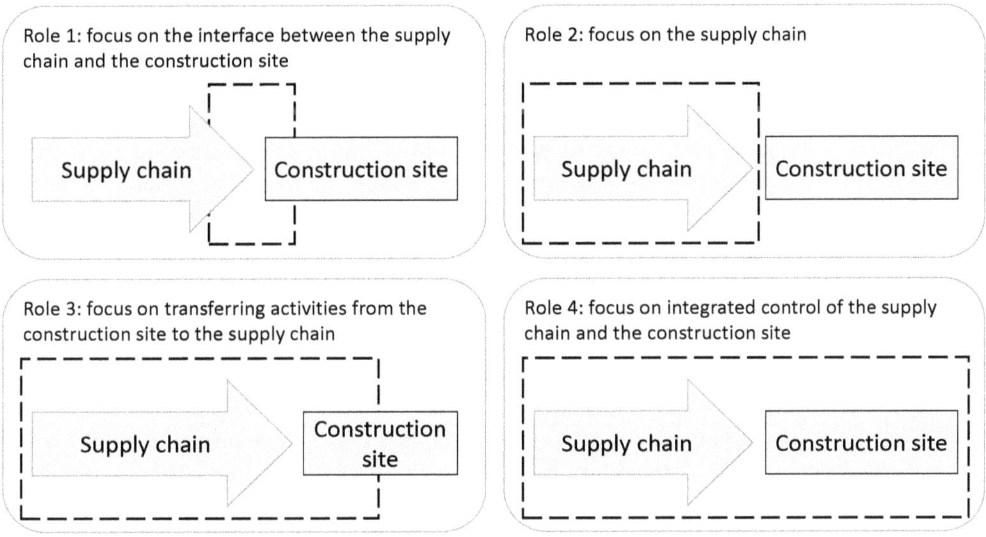

Figure 13.2 Four basic roles of supply chain management in construction (adapted from Vrijhoef and Koskela, 2000)

between the site and direct suppliers. The contractor, whose main interest is in site activities, is in the best position to adopt this focus;

2. Secondly, the focus may be on the supply chain itself, with the goal of reducing costs, especially those relating to logistics, lead time and inventory. Material and component suppliers may also adopt this focus;
3. Thirdly, the focus may be on transferring activities from the site to earlier stages of the supply chain. This rationale may simply be to avoid the basically inferior conditions on-site, or to achieve wider concurrency between activities, which is not possible with site construction with its many technical dependencies. The goal is again to reduce the total costs and duration. Suppliers or contractors may initiate this focus;
4. Fourthly, the focus may be on the integrated management and improvement of the supply chain and the site production. Thus, site production is subsumed into CSCM. Clients, suppliers or contractors may initiate this focus.

13.4 Role of logistics in CSCM and impact on environmental performance

The emergence of SCM is due to the same shift in theoretical concepts as the emergence of Lean production and Lean logistics (Lamming, 1996). The traditional view on managing the supply chain is based, to a large extent, on a transformation view of production, whereas SCM is primarily based on a flow view of production and relationships delivering value from supply base to end customer (Christopher, 2016). The transformation view suggests an independent control of each stage of production, whereas the flow view suggests a focus on the control of the total flow of production (Koskela, 1999). Related to this is the concept that the supply chain can be seen as a 'logistical factory'; an organisational view of linkages between upstream and downstream processes and activities producing products, services and value in the hands of the end customer (Christopher, 2016). Thus, the same intra-organisational principles and methods that have been used to develop actual factories can also be used inter-organisationally to improve supply chains as 'virtual factories' (Luhtala, Kilpinen and Anttila, 1994). On the other hand, practices particular to quality control in SCM have a third basic conceptual basis, which is the view of value generation (Koskela, 2000).

Logistics management connects to SCM as the part that plans, implements and controls the effective and efficient flows and stocks of materials, equipment, capacity and information needed 'from the point of origin to the point of consumption in order to meet customers' requirements' (Lambert, Pugh and Cooper, 1998). In a traditional view, construction logistics processes are often organised as a linear system in a straightforward sequence without controlled reliability buffers between the supply base and the construction site (Figure 13.3).

In addition to the organisational and economic aims of SCM and logistics management, environmental aims have more recently been added in terms of transport logistics (Vrijhoef, 2015). For instance, in the last few years governments have aimed to reduce the effects of transportation on the climate, and particularly cities have restrained the number of polluting vehicles from entering city centres in order to improve the inner-city climate and air quality in general. Notably, construction transport has been found to be relevant to this aim, while typically 30 to 40% of all business transport is related to construction (Transport and Travel, 2010). This represents a large portion of vehicle emissions and road congestions. On the other hand, studies on load factors indicate a need to act; while these tend to remain structurally under 50%, in a few cases they are down to 15%, and lag far behind other sectors of transport (Sullivan, Barthorpe and Robbins, 2010).

13.4.1 Consolidated logistics for improved logistics performance

Logistics management techniques, such as 'decoupling' the delivery process by introducing a reliability buffer between the supply base on the one hand, and the construction site on the other, are seen by various stakeholders involved in the supply chain as a potentially viable solution. This solution should contribute to organising construction logistics and transportation in a more effective, efficient and sustainable way, particularly within urban areas (Constructing Excellence, 2007).

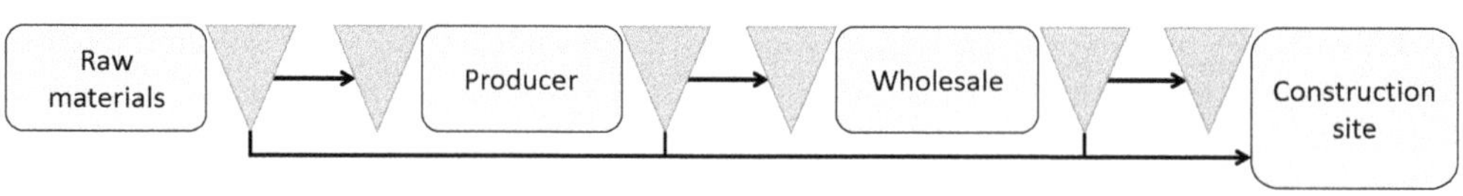

Figure 13.3 Traditional logistics process to a construction site (adapted from Vrijhoef, 2015)

In the UK and notably in and around London, including Heathrow Airport, local government and the industry have undertaken recent action to reduce construction transport movements, with demonstrable results (Transport for Transport for London, 2008). In particular, the application of so-called construction consolidation centres (CCC) or logistics 'hubs' has had quite an impact on reducing transport movements, emissions and safety issues (Sullivan, Barthorpe and Robbins, 2010) (Figure 13.4). Other scholars have echoed the merits of applying logistics centres for construction projects (e.g. Elfving, Ballard and Talvitie, 2010; Hamzeh et al., 2007).

The London Construction Consolidation Centre (LCCC) claims that the number of construction vehicles entering the City of London, and delivering to the sites being served by the LCCC, has reduced by 68% (Brett, 2007), while supplier journey times have been reduced by two hours (MVA, 2006). The delivery performance of goods being delivered correctly the first time round is 97% (Constructing Excellence, 2007). LCCC claims that a reduction of materials waste generated is up 15% through reduced damages, fewer materials that are lost, and less theft (Transport and Travel, 2010). Also, a reduction of materials and packaging on-site has been reported (Department for Transport, 2007). As a result of the reduction in vehicles and movements, a reduction of greenhouse gas emissions has been reported at around 75% (Transport for Transport for London, 2008).

Besides consolidation, decoupling is another important aim of a logistics hub. The hub as decoupling point has played an important role in logistics management, as well as in other sectors, e.g. in manufacturing and the food industry, separating the demand side of the supply chain from the supply side (Hoekstra and Romme, 1992). Associated with the decoupling point is the issue of postponement. The principle of postponement can be subdivided into three generic types: form, time and place postponement (Bowersox and Closs, 1996; Bowersox, Closs and Cooper, 2002). Form postponement entails delaying final production activities that determine the end form or late assembly of materials and components. Time postponement means delaying the forward movement of materials until customer orders have been received. Finally, place postponement refers to the placing of inventories on controlled locations to postpone the forward movement of

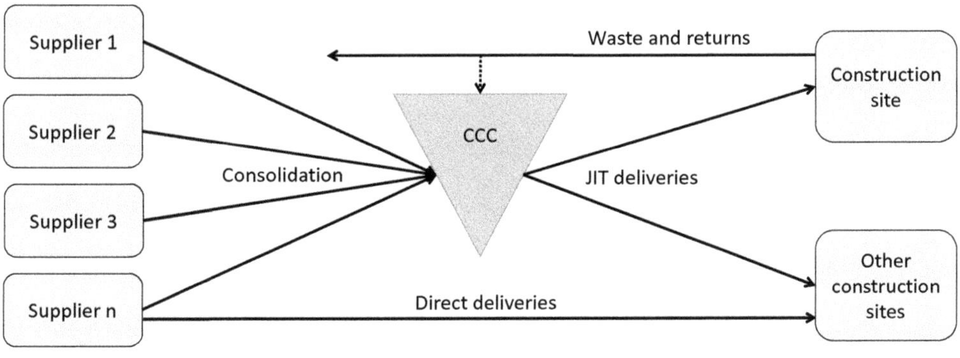

Figure 13.4 Construction logistics process using a CCC, i.e. logistics hub (adapted from Vrijhoef, 2015)

materials further in the production and delivery process. Typically, time and place postponement relate to the concept of reliability buffering (Park and Peña-Mora, 2004).

In terms of the three types of postponement, the hub functions as a decoupling point between supplier and construction site for temporary storage at or near the construction site, where final prefabrication or preassembly of materials and components can take place before being shipped to site, as well as the conditioned storage of materials. The hub then functions as a decoupling point at which supplies of materials from suppliers to the hub are controlled in either a 'push' or a 'pull' way based on the project planning, or either a pull planning and then paused and stored at the hub. Later materials are called off from the hub to the site based on real and actual progress on-site and thus controlled in a 'pull' way and delivered just in time to site.

There is also a possibility that supplies and transports are halted and cross-docked at the hub for further transport, controlled by needs and time slots agreed upon with the site. By organising the supply in this way, actual shipments and supplies to site are more time-independent and can be controlled just in time, following the speed and progress on-site, as opposed to direct and separate shipments traditionally controlled by each supplier based on an individual transport planning.

To conclude, a hub can also handle return flows from the construction site in the same haul including waste, packaging and equipment for further handling, recycling and returns. Studies in different local settings show that applying this kind of 'reverse logistics' technique should reduce empty return runs by up to about 35% as a result of vehicles leaving construction sites loaded with construction and demolition waste (CDW) and other return shipments (Shakantu et al., 2008).

13.4.2 Urban logistics aimed at improved environmental impact

The aim of the case studies presented in the case descriptions of this chapter is to demonstrate the intended applications of SCM on improved logistics, and the advantages for both the site and the surroundings. The dominant measure taken in all of the case studies includes the use of a construction logistics hub as explained above. In addition, in this case all materials and equipment flows, and the construction site, were managed by a logistics service provider (LSP). This particular firm managed the inventory at the hub and issued delivery time slots; their own material-handling crews brought materials from the arriving vehicles to the point of installation at night before subcontractors would start the installation. In the same haul, they carried out the waste and trash of the day before and transported this out, back to the hub on the same vehicles. The cases have all aimed at achieving effects both on the site and in the supply chain in terms of efficiencies, as well as mitigating effects for the environment and the surroundings.

The logistics hub in the case of the City of Utrecht reported here is based on the concept presented above. In addition, it includes reverse logistics by bundling, sorting and recycling CDW at the hub and further reducing total transportation as a result. In this particular case, this is important as it contributes to the objectives of the City of Utrecht, as with other cities, to try to become carbon and resource neutral, complying with the objectives of the 'circular economy' in the next decades (Van Buren et al., 2016). At the same time, the city aims to create jobs for low-skilled labour, and opportunities for creative companies and products that create new products with waste, particularly in the local built environment (Gemeente Utrecht, 2015). These societal side effects have been found to be of great importance for government and industry alike.

On a larger city scale, the consolidation hub in Utrecht is located in an industrial area aimed at channelling materials entering the city and being distributed efficiently via the hub to local construction sites. The area will be a grant hub location and a central point for city distribution connected to many sites and firms in the city. These connections are aimed at increasing the levels of transport efficiencies and waste through the area, and thus increasing the value and sustainability of the business case of the hub for the city as a whole.

13.5 Four factors of influence on construction logistics performance

Construction logistics and transport, as well as their effect on process efficiency and the environment, can be influenced by different factors. Based on contemporary literature, four factors of influence have been identified including corresponding concepts and measures that influence construction logistics and the performance of construction (based on Lundesjo, 2015; Sobotka, Czarnigowska and Stefaniak, 2005; Sullivan, Barthorpe and Robbins, 2010), discussed below.

First, transport can be influenced by the way the logistics are managed. More specifically, this concerns the extent to which the logistics processes can be decoupled and postponed to support just-in-time deliveries, and load factors of vehicles to be maximised by means of consolidation. In all case studies, reliability buffering and, in most cases, the logistics hub have played a major role in improving the logistics performance and the environmental impact of the particular projects.

Second, the information management of the construction process and the logistics in particular influence the efficiency of construction transport. On the one hand, in the design and project preparation phase this applies to advanced planning and modelling systems such as 4D BIM applications to support the logistics planning. In the project execution and materials supply management, warehouse management systems (WMS) and transport management systems (TMS) online software applications, for example, support the inventory and transport management.

Third, on the material level, the packaging and integration of materials into bundles, material kits or prefabricated modules that can be shipped and moved in a smart way to the place of installation on-site influence the efficiency of transport and handling on-site. In all case studies, the daily kits of materials was a measure taken. In essence, the kits included exact amounts of materials per room or location in the building, to be processed per day or in a few days. The kits were compiled and shipped at the hub. In a few cases, materials were also preassembled or prefabricated at the hub before being shipped to site.

Fourth, the extent to which project tendering and purchase contracts consider transport and the effects of it, such as GHG emission as variables or criteria, influences the effort that contractors, suppliers and LSPs put into transport efficiency. In two of the case studies, such contracts were in place and contractors needed to comply with contract implications and vehicle registration to prove that the emission levels were not exceeding the contractual binding levels. Intermediate control and compliance could lead to extra logistics measures to improve logistics performance as needed.

Below, the four factors of influence of logistics performance are further explained as applied to the four case studies presented in the case descriptions below. Each of the four factors includes various logistics measures and interventions to improve the logistics management and the environmental performance as a result. In fact, the four cases presented below were experimental with regard to most of the logistics measures that were taken in the four respective case projects (Table 13.1).

Table 13.1 Logistics factors and measures taken in case study projects

	Project A	*Project B*	*Project C*	*Project D*
Logistics management	Logistics hub, truck holding, consolidation of deliveries, waste disposal in same haul, group transport of workers, transported equipment	Truck holding areas close to site	Logistics hub, truck holding, consolidation, waste disposal in same haul	Logistics hub, truck holding, consolidation, waste disposal in same haul, public and group transport of workers, transported equipment
Information management	Logistics data export from engineering phase via Building Information Model into TMS of the hub	Online transport call-offs per subcontractor, integrating call-offs into full truck loads at hub, WMS for online insight in stock, WMS directly connected to TMS	Online transport call-offs per subcontractor, integrating call -offs into full truck loads at hub, WMS for online insight in stock, WMS directly connected to TMS	Printed 'transport tickets' for planning call-offs of deliveries from hub to site
Prefabrication, kitting	Prefabricating, e.g. of rebar modules at hub, kitting at hub of daily materials batches per room	Kitting at hub of daily materials batches per room, preinstalling materials per room	Kitting at hub of daily materials batches per room	Kitting at hub of daily materials batches per room
Procurement, purchasing	n/a	Balanced score tender aimed at reducing vehicle movements and GHG	n/a	GHG criteria in balanced score tender aimed at reducing vehicle movements

13.5.1 Logistics management: decoupling and consolidation

Decoupling and consolidation are typically organised via logistics hubs normally on the city border, often close to motorways. These hubs function as a temporary storage, i.e. reliability buffer at or near the construction site, and as a holding point at which large deliveries from suppliers are received and stored for a short time. Based on a pull planning and actual progress on-site, materials are called off by subcontractors on a daily basis. Based on the call-offs, materials are shipped from the hub to the construction site just in time, in full truckloads and bundled material kits, and then put in place per installation location in the building.

Materials that are too bulky to be consolidated and repacked in material kits are held at the hub until they are called off by the site in order to prevent congestion on and around the site. For instance, this is the case for large quantities or bulk materials that are processed on-site directly from the truck, such as concrete elements or ready-mixed concrete, full truckloads of heavy material, and other supplies that require direct handling on-site. Dependent on the type of truck, on the return trip it takes return flows from the construction site including waste, packaging and used equipment.

13.5.2 Information management: integrated planning and delivery

The planning of the delivery to the construction site is done by the contractor or by a specialised LSP in collaboration with the subcontractors on-site. Also, the suppliers are active in planning the deliveries, particularly the time-critical products and the direct deliveries. The overall planning is finalised and controlled by the main contractor at the construction site.

The resources needed at the construction site are called off at the hub by the responsible person at the construction site through the WMS functioning at the hub. This request for resources is at minimum one day before the goods are needed at the construction site. The goods can be picked based on this signal and be transported in consolidated shipments to the construction site controlled by the connected TMS.

The balancing of supply chain resources with supply chain requirements is done by the site manager. Orders that are placed come from the project planning. The information in an order generally contains an article code and name, construction part, the planning, and location on-site. To organise the information about goods to the construction site in a situation with the hub, a so-called 'building ticket' is used. This ticket is a form or card with which the subcontractor can call off materials, comparable to a Kanban card. Based on the tickets of all subcontractors, the logistics coordinator can plan integrated deliveries of full truckloads to the construction site.

13.5.3 Preassembly: off-site preparation and materials kitting

Particularly on confined construction sites in busy surroundings, there is often a lack of space to stock and assemble materials. Therefore, a solution to this can be to preassemble materials or put materials off-site and deliver materials in precise quantities to site. Subcontractors can preassemble materials at the hub and also have them shipped just in time via the hub to the site for final installation. At the hub, materials are pulled from stock, taken to the working area at the hub where the subcontractor does the assembly, after which the assembled product is put back to stock, waiting to be called off by the site. Based on the call-off schedule, preassemblies are then optimally planned and shipped to the construction site in bundled shipments.

In case of bulky preassemblies or larger prefabricated modules or material kitting taking place at suppliers, a decision can be made that those modules or kits are shipped to the construction site directly from the suppliers. This means that the modules or kits do not pass the hub and are not put in stock to be called off from there. This particularly concerns large engineered-to-order modules, large customised subassemblies, large bulky materials, or large material kits to finish larger spaces in the building, for instance. These deliveries usually go to site directly, while extra handling at the hub is basically not needed, unpractical or expensive. Those deliveries may travel quite a distance from various suppliers' locations. The planning and the time of arrival of those transports need to be accurate and planned well ahead so as to avoid long waiting times once arriving at the site. Vehicle marshalling and digital planning platforms are means to achieve this, in particular. If needed, the transports are held outside city limits or at the hub before being called in to the site.

13.5.4 Procurement: MEAT tender and strategic purchasing

Local government has installed restrictions to enter cities and wants to keep vehicles out. Clients who care about sustainability tender via innovative routes, and include and reward GHG reduction as a MEAT criterion (Most Economically Advantageous Tender) in contracts. Requirements may also include a reduction of hindrance and ensuring that roads around the site are not obstructed. Green procurement and corresponding contracts force contractors to comply with those requirements in their bids in order to win (Varnäs, Balfors and Faith-Ell, 2009). In addition, local governments may require transport and deliveries outside rush hours or stimulate transport and deliveries at night.

In turn, some contractors stress the importance and stimulate their suppliers and transport firms to achieve logistics efficiencies. Some studies on supply chain costs rate logistics costs at 10 to 20% of sales prices (www.instituteforsupplymanagement.org/). However, adversarial procurement by clients tends to stimulate contractors to adversarial purchasing of subs and suppliers at the lowest price, thus affecting the opportunity to better organise logistics. In turn, subs have generally included logistics costs in their prices and contractors have often bargained a good purchase price from suppliers before the discussion about logistics costs would normally begin in projects. This discussion often comes too late, though, for logistics to be discussed properly. Therefore, contractors often shift to strategic and long-term kinds of purchasing relations with subs and suppliers to drive down logistics costs on a structural basis collaboratively.

13.6 Empirical research on advancing construction logistics

This chapter has been partly based on an applied research project in the Netherlands, which aimed at advancing construction logistics and demonstrating the effects in construction practice. Four construction projects have been observed in close collaboration with the firms and researchers involved. In the projects, various kinds of selected solutions have been applied and tested. The effects have been measured via a set of Key Performance Indicators (KPIs) presented below. The research project had aimed at gaining academic insights and advancing logistics in practice, applying solutions such as the use of hubs, leading to increased production efficiency on-site as well as reducing construction transport to site.

13.6.1 Ethnographic participatory action research approach

The research used a combined application of independent and dependent variables of construction supply chain and logistics management. The insights of prior research were used to define the independent variables and how they occurred in the cases (e.g. Vrijhoef, 2011). Further, the SCOR dimensions of the Supply Chain Council were taken as dependent output variables and measured how cases actually performed (SSC, 2010). A participatory action research approach was applied in which the researchers took part in the project teams to shape the interventions, and observed the effects of them within the empirical context (Denzin and Lincoln, 2008).

The researchers collected data by participating in the case projects for an extended period of time (Bryman, 2008). The cases took place during a period of two years. In the two years, the researchers visited and gathered data based on the opportunities the projects offered during construction. Also, the hub that was in place for most of the case projects was visited for observations and data gathering, e.g. data from digital warehouse and transport systems. There was extensive interaction between researchers and practitioners in the cases, to discuss adjustments and take further action to implement the logistics measures to be researched. Due to this interaction, the research can therefore be considered to be an ethnographic research, i.e. 'interactive-inductive research involving direct and sustained contact with human agents, within the context of their daily lives, watching what happens, listening to what is said, asking questions, and producing a richly written account' (Pink, Tutt and Dainty, 2013).

Interviews, document studies and observations took place every two to four weeks, which included meetings with project and logistics personnel, analysing planning documentation and documented delivery registrations, time measurements on-site, and attending site meetings. Researchers and students took part in those activities. In addition, meetings took place every two months with senior researchers and the management of the firms to discuss new experiments and interventions in the projects, modelling exercises of 4D planning using BIM, and effect measurements of logistics measures taken. For the measurement, the below KPI framework was used, and the outcomes were then validated with practitioners in the case projects and confirmed with external experts.

13.6.2 Case selection

Four cases were analysed with sufficient similarities, but also differences to increase the level of understanding and uncover areas for further application and research (Table 13.2). The projects differed in type, size and location. This enabled conformities to show how logistics solutions were implemented in general and how those led to similar improvements. The differences gave insight into how project characteristics affected the dimensions of logistics and the specific effects of the solutions applied per project.

Table 13.2 Characteristics of the case study projects

	Project A	*Project B*	*Project C*	*Project D*
Type of project	Newly built housing blocks, including commercial spaces and parking basement	Newly built hotel, including parking basement	Newly built multifunctional expansion of shopping centre, including hotel and apartments, and parking basement	Internal refurbishment of office buildings of a larger trade centre including the entrance of the centre
Location	Near city centre	Urban area close to motorway	In city centre	In city centre
Duration of the project	2014–2016	2013–2015	2013–2019	2014
Size of the building	255 small apartments, 1,500 m^2 commercial space, parking basement	18,500 m^2 hotel, 4,000 m^2 parking for 136 spaces	35,000 m^2 shopping, 5-level parking basement for 1,300 spaces	65,000 m^2

13.6.3 Case study organisation and data collection

The researchers were present as observers in the project teams. During construction, the LSP coordinated the logistics activities based on the agreements that were set up during the purchasing processes with suppliers and subcontractors. During construction, the main contractor made a six-week lookahead planning. This planning was shared with the LSP at the hub so that they could keep a sufficient inventory of materials to respond to call-offs at the site.

Next, a detailed weekly planning was set up. This planning gave insight into the day-to-day operation. Also, this planning was shared with the LSP. Based on this planning, delivery plans were made for daily deliveries of materials to the construction site. Also, this planning led in the planning of call-offs of materials by the contractor and subcontractors on-site. Deliveries of the materials were then planned in full truckloads from the hub to the construction site. In the event that materials were to be shipped directly from suppliers to the construction site, these materials were called off by the hub at the suppliers and went straight to the site.

Based on the KPI framework, the personnel of the firms involved and the researchers measured and reported data of the deliveries and the material movements both taking place at the hub and on-site. These measurements and reports have systematically been put in a database and thus produced the input data to calculate the formulas of the KPIs.

13.7 Case descriptions and analysis

Observations within and between the four projects have been analysed using the KPI framework as a guide (Table 13.3). The findings are partly qualitative and partly quantitative. The descriptions of the findings partly focus on elements that are not present in the framework or that might deepen an understanding of how to operationalise its dimensions. Within the research reported, the KPI framework was subject to further insight, which was also true for the projects and solutions applied. The fieldwork carried out at the projects was therefore input for both testing and further advancement of the logistics solutions, and the measurement of effects via the KPIs.

13.7.1 KPI framework

The KPI framework that was developed for the case studies was based on the Green-SCOR model (SCC, 2010) adapted to a construction context. The indicators and measurements were further improved based on practical examples in the UK and the Netherlands applying advanced logistics solutions. The KPIs were specifically formulated to be able to monitor the construction and logistics processes of the case studies, and also to gain data to measure the effects of the logistics measurements taken.

13.7.2 Case study project A

This case project applied a hub as a consolidation centre. The project was located at a canal and thus the potential of water transport had been studied, as well as the consequences for costs, emission and planning. It remained unclear what exactly the extra costs and time use would be of the transfer to water transport. CO_2 would reduce, but other GHG emissions such as NO_x and SO_2 would increase because of the fuel used by

Table 13.3 KPI framework: indicators and their measurement

	Indicators	*Measurement*
Reliability	Percentage deliveries on time, and complying to quality and requirements	Amount of deliveries on time and complying to requirements, and reason why not if not, and consequences/costs to fix
Responsiveness	Offloading and waiting times of deliveries on site	Waiting time of trucks before being offloaded on site, and waiting time of personnel and equipment on site before being able to offload
Agility	Lead times of orders delivered from hub to site, and from suppliers to hub	Time needed to place and process an order from the hub and suppliers, and being delivered to the site
Costs	Productivity gains on site, purchase price reductions, transport costs reductions	Less time wasted by workers on site, eliminating logistics cost elements by suppliers, reduction of travel distance and time or bundling of deliveries of suppliers
Assets	Transport efficiency, load factor of trucks, consolidation factor of hub, days of stock at hub	Transport volume and weight of trucks to and from site; amount of trucks from supplier to hub vs hub to site
Environment	Environmental impact, GHG emission, waste reduction, noise, hindrance, health, safety	Amount of fuel use, amounts of waste, packaging, debris, numbers of complaints, incidents, accidents, traffic jams caused

the boats. Also, the materials needed to be packed in specific ways in order for them to be transported by boat. As a result, transport via water was not further applied, but trucks were used. On-site, specific equipment had been developed and tried, such as floor extensions, to be able to easily pull in material kits from the building elevator onto the floors, and further to the work spots on the floors.

For the information management and support of optimised transport, all data needed from suppliers was analysed and validated in the BIM model of the project. This data contained all product-specific details, location, time and transport, and handling equipment. This data was connected to the planning software on-site and the TMS of the transport firm. The data was printed on 'building tickets' that functioned as an order form from site to the hub and suppliers, including QR codes. Tagging via sensor technology such as Bluetooth and RFID had been studied, but applied in the next project.

13.7.3 Case study project B

In this case, the building of a hotel, the logistics management had been concentrating on the tower crane efficiency: the materials supply onto floors, and carrying off debris from the floors. The aim was to balance and increase the efficiency of the arrival of deliveries on-site, subsequent offloading by the crane, storage on-site, the crane taking the materials up to the floor, and installing the materials on the floor. This project did not work with a hub, but was supplied directly from the supplier's plants, particularly concrete, rebar and formwork. This was followed by the HVAC firm, whose factory functioned as a place where the kitting of all materials per hotel room took place. Also, all ductwork and finishing materials were put in the material kits.

The site used an online TMS and 'building tickets', including registration of all arriving deliveries from the suppliers and the HVAC firm's location. This led to a continuous information flow to be able to manage the deliveries more accurately, and plan the tower crane's capacity more efficiently. This also reduced the amount of waiting vehicles and traffic jams around the site. As a safeguard, once trucks approached the site they were able to use a buffering location if needed, in case they should arrive untimely.

13.7.4 Case study project C

This case is logistically comparable to case B, although in this case a hub was used. Since the project is very large, there were dedicated logistics personnel on-site as well as at the hub. These personnel were equipped with portable computers with applications to access the online TMS and WMS platform, including the planning of all cranes, and all entrance and storage locations on-site. These integrated systems enabled suppliers and transport firms driving up and down from the site to the hub to deliver materials and carry off debris in a highly accurate manner. However, the size of the project and the number of suppliers and subcontractors caused quite a bit of improvisation and deliveries that evaded the online system, and still caused traffic jams and inefficiencies on the site and around.

Since the project site is located in the very city centre next to the central railway station, all disruptions caused considerable congestion. The city is keen to prevent this and reduce emissions caused by vehicles. As an extra measure, most of the deliveries were planned outside rush hours and particularly in the early mornings and in the evenings. As a consequence, these deliveries took less time and produced less emission.

13.7.5 Case study project D

This project had been contracted as a MEAT tender applying CO_2 emissions of deliveries, travel of personnel and carrying off debris as criteria to the bids of contractors and rewarded in the appraisal of the bids, alongside the price, planning and quality. The winning bid managed to offer the lowest CO_2 emission. In the tender, a traditional calculation of the emission was given of 42,000 kg which the project would produce when executed by applying traditional logistics management. The winning bid offered to do the project by applying a hub for all deliveries and carrying off debris, and group and public transport of personnel, resulting in 17,000 kg emission, i.e. a reduction of 60%. Based on the registration of all vehicle movements, the real emission of the project appeared to be 22,500 kg, i.e. a reduction of 46%. This was caused by displacement of the hub at a further distance from the site, and lower load factors than were offered in the bid.

13.7.6 Cross case analysis

Looking across the case studies, following the four factors respectively, four corresponding analyses can be made. These analyses indicate four provisional directions for extending the understanding of CSCM and its further interpretation related to improved logistics, notably aimed at environmental performance (Table 13.4).

First, CSCM could benefit from more explicit refocus and by the adding of elements of logistics management to its domain. Contemporary CSCM appears to have lost

Table 13.4 Empirical findings and results of case study projects

	Project A	*Project B*	*Project C*	*Project D*
Reliability	Most deliveries arriving on time from hub to site	More than traditional deliveries arriving on time from suppliers to site	With online transport ticket deliveries within 10 min time frame; without ticket 1.30 hrs	Most deliveries arriving on time from hub to site
Responsiveness	TMS gives insight into deliveries, taking less time for site personnel to offload	Offloading of all deliveries strictly planned via online tickets and followed up by personnel	Offloading of all deliveries strictly planned via online tickets and followed up by personnel	In most cases site personnel are aware of deliveries planned and take less time for offloading
Agility	Orders from site to hub come a few days to a week in advance based on online planning	Online WMS allows orders from site to hub up to two days in advance	Online WMS allows orders from site to hub up to two days in advance	Local WMS at hub allows orders from site a few days in advance
Costs	Prefabrication of rebar and kitting resulted in faster working, less time wasted on site	Bundled deliveries, kitting resulted in faster working, less time wasted on site	Bundled deliveries, kitting resulted in faster working, less time wasted on site	Lower bid caused by lower transport costs and higher productivity based on firms' previous experiences
Assets	Fewer vehicles to and from site than traditional, bundling, debris in same loop	Fewer vehicles to and from site than traditional, bundling, debris in same loop	Consolidation of deliveries at hub leading up to 60% fewer vehicles from hub to site	Consolidation of deliveries at hub leading up to 70% fewer vehicles to site and no return
Environment	Less emission due to fewer vehicles, less annoyance within neighbourhood of workers not taking parking space	Less emission due to fewer vehicles, less hindrance around site due to stricter planned deliveries and faster offloading	Deliveries avoiding rush hours took 25% less travel time, 11% less emission as a result	Reduced CO_2 emission 22,500 kg during project, i.e. 46% less than traditional as based on contract

aspects of its logistics management and transportation origin. However, contemporary logistics management, including concepts of the logistics hub and additional advancements for the control, decoupling, postponement and reliability buffering of logistics processes can support the aims of CSCM and Lean thinking to balance material flows and reduce variability in the supply chain. The resulting improvement of the logistics performance will also improve supply chain performance and flow, and reduce negative impacts such as process waste and environmental effects alike.

Second, the information management elements supporting construction logistics and construction transport are the same that are supporting CSCM to a large extent. In particular, design and project preparation phase ICT systems, including advanced planning and modelling systems such as location-based planning, takt time planning, last planner and 4D modelling applications, serve Lean CSCM as well as logistics planning.

Third, the approach to delivery of materials and components to site via smart bundling of materials and increased prefabrication and preassembly serve both Lean CSCM and logistics management aims. The just-in-time approach to delivery and the exact planning of time, quantities and fitting of materials on-site is central in logistics and will serve Lean CSCM to achieve flow and reduce waste, as well as improve takt time approaches and productivity on-site.

Fourth, with regard to logistics management, CSCM could benefit from more refocus and by reintroducing elements of its origins in strategic purchasing and incentivised procurement. Effective indicators could be included in project tendering and purchase contracts, and thus stimulate and reward logistics performance improvement, further efficiency improvements, and mitigating effects on the environmental impact of construction projects and transportation, such as GHG emissions.

13.8 Discussion: extending the roles of construction supply chain management

Based on the empirical analyses of the case studies above and the contemporary body of knowledge of CSCM, limitations and corresponding additions to each of the existing roles of CSCM can now be discussed. Limitations may exist in the prior formulations of the roles as such, when they are compared to new knowledge and modern conceptualisations of SCM. Limitations may also occur when assessing the roles based on external views in society and the meaning the roles have in the contemporary understanding of construction. Particularly in this chapter, recent demands from society and governments have caused the industry to reformulate and extend the existing roles of CSCM, pushing the industry to move towards higher levels of efficiency and environmental awareness, notably from a logistics perspective. Based on the limitations of the existing roles on the one hand, and new demands on the other, directions for extending the four roles of CSCM are discussed below.

13.8.1 Extending role 1: improving the interface between the site and the supply chain

In the basic approach, both CSCM and logistics initiatives have stressed cost and time benefits particularly, and have failed to address the impact of supply chain complexity on-site assembly variability, and the dynamics of connecting between the site variabilities, as well as the supply chain complexities including the need for responsiveness of transports based on progress on-site.

What was found in the case studies in this regard was that pull planning provides a solution, including possibilities for the site to call off materials flexibly per day from the hub one or two weeks ahead. Including all logistics and ICT systems in place, this enabled real-time and actual just-in-time deliveries based on pull demand from site.

The logistics extension applies particularly to putting a hub facility on the supply chain to buffer changing situations on-site and respond efficiently and flexibly. In addition, it enables the site to supply accurately and support the efficiency of site work. This yields positive effects for both productivity, traffic situations and safety around construction sites, notably in dense urban areas.

13.8.2 Extending role 2: improving the efficiency of the supply chain

Besides resolution of issues in the interface, problems at both ends of the information and delivery processes within the supply chain itself may persist. The basic role of CSCM is to prevent problems in product definition, product manufacture and, in the end, presupposing the timely delivery to site for installation.

In the case studies, suppliers from materials to the hub, as well as direct shipments to site, were orchestrated by the LSP at the hub, which was in close communication with the contractor on-site. The decoupling and consolidation functionalities of the hub enabled the site to optimise the supply side to the maximum independently from variabilities on the side of the construction site.

The logistics extension implies improved shielding of the supply chain independently from the site. This enables the management of the supply chain and the logistics to run efficiently. Robust logistics management in the supply chain, including information support systems often controlled by suppliers and LSP, lead to higher levels of transport efficiency and lower levels of GHG emissions and other potential negative effects and risks of transport movements, particularly in urban contexts.

13.8.3 Extending role 3: transferring activities from the site to the supply chain

Transferring activities off-site yields objective and well-known benefits. Industrialisation as such leads to higher levels of accuracy and error correction. As a result, the productivity of manufacture increases compared to site construction. However, the total process of industrialised construction tends to be more complex and vulnerable to variability, even if part of the process on-site becomes less complex. The transfer of activities off-site also results in more complexity in the supply chain, which should be managed.

In the case studies, materials suppliers had been preassembling and prefabricating subassemblies and prefab modules at their plants. Also, the hub location provided work space and equipment to do additional preassembly and prefabrication during the last stage before materials and subassemblies were actually being called off and shipped to site. Besides this, as a basic logistics measure, the hub provided for the bundling of daily kits of the exact required materials per work location in the building. These measures taken collectively supported takt time planning, flow and productivity on-site.

The extended role of logistics here is to add to the controllability of the complexity and shielding against the increased vulnerability to variability. In addition, off-site construction generally leads to shipments of larger components to site that can be installed more efficiently, but they also require greater logistics accuracy. As well as this, the hub locations offer opportunities and accommodation for value-added logistics, e.g. flexible just-in-time prefabrication and preassembly of materials and components just in the right time and quantities, and adaptive to work speed on the construction site.

13.8.4 Extending role 4: integration of the site and the supply chain

Besides transferring activities off-site, integration of both off-site and on-site activities leads to stable supply chains that are integrated with site control with regard to products and information. This largely serves the aim of increased agility of logistics in the supply chain, leading to increased interaction and integration between the supply chain and the site.

In the case studies, two developments particularly supported the integrated control of the site and the supply chain, namely the design-for-logistics approaches and the procurement and purchasing routes. In particular, BIM enabled the planning of logistics provisionally in the design phase and conveying of logistics requirement to suppliers in the early stages, and enabled suppliers to plan and embed this in the manufacture of materials and components. In most cases, this was added to the purchasing discussions and contracts with suppliers. This was particularly the case if there was pressure from a dominant role for logistics improvement in the procurement between client and contractor, for instance by incentivised environmental issues, such as reduced GHG.

Further to value-added logistics, consideration of improved logistics and constructability become design criteria. This kind of wider integration in the supply chain results in further efficiency improvements in the integrated logistics and management of the entire process from the client's procurement, the design, the manufacture, the transportation, and the final installation on-site. This efficiency will influence the performance of the supply chain and the construction, and improved effects on the context and environment.

13.9 Conclusion

In this chapter, we observed four project cases applying advanced logistics solutions. The effects of these applications were assessed based on the KPI framework which was constructed for the research project reported. In the six dimensions of the framework, all four projects performed better logistically than what the projects would do traditionally. This implied improved transport efficiency and reduced environmental impact of the transport. The observations supported the applicability and usefulness of the framework and the benefits of improved transport and logistics management in construction.

The KPI framework appeared to be quite useful in describing the level of performance of the projects and the four factors of influence on the construction transport of the projects. Although the scope of the performance and the factors covered a wider range of issues than the framework could assess directly, the framework was able to accommodate those influences on the projects and the construction transport.

The application of a logistics hub and other efforts to decouple and consolidate construction transport have been found to play a key role in the logistics solutions and to achieve the effects reported. Second, applications of information management, notably WMS and TMS, are playing an increasingly important role in coordinating the stocks at the hub and ensuring accurate deliveries from the hub to site. The coordination is starting to extend to the design phase via BIM applications in order to determine the efficient delivery of daily materials packages to construction sites. Third, modularisation and the kitting of materials have appeared to be logistics solutions benefiting productivity on-site. Fourth, tendering and purchasing based on MEAT criteria rewarding GHG reductions by clients and governments, and also construction firms towards suppliers and transport firms, will likely encourage further efforts to improve transport efficiencies and reduce the environmental impact of construction transport.

Notwithstanding the benefits and potential, the scope of these improvements focuses mainly on the 'last mile', notably the transport from the hub to the site. In most cases, however, the deliveries from suppliers to the hub are not acted upon, although there is an improvement potential here, too. Further, and this goes for the application of the KPI framework as well as for ICT instruments mentioned, firms and individuals along the supply chain must be aware and systematically use and keep up those tools in order to be effective and cause the desired impact.

Based on the new empirical understanding and the previously introduced four basic roles of CSCM, extensions in each role have been observed and discussed. Regarding the extension of each of these roles, not only limited business objectives of firms in the supply chain are served, but also the wider interests of the public domain in which firms are developing their activities on construction sites and in supply chains.

To conclude, it can be stated that the generic body of knowledge accrued in the framework of CSCM has led to a basic understanding of the nature of construction supply chain problems, and provides direction for action and improvement, although with limited impact. However, current developments and demands in society and in cities in particular have called upon more comprehensive approaches to achieve wider impact when applying extended conceptualisations of CSCM. This should not only benefit the construction industry itself but also the environment, and improve the sense of well-being in society in which firms operate.

Acknowledgements

The empirical findings presented in this chapter are partly based on a joint research project of the NWO-financed 4C Construction Control Tower consortium of firms, governments and universities, including Delft University of Technology and Utrecht University of Applied Sciences. The project was aimed at the advancement of construction logistics in the Netherlands. The project has indicated directions for projects, firms and governments on how to advance CSCM and construction logistics management leading to improved logistics performance, more efficient construction projects, and reduced environmental impact.

References

Ala-Risku, T. and Kärkkäinen, M. (2006) 'Material delivery problems in construction projects: a possible solution', *International Journal of Production Economics*, *104*(1), pp. 19–29.

Bowersox, D.J. and Closs, D.J. (1996) *Logistical management: the integrated supply chain process*. Edition 3, McGraw-Hill, New York, 730.

Bowersox, D.J., Closs, D.J. and Cooper, M.B. (2002) *Supply chain logistics management*. McGraw-Hill, Boston.

Brett, P. (2007) 'Construction consolidation centres; an assessment of the potential for London wide use', *Project Ref: 17321/004*.

Bryman, A. (2008) 'Of methods and methodology', *Qualitative Research in Organizations and Management: An International Journal*, *3*(2), pp. 159–168.

Chan, A.P., Scott, D. and Chan, A.P. (2004) 'Factors affecting the success of a construction project', *Journal of Construction Engineering and Management*, *130*(1), pp. 153–155.

Christopher, M. (2016) *Logistics and supply chain management*. Pearson, Harlow.

Constructing Excellence (2007) London construction consolidation centre. Interim Report – May 2007. Transport for London. In partnership with Bovis Lend Lease, Constructing Excellence, Stanhope.

Cooper, M.C., Lambert, D.M. and Pagh, J.D. (1997) 'Supply chain management: more than just a new name for logistics', *The International Journal of Logistics Management*, *8*(1), pp. 1–13.

Cox, A. and Ireland, P. (2002) 'Managing construction supply chains: the common sense approach', *Engineering, Construction and Architectural Management*, *9*(5/6), pp. 409–418.

Dahlgaard, J.J., Khanji, G.K. and Kristensen, K. (2008) *Fundamentals of total quality management.* Routledge, London.

Dainty, A.R., Millett, S.J. and Briscoe, G.H. (2001) 'New perspectives on construction supply chain integration', *Supply Chain Management: An International Journal, 6*(4), pp. 163–173.

Denzin, N.K. and Lincoln, Y.S. (2008) *Strategies of qualitative inquiry.* (Vol. *2*). Sage, Thousand Oaks, CA.

Department for Transport (2007) *London construction consolidation centre.* Freight Best Practice. Department for Transport.

Elfving, J.A., Ballard, G. and Talvitie, U. (2010) 'Standardizing logistics at the corporate level towards lean logistics in construction', *Proc. 18th Annual Conference of the International Group for Lean Construction*, Haifa, Israel, pp. 222–231.

Fellows, R. (2009) 'Culture in supply chains', in Pryke S. (Ed. *Construction supply chain management: concepts and case studies*, Blackwell Publishing Ltd, Chichester, pp. 42–72.

Gemeente Utrecht (2015) Actieplan goederenvervoer 2015–2020: slimmer, schoner, veiliger en efficiënter goederenvervoer (In Dutch).

Hamzeh, F.R., Tommelein, I.D., Ballard, G. and Kaminsky, P.M. (2007) 'Logistics centers to support project-based production in the construction industry', *Proc. 15th Annual Conference of the International Group for Lean Construction*, EUA, East Lansing, MI, pp. 181–191.

Harland, C. (2015) *Supply chain dynamics.* Wiley Encyclopedia of Management, 1–2. John Wiley & Sons Inc, Hoboken, NJ.

Hoekstra, S. and Romme, J. (1992) *Integral logistics structure: developing customer oriented goods flow.* McGraw-Hill, London.

Koskela, L. (1999) 'Management of production in construction: a theoretical view', *Proc. 7th Annual Conference of the International Group for Lean Construction*, Berkeley, USA, pp. 241–252.

Koskela, L. (2000) 'An exploration towards a production theory and its application to construction', PhD Thesis, VTT Technical Research Centre of Finland.

Koskela, L. and Leikas, J. (1997) 'Lean manufacturing of construction components', in Alarcón L Ed. *Lean construction*, AA Balkema, Rotterdam, pp. 263–271.

Lambert, D.M., Pugh, M. and Cooper, J. (1998) 'Supply chain management', *International Journal of Logistics Management, 9*(2), pp. 1–19.

Lamming, R. (1996) 'Squaring lean supply with supply chain management', *International Journal of Operations and Production Management, 16*(2), pp. 183–196.

London, K. (2008) *Construction supply chain economics.* Taylor and Francis, London.

Luhtala, M., Kilpinen, E. and Anttila, P. (1994) *LOGI: managing make-to-order supply chains.* Helsinki University of Technology, Espoo.

Lundesjo, G. (2015) *Supply chain management and logistics in construction: delivering tomorrow's built environment.* Kogan Page, London, 288.

Mentzer, J.T., DeWitt, W., Keebler, J.S., Min, S., Nix, N.W., Smith, C.D. and Zacharia, Z.G. (2001) 'Defining supply chain management', *Journal of Business Logistics, 22*(2), pp. 1–25.

Monden, Y. (2011) *Toyota production system: an integrated approach to just-in-time.* Productivity Press, New York.

MVA (2006) Overview on consolidation centres. *MVA Consultancy*, 20 October 2006.

O'Brien, W.J. (1998) *Capacity costing approaches for construction supply-chain management.* PhD Thesis, Stanford University, Stanford, CA.

Park, M., and Peña-Mora, F. (2004) 'Reliability buffering for construction projects', *Journal of Construction Engineering and Management, 130*(5), pp. 626–637.

Pink, S., Tutt, D. and Dainty, A. (2013) *Ethnographic research in the construction industry.* Routledge, Abingdon.

Saad, M., Jones, M. and James, P. (2002) 'A review of the progress towards the adoption of supply chain management (SCM) relationships in construction', *European Journal of Purchasing & Supply Management, 8*(3), pp. 173–183.

SCC. (2010) *COR supply chain operations reference (scor®) model overview – version 10.0.* Supply Chain Council, Cypress, TX, USA, Available at: supply-chain.org. Supply Chain Council, Cypress, TX.

Shakantu, W.M., Muya, M., Tookey, J.E. and Bowen, P.A. (2008) 'Evaluating truck empty running in construction: a case study from Cape Town, South Africa', *Construction Economics and Building, 8*(2), pp. 41–49.

Sobotka, A., Czarnigowska, A. and Stefaniak, K. (2005) 'Logistics of construction projects', *Foundations of Civil and Environmental Engineering, 6*, pp. 203–216.

Sullivan G., Barthorpe S. and Robbins S. (2010) *Managing construction logistics*. Wiley-Blackwell, Chichester.

Transport and Travel (2010) Freight consolidation centre study. *Final Report*. Prepared for Department for Transport, Transport and Travel Research Ltd, Scott Wilson Ltd, Version 1.0, 14th July 2010.

Transport for London (2008) London construction consolidation centre. Final Report, Transport for London in partnership with Bovis Lend Lease, Constructing Excellence, Stanhope, Wilson James and Transport Studies Department from University of Westminster.

Transport for London. (2017) *Construction logistics plan guidance*. Version 3.0 July 2017 Transport for London, London.

UN (2017) Towards a zero-emission, efficient, and resilient buildings and construction sector. UN Environment and International Energy Agency, Global Status Report 2017.

Van Buren, N., Demmers, M., Van der Heijden, R. and Witlox, F. (2016) 'Towards a circular economy: the role of Dutch logistics industries and governments', *Sustainability, 8*(7), p. 647.

Varnäs, A., Balfors, B. and Faith-Ell, C. (2009) 'Environmental consideration in procurement of construction contracts: current practice, problems and opportunities in green procurement in the Swedish construction industry', *Journal of Cleaner Production, 17*(13), pp. 1214–1222.

Vrijhoef, R. (2011) *Supply chain integration in the building industry: the emergence of integrated and repetitive strategies in a fragmented and project-driven industry*. PhD Thesis, Delft University of Technology, IOS Press, Amsterdam.

Vrijhoef, R. (2015) 'Reducing the environmental impact and improving the efficiency of construction transport', *Proc. CIB Joint International Symposium Going North for Sustainability: Leveraging Knowledge and Innovation for Sustainable Construction and Development*, South Bank University, London, UK, IBEA Publications Ltd., pp. 363–375.

Vrijhoef, R. and Koskela, L. (2000) 'The four roles of supply chain management in construction', *European Journal of Purchasing and Supply Management, 6*(3–4), pp. 169–178.

Wegelius-Lehtonen, T. (2001) 'Performance measurement in construction logistics', *International Journal of Production Economics, 69*(1), pp. 107–116.

Womack, J.P. and Jones, D.T. (1996) *Lean thinking*. Simon and Schuster, New York.

14 Location-based management system now and in the future

Olli Seppänen

14.1 Introduction

A location-based management system starts from the assumption that the best way to plan and monitor progress of work in a construction project is based on physical locations. Locations are fixed and their status can be easily monitored (Kenley and Seppänen, 2010, p. 123). Tracking crews and operations is more complicated because they are, by nature, mobile, although real-time approaches in tracking crew flows are being developed in order to monitor workers (e.g. Olivieri, Seppänen, and Peltokorpi, 2017). From the point of view of Koskela's Transformation, Flow and Value (TFV) theory (Koskela, 2000), location-based management focuses on transformation processes within locations, the flow of work in a location and flows of workers through multiple locations.

Location-based planning methods can be compared and contrasted with activity-based planning methods, such as the critical path system. Although locations are often included in the Work Breakdown Structure of a Critical Path Method schedule, activity-based systems do not enforce the use of the same locations everywhere or maintain a consistent hierarchy of locations. Locations can be used to filter activities, for example based on location code, but they cannot be used to automate logic generation or plan continuous work. As a result, activity-based systems of a real construction project include hundreds or thousands of activities and their logic can be very complicated (Kenley, 2005).

In contrast, location-based planning systems use location as the basic unit of planning and control. Tasks are assumed to flow through locations. Logic is assumed to repeat in each location where the same two tasks exist, which means that the number of logic dependencies is greatly reduced in location-based systems. Kenley (2005) wrote that the efficiency of combining similar work in different locations solves the complexity problem of activity-based schedules. Activity-based approaches are mostly focused on Transformation and ignore several aspects of flow, such as the movement of crews through locations (Koskela, 2000).

In addition to decreased complexity, tasks flowing through locations (or operation flow as defined by Sacks, 2016) can be used to plan continuous work. Indeed, the emphasis of location-based planning is to plan for productivity. Continuous work means that the same crew is able to work on the same task continuously without breaks from location to location. The benefits of continuous work include increased learning effects, increased productivity, clear directions to crew members and a smaller risk of subcontractor crews leaving the project or charging for waiting time. Discontinuous work or 'Starts and stops' are a very important factor for subcontractor profitability and impact their decisions about which projects to prioritise (Sacks and Harel, 2006).

The Location-based Management System (LBMS) uses these concepts of locations and tasks flowing through locations to augment the traditional Critical Path Method (CPM) with concepts enabling workflow and using locations to automate the planning of logical relationships. In LBMS, the work is continuous by default and it is a planning decision to break continuous flow. In that sense, LBMS is an improved CPM algorithm. The term LBMS also refers to a method of planning using the LBMS algorithm emphasising the continuous workflow of crews and schedule optimisation by synchronising production rates and removing float between tasks.

These LBMS planning guidelines have been combined with the social process of Last Planner System (Seppänen, Ballard, and Pesonen, 2010) to make the planning process collaborative. LBMS as a planning process includes defining the LBS of the project, defining tasks and their quantities by location, defining relationships between tasks, aligning production rates and optimising the schedule. Finally, buffers can be inserted between tasks to account for variability and decrease the risk (Kenley and Seppänen, 2010). Takt planning is another set of location-based planning guidelines. Takt planning and control can use the same LBMS algorithm for calculations, but the planning guidelines emphasise exactly aligned production rates and, rather than using time buffers between tasks, risk management is done by underloading resources (Frandson, Seppänen, and Tommelein, 2015) or by adding buffers between work stages or at the end of the project.

In contrast with activity-based methods, LBMS emphasises production control during the execution phase. A traditional controlling approach in activity-based systems is based on a thermostat model of project control, in effect reacting to deviations on the critical path after they have happened (e.g. Meredith and Mantel, 1995). Production control in LBMS emphasises real-time information and forecasting problems before they happen. Seppänen (2009) defined how production can be forecast to warn of upcoming production problems at least two weeks before they happen. Armed with this information, production control becomes proactive, aiming at preventing problems before they happen by making adjustments to production rates and sequences. It can be said that LBMS puts more emphasis on production control than production planning and, in that sense, is clearly a Lean technique based on pull controlling (Seppänen, 2009). While LBMS is a more centralised strategy, takt controlling places more responsibility on the workers and trades themselves by making clear the targets and making the controlling process more transparent.

This chapter starts with a very short history of location-based planning. Next, the location-based planning system is presented. Although the differences between CPM and LBMS algorithms are described briefly, most of the emphasis is on discussing the production system risk and how to plan optimal schedules using LBMS or takt planning methods. Then, the location-based controlling system is presented. Again, the mathematics are briefly described but most of the section is devoted to reviewing assumptions of production control, a location-based controlling process and empirical results related to production control. The proposed process to combine the Last Planner System and LBMS is described next. The chapter concludes with a discussion of currently open research questions and potential future research directions. It should be noted that other methods, such as cost-loading schedules, management of detailed design, scheduling of deliveries, safety aspects, etc., are also heavily dependent on location. Many of these additional methods have been discussed by Kenley and Seppänen (2010, pp. 163–200), but they will not be discussed within the scope of this chapter.

14.2 A short history of location-based planning

Location-based planning and control methods have a long history. The earliest documented case study using location-based planning was the Empire State Building (built 1930–1931), which was built by Starrett Brothers. They completed the 102-storey building in record time, in 18 months, from sketch designs to opening, completed the structure at the speed of one floor per day and completed under budget and with a high safety record (for the time). The management of the project was based on repetition, continuous flow and trying to achieve an assembly line of production (Willis and Friedman, 1998). Shreve (1930) first introduced the concept of cascading delays and stated that, to achieve high speed, they needed to disconnect the different portions of the work as much as possible. This concept of buffers is an important part of current LBMS methodologies. However, the location-based approach of Starrett Brothers did not have an analytic method based on calculations and was more of a method of presentation.

Line-of-balance was the first more analytic method. It originated from the US Navy in the 1940s where it was used as a planning and control tool (Lumsden, 1968). Lumsden (1968) described that the technique is a way to model repetitive construction. Repetitive units were modelled with their own CPM network and two lines were drawn in a line-of-balance diagram: one for the start of sub-network and one for the end of sub-network. The vertical axis showed the number of produced repetitive units. Line-of-balance also included balancing production rates by changing the number of crews (Lumsden, 1968). The line-of-balance method relied heavily on having exactly repeating locations and was mainly used for housing schemes of repeating units. The line-of-balance approach was expanded to have more flexibility later by Arditi, Tokdemir, and Suh (2002).

The flowline approach by Mohr (1979) is based on the work of Selinger (1973, 1980)) and Peer (1974). Flowlines specifically show crew movements. Each task is represented as a single line, rather than the dual lines of line-of-balance. Rather than having the number of repetitive units on the vertical axis, the flowline method was based on discrete locations. However, the flexible location breakdown structures of LBMS, where locations can be defined on multiple hierarchy levels and can differ in size, were not considered in the flowline method and the method was still largely a visualisation technique. Flowline visualisation (Figure 2) is still being used as the primary schedule visualisation method of LBMS.

Several location-based methods can be considered *integrated* methods in the sense that analytic CPM methods are integrated with location-based methods. As an example, Russell and Wong (1993) first tried to solve the complexity problem of activity-based schedules with a system they called *representing construction.* They created a classification of logic types that could be automated based on locations. These logic types are very similar to the layered logic used in LBMS. The Repetitive Scheduling Method is another attempt to integrate CPM and location-based methods (Harris and Ioannou, 1998).

The LBMS builds on the earlier work and is based on an augmented CPM algorithm that incorporates *layered logic* (related to Russell and Wong's (1993) work) and *continuity heuristics* to plan for continuous work. The planning and controlling methodologies and processes are heavily based on the work of Kankainen and Kiiras from Helsinki University of Technology (Kiiras, 1989; Kankainen and Sandvik, 1993). The controlling methods and calculations have been developed by Seppänen (2009). The controlling methodologies have been improved over the years by empirical studies (Seppänen, 2009; Kenley and Seppänen, 2010; Seppänen,

Evinger, and Mouflard, 2014). The system has been presented in numerous IGLC conferences (first appearance Kankainen and Seppänen, 2003). Recent development has focused on the combination of Last Planner System and LBMS (Seppänen, Ballard, and Pesonen, 2010; Seppänen, Modrich, and Ballard, 2015; Dave, Seppänen, and Modrich, 2016) and comparing the LBMS and takt time planning methods (Seppänen, 2014; Frandson, Seppänen, and Tommelein, 2015).

14.3 Location-based planning system

The LBMS requires as its foundation a location-based plan. The location-based planning system is described in this section. It is composed of a technical system based on the LBMS algorithm, flowline visualisation, and guidelines and best practices for planning and optimising a schedule and analysing its feasibility and risk levels. This section starts by describing the various components of the location-based plan, then the flowline visualisation of a plan is described. Logic and calculations related to the LBMS algorithm are briefly described. Risk management and buffers are an important part of location-based planning and they are described next. Finally, the guidelines for optimising a plan are described from two alternative viewpoints: location-based planning system and takt planning. These viewpoints are then compared and contrasted.

14.3.1 Location breakdown structure

The Location Breakdown Structure (LBS) is one of the most important up-front planning decisions in LBMS. LBMS is the first location-based planning method that allows for a hierarchical LBS with unlimited hierarchy levels. For example, the project can first be divided into buildings, buildings can be subdivided into structurally independent sections, which can be divided into floors and then to interior zones. Different construction phases can have a different breakdown. For example, exterior work can ignore floors and be divided based on the side of the building, and structural work can be divided based on pour areas. The most important thing is to have the same LBS for all tasks of the same phase because sharing the same LBS decreases complexity and increases the power of the system (Kenley and Seppänen, 2010). For this reason, when combining LBMS and LPS, the phase scheduling sessions begin by defining a common LBS for all tasks of the phase (Seppänen, Ballard, and Pesonen, 2010).

Because projects have widely varying characteristics, it is difficult to present general guidelines for how to create locations. However, for building construction projects, Kenley and Seppänen (2010) proposed that the highest level division should be based on structural independence (for example, different buildings or structurally independent areas based on movement joints, etc.). The middle levels include a floor division to model physical constraints. Finally, the lowest hierarchy level should always have small locations so that only one subcontractor can work efficiently in the area at one time (Kenley and Seppänen, 2010). Linear construction projects often have several roadlines, intersections, tunnels, bridges and so on. These form the highest level of division. At the lowest level, linear projects can be scheduled based on chainage of the project and calculations can be made continuously on a meter level of accuracy rather than based on 'discrete' locations such as floors of a building (Kenley and Seppänen, 2010).

14.3.2 Tasks, location-based quantities and duration calculation

In LBMS, tasks are packages of work, which can be completed in a location by the same crew with no breaks and share the same external dependencies to other tasks. A task usually contains work in several locations. This is a key difference to CPM, where activities are always located in one location. In a way, tasks are collections of CPM activities. The basic assumption of LBMS is that tasks are performed continuously, without breaks from one location to the next (Kenley and Seppänen, 2010).

The work content of a task can be based on quantities. There are two ways to achieve this. Firstly, if the project's quantity take-off has been created by location, for example by using BIM tools (e.g. Kala, Seppänen, and Stein, 2010), it can be used as the basis for scheduling. Alternatively, tasks can be determined first, for example by integrating the collaborative Last Planner System™ phase scheduling process and LBMS, and then quantities can be estimated for each collaboratively determined task separately. In each case, one or more quantity items will be assigned to a task. Quantities describe the scope that will be accomplished when a task is finished in the location and make it easier to evaluate whether the task is complete. For example, the same drywall crew can install different types of drywall in a location before moving to the next location. The quantity items could include installing full-height drywall for corridors, water-resistant bathroom wall, double board wall for meeting rooms, etc. Each quantity item can have its own resource consumption, measured as man-hours/unit (for example, 0.46 man-hours/m^2 for standard living room wall). By multiplying each quantity by its labour consumption, the total number of man-hours in each location can be calculated.

Duration calculations of LBMS are based on these total man-hours. To calculate the duration, more planning input is required related to crews, shift length and the difficulty factor of a location. The basic assumption of LBMS is that tasks have an optimum crew composition which will most efficiently complete the work. However, duration can be changed by increasing or decreasing the number of crews of optimal composition (Arditi, Tokdemir, and Suh, 2002). If the locations are big enough to accommodate multiple crews, the assumption of LBMS is that adding crews will not impact productivity (but will increase risk as described later in the risk management section). Therefore, the number of optimal crews becomes a critical planning decision. The duration in number of shifts can be calculated using the following steps (Kenley and Seppänen, 2010, p. 133):

1. Quantity of man-hours needed to complete the location;
2. Divide by the total number of crew members (duration in hours);
3. Divide by the shift length (duration in shifts);
4. Multiply the duration in shifts by the difficulty factor.

In addition to duration calculation, the quantities can be used in several ways in other parts of the LBMS. For example, deliveries can be planned when the need time for each quantity item is known. On the other hand, quantities can link together the production schedule and procurement schedule or be used for cost loading the schedule. These additional methods are not within the scope of this chapter but are described in Kenley and Seppänen (2010, pp. 163–193).

14.3.3 Flowline visualisation

Quantities determine the locations where each task is located and the duration of these tasks. This information can be used to plot the flowline of a task. In a flowline figure, the LBS is shown on the left and the time is shown horizontally going to the right. Each task is shown as a diagonal line. The slope of the line signifies the production rate of the task. Assuming that the difficulty factor and crew size of each location are the same, the flowline slope reflects quantity variation between locations. When multiple flowlines are shown together, work sequence can be read horizontally. Optimisation opportunities and wasted time can be seen in the schedule by looking at empty areas between tasks. Figure 14.1 shows a sample flowline figure of three tasks.

Compared to Gantt Charts, flowline figures are a very efficient way of showing information and they enable seeing the big picture. It is very easy to see whether the flowline schedule has been optimised by looking for empty areas. In contrast, it is quite difficult to evaluate the quality of schedules from a Gantt chart. Large Gantt charts can include thousands of activities on dozens of pages. In contrast, even large construction phases

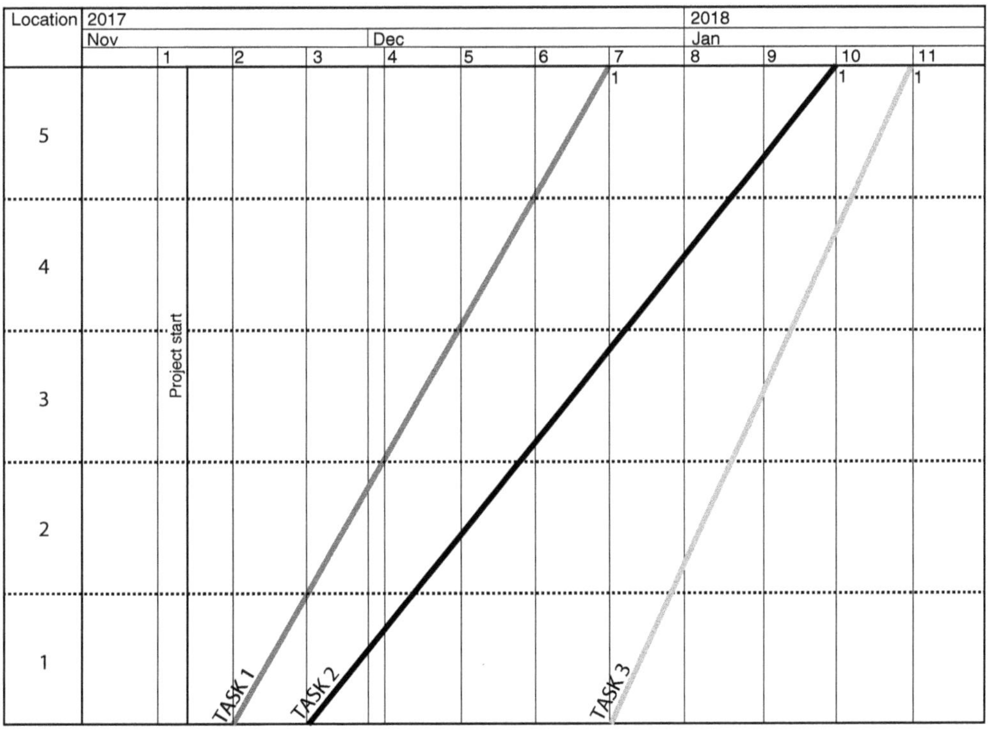

Figure 14.1 A flowline figure of three tasks. All tasks are performed continuously because there are no gaps between locations. It is possible to see optimisation opportunities by looking at empty space between the three flowlines, work sequence by reading horizontally from left to right and the flow of operations by following flowlines diagonally

can be represented in one flowline diagram. However, visualisation is not a key difference between the different systems, because even a CPM schedule which has locations and dates can be visualised as a flowline diagram (Olivieri, Seppänen, and Granja, 2018). We will next move to the LBMS calculations and methodologies, which are the real difference between activity-based and location-based approaches.

14.3.4 Layered CPM logic in location-based schedules

LBMS uses the locations to automate the creation of logic between tasks. The *layered logic* of LBMS includes *five layers* which use locations or hierarchy levels in a different way to do this. These logic layers are described next and, finally, two flowline figures illustrating all the different layers are presented (Kenley and Seppänen, 2010, pp. 133–142).

14.3.4.1 Layer 1: external logic relationships between activities within locations

In this logic layer, a relationship applied between two tasks will be applied in each location where both tasks exist. For example, a relationship stating that **painting** (a task) must happen after **drywall** (a task) **on each floor** would be a layer 1 logic link (Figure 14.2).

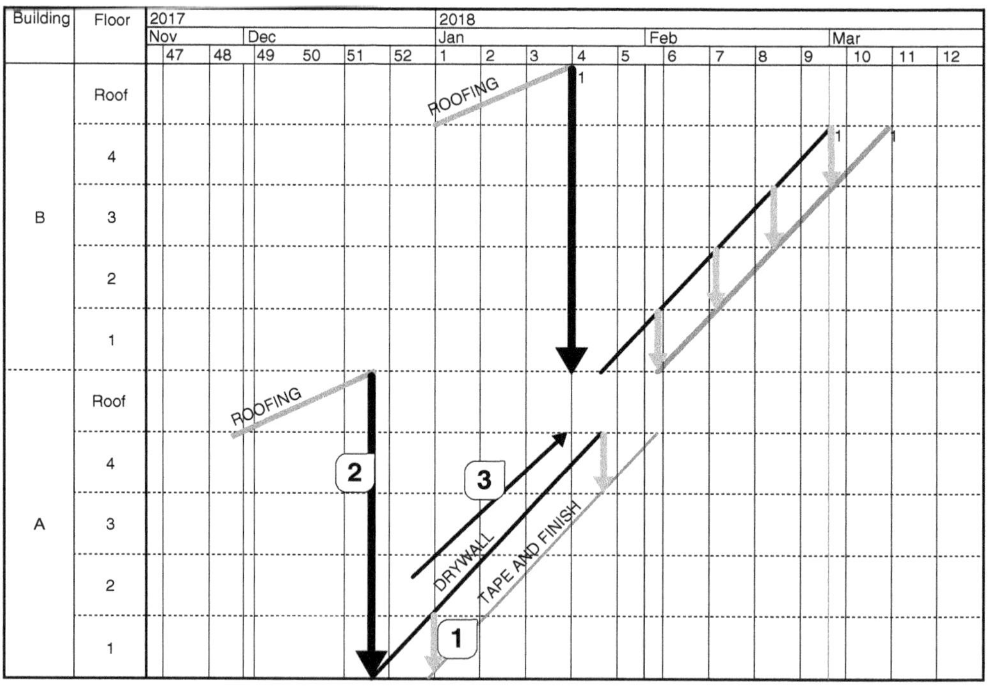

Figure 14.2 A flowline figure illustrating logic layers 1–3. The figure calls out layer 1 links between Drywall and Tape and Finish. Roofing has a layer 2 link to drywall. All tasks have layer 3 links to model internal location sequence (Drywall shown in the figure)

14.3.4.2 Layer 2: external logical relationships driven by different hierarchy levels

Layer 2 extends Layer 1 logic by allowing a different hierarchy level of the LBS to determine the logic link. For example, a relationship stating that **roofing** (a task) must precede **drywall** (a task) **in each building**, would be a layer 2 logic link (Figure 14.2).

14.3.4.3 Layer 3: internal dependency logic between locations within tasks

Layer 3 links are unique to LBMS. They are used to model the movement of crews through locations (the flow of operations). The basic assumption of LBMS is that a crew completely finishes a location before moving to the next location. The links are generated based on a task's location sequence, which can be planned individually for each task or for several tasks at once. For example, **drywall** (a task) can be planned to proceed from Building A, first floor, up through the building and then to Building B, first floor and up through Building B (Figure 14.2).

14.3.4.4 Layer 4: additional location-based links

Layer 4 links account for location lags in external logic. This is similar to layer 1 logic, but includes a location lag which can be positive or negative. For example, in a cast-in-place structure, **the pouring of horizontal concrete** (a task) precedes the **formwork** of the floor above with a location lag of one floor. It also precedes **masonry walls** with a negative lag of two floors (Figure 14.3).

14.3.4.5 Layer 5: standard CPM links between any tasks and different locations

Finally, layer 5 allows for any task and any location to precede any other task in any location. This is the only layer of logic in the standard CPM. In LBMS, layer 5 links are typically used to tie different construction phases together because construction phases often do not have the same locations. For example, fireproofing could be the last task of the structural phase and kick off the interior rough-in phase, which divides the locations differently. In this case, fireproofing would need to be linked to the first task of interior rough-in with layer 5 links.

Layered logic reduces the complexity of schedules because the same project can be modelled with much fewer links (Kenley, 2005). The benefit of LBMS increases with more tasks and more locations. Olivieri, Seppänen, and Granja (2018) modelled three schedules of real projects in CPM and LBMS and the number of links required to model the schedules was reduced between 33% and 44%.

14.3.5 Differences between LBMS algorithm and CPM algorithm

LBMS algorithm is similar to CPM algorithm with a few important subtle differences. The algorithm is not presented in detail here, and interested readers can refer to Kenley and Seppänen (2010, pp. 147–156). However, a few key differences are worth mentioning here. In schedule planning, the difference of LBMS and CPM calculation relates to planning continuous work and to float and criticality calculations when continuous work has been planned.

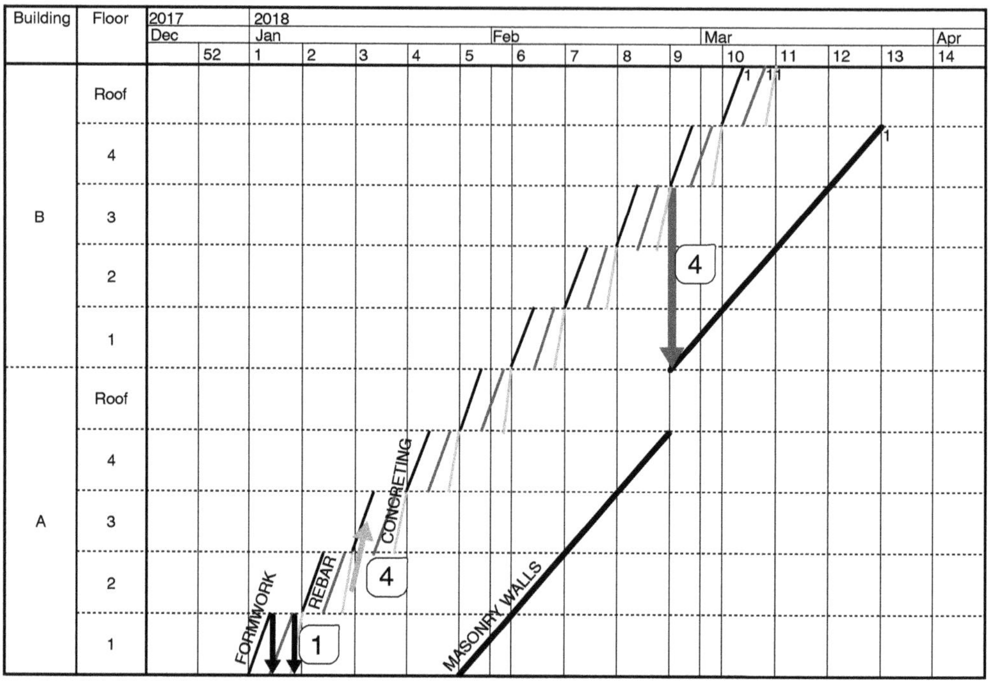

Figure 14.3 Two examples of layer 4 links are shown – concrete of the previous floor must be completed before formwork on the next floor and the concreting must be ready two floors above before masonry walls can start in a location. Additionally formwork, rebar and concreting on the same floor follow layer 1 dependencies

Figure 14.4 illustrates four tasks: a task of standard production rate, a slower task, a faster but continuous task, and a faster but discontinuous task. The third task, faster and continuous, is only possible with the LBMS algorithm. The earlier locations of the task are 'pulled' by the later locations enabling continuous work. In standard CPM, faster tasks are always discontinuous due to the lack of this continuity heuristic. This is a critical difference because forcing work to be continuous enables schedule optimisation by synchronising production rates. By default, all tasks in LBMS are continuous and it is a planning decision to make them discontinuous.

14.3.6 Risk management and buffers

One of the main goals of LBMS is to decrease the risks related to schedules. There are several types of uncertainties that can impact production, for example uncertainties related to environment and prerequisites of production. The most important ones handled directly by LBMS include uncertainties related to adding resources, resource availability, productivity rates and locations (Kenley and Seppänen, 2010, p. 181).

Every time a new mobilisation is called for by the schedule, there is a risk that the resources will not be available when needed. This risk applies for the first mobilisation as well as for any subsequent ones if the work is discontinuous or additional resources are

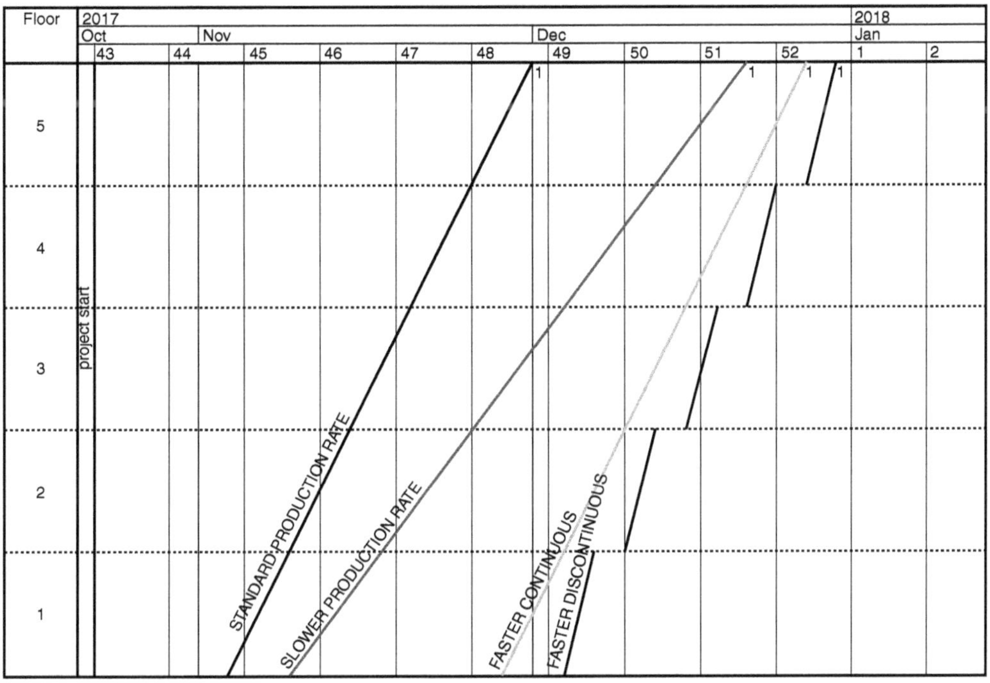

Figure 14.4 The main difference of LBMS and CPM algorithms illustrated with a flowline diagram. The start dates of tasks can be pulled by the later locations in the LBMS algorithm to make work continuous. In this figure, the start date of task 'Faster continuous' has been pulled to achieve continuous flow. Ordinary CPM would leave faster tasks discontinuous (like task 'Faster discontinuous')

required in the schedule (Kenley and Seppänen, 2010, p. 182). This risk can be minimised by planning continuous work and protecting the continuous workflow from variability by adding buffers.

Resource availability is always a risk. It is possible that the subcontractor does not have enough crews available, or could have more important projects which are delayed at the time when task commences. Similarly, a subcontractor could have too many resources available and mobilise with too large a crew if other projects have lower demands (Kenley and Seppänen, 2010, pp. 182–183). This risk can be mitigated by collaborative phase scheduling where commitments are made to resource levels required to achieve the phase schedule.

Productivity rates used to plan a schedule are always based on averages and set a good target productivity. However, there are huge individual differences in productivity and even the same individual can have different productivity over time (Kenley and Seppänen, 2010, p. 182). This risk can be mitigated by first run studies (first brought to construction by Parker and Oglesby, 1972) and by active production control during construction.

All the different types of uncertainty lead to variability. Lean Construction aims to minimise variability. However, some variability will always remain in the production system. LBMS protects against the remaining variability by including buffers in the schedule. Buffers are inserted between critical tasks to protect their continuous flow. Buffers delay the start date of the succeeding task from the earliest possible start date. They can be absorbed during production if predecessors get delayed. This gives time for control actions before the predecessor task impacts the successor task (Seppänen, 2009). In ordinary CPM, the buffers are included within task durations. In LBMS, the buffers are taken out of durations and made explicit between the tasks.

It should be noted here that the main difference between takt planning and LBMS approaches is the type of buffer used. Takt approaches protect against variability by underloading resources (i.e. having more resources available than needed to do the work). If the workers run out of work in takt planning, they will work on other, non-repetitive backlog work in the building (Frandson, Seppänen, and Tommelein, 2015). Additionally, buffers can be placed at the end of a project. Currently ongoing project work utilising the takt approach suggests that removing buffers from between tasks creates a continuous urgency in takt projects, establishes social interaction between the tradesmen who work very close to each other, and allows the buffer to be used up only when required by the project. The LBMS approach decreases cycle times compared to the CPM approach, but takt seems to be able to achieve even shorter cycle times because the buffer is only used when required. Empirical results related to these observations have not yet been published.

14.3.7 Schedule optimisation using location-based planning techniques

The starting point for schedule optimisation in LBMS is a schedule where resources have been determined for each task separately (for example, by discussing with subcontractors or by using one crew for all tasks). In the initial schedule, all tasks are continuous. This will result in some trades flowing through the building at a slower rate than others. The continuity requirement pulls the start dates of faster tasks and leaves empty space between tasks (Kenley and Seppänen, 2010, p. 221). An example schedule of a six-storey building with unsynchronised production rates is shown in Figure 14.5.

The optimisation process focuses on aligning the schedule in such a way that the empty spaces are eliminated. In Figure 14.6, there are empty spaces before Install Ductwork (slower than predecessor), Framing (faster than predecessor) and Finishing work (slower than predecessor). Empty spaces can be eliminated by changing resources (for example, adding a larger crew to install ductwork), changing scope (for example, having Drywall Install crew do some Finishing work), changing location sequence (does not apply to this example), splitting tasks (for example, performing floors 1–3 continuously and then having a break before floors 4–6), or switching to discontinuous work (Framing crew leaving the site after each floor) (Kenley and Seppänen, 2010, pp. 221–222). In this example, the schedule is aligned by changing the number of crews.

Figure 14.6 shows the same schedule after the resources from Framing have been decreased from two crews to one crew. The total scheduled duration decreased from 44 weeks to 40 weeks. In location-based planning, improving the alignment of schedules will shorten project durations. The ability to complete the building earlier by decreasing manpower is called the *location-based planning paradox*.

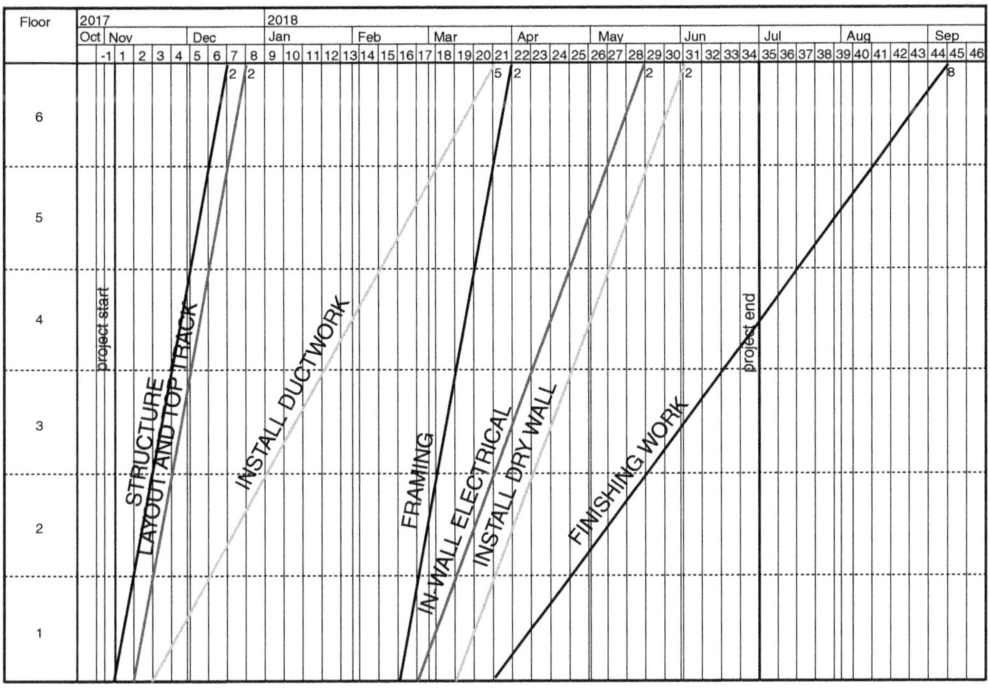

Figure 14.5 A flowline of a six-storey building with unsynchronised production rates

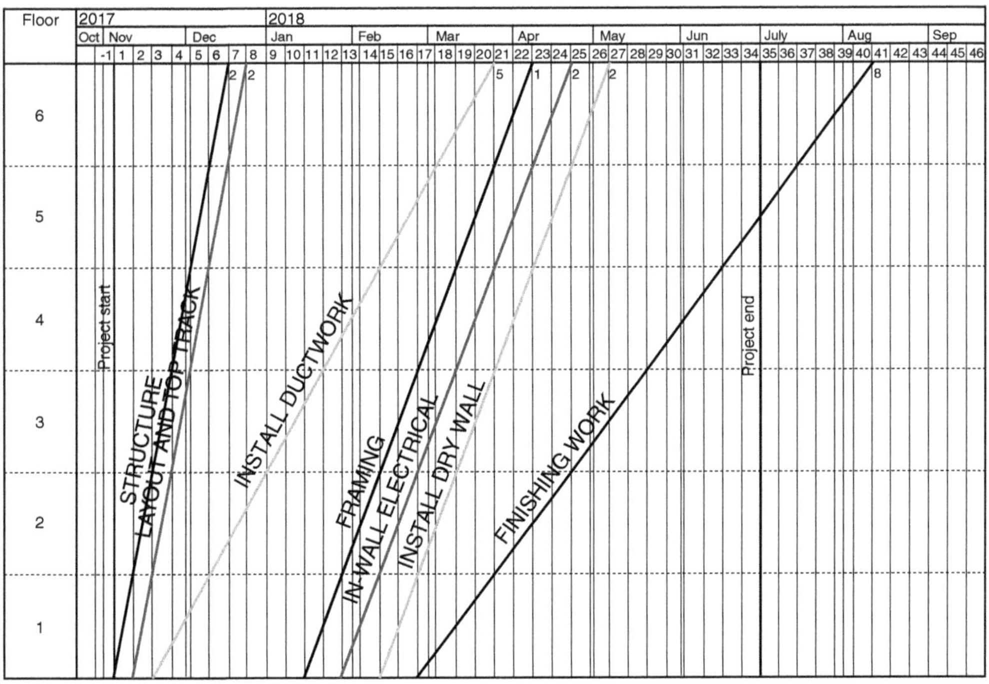

Figure 14.6 Flowline figure of the six-storey building, one crew decreased from Studs

As the final alignment step in this simple example, more resources are required to Overhead MEP install and Finishes tasks. Resource constraints should be reviewed with the subcontractors when increasing resources. Increasing Ductwork crews from 5 to 8 (optimal crew of 2) and Finishes crew size from 8 to 16 would perfectly synchronise all tasks from Overhead MEP to Finishes. Figure 14.7 shows the results of these two changes. The project duration has decreased to 22 weeks, which is 12 weeks earlier than the deadline (vertical line in the figure).

Up to this point, LBMS and takt approaches are similar. They both aim at perfectly aligned production rates. However, the schedule of Figure 14.7 is more risky than the original because it does not include any buffers between tasks and it requires more resources for the Ductwork and Finishes crews. Any delay along the way would cascade through the project. To prevent these cascading delays, the LBMS method would add buffers between the tasks, while considering the risks and variability related to each task.

In contrast, the takt approach would underload the resources to make sure that each crew will be able to finish on time in each location, most of the time. If there is unused capacity, it can be consumed in 'off-takt' workable backlogs (Frandson, Seppänen, and Tommelein, 2015). Because the total planned duration with takt approach is shorter, there is also a buffer at the end of the project (weeks 22–34), as in our example above. This buffer can be used when required. More research is required on how these different types of buffer actually impact production. Anecdotal evidence suggests that time buffers of LBMS make workers and management lose the sense of urgency, and buffers are consumed even if there are no problems. In contrast, takt buffers seem to lead to a continued sense of urgency, which may lead to higher performance during production phase.

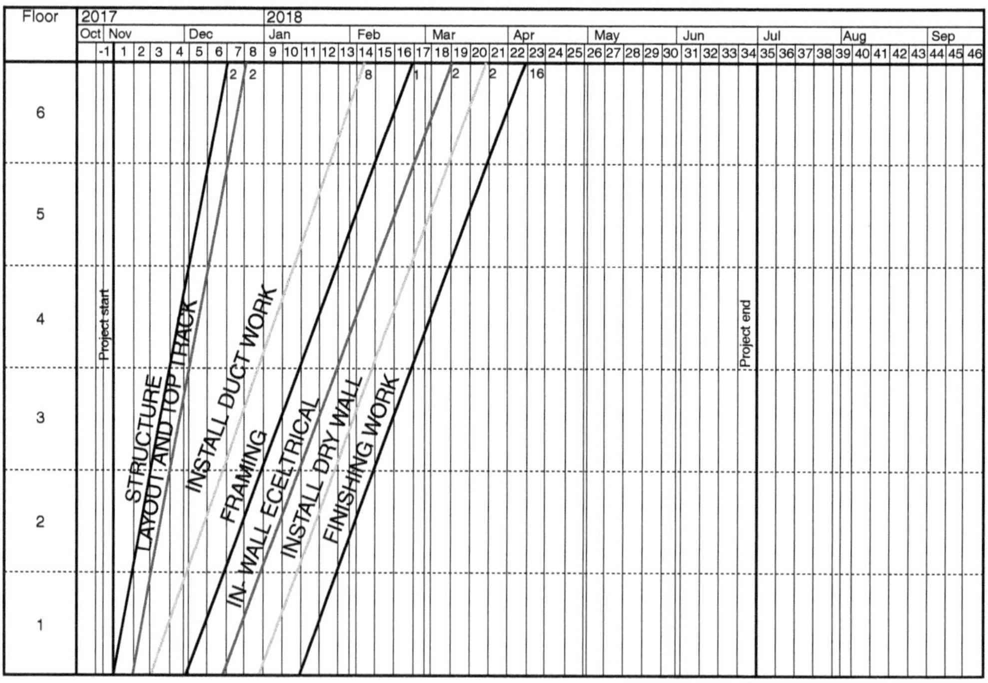

Figure 14.7 Aligned schedule

14.4 Location-based controlling system

In LBMS, controlling is given more weight than planning. Plans are always based on assumptions and the best way to control the project is to collect as much real-time information as possible, react to any deviations and proactively make things happen according to plan. This is in contrast to the 'after-the-fact' approach of controlling, which is at the core of the CPM controlling model (Meredith and Mantel, 1995). Koskela and Howell (2001) called the activity-based model the *thermostat* model of controlling, which is overly simplistic.

Controlling in LBMS includes monitoring the status of locations and labour on-site to calculate actual productivity, visualising status in control charts and flowlines, forecasting progress based on actual production rates and giving alarms to warn of upcoming problems to enable proactive control.

14.4.1 Location-based status monitoring

The basic progress monitoring in LBMS focuses on four aspects (Seppänen and Kenley, 2005):

1. Actual start and finish dates and interruptions;
2. Actual quantities;
3. Actual resources;
4. Actual shift length and days off.

Actual start and finish date of each location is a basic requirement for tracking and is required for all downstream calculations and visualisation. If actual start dates and finish dates are known, it is possible to show the status in a control chart or plot progress in a flowline diagram. It is also possible to calculate actual production rates (quantities/shift). However, actual production rates get distorted if any interruptions longer than a day are not recorded. Because the goal of LBMS is to minimise interruptions, the number of interruptions is an important metric in its own right (Kenley and Seppänen, 2010, p. 273).

Tracking actual quantities for each location enables the detecting of quantity deviations, which can become critical if they repeat in other locations. These can be caused by measurement errors, undocumented change orders, or an attempt by a subcontractor to invoice for work outside their scope (Kenley and Seppänen, 2010, p. 272). If quantity deviations are not detected, any attempts to calculate production rates (units/shift) or resource consumption (man-hours/unit) will be based on incorrect quantities and are not usable for estimating future work. Actual quantities can be easily measured by using BIM tools, assuming that the model reflects as-built conditions.

Actual resources are important for the calculation of actual resource consumption and can be useful to detect root causes of deviations. For example, a poor production rate may be caused by higher labour consumption than planned (lower productivity), indicating incorrect estimates or problems with production. Fewer resources than planned could also be the cause. It is particularly interesting to monitor changes in resource consumption because this can indicate problems in production. Actual resources have traditionally been hard to track and are often based on self-reporting by subcontractors. However, it is common that subcontractors report the information only as total workers on-site and it is hard to figure out where the resources were working

(Kenley and Seppänen, 2010, pp. 272–273). In the future applications of automatic tracking technology based on indoor positioning (e.g. Olivieri, Seppänen, and Peltokorpi, 2017), labour can be tracked automatically, and these problems involving inadequate production control data can be resolved.

14.4.2 Forecasting and alarms

The actual progress can be used to calculate forecasts. Forecasts are based on the actual resource consumption (if available) or on actual production rates. The assumption is that the task will continue with the same resources and the same productivity unless control actions are taken. Alarms are generated when a predecessor is about to interfere with the successor. The goal of proactive production control is to prevent the alarms from turning into actual production problems, which can start a chain of cascading delays. The mathematics of calculating forecasts were described by Seppänen (2009, pp. 113–115). Figure 14.8 shows a flowline figure with planned schedule, actual progress, forecasts and alarms.

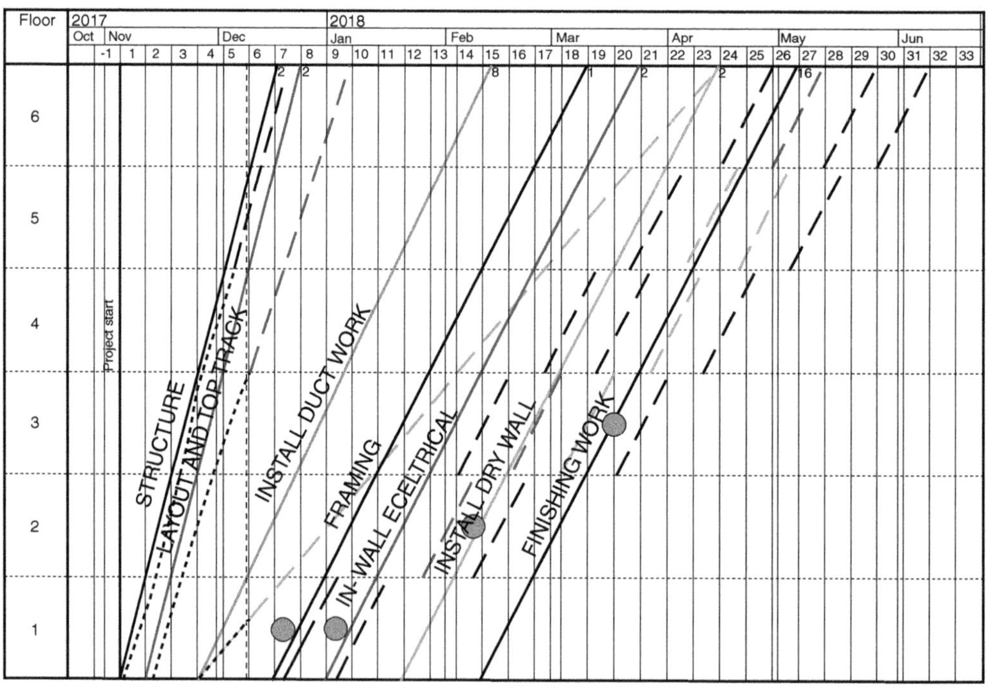

Figure 14.8 Flowline figure with planned lines (solid), actual lines (dotted) and forecasts (dashed). Tasks Structure and Layout/Top Track are slightly behind schedule, but have not caused any alarms. Ductwork installation is one week behind, but is going slow. It will impact Framing in two weeks and cause a cascading delay if the production rate is not corrected

14.4.3 Planning control actions

Control actions are taken to recover from a deviation in order to prevent interference with other tasks or project delay. In LBMS, plans (solid lines in flowline) are not updated. Rather, a control action is defined with the specific goal of preventing interference. Control actions adjust the forecast (the dashed lines in flowline). Examples of possible control actions include: improving productivity by reducing waste, changing the number of resources, working overtime or on weekends, changing sequence, delaying successor task, etc. (Kenley and Seppänen, 2010, p. 283).

For example, in the situation of Figure 14.8, the control actions could target the task causing the problem: Install Ductwork. First, the root cause of the problem would be identified. Then, an action plan would be documented (possibly as an A3 report) and its likely impact reflected in the forecast to see whether the plan solves the problem. If the root cause was an understaffed crew, a control action could be mobilising additional resources (assuming the same productivity).

14.4.4 Cascading delays in construction

Seppänen (2009) investigated the impact of production problems on production. A production problem was defined as a start-up delay, discontinuity, or slowdown caused by interference from other tasks. Production problems were found to cause downstream problems via multiple mechanisms. Cascading delay chains were started by complex combinations of resource issues, production management decisions and out-of-sequence work. These led to multiple contractors working in the same location, resulting in slowdowns and demobilisations with the associated return delays. However, LBMS was able to create alarms before they happened; and, in the study, the LBMS forecasting method was further developed by adding more information about resource availability to generate alarms even earlier (Seppänen, 2009, p. 162). Figure 14.9 shows an example of cascading delays from the interior phase of an office project. The contractor providing the vinyl floor covering had resource problems during the summer holiday period and their slow initial progress caused a cascade of problems that lasted until commissioning. Several return delays where contractors had to leave the site and did not come back immediately when the predecessor was completed can be seen from Figure 14.9.

14.4.5 Empirical results of location-based control

Several empirical studies have been carried out related to location-based control. A series of 30 master's theses based on action research were done in Finland when the control methods were first being developed in the 1980s and 1990s. Each thesis brought some improvements to the controlling techniques. The overall results were documented in Finnish in a handbook of production control, published in three editions (Kankainen and Sandvik, 1993; Kolhonen, Kankainen, and Junnonen, 2003). The research method on all these was action research.

Empirical research based on six projects, including four residential projects, a business park and a school building, was conducted by Seppänen and Kankainen (2004). The analysis was based on archival analysis where the planned and actual flowlines were analysed for start-up delays, production rate deviations, interruptions and final delays. The findings included serious problems related to the controlling of interior work. In general,

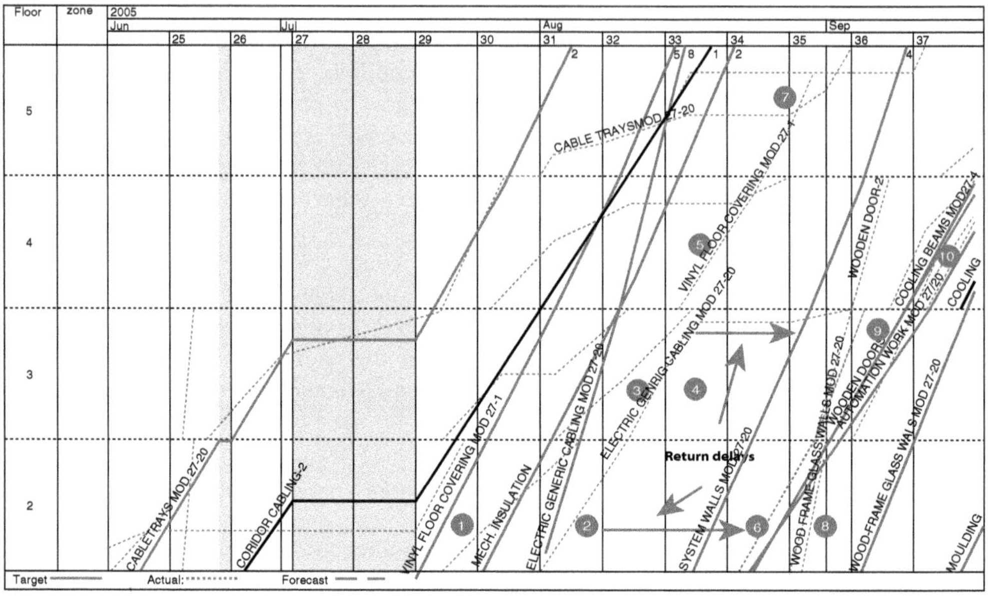

Figure 14.9 Cascading delays in an office project, caused by slow start of vinyl floor covering work (problem 1). Return delays were caused to successor trades (problems 2 and 4), which caused overcrowding in locations (problems 9–10) and general slowing down of work (adapted from Seppänen, 2009)

tasks started on time but had interruptions or slower than planned production, resulting in a typical delay of 2–4 weeks for each interior task. Overall, 71% of tasks were planned to be continuous, but only 33% of tasks were actually continuous. Buffers between tasks were found to have a statistically significant negative correlation to interruptions (i.e. more buffers, fewer interruptions). The study also calculated the prevalence of production problems. Tasks which started late had fewer slow-downs. The conclusions of the research are that simply planning continuous work is not enough, that controlling is a critical process, that discontinuities are the hardest deviation type to recover from and that starting too early results in slowdowns.

This study led to the doctoral thesis of Seppänen (2009) where the goal was to find out how production control on-site works, how reliable are the production plans, which factors explain the success or failure of the plans, and attempts to improve the LBMS controlling system to give better information to decision makers. The study was based on archival review but also detailed observations of process on-site to find production problems. The factors related to the success and failure of the plans were found to be cascading delay chains, resource problems, detailed planning process, not taking control actions, and uninformed production management decisions including push controlling start dates. However, it was possible to forecast many of the problems before they happened. All three case studies had a lot of cascading delays and production problems but still managed to achieve substantial completion on time, although in each case the time for hand-off and self-commissioning activities were compressed. One of the projects was able to achieve a duration compression of 10% through the use of schedule optimisation, while others achieved durations comparable to other projects of the same type.

Kala, Mouflard, and Seppänen (2012) evaluated the production control approach from three perspectives on a large hospital project in California: (1) time used running LBMS technical system compared to keeping a CPM schedule up to date, (2) quality of information for decision making from LBMS, (3) reliability of the planning process. The researchers found that the CPM schedule required more man-hours to operate the technical system than the LBMS process. LBMS was also able to provide better information for superintendents for decision making to enable proactive control. In contrast with Seppänen's (2009) study, start dates had delays of three weeks on average, but the production rates were very close to planned. It could be that the improved forecasting ability of LBMS based on Seppänen (2009) and the actual use of forecasts to guide decisions helped to keep the production rates in control. An interesting result related to man-hour estimates by subcontractors was found. Subcontractors were over-estimating their resource consumption by 30–40% on average. This could either reflect considerable capacity buffers (in addition to time buffers) or substantially better than average productivity on this project.

Evinger, Mouflard, and Seppänen (2013) evaluated the productivity effects of following the standard practice of starting as soon as possible. They used two case studies where one was performed with a CPM schedule and traditional management practices, and one had a mix of CPM and LBMS strategies. The most interesting results were in project two, where similar work was carried out on different floors based on CPM or LBMS. The floors were patient floors and had identical scope. CPM floors had on average 18% higher labour consumption (poorer productivity). On average, CPM floor tasks had 10% lower production rates than LBMS floors. The authors concluded that running the project with a CPM methodology resulted in lower productivity and lower production rates than using LBMS methodology.

Seppänen, Evinger, and Mouflard (2014) evaluated the impacts of LBMS on production rates and productivity. They tracked the production alarms generated by LBMS and what kind of control actions were taken by the production team in three projects. Of the LBMS alarms, 39% resulted in control actions agreed with the team; 65% of actions related to improving production rates were actually able to achieve a higher production rate and the increase was on average 37%; and 50% of the actions were able to prevent a production problem. In several cases, the number of resources did not increase correspondingly, but the productivity of the subcontractor increased. The authors suspected that production rate increases were mostly achieved by actions targeting productivity rather than requiring more resources. The authors concluded that the General Contractor can impact subcontractor production rates with active control based on LBMS principles and, thus, decrease project durations.

Olivieri, Seppänen, and Granja (2018) compared the results of CPM and LBMS planning processes using three case projects. Their goal was to achieve the same duration but with more continuous flow of operations. By optimising the schedules created by CPM experts, it was possible to achieve the same total duration with lower resource use and more continuous use of resources. The optimisation could be achieved with a low number of operations because the number of planning elements in LBMS (tasks as opposed to activities) was a fraction of the activities required to model the schedule in CPM (reduction of 95–96%).

In recent years, several studies have been carried out related to takt planning and its relationship to LBMS (e.g. Seppänen, 2014; Frandson, Tommelein, and Seppänen, 2015). There is limited empirical evidence of results using takt time planning because most of the studies have focused on planning or individual construction phases. However, Heinonen and Seppänen (2016) presented a case study of a cruise ship refurbishment project where project durations were decreased by 73% in three years of development and after several projects. This indicates that takt planning is a more aggressive strategy and can result in substantially larger time savings. The psychological impact of having crews close together and solving problems and establishing trust at crew level seems to also alleviate the concerns related to cascading delays. It can be that a mix of these two approaches is required for optimal results, depending on uncertainty of the project. More empirical research is certainly required in order to understand the best strategy for each project.

14.5 Location-based controlling process based on the combination of LBMS and LPS

LBMS controlling is focused on preventing cascading delays caused by interference between trades (Seppänen, 2009). LPS focuses on the social process, constraint screening and commitment. Both of these views are critical, so the systems are complementary and raise different issues for discussion and resolution (Seppänen, Modrich, and Ballard, 2015). The process combining LBMS and LPS includes the following steps (Seppänen, Modrich, and Ballard, 2015):

- Identify tasks and locations in the look-ahead window (LPS/LBMS);
- Break down tasks and locations to operations (LPS);
- Identify, assign and remove constraints (LPS);
- Review actual production to identify ongoing production problems (LBMS);
- Review forecasts and alarms to identify future production problems (LBMS);
- Root cause analysis for problems (LPS);
- Re-plan to address current and upcoming problems (LPS/LBMS);
- Release constraint-free operations, tasks and locations to workable backlog (LPS);
- Prepare for upcoming operations (LPS).

LBMS supports the process by providing data on ongoing and future problems and the system where the phase schedules and look-ahead schedules are stored. Root cause analysis of LPS tackles all the problems identified based on constraint identification, constraint removal or actual production. LBMS provides numerical support, for example actual production rates and labour consumptions, which can be used to drive LPS discussions. Any agreed actions that impact the crew size or future productivity of an operation will trigger recalculation of the LBMS forecast to determine whether the actions are sufficient to fix the problem. The control chart can visualise the workable backlog by colour-coding tasks in locations based on constraints and status (Seppänen, Modrich, and Ballard, 2015). For example, any tasks with constraints can be greyed out (Kenley and Seppänen, 2010, p. 329).

Based on the ideas above, a weekly production control routine, including collecting and reporting progress, analysing LBMS and LPS data, and superintendent and subcontractor meetings, can be defined. The routine should be based on regular superintendent look-ahead meetings, daily huddles, weekly production planning and commitment meetings, and phase scheduling meetings for up-coming phases. The actual routine will be defined based on project size, organisation and requirements, and it is impossible to define a routine that would work in all cases. Some parts of the process related specifically to LBMS are elaborated below.

The weekly routine including LBMS always includes progress data collection and reporting. There are several ways to approach data collection. Data collection can be *centralised*, which requires the person responsible for monitoring status to tour every location to observe and record the status of work (Kenley and Seppänen, 2010, p. 337). This approach can give an accurate snapshot of status on the status date, but does not get accurate actual resources or what happened within the monitoring period. It is also very labour-intensive. It is possible to *decentralise* information collection, where subcontractors or superintendents self-report work status. This method presents the risk that the status data may be incorrect, but decreases the workload of any one actor (Kenley and Seppänen, 2010, pp. 337–338). Digital tools make the data collection easier because *decentralised* strategies may be implemented with mobile tools with subcontractors recording progress using a mobile phone (e.g. Dave, Seppänen, and Modrich, 2016). Regardless of the technology used, in most documented case studies the General Contractor verifies the information provided by others (e.g. Seppänen, Evinger, and Mouflard, 2014). In the future, capturing progress data will be automated as a combination of sensor data, indoor positioning (e.g. Olivieri, Seppänen, and Peltokorpi, 2017) and image recognition technologies. A more objective status picture will benefit production control in both LBMS and takt environments.

After data collection, the LBMS software (for example, Trimble Schedule Planner) is updated and forecasts and alarms are reviewed. In larger projects, this is typically done by a production engineer who can be a full-time resource or a person spending some work time on LBMS-related tasks (Seppänen, Evinger, and Mouflard, 2014). Their task is to identify either ongoing (based on actual progress) or upcoming (based on forecasts) issues and prepare reports and recommendations to review with the project team. Typical reports reviewed by project teams to address issues include just two or three flowlines with plan, actual and forecast to make it easy to understand for subcontractors. Any agreed control actions are reflected back in the forecast and logged in a control action log (Seppänen, Evinger, and Mouflard, 2014).

Seppänen, Modrich, and Ballard (2015) concluded that a critical part of production control is to reveal as many problems as possible and as early as possible. By combining the LBMS progress data, forecasts and alarms to LPS constraints and weekly commitments and verifying that weekly commitments meet the requirements of LBMS production rate, more problems can be raised earlier for discussion. The authors presented a hypothesis that the amount of problems identified increases and the information from the combined system can help in resolving the problems (Seppänen, Modrich, and Ballard, 2015). This hypothesis has not yet been tested empirically, although several practical implementations are ongoing.

14.6 Limitations and implementation issues

From a technical perspective, LBMS involves simply adding additional concepts based on production and locations on top of the CPM algorithm. For practitioners used to CPM, these additional layers of logic require a mindset change. For many, it is a new idea that schedules can be optimised and followed. Traditional CPM was always more of a contract document and a way to illustrate changes rather than a production management tool. The practical difficulty of this is demonstrated by low levels of implementation outside of Finland. Most of the case studies reported in literature are either in Finland or California.

Especially in the United States, the practitioners have been hesitant to adopt the system because the legal system does not have previous cases with LBMS schedules. The additional layers of logic have not been tested in courts and, if an LBMS schedule is analysed with CPM tools, the results may pose a legal risk to the contractor. Development of processes to use LBMS, together with CPM to decrease this risk, have started (e.g. Olivieri, Seppänen, and Granja, 2016). However, a lot of work is still required to address the issues related to delay management and legal risks.

Key issues seem to be the separation between Project Management and Project Production Management systems which tackle different problems. CPM is often a contract requirement and is used for managing contracts and for critical path analysis. LBMS is primarily a production management system and mostly associated with production control benefits (Olivieri et al., 2019). The use of LBMS for contract or delay management is problematic because LBMS forecasts the future based on actual production rate of the past. If this forecast is used for delay analysis, the problems of the past are assumed to continue in the future, and the party responsible for poor production rate is also blamed for delays that have not yet occurred. Therefore, CPM calculations are a better choice to use for contract and delay management purposes. In contrast, for production management purposes, it is critical to understand where current production rates will take the project if no action is taken. At the moment, software packages that can perform both calculations easily do not exist and workarounds have to be used. Future development of processes and tools in this regard is important because project management functions are also important, and lack of clarity of project management versus production management is hindering the adoption of location-based methods.

14.7 Future research directions

Although some attempts have been made to compare LBMS and takt approaches, the psychological impacts of different types of buffers have not been thoroughly investigated. Anecdotal evidence suggests that the workable backlog, capacity buffers and end-of-project buffers of takt result in better performance than time buffers of LBMS. Takt also includes several controlling methods to protect against cascading delays (Binninger et al., 2017). In real applications of takt, based on discussions and interviews by the author, the crews rarely experience any waiting time. It seems that the cascading delays can be prevented at crew level. However, sometimes the workable backlog becomes critical, necessitating the formal treatment of these backlogs. Although the takt process seems to have many advantages over the LBMS process, the technical system of LBMS may still be required to comply with contractual obligations. LBMS software should be adjusted to also enable the calculation of standard CPM algorithms for project management purposes. An important future research direction is to attempt a synthesis between the LBMS technical method and the takt process, integrating Last Planner concepts. The synthesis must address the formal treatment of workable backlogs, how to use forecasts and alarms within the takt time, and how to handle delay management for contractual purposes.

Another important future research angle is to integrate logistics and supply chain management based on locations. Logistics becomes a critical part of the production system when takt times are compressed and buffers are taken out from between the operations (e.g. Heinonen and Seppänen, 2016). Logistics can be used as a control action because typically 20% of a tradesman's time is spent on logistic-supporting activities. Therefore, just by adding or subtracting logistics from the scope of a trade, a contractor can increase or decrease production rates by 20% without any changes in manpower.

Progress with automated tracking of labour, materials and equipment, and automatic recognition of quality and progress will enable the automation of many manual processes related to production control. Research using these technologies is currently ongoing on both LBMS and takt projects to see how the different flows are impacted by different buffering strategies. For the first time, it is becoming possible to get objective, real-time data about the status of production (Zhao et al., 2019). In future, this could mean that alarms of LBMS could be calculated in real time in a takt process, thus giving the same proactive control as in LBMS without requiring time buffers between tasks.

14.8 Conclusion

The LBMS is composed of a planning and controlling system. Both systems include a technical system, and best practices or guidelines on how to best use them to improve planning and controlling results. The differences compared to other planning systems are mathematical and process related. In terms of the LBMS algorithm, locations can be used to greatly simplify the planning. On the other hand, the LBMS algorithm enables continuous work. In contrast with CPM, more emphasis is placed on controlling than planning.

LBMS controlling includes collecting detailed location-based progress data and refining it by calculating progress metrics, forecasts and alarms. These can be utilised to raise problems for resolution, for example by using the social process of LPS. Empirical data of the results has been reported and there is convincing evidence that LBMS outperforms CPM.

Takt planning is a related location-based technique. It can use the same technical system but emphasises different things in terms of guidelines. In a closely related industry of cruise ship refurbishment, impressive benefits have been documented. Based on anecdotal evidence, it seems that the takt process and buffering outperforms the LBMS process because buffers are only used when required.

It can be concluded that location-based systems outperform CPM in both planning and controlling, but more research is required to determine the best location-based planning and controlling strategy. However, CPM still has a role to play in project management functions. It is likely that the best strategy is a mix of LBMS and takt planning approaches and depends on project (or phase) characteristics. New technology development related to real-time tracking and reality capture has the potential of real-time production control and allows the use of LBMS controlling system in a takt environment.

These conclusions highlight several needs for future research. An LBMS technical system could be used as a starting point to develop a technical system for takt. Several calculations need to be adjusted in order to account for capacity buffers (currently not considered in LBMS). The technical system development should also take into account the real-time production control possibilities allowed by the latest technology development. The technical system should allow for multiple types of buffers, including buffers incorporated into takt durations (capacity buffer), and allow for duration calculation based on takt approach (resources and buffers adjust, durations stay constant), or LBMS approach (resources and buffers stay constant, durations adjust). These decisions could be made separately for each process. The LBMS controlling system should be adjusted likewise and forecasts should absorb both capacity buffers (when planned using takt approach) or time buffers (when planned with LBMS approach). Expanding the LBMS technical system would make sense because it already incorporates CPM functions and could thus be used for project management purposes.

In addition to these currently unaddressed technical questions, several process-related questions remain. Takt seems to compress repetitive work processes, but non-repetitive parts are considered part of workable backlog. It is entirely possible and even likely in some project types that non-repetitive parts determine the project duration; for example, in office buildings or hotels, often the public areas are the last areas to finish, and rushing the offices or hotel rooms does not expedite the project. What are the optimum approaches to expedite the non-repetitive parts of a project? Should we aim for processes in takt to be critical or non-critical? Should takt time for repetitive parts be determined in part based on project criticality? Which parts of non-repetitive work are workable backlog and which are critical project deliverables? How do we manage the difference? Removing time buffers between tasks increases the importance of supporting operations, such as design management, procurement, prefabrication, supply chain management and logistics. The production system becomes less forgiving for any failures in these aspects of production system. How can we make all supporting processes optimally support takt? A great deal of research is being done on these individual aspects of production system, but the work should be reviewed from the takt point of view.

References

Arditi, D., Tokdemir, O. and Suh, K. (2002) 'Challenges in line of balance scheduling', *ASCE Journal of Construction Engineering and Management*, 128 (6), pp. 545–556.

Binninger, M., Dlouhy, J., Steuer, D. and Haghsheno, S. (2017) 'Adjustment mechanisms for demand-oriented optimisation in takt planning and takt control', *Proc. 25th Annual Conference of the International Group for Lean Construction*, Heraklion, Greece, pp. 605–612.

Dave, B., Seppänen, O. and Modrich, R.-U. (2016) 'Modeling information flows between Last Planner and Location Based Management System', *Proc. 24th Annual Conference of the International Group for Lean Construction*, Boston, MA, USA, USA.

Evinger, J., Mouflard, C. and Seppänen, O. (2013) 'Productivity effects of starting as early as possible in hospital construction', *Proc. 21st Annual Conference of the International Group for Lean Construction*, Fortaleza, Brazil.

Frandson, A., Seppänen, O. and Tommelein, I. (2015) 'Comparison between location based management and takt time planning', *Proc. 23rd Annual Conference of the International Group for Lean Construction*, Perth, Australia, pp. 28–31.

Harris, R.B. and Ioannou, P.G. (1998) 'Scheduling projects with repeating activities', *ASCE Journal of Construction Engineering and Management*, 124 (4), pp. 269–278.

Heinonen, A. and Seppänen, O. (2016) 'Takt time planning: lessons for construction industry from a cruise ship cabin refurbishment case study', *Proc. 24th Annual Conference of the International Group for Lean Construction*, Boston, MA, USA, pp. 23–32.

Kala, T., Mouflard, C. and Seppänen, O. (2012) 'Production control using location-based management system on a hospital construction project', *Proc. 20th Annual Conference of the International Group for Lean Construction*, San Diego, CA, USA.

Kala, T., Seppänen, O. and Stein, C. (2010) 'Using an integrated 5D and location-based management system in a large hospital construction project', *Lean Construction Journal*, pp. 102–112.

Kankainen, J. and Sandvik, T. (1993) *Rakennushankkeen Ohjaus. (in Finnish) (Controlling a construction project)*. Helsinki, Finland: Confederation of Finnish Construction Industries, Rakennustieto Oy.

Kankainen, J. and Seppänen, O. (2003) 'A line-of-balance based schedule planning and control system', *Proc. 11th Annual Conference of the International Group for Lean Construction*, Blacksburg, VA, USA, pp. 22–24.

Kenley, R. (2005) 'Dispelling the complexity myth: founding lean construction on location-based planning', *Proc. 13th Annual Conference of the International Group for Lean Construction*, Sydney, Australia.

Kenley, R. and Seppänen, O. (2010) *Location-Based Management in Construction: Planning, Scheduling and Control*. London, UK and New York, USA: Spon Press.

Kiiras, J. (1989) *OPAS ja TURVA, Erityiskohteiden Työnaikaista Ohjausta Palveleva Aikataulu- Ja Resurssisuunnittelu. (in Finnish) (A schedule and Resource Planning System for the Implementation Phase of Unique Projects)*. Espoo, Finland: Helsinki University of Technology, Construction Economics and Management Publications. 217.

Kolhonen, R., Kankainen, J., and Junnonen, J-M. (2003) *Rakennushankkeen Ajallinen Hallinta. (in Finnish) (Time management of construction projects)*. Espoo, Finland: Helsinki University of Technology, Construction Economics and Management Publications. 217.

Koskela, L. (2000) 'An exploration towards a production theory and its application to construction', PhD Thesis, VTT Technical Research Centre of Finland.

Koskela, L. and Howell, G. (2001) 'Reforming project management: the role of planning, execution and controlling', *Proc. 9th Annual Conference of the International Group for Lean Construction*. **N**ational University of Singapore, Singapore.

Lumsden, P. (1968) *The Line of Balance Method*. Oxford, UK: Pergamon Press.

Meredith, J. and Mantel, S. (1995) *Project Management: A Managerial Approach*. New York, USA: John Wiley & Sons.

Mohr, W. (1979) *Project Management and Control (in the building industry)*. 2nd edition. Department of Architecture and Building, Parkville, VIC: University of Melbourne.

Olivieri, H., Seppänen, O. and Granja, A. (2016) 'Integrating LBMS, LPS and CPM: a practical process', *Proc. 24th Annual Conference of the International Group for Lean Construction*, Boston, MA, USA.

Olivieri, H., Seppänen, O. and Granja, A.D. (2018) 'Improving workflow and resource usage in construction schedules through location-based management system (LBMS)', *Construction Management & Economics*, 36 (2), pp. 109–124.

Olivieri, H., Seppänen, O. and Peltokorpi, A. (2017) 'Real-time tracking of production control: requirements and solutions', *Proc. 25th Annual Conference of the International Group for Lean Construction*, Heraklion, Greece.

Olivieri, H., Seppänen, O., Alves, T. D. C., Scala, N. M., Schiavone, V., Liu, M., & Granja, A. D. (2019) Survey Comparing Critical Path Method, Last Planner System, and Location-Based Techniques, *Journal of Construction Engineering and Management*, *145* (12), p. 04019077.

Parker, H. and Oglesby, C. (1972) *Methods Improvement for Construction Managers*. New York, USA: McCraw-Hill.

Peer, S. (1974) 'Network analysis and construction planning', *ASCE Journal of the Construction Division*, 100 (CO3), pp. 203–210.

Russell, A.D. and Wong, W.C. (1993) 'New generation of planning structures', *ASCE Journal of Construction Engineering and Management*, 119 (2), pp. 196–214.

Sacks, R. (2016) 'What constitutes good production flow in construction?', *Construction Management and Economics*, 34 (9), pp. 641–656.

Sacks, R. and Harel, M. (2006) 'An economic game theory model of subcontractor resource allocation behavior', *Construction Management & Economics*, 24 (8), pp. 869–881.

Selinger, S. (1973) *Method for Construction Process Planning Based on Organisation Requirements.* PhD Thesis, Technion-Israel Institute of Technology, Haifa, Israel.

Selinger, S. (1980) 'Construction planning for linear projects', *ASCE Journal of the Construction Division*, 106 (CO2), pp. 195–205.

Seppänen, O. (2009) *Empirical Research on the Success of Production Control in Building Construction Projects.* PhD Thesis, Department of Structural Engineering and Building Technology, Faculty of Engineering and Architecture, Helsinki University of Technology, Espoo, Finland.

Seppänen, O. (2014) 'A comparison of takt time and LBMS planning methods', *Proc. 22nd Annual Conference of the International Group for Lean Construction*, Oslo, Norway, pp. 23–27.

Seppänen, O., Ballard, G. and Pesonen, S. (2010) 'The combination of last planner system and location-based management system', *Lean Construction Journal*, 10, pp. 44–54.

Seppänen, O., Evinger, J. and Mouflard, C. (2014) 'Effects of the location-based management system on production rates and productivity', *Construction Management and Economics*, 32 (6), pp. 608–624.

Seppänen, O. and Kankainen, J. (2004) 'An empirical research on deviations in production and current state of project control', *Proc. 12th Annual Conference of the International Group for Lean Construction*, Helsingor, Denmark.

Seppänen, O. and Kenley, R. (2005) 'Performance measurement using location-based status data', *Proc. 13th Annual Conference of the International Group for Lean Construction*, Sydney, Australia.

Seppänen, O., Modrich, R-U. and Ballard, G. (2015) 'Integration of last planner system and location-based management system', *Proc. 23rd Annual Conference of the International Group for Lean Construction*, Perth, Australia, pp. 123–132.

Shreve, R. (1930) 'The Empire state building organization', *The Architectural Forum*, LII (6), pp. 771–778.

Willis, C. and Friedman, D. (1998) *Building the Empire State Building.* New York, USA: W.W. Norton and Co.

Zhao, J., Seppänen, O., Peltokorpi, A., Badihi, B., & Olivieri, H. (2019) Real-time resource tracking for analyzing value-adding time in construction, *Automation in Construction*, *104*, 52–65.

15 Relating construction production design and planning activities with location-based scheduling techniques

Clarissa Biotto and Mike Kagioglou

15.1 Introduction

Lean Construction is a relatively new production management philosophy that uses adapted concepts, tools and techniques from Lean manufacturing in order to design, plan, control and improve construction production systems. During the last 25 years, many terms have emerged in the Lean Construction literature to refer to the act of designing and planning a production system, namely: Production System Design (PSD); Phase Scheduling (PS); and Work Structuring (WS). In parallel, with the advent of Location-based Scheduling (LBS) techniques, Lean Construction has increased the number and diversity tools available to schedule activities defined by the PSD and plan. In many instances, the scope, capabilities, limitations and contexts of use of these terms and techniques were not completely understood by the Lean Construction field. This confusion led to the inconsistent application of LBS on Lean projects, which still carries on to this day.

There are three main reasons why successful Lean deployments are limited. Firstly, the activities of designing and planning a production system have emerged in Lean Construction literature quite erratically. In 1999, the design of a production system was introduced by Ballard (1999) called Work Structuring. It referred to the design of a product in alignment with its production processes and operations (Ballard, 1999). Ballard (2000a) expanded the scope of WS including the structuring of the 'supply chains, the allocation of resources, and design-for-assembly efforts' as part of the Lean Project Delivery System (Ballard, 2000a).

In 2001, the term 'Production System Design' was used by Ballard et al. (2001a) to specify the process of design and to make products based on Koskela's theory of Transformation-Flow-Value (Koskela, 2000). That is, the Lean production system should be designed in order to deliver the project while maximising the value for clients and minimising waste (Koskela and Ballard, 2003). Still in 2001, Ballard et al. (2001b) revisited the term WS, equalling it to PSD. In these cited works, both PSD and WS scope focus on structuring the production of a Lean Construction project.

In 2004, Schramm (2004) developed a model to implement PSD in construction companies. In his model, decisions are made not only to structure the production operation, but also to devise it according to the company's strategy, showing that the PSD scope is broader than the WS. Although PSD occurs before the construction act (Koskela and Ballard, 2003), it is known that, during the production system operation, some production planning activities also take place in Lean projects. Whether the Last Planner System™ (Last Planner System is a trademark of the Lean Construction Institute, see Chapter 3 for a detailed description)

(Ballard, 2000b) is being used or not, PS should be undertaken to establish which processes should be executed in a particular project phase, how and by whom.

Secondly, another reason for ineffective Lean Construction deployment has to do with the fact that Lean Construction practitioners and academics do not fully understand the advantages and disadvantages of different types of LBS techniques. Since the 1950s, the variety of LBS techniques has grown considerably. Since that time, Line of Balance (LOB), Linear Schedule (LS), Linear Scheduling Method (LSM), Vertical Production Method (VPM), Flowline (FL), Location-based Management System (LBMS) and Takt-Time Planning (TTP) have been deployed in construction projects. These tools use locations of a building as the basic unit for production planning and control. Despite the similarity among different types of LBS techniques, their concepts, visual representation and context of use vary. Careful consideration of the design and plan of the construction production system is required to support an informed LBS technique selection.

Thirdly, there is a complete absence of a conceptual framework, in which the PSD and planning activities are related with the LBS techniques. The LOB, FL and TTP have proven to be very helpful for the production schedule, and therefore similar tools are needed to assist the designer of construction production systems. Some seminal contributions compared location-based planning tools with networks techniques, such as Critical Path Method (CPM) and Program Evaluation and Review Technique (PERT), in linear repetitive construction projects (Peer, 1974; Arditi and Albulak, 1979; Birrell, 1980).

Along with the improvement of the LBS techniques, comparisons were made with network methods, namely LOB (Lucko and Gattei, 2016), Linear Schedules (LS) (Kallantzis, Soldatos and Lambropoulos, 2007), Linear Scheduling Method (LSM) (Yamin and Harmelink, 2001), VPM (O'Brien, Kreitzberg and Mikes, 1985), FL and LBMS (Kenley and Seppänen, 2010). The LBS techniques were also compared among themselves, for instance between LSM and LOB (Mattila and Abraham, 1998; Su and Lucko, 2015; Lucko and Gattei, 2016), in which the authors analysed the historical evolution, the type of network and the analytical capabilities of tools.

Although contrasting the LBS techniques is a common research, none of the papers mentioned above compared the most used LBS tools in Lean projects, such as LOB, FL and TTP, or connected them to the Lean PSD activities. As a result, this chapter is making an original contribution in presenting a conceptual framework for the use of LOB, FL and TTP in the context of PSD, PS and WS. This framework aids Lean Construction practitioners and academics to make informed decisions about the LBS techniques to be used in the PSD and planning based upon the visual representation, use of buffers, activities balancing and context of use regarding the collaboration of subcontractors and the project uncertainty. As such, it is of some significance to practice as well as articulating complex theoretical positions in relation to LBS.

15.2 Research method

The chapter is organised in two sections. The first section is a literature review on current research about PSD, WS and PS, followed by a comparison among these terms according to the focus of designing activities, the collaboration for decision-making, project stage when it is developed, and the output for the production planning and control.

In the second section, a literature review was conducted about LBS techniques, such as LOB, FL and TTP. Next, the authors made a comparison among the LBS techniques, focusing on capabilities to visually represent the plan, to employ different activity networks and define pace; to use different types of buffers and balance the activities; and to define in which context the LBS techniques should be implemented.

For the literature review, papers were identified from the International Group for Lean Construction conferences, *Lean Construction Journal*, Lean Construction Institute white papers, and many other relevant conference and journal papers utilising the main terms identified in the research design phase, including PSD, WS, PS, LBS, etc. The search for the papers covered some seminal works about location-based tools from a theoretical as well as a practical (practice-based) perspective with applications and case studies.

Further, cross comparisons were made between the production design activities and the LBS techniques to establish their potential combined use in Lean projects. As a result of this comparison, a conceptual framework was devised representing the relationship between the LOB, the FL and the TTP with the PSD, PS and WS.

The conceptual framework is an assumption based on the literature review and the varied professional experience of the authors in Lean Construction implementation, which requires further investigation and validation.

15.3 Production System Design (PSD)

The PSD is the first managerial activity before the beginning of any productive effort (Koskela and Ballard, 2003). It should be part of the design stage discussions, before the construction stage (Schramm, Costa and Formoso, 2004). The PSD enables the discussion and translation of the desired production strategy in a set of decisions about the production system that will be managed during the system operation (Schramm, 2004; Schramm, Rodrigues and Formoso, 2006).

The PSD based on Lean Construction principles has three main goals (Koskela, 2000). These goals are: (1) Deliver the project; (2) Maximise value; and (3) Minimise waste. This represents the most basic form and opportunity for minimising the effects of variability on production, as long as it contributes to achieving the major project goals (Ballard et al., 2001a).

The PSD extends from the global organisation of the company until the project operations, defining who should be involved in the roles for decisions, and how the physical work should be performed (Ballard et al., 2001a). Although it focuses on the operational aspects of production, especially because it is traditionally developed by the production manager, the PSD should reflect the strategical decisions of the project and the construction companies (Schramm, Costa and Formoso, 2004). For instance, when a construction company decided to offer customisation of the apartments for clients of a low-income residential project, they needed to redesign the established PSD. In order to support the new company's strategy, the company had to reduce the production batch size and variability, and increase the process flexibility and transparency (Schramm et al., 2009).

The main decisions made in the PSD are (Schramm, Costa and Formoso, 2004): definition of production batch size and transfer batch size; definition of work packages; definition of work packages dependencies and sequence; level of vertical integration; estimating the production system capacity related to equipment and manpower; study of project workflows; construction strategy; site layout; and critical process identification and design.

Decisions made at this stage are interdependent; therefore, if one is changed, the others will be affected (Meredith and Shafer, 2009). That is why the PSD based on Lean Construction principles recommends a collaborative process of decision-making in which designers and builders align their interests with the project goals. Adding to this, because the PSD occurs before the construction stage and major subcontractors are still unknown, the level of uncertainty is high. The project complexity imposes challenges on decision-making for the participants of the PSD. Therefore, to overcome these difficulties and support informed decisions, some researches apply computational simulations on a construction phase or activities to measure the impacts of variability, different production batch sizes, and the crew's skills on the production system performance (Alves, Tommelein and Ballard, 2006; Esquenazi and Sacks, 2006; Schramm and Formoso, 2007; Schramm et al., 2008).

Whether deterministic or probabilistic, the PSD produces as output the project master plan (Schramm, Costa and Formoso, 2004) that should be used during the construction stage. Due to previous definitions of production pace, capacity and deadlines in the PSD, it becomes simpler to devise the master plan (Schramm, Costa and Formoso, 2004).

15.4 Phase Scheduling (PS)

The term Phase Scheduling emerged in the Lean Construction literature in Ballard's (2000c) white paper. PS is also known as pull planning or reverse PS. It is a collaborative production design activity to structure the work of a project phase (Ballard, 2008). PS occurs during the production system operation due to information being available and accurate for planning when the subcontractors are hired.

PS was incorporated into the Last Planner System (LPS) (Ballard, 2000b) to bridge the gap between the master plan and the lookahead plan. For that, the PS participants use a mix of push and pull flows for planning the work. The construction phase's milestones that were set up at the project's master plan are pushed to the phase planning. Next, the phase's activities are broken down into tasks and handoffs. The network and duration of tasks are defined by the contractors of the phase using sticky-notes (among other means) on a wall (or other physical and digital media). Then, a reverse plan of the phase's tasks is devised, pulling the tasks from the phase deadline towards the phase start date (Alarcon, Betanzo and Diethelm, 2004). The contractors define collaboratively the handoffs between the crews and project phases, insert buffers and guarantee the completion of the work on time (Ballard and Howell, 2003; Alarcon, Betanzo and Diethelm, 2004; Ballard, 2008).

One of the outputs of the PS is the plan of the project's phase (Ballard, 2008). The plan can be scheduled using traditional tools, such as a Gantt chart (Knapp, Charron and Howell, 2006), or LBS techniques, such as LOB (O'Brien, Kreitzberg and Mikes, 1985[1]), FL (Seppänen, Ballard and Pesonen, 2010) and TTP (Fiallo and Howell, 2012). Moreover, computational simulations can be used to support the decision-making process by the phase's participants (Tsao, Draper and Howell, 2014).

As PS is a transparent and collaborative process of decision-making, it promotes the 'team work, awareness of the impact of individual actions on all participants', which enhances the subcontractors' commitments (Alarcon, Betanzo and Diethelm, 2004).

15.5 Work Structuring (WS)

The term Work Structuring was introduced in construction by Ballard (1999) to mean 'the development of operation and process design in alignment with product design, the structure of supply chains, the allocation of resources, and design-for-assembly efforts' (Ballard, 1999). Further, it was related to the PSD in Ballard et al.'s (2001b) white paper. Despite the change of the name from WS to PSD, the scope from the perspective of the production organisation has remained the same. However, it is essential to highlight that the PSD extended the borders of decision-making, including the organisational structure and the alignment of strategies among stakeholders (Ballard et al., 2001b).

When applied for a narrow scope, considering only a task, the WS has been known as task planning (Junnonen and Seppänen, 2004). In the latter, task planning is defined as the process of planning one task before its execution, considering the time, cost and quality specified in the master schedule (Junnonen and Seppänen, 2004).

The WS aims to make the workflow smooth and reliable, reducing wastes, while delivering value to the customer (Ballard, 1999; Milberg, 2007). In order to do that, the WS breaks down the product and the process into parts, sequences and assignments based on three main elements (Tsao, 2005). These elements are (1) **Production unit** which is the gang assigned to perform the work; (2) **Work chunk** which is the work package that can be handed off to the next gang; (3) **Handoff** which is the completed work chunk that is released to the subsequent gang after its acceptance. The relationship between these three elements is presented in Figure 15.1. Then, in order to define the structure and the flow of the production, the following questions need to be answered (Ballard, 1999):

- In what chunks will work be assigned to specialists?
- How will work chunks be sequenced?
- How will work be released from one production unit to the next?
- Where will decoupling buffers be needed and how should they be sized?
- When will the different chunks of work be done?

In reviewing the literature, the WS is used before the construction stage, but it is applicable along the construction (Ballard, 1999), or in any critical activity of the project. For instance, Tsao et al. (2004) used the WS to support decision-making in the design development of a stone-on-truss curtain wall, which was a critical element of the project.

Both PSD and PS deploy the WS while breaking down the work in work chunks or work packages, handoffs and production units or crews, to make the workflow smooth.

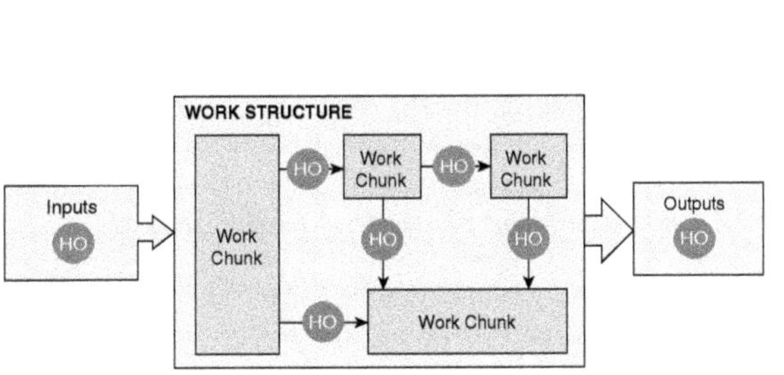

Figure 15.1 Relationship between work chunks and handoffs (based on Tsao, 2005)

15.6 Comparison among the different production system design activities

To compare the PSD, PS and WS, four aspects were analysed: (1) The focus of the PSD; (2) The stakeholders' collaboration for decision-making; (3) The project stage when it was deployed; and (4) The output of the design process.

15.6.1 Focus of the production system design

The PSD focuses on strategical decisions about the construction project and concerns the project's viability, budget and lead time, which are consequences of the production system organisation (Ballard et al., 2001a, 2001b; Schramm, Costa and Formoso, 2004; Schramm, Rodrigues and Formoso, 2006; Mota, Mota and Alves, 2008). On the other hand, the PS tries to ensure that phase activities are clearly defined in handoffs for sub-contractors and that the phase's lead time fits into the master schedule (Ballard and Howell, 2003). While the WS focuses on the process view, it is used for single tasks (Junnonen and Seppänen, 2004; Tsao et al., 2004; Alves, Tommelein and Ballard, 2006), for phase's tasks, and for the whole PSD (Schramm, Rodrigues and Formoso, 2006).

15.6.2 Collaboration for decision-making

Whereas the PSD focuses on the wider scope of the project's strategies, it requires the involvement of the main stakeholders, including people from departments not directly connected to the construction, such as the companies' board, marketing and purchasing departments (Schramm, 2004). However, as information is not accurate at the beginning of the project, and the main contractors are not yet on board, the decisions are detailed along the project execution. Then, collaboration among all stakeholders is desirable, but not always possible.

In contrast with the PSD, the PS necessarily involves the phase's contractors to make collective decisions and visualise their impacts on the workflow. In contrast to the PS, the WS is opportunistic, i.e. when used to plan the PSD, it can be less collaborative, but when planning a specific phase or task it requires accurate information from the participants.

A direct relation can be stated here: the higher the level of detail of a task, the higher is the necessity of involvement of stakeholders in the decision-making process.

15.6.3 Project stage

All PSD activities need to take place before the beginning of the referred project, phase or task. The PSD's decisions should be made before the construction stage, mainly because of their impact on the many participants. As the PS designs the project's phase, it should anticipate its starting. The same is applicable for the WS when used for a particular task.

15.6.4 Output for the production planning and control

The output of the PSD is the project master plan that should be used by the LPS levels in Lean projects (Schramm, Costa and Formoso, 2004). This master plan is then used by the PS, which produces the phase plan. Further, it can be detailed in the lookahead planning through the WS deployment. It will then produce the task plan for the weekly planning.

Table 15.1 Comparison among the production system design activities

	Focus	*Decision-making*	*Stage*	*Output*
Production System Design	Strategical decisions about the whole production system, including organisational levels and product design	Collaboration is desirable	Before construction stage	Project master plan
Phase Scheduling	Project's phase activities, handoffs between subcontractors	High collaboration between contractors	During construction stage, before project phase	Project phase plan
Work Structuring	Project activities, product design, assemblage, handoffs	High collaboration between design, manufacturing and contractors	During construction stage, before activities	Activities plan

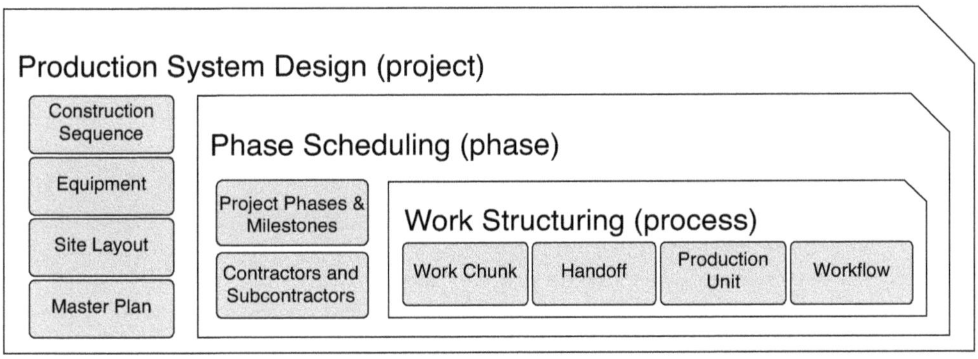

Figure 15.2 Work structuring is part of the decision scope of production system design and phase scheduling

Table 15.1 summarises the comparison among the three PSD activities discussed in this section, whereas Figure 15.2 represents their relationship.

15.7 Location-based Scheduling (LBS) techniques

The second section of this chapter presents the most common location-based techniques used in Lean projects. The term Location-based Schedule was proposed by Kenley (2004) to designate the techniques that use the location or unit as a basis for the production planning and control. LBS techniques, such as LOB, FL and TTP, were originally developed in manufacturing and have been adapted for construction. The adaptation occurred by changing the vertical axis: from units produced to location units (Henrich, Tilley and Koskela, 2005; Kenley and Seppänen, 2010).

The LBS techniques began with the advent of the LOB and the FL. Moreover, relevant similar techniques were developed with different names, which may have caused confusion regarding the conceptual basis, the contextual use, visual representation and limitations. Some of these techniques are variations of the LOB and the FL.

The LBS techniques are diagrams of two axes (vertical and horizontal) of information: one for quantity produced or unit or location, and the other for time. The activities are represented by slope lines. The visualisation of location units is an important factor for production planning since the construction is characterised by physically fixed products (location units) which are processed by crews, tools and equipment that move through the locations. This production system characteristic is normally found in ship building, aerospace manufacture and assembly, among others. These techniques can be applied on construction PSD activities described in the previous section to represent and support the decisions made by the project team.

The LBS techniques are preferred by the Lean Construction community to the methods based on activity. Examples of methods to plan construction based on activity are the well-known CPM and PERT. Both methods are frequently criticised by Lean researchers due to their inability to deal effectively with construction complexity (Peer, 1974; Birrell, 1980; Koskela and Howell, 2002; Koskela and Ballard, 2006; Koskela et al., 2014; Dave et al., 2015). CPM produces a plan that is quickly out of date and therefore ignored by operational professionals, resulting in inconsistencies and limited coordination. Additionally, it does not support continuous workflow and clear handovers for them (Peer, 1974; Arditi and Albulak, 1979; Birrell, 1980).

It is important to highlight that the aim of using LBS techniques is to design a production system with continuous workflow and uninterrupted flow for crews throughout the location units (Moura, Monteiro and Heineck, 2014). To make the workflow smoother and reduce the work in progress, the activities should be planned in only one rate, i.e. in parallel lines (Mendez and Heineck, 1998). The achievement of the same delivery rate is not always possible due to the different amounts of work executed by crews and/or different areas of the location units; when this occurs, the balancing process will guarantee the achievement of similar paces among activities.

Through the LOB technique and, generally, through the LBS techniques, it is possible to visualise the activities sequenced along the time. The LOB contains information such as: the delivery rate; activities synchronism, parallelism and interferences; distribution of workers' and crew's workflow; strategy of construction execution; buffers; production and transfer batches; activities cycle time in a batch and activity lead time (Moura, Monteiro and Heineck, 2014) – Figure 15.3.

Before presenting the tools, it is important to stress here some concepts frequently discussed in a production system:

- **Work in Progress (WIP):** 'items of work between processing steps' (Marchwinski and Shook, 2003). The importance of visualisation and control of WIP has increased in construction due to the popularity of location-based tools (Faloughi et al., 2015). In this context, WIP is defined as the amount of time that location units contain unfinished work, i.e. do not receive any transformation activity. Graphically, WIP is represented in Figure 15.3;
- **Buffer:** used to protect a production system against the variability. It can be buffer of inventory, capacity and time (Hopp and Spearman, 2011). In the LBS tools, such as LOB and FL, it is possible to visualise the **time buffers** (time gap between the tasks) and the inventory buffers, or **work buffers**, as described by Lucko and Gattei (2016)

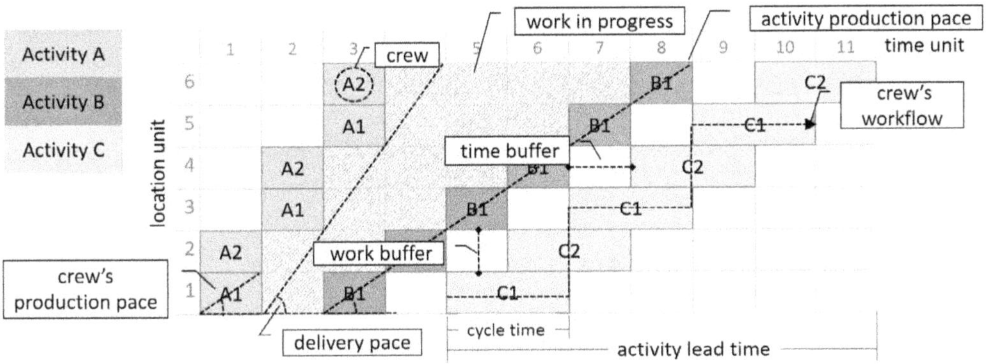

Figure 15.3 Visual information on location-based tools

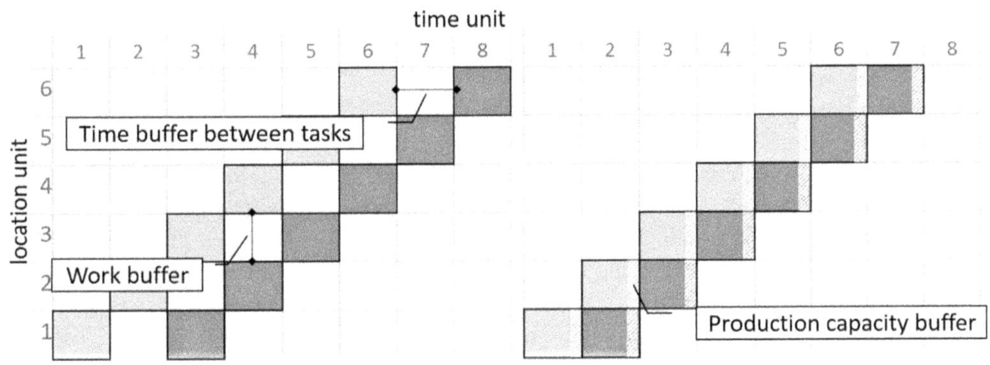

Figure 15.4 Three types of buffers used by location-based tools

(location/unit gap between the tasks). A type of buffer commonly explored by the TTP is the **production capacity buffer** (Frandson, Seppänen and Tommelein, 2015), which underloads the crews' capacity to perform an activity. In order to facilitate the understanding of buffers, the authors made a schematic representation of the three types of buffers' capacity inside a task, exemplified in Figure 15.4.

15.7.1 Line of Balance (LOB)

LOB is a planning technique developed by the Goodyear Company in the 1940s and used in manufacturing for repetitive processes. It was originally developed for an industrial programme, by the US Navy, in the 1950s, and the LOB technique was published by the Office of Naval Material in 1962. The LOB papers emerged in the 1970s presenting another option, rather the CPM, for LSM in construction (Schoderbek and Digman, 1967; Carr and Meyer, 1974; Peer, 1974; O'Brien, 1975; Kleinfeld, 1976).

The LOB is used by the construction industry, especially in repetitive projects such as houses (Lumsden, 1968), high-rise buildings (Carr and Meyer, 1974; O'Brien, Kreitzberg and Mikes, 1985; Mendez and Heineck, 1998; Arditi, Tokdemir and Suh, 2001; Lucko, Alves and Angelim, 2014), roads (Arditi and Albulak, 1986), but also for resource levelling (Damci, Arditi and Polat, 2013, 2016) and so on. The use of the LOB is recommended for repetitive projects due to the advantage of continuity of work (Mendez and Heineck, 1998). However, it can be applied in non-repetitive projects, such as leisure areas (Valente et al., 2014).

15.7.1.1 Visual representation

The LOB has two visual representations. In the seminal papers, activities were represented as dual parallel lines (Lumsden, 1968), which the visualisation of crew's workflow along the production units is considered not to be explicit (Frandson, Seppänen and Tommelein, 2015). However, the second visual representation of the LOB shows the activities represented by coloured boxes with the crew's label (Kemmer, Heineck and Alves, 2008), used in some countries, such as Brazil.

15.7.1.2 Network method and pace representation

The LOB is based on activity-on-arrow (AOA) networks (Lumsden, 1968), i.e. the activity is represented by an arrow between two nodes (events) (Figure 15.5). In the arrow, the tail represents the start date and the head represents the finish date of the activity (Lumsden, 1968; Su and Lucko, 2015). This network methodology focuses on the events; for this reason, the activity in the LOB is enveloped by two parallel lines (the events). The first line is the start date, and the second line is the delivery of the production unit. Because this technique is focused on the delivery of completed units, the delivery rate begins counting 'when the first unit has been finished' (Seppänen, Ballard and Pesonen, 2010; Su and Lucko, 2015). Accordingly, the line's slope represents the delivery rate that 'counts how fast work units are finished' (Su and Lucko, 2015).

15.7.1.3 Use of buffers and balancing

In the LOB technique, practitioners are combining different types of buffer, such as crew's production capacity buffers and time buffers. Although Frandson, Seppänen and Tommelein (2015) state that the LOB uses only time buffers, there are cases of use of capacity buffers, for example in Valente et al. (2014). In this latter paper, the authors round up some tasks' duration to reach the 5-day pace, yet not balancing the workers' operational time.

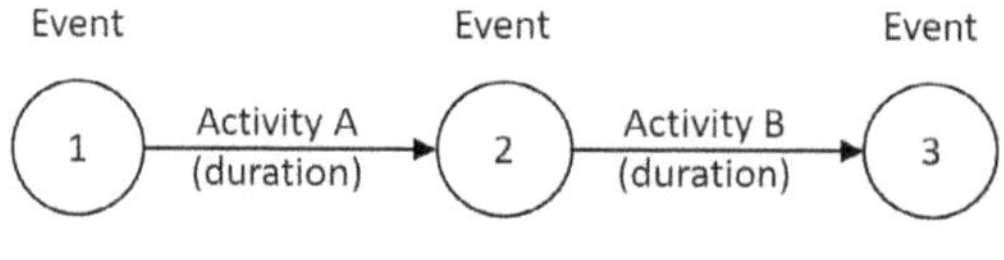

Figure 15.5 Example of activity-on-arrow (AOA) network

The balancing process in the LOB tackles the crew continuity and the productivity alignment (Lucko and Gattei, 2016). To achieve a single delivery pace for all activities, it is possible to vary the number of crews performing the same task, and/or assign more or fewer resources, which affects the productivity, i.e. the duration of the task (Lucko and Gattei, 2016).

To exemplify a balancing process using LOB, Figure 15.6 illustrates four scenarios of plan, where the number of crews or the crew's composition change:

a) If there is an activity with 2-time units' duration (2 weeks, for example), and the desired pace is to deliver 1 location unit per time unit (one week), it is possible to increase the number of crews for 2;
b) Another option is to increase the number of workers inside a crew and reduce the duration of the activity to 1-time unit. This strategy is the one adopted by the TTP;
c) Or, it is possible to deliver 2 location units per 2-time units if both crews start working in parallel;
d) And, if there is no possibility to increase the number of crews or workers, the delivery pace will be 1 location unit per 2-time units.
e) It is interesting to note that, even when the delivery pace is the same, the activity production pace may vary, as it occurs in example (a), in which the delivery pace is the same as (b) and (c). However, the activity production pace is different, i.e. 4 location units in 5-time units. Then, different production organisations can be used to achieve the same delivery pace, which means more flexibility for using the resources.

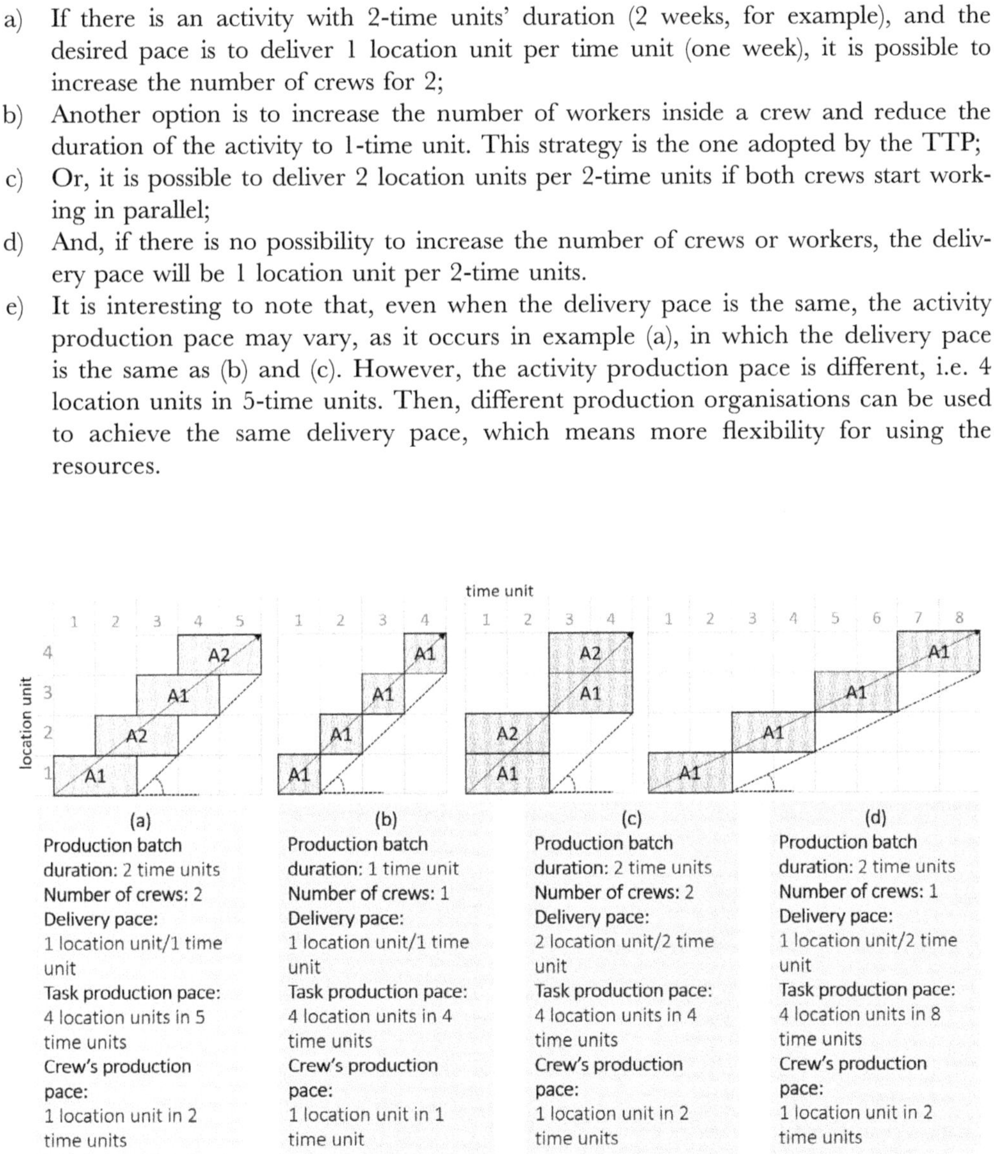

Figure 15.6 Possibilities for balancing the lines in Line of Balance technique

When comparing option (a) with (b), it is necessary to think about the physical flow in the location unit. In option (b), a higher number of workers will occupy the same space at the same time compared to option (a), in which a lower number of workers will be sharing the same work space. Despite this, increasing the crew size does not increase the productivity in the same proportion (Damci, Arditi and Polat, 2013). When comparing option (a) with (c), the idea of a lower number of workers sharing the same work space is equal; however, in option (a), the amount of materials to be supplied for one location unit is distributed in 2-times units, reducing the workload of transport equipment.

15.7.1.4 Context of use in Lean Construction

The LOB technique is used to devise the production system design, as well as the master plan of construction projects (Schramm, Costa and Formoso, 2004; Schramm, Rodrigues and Formoso, 2006; Kemmer, Heineck and Alves, 2008; Biotto, Formoso and Isatto, 2015). Moreover, the LOB can be used with the LPS to facilitate the development of other levels of planning, such as the lookahead and the commitment plan (Barbosa et al., 2013; Moura, Monteiro and Heineck, 2014; Olivieri, Granja and Picchi, 2016).

Due to the LOB being used in the PSD at the very beginning of the project, there is a high level of uncertainty, which, according to Laufer and Tucker (1988), the longest is the horizon of planning, while the lowest is the level of detail of activities. For this reason, added to the fact that many subcontractors are not on board at the moment of the PSD or master plan deployment, the LOB has also been used in less collaborative environments. However, when used in the integration of the LPS, the activities are progressively detailed with the collaboration of trades. The LOB is a flexible tool to be deployed in construction projects, mainly for using different strategies of balancing and buffers.

15.7.2 Flowline (FL)

FL is a term coined by Mohr (1979). However, the technique has been reported to have been used in the 1950s in Eastern Europe for the mass production of housing, industrial buildings, power plants, roads, rail and other projects (Slipchencko, 1966, cited in Lucko and Gattei, 2016). The method includes variations known as work-time diagrams, cyclogram or velocity diagrams and LSM, time space scheduling method, and repetitive scheduling method (see Lucko and Gattei, 2016). Notwithstanding this, the FL was also developed by Selinger (1973) and Peer (1974).

15.7.2.1 Visual representation

The FL is a derived technique from the LOB. However, the activity is represented by a single line, which Kenley and Seppänen (2010) consider a much cleaner representation compared to the LOB. In order to visualise the crews who are executing the same work package, the activity's FL can be broken down into crews' lines (Kenley and Seppänen, 2010; Lucko and Gattei, 2016).

The activity's line is drawn from the start date (X-axis) of the first unit location (Y-axis) to the finish date (X-axis) of the last unit location (Y-axis) (Kenley and Seppänen, 2010).

15.7.2.2 Network method and pace representation

The FL is rooted in the activity-on-node (AON) network, in which the activities are the nodes and the connections between them are the logical links (Figure 15.7) (Su and Lucko, 2015). In the FL plan, the line represents the start and end of an activity (Su and Lucko, 2015). The slope of the activity's line represents the production rate, which is the total quantity of units divided by the total activity duration (Su and Lucko, 2015).

15.7.2.3 Use of buffers and balancing

In the FL, production systems are designed using two types of buffers: time and work buffers. The FL balancing process is similar to the LOB, i.e. a single production pace among all activities is achievable by varying the number of crews performing the one same task, and/or varying the crew size that varies the duration (Lucko and Gattei, 2016).

15.7.2.4 Context of use in Lean Construction

The FL is commonly used in repetitive construction projects. However, it can also be deployed for complex construction projects by breaking down the project into equal-sized locations or equal amounts of work content (Kenley and Seppänen, 2010).

In contrast to the LOB, which does not have a specific software to draw it, the FL method can be devised using the Vico Software developed in Finland. The software uses the LBMS derived from the FL and LOB (Kenley and Seppänen, 2010) to plan and control the construction activities.

Similarly, the FL is used at the PSD, and the LBMS can be integrated with the LPS (Seppänen, Ballard and Pesonen, 2010), which means that it is also a flexible tool to progress the detail of activities planning according to the availability of subcontractors and information.

15.7.3 Takt-Time Planning (TTP)

TTP in construction is derived from the takt time used in Lean manufacturing. It is used to plan the production system by setting its rates according to the demand rate. Frandson, Berghede and Tommelein (2013) define takt time as the 'unit of time within which a product must be produced (supply rate) in order to match the rate at which that product is needed (demand rate)'.

The main aim of the TTP is to design the production system for continuous workflow, keeping the trades at a balanced pace of work (that matches the demand rate) through a sequence of zones (Frandson, Berghede and Tommelein, 2014; Frandson, Seppänen and Tommelein, 2015). The zones are 'physical and clearly defined locations' to avoid ambiguity about location boundaries, as in the LOB (Frandson, Seppänen and Tommelein, 2015).

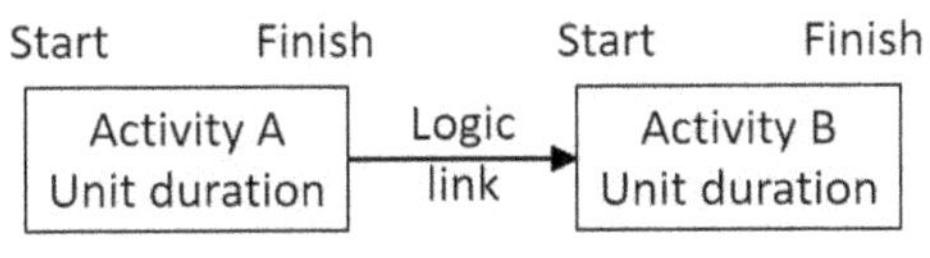

Figure 15.7 Example of activity-on-node (AON) network

To design a production system using TTP it is necessary to (Frandson, Berghede and Tommelein, 2013): (1) gather data about the exact work to be done, in what amount, by whom, where and in what sequence; (2) define zone and takt time definition; (3) identify trade sequence; (4) determine individual trade durations; (5) balance workflow; and (6) finalise the production schedule.

15.7.3.1 Visual representation

The TTP represents the activities using coloured boxes per zone. The boxes represent the takt time and the crew assigned to perform the task.

15.7.3.2 Network method and pace representation

The network methods have not yet been widely explored in TTP. However, by analysing the studies with the deployment of TTP, it is possible to assume that the network method used is the AON, especially because the focus of LBS is the logical link among activities and duration.

In a production plan devised using the TTP method, the trades must complete their work in the assigned zone within an amount of time set by the takt time (Frandson, Seppänen and Tommelein, 2015). Then, the pace of all activities should follow a single takt. In contrast to the LOB and FL, in the TTP the production and delivery paces are always the same, which is equal to the takt time. This occurs as only one crew is assigned to perform a task in the zone, which means there are no parallel crews.

15.7.3.3 Use of buffers and balancing

The TTP uses only production capacity buffers, in which the trades must underload their production capacity of around 70 to 80% (Frandson, Seppänen and Tommelein, 2015). This means that, when calculating the task's duration, the planners use a lower crew's productivity; as a result, the duration will be longer, but the possibility of the crews completing the tasks earlier is higher. However, due to the fact that there are no time buffers between tasks, if the crews complete the work later than the takt time, the risk of a cascade delay through the whole production system is increased (Seppänen, 2014). In order to avoid project delays, Frandson and Tommelein (2016) considered the possibilities of using time buffers between some trades due to non-field-related works that can hamper the tasks to start on the planned date.

It is worth mentioning that, although it is not a work buffer as represented in the LOB and the FL, the TTP uses workable backlogs as buffers to prevent the crews' idleness. This type of buffer is necessary to maintain the continuous workflow of the crews that completed their work earlier than the takt time (Linnik, Berghede and Ballard, 2013).

In order to plan and operate the production system with all trades in the same takt time, the balancing process in the TTP adjusts the amount of work in a work chunk of a particular zone, the crew size and capabilities, and the means and methods used by this crew. All these three aspects define the 'work density' (Frandson and Tommelein, 2016; Tommelein, 2017).

For instance, while operating the production system, if a work chunk is being executed longer than the takt time, then it is necessary to increase the production capacity. When the opposite occurs, and the work chunk is executed faster than the takt time, then the crew size should be reduced. Another alternative is to change the amount of work for the crew by changing the zone size and/or redistributing the tasks of the work chunk between similarly skilled crews.

The TTP achieves shorter schedules compared to the LOB and LBMS (Seppänen, 2014), especially because it suppresses the buffers between activities and eliminates work in progress derived from differences in task paces. However, a higher number of workable backlogs is necessary to keep the trades on-site, thus avoiding their demobilisation (Seppänen, 2014).

15.7.3.4 Context of use in Lean Construction projects

The use of TTP in construction started with its application in the development of the PSD – or, more specifically, the PS (Fiallo and Howell, 2012; Frandson, Berghede and Tommelein, 2013; Linnik, Berghede and Ballard, 2013; Yassine et al., 2014; Frandson, Seppänen and Tommelein, 2015; Dlouhy et al., 2016; Frandson and Tommelein, 2016; Tommelein, 2017). There are some applications of TTP for planning the refurbishment of cruise ship cabin works (Heinonen and Seppänen, 2016).

Few works implement the TTP in the PSD and master planning. Some assumptions as to the reason for this are that all steps of the TTP require intensive collaboration between trades and the general contractor in an iterative fashion, mainly because the decisions are made by exploring production system alternatives (Frandson, Seppänen and Tommelein, 2015). The participation of contractors and subcontractors in planning may be achieved only after the design of the production system and definition of master plan.

The TTP is used in the LPS (Emdanat, Linnik and Christian, 2016); therefore, the PS is part of it. The TTP defined for the project's phase is used by the lookahead, make ready, weekly and daily planning (Frandson and Tommelein, 2016).

More than one phase can be planned using TTP. For each phase, different production batch sizes and takt time may be defined, which may cause difficulties when visualising the WIP between phases (Faloughi et al., 2015).

15.7.4 Comparison among location-based scheduling techniques

When comparing the three LBS techniques for construction planning, there is a main similarity among them: achieving continuous workflow simultaneously sets a unique production or delivery rate among activities in order to reduce the WIP. However, as visual tools, they have different graphical representations of activities (see Figure 15.8). In the LOB technique, one activity is visualised by dual parallel lines. The crew's workflow becomes clear in the current LOB, through the use of boxes with the crew's label. In turn, the FL represents an activity by a single line starting at the beginning of the first day and finishing at the end of the last day. And, in the TTP, an activity is illustrated by coloured boxes.

Both FL and TTP are based on AON networks, which focus on defining the logical link among the activities. In contrast, the LOB plans are developed based on AOA networks that explore the events that mark the beginning and end of tasks. That is the reason why the pace visualised in the LOB plans is the delivery pace (counted at the end event of an activity), while the activities' slopes in the FL plan illustrate the production

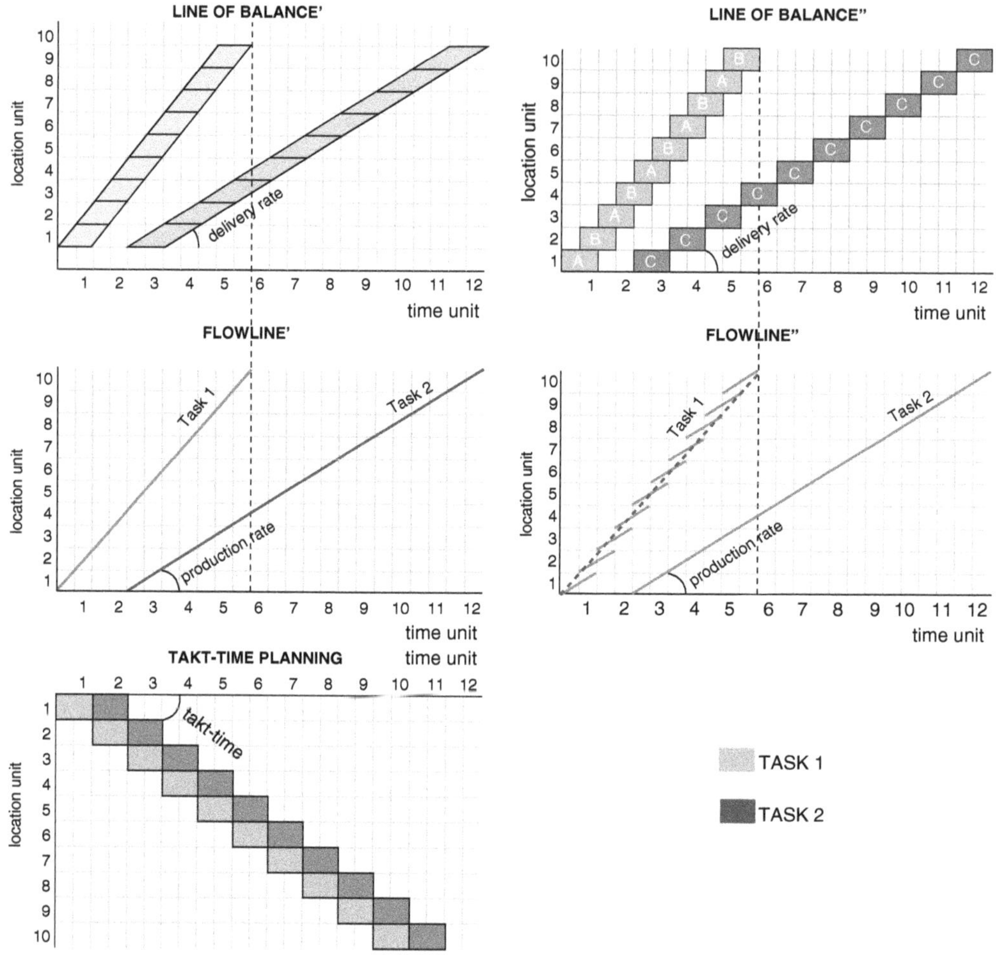

Figure 15.8 Different construction planning techniques based on location

pace of a whole task or crew (when detailing the crew's workflow). The TTP focuses on keeping the production pace equal to the delivery pace, and both match the takt time established for the plan.

Buffers are also used in different ways: the LOB uses time and work buffers between activities and production units respectively, similar to the flowline. At the same time, the TTP incorporates buffers in the crew's production capacity (i.e. the activity's cycle time is shorter than the takt time) (Frandson, Seppänen and Tommelein, 2015). Also, the TTP employs workable backlog to avoid trades' idleness when the work is finished earlier than planned. One criticism of the authors regarding this type of buffer is that, in construction projects where the manpower is specialised, it is more difficult to plan workable backlogs because the amount of work is already forecasted in the contract and in the construction plan. Usually, in the TTP, the workable backlogs are not visualised graphically in the plan, thus becoming a peripheral plan.

Regarding the balancing process of the activities' pace, it is possible to observe that the LOB and FL are flexible techniques that study adjustments in the crews' composition, number of crews and amount of work in a work package to achieve a common pace. In contrast, in the TTP the planners can study the 'work density' to meet the takt time, by modifying not only the crews' size, but also the production batch size, the amount of work in the work chunk and the means and methods used by the crews to perform the tasks. It is worth highlighting that, in the TTP, only one crew is assigned to execute a task in a particular zone.

The context of the construction project uncertainties also influences the way that these methods are implemented. In low complexity projects, or in projects with intense collaboration of partners, the uncertainty is lower, and therefore the interdependencies are known. In this scenario, buffers between activities can be reduced, and TTP can be applied, in most of the cases, in the PS. However, in scenarios where the project has high uncertainty, it is recommended to protect the production from cascading delays by allocating time and work buffers between activities, such as the LOB and FL.

The LOB and FL are usually devised for the whole construction project, which allows the visualisation of WIP in the early stages of planning. On the other hand, in projects that apply TTP, the master plan is devised by using CPM and a Gantt chart, i.e. traditional methods that do not allow the visualisation of WIP on locations. Based on this, the use of TTP focusing on a unique construction phase may hinder the visualisation of WIP between project phases (unfinished location units from the end date of a phase until the start date of the next phase), and the takt time required for one phase may not be the same for the next one, which nullifies the idea of phase demand rate used to plan the activities (Faloughi et al., 2015).

As mentioned previously, the TTP may use different production batch sizes between phases. On the other hand, the LOB and FL are more flexible regarding the production and transfer batch sizes, according to the example in Figure 15.9. The use of a common location breakdown structure for the whole project is useful for the visualisation of activity interferences and WIP.

A comparison among the LBS techniques is presented in Table 15.2.

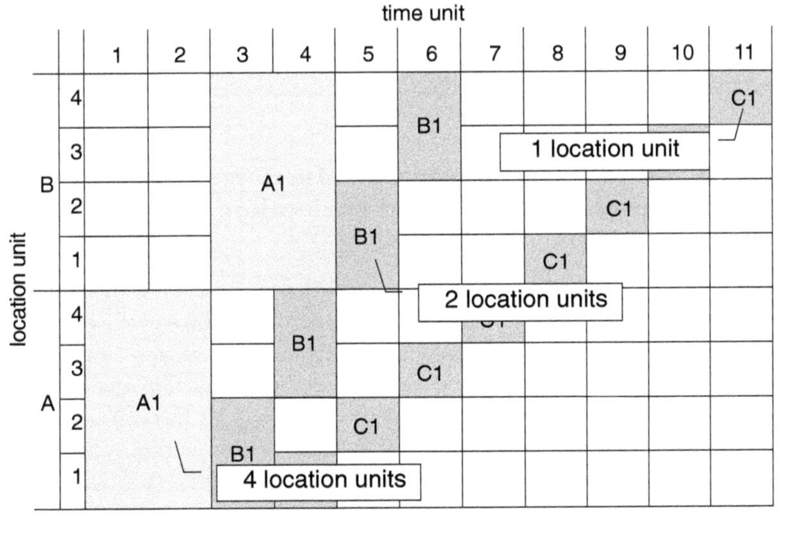

Figure 15.9 Example of different production batch sizes in the line of balance

Table 15.2 Comparison among the Lean LBS techniques for construction planning

	Line of Balance	*Flowline*	*Takt-Time Planning*
Tasks are represented by	Formerly: two parallel lines; Currently: boxes	One line	Boxes
Tasks lines or boxes are represented by	Formerly: Start and finish dates of first and last units; Currently: box at the start and finish dates per each unit	Start date at the first unit and finish date at the last unit	Box at the start and finish dates per each unit
Slope of line represents	Delivery pace	Production pace	Takt time: available production time divided by demand
Type of buffers	Production capacity buffers (inside the work package duration per unit); Buffers between activities	Production capacity buffers (inside the task duration); Buffers between activities	Production capacity buffers (inside the work package duration): difference between takt time and cycle time
Pace achievement (balancing the lines)	Adding or reducing the number of crews to execute an activity; Changing the crews' composition and amount of service inside the work package	Changing the crews' composition	Changing the crews' composition and amount of services inside the work package; Distributing the workload among crew's members; workable backlogs
Level of planning detail	Flexible, mostly used in the PSD and Master Planning	Flexible, commonly use at PSD and Master Planning	High, mostly used in Phase Scheduling
Collaboration	Varies according to the level of planning detail	Varies according to the level of planning detail	Highly necessary due to high level of planning detail

15.8 Crossed comparison between production system design activities and location-based scheduling techniques

After explaining the location-based tools characteristics, it is possible to relate them to the PSD design activities. Some factors, such as availability of information, of contractors and subcontractors, and their collaboration, impose different barriers and opportunities for the location-based tools. That is, the PSD occurs before the production operation, when strategical decisions for the project are made, commonly at a high level of uncertainty. In this scenario, it is suitable to use the LOB or the FL to plan the production system, because these tools are flexible regarding the level of detail, buffers and paces. In fact, after concluding the PSD, these location-based tools become the project master plan, and their information will be used to operate the production system.

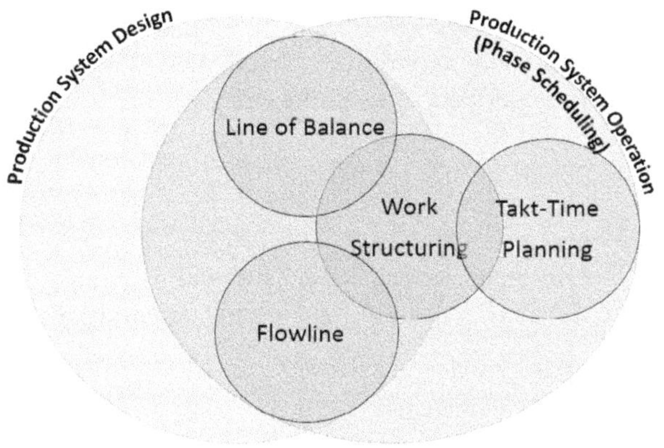

Figure 15.10 Position of Lean tools in relation to planning activities in Lean Construction management

Throughout the system operation, the level of uncertainty may be reduced considerably, compared to the PSD. Here, the main contractors and subcontractors are onboard and may take an important role in the production planning and control. In this scenario, the TTP is the suitable location-based tool to plan the work; the demand is known, and the work can be structured precisely, supressing the buffers between activities. Thus, TTP is suitable to be deployed in the PS.

The WS is the basis of any PSD effort. It must be used before and during the production system operation. What differs between its use in the PSD and PS is the availability of information at the moment of the decision-making. The higher the certainty, the more accurate the decisions made. Besides, every activity of PSD and planning has an opportunistic character regarding information.

The potential applications of LBS techniques in the PSD activities described in this section are presented in the conceptual framework in Figure 15.10.

15.9 Conclusion

In this chapter, the authors compared the Lean Construction activities of PSD, PS and WS, and the LBS tools of LOB, FL and TTP.

Comparing the three location-based tools, it was found that the LOB and FL are very similar tools, and both have evolved in construction to become flexible enough to be applied in any level of uncertainty of the WS. Both protect the production system against variability and delays through the allocation of buffers between activities. Whereas the LOB tries to achieve the same delivery rate, the FL tackles the production rate. Both tools do not force the balancing of activities as the TTP does. The latter is the most recent method applied in construction, and it requires planning workable backlogs in order to avoid the crews' idleness or demobilisations. Crews must work at the same pace or be quicker than the takt time. No buffers are allocated between activities, which increases the risk of cascading delays in projects. TTP has been used in the PS, in which a collaborative participation of contractors is highly necessary.

Differences among the location-based tools were identified, as well as the PSD activities. Therefore, a conceptual framework was developed representing the relationship between the location-based tools and the PSD activities. The framework may support Lean Construction practitioners and scholars in choosing the most suitable technique to deploy in their projects.

Acknowledgments

The authors would like to thank the CNPq, National Council of Technological and Scientific Development, Brazil, for sponsoring the PhD research of the first author.

Note

1 O'Brien, Kreitzberg and Mikes (1985) present the Vertical Production Method (VPM) that combines CPM network with location for repetitive work in a high-rise building, i.e. a prototype of the line of balance. The method was applied to plan many construction phases of a 32-storey high-rise building.

References

Alarcon, L. F., Betanzo, C. and Diethelm, S. (2004) 'Reducing Schedule in Repetitive Construction Projects', *Proc. 12th Annual Conference of the International Group for Lean Construction*, Helsingør, Denmark.

Alves, T. D. C. L., Tommelein, I. D. and Ballard, G. (2006) 'Simulation as a Tool for Production System Design in Constructioned', *Proc. 14th Annual Conference of the International Group for Lean Construction*, Santiago, Chile, pp. 341–353.

Arditi, D. and Albulak, M. Z. (1979) 'Comparison of Network Analysis with Line of Balance in a Linear Repetitive Construction Project', *Proc. Sixth INTERNET Congress*, Garnish-Partenkirchen, Germany.

Arditi, D. and Albulak, M. Z. (1986) 'Line-of-Balance Scheduling in Pavement Construction', *Journal of Construction Engineering and Management*, 112(3), pp. 411–424.

Arditi, D., Tokdemir, O. B. and Suh, K. (2001) 'Scheduling System for Repetitive Unit Construction using Line-of-Balance Technology', *Engineering, Construction and Architectural Management*, 8(2), pp. 90–103.

Ballard, G. (1999) 'Work structuring', in *White Paper*, Las Vegas, NV, USA: Lean Construction Institute.

Ballard, G. (2000b) *The Last Planner System of Production Control*. PhD Thesis, The University of Birmingham, Birmingham, UK. Available at: http://leanconstruction.org.uk/media/docs/ballard2000-dissertation.pdf.

Ballard, G. (2000a) 'Lean Project Delivery System—Revision 1', *LCI White Paper*, Lean Construction Institute, 8, pp. 1–6.

Ballard, G. (2000c) 'Phase Scheduling', *LCI White Paper*, 7.

Ballard, G. (2008) 'Phase Scheduling', *P2SL Research Workshop*.

Ballard, G. and Howell, G. A. (2003) 'An Update on Last Planner', *Proc. 11th Annual Conference of the International Group for Lean Construction*, Blacksburg, VA, USA.

Ballard, G., Koskela, L., Howell, G. and Zabelle, T. (2001a) 'Production System Design in Construction', *Proc. 9th Annual Conference of the International Group for Lean Construction*, 9.

Ballard, G., Koskela, L., Howell, G. and Zabelle, T. (2001b) 'Production System Design: Work Structuring Revisited', *White Paper*, 11.

Barbosa, G., Andrade, F., Biotto, C. and Mota, B. (2013) 'Implementing Lean Construction Effectively in a Year in a Construction Project', *Proc. 21st Annual Conference of the International Group for Lean Construction*, Fortaleza, Brazil, pp. 1017–1026.

Biotto, C. N., Formoso, C. T. and Isatto, E. L. (2015) 'Uso de modelagem 4D e Building Information Modeling na gestão de sistemas de produção em empreendimentos de construção', *Ambiente Construído*, 15(2), pp. 79–96.

Birrell, G. S. (1980) 'Construction Planning—Beyond the Critical Path', *Journal of the Construction Division*, 106(3), pp. 389–407.

Carr, R. I. and Meyer, W. L. (1974) 'Planning Construction of Repetitive Building Units', *Journal of the Construction Division*, 100(3), pp. 403–412.

Damci, A., Arditi, D. and Polat, G. (2013) 'Multiresource Leveling in Line-of-Balance Scheduling', *Journal of Construction Engineering and Management*, 139(9), pp. 1108–1116.

Damci, A., Arditi, D. and Polat, G. (2016) 'Impacts of Different Objective Functions on Resource Leveling in Line-of-Balance Scheduling', *KSCE Journal of Civil Engineering*, 20(1), pp. 58–67.

Dave, B., Hämäläinen, J. P., Kemmer, S., Koskela, L. and Koskenvesa, A. (2015) 'Suggestions to Improve Lean Construction Planning', *Proc. 23rd Annual Conference of the International Group for Lean Construction*, Perth, Australia, pp. 193–202.

Dlouhy, J., Binninger, M., Oprach, S. and Haghsheno, S. (2016) 'Three-Level Method of Takt Planning and Takt Control – a New Approach for Designing Production Systems in Constructioned', *Proc. 24th Annual Conference of the International Group for Lean Construction*, Boston, MA, USA.

Emdanat, S., Linnik, M. and Christian, D. (2016) 'A Framework for Integrating Takt Planning, Last Planner System and Labor Trackinged', *Proc. 24th Annual Conference of the International Group for Lean Construction*, Boston, MA, USA, pp. 53–62.

Esquenazi, A. and Sacks, R. (2006) 'Evaluation of Lean Improvements in Residential Construction Using Computer Simulationed', *Proc. 14th Annual Conference of the International Group for Lean Construction*, Santiago, Chile, pp. 137–149.

Faloughi, M., Linnik, M., Murphy, D. and Frandson, A. G. (2015) 'WIP Design in a Construction Project Using Takt Time Planning', *Proc. 23rd Annual Conference of the International Group for Lean Construction*, Perth, Australia, pp. 163–172.

Fiallo C. M. and Howell, G. (2012) 'Using Production System Design and Takt Time To Improve Project Performance', *Proc. 20th Annual Conference of the International Group for Lean Construction*, San Diego, USA.

Frandson, A., Berghede, K. and Tommelein, I. D. (2013) 'Takt Time Planning for Construction of Exterior Cladding', *Proc. 21st Annual Conference of the International Group for Lean Construction*, Fortaleza, Brazil, pp. 527–536.

Frandson, A., Berghede, K. and Tommelein, I. D. (2014) 'Takt-Time Planning and the Last Planner', *Proc. 22nd Annual Conference of the International Group for Lean Construction*, Oslo, Norway.

Frandson, A.G., Seppänen, O. and Tommelein, I.D. (2015) 'Comparison Between Location Based Management and Takt Time Planning', *Proc. 23rd Annual Conference of the International Group for Lean Construction*, Perth, Australia.

Frandson, A. G. and Tommelein, I. D. (2016) 'Takt Time Planning of Interiors on a Pre-Cast Hospital Projected', *Proc. 24th Annual Conference of the International Group for Lean Construction*, Boston, MA, USA, pp. 18–24.

Heinonen, A. and Seppänen, O. (2016) 'Takt Time Planning in Cruise Ship Cabin Refurbishment: Lessons for Lean Constructioned', *Proc. 24th Annual Conference of the International Group for Lean Construction*, Boston, MA, USA.

Henrich, G., Tilley, P. and Koskela, L. (2005) 'Context of Production Control in Construction', *Proc. 13th Annual Conference of the International Group for Lean Construction*, Sydney, Australia, pp. 189–198.

Hopp, W.J. and Spearman, M.L. (2011) *Factory Physics*. Long Grove, IL: Waveland Press.

Junnonen, J. and Seppänen, O. (2004) 'Task Planning as a Part of Production Control', *Proc. 12th Annual Conference of the International Group for Lean Construction*, Helsingør, Denmark, pp. 183–193.

Kallantzis, A., Soldatos, J. and Lambropoulos, S. (2007) 'Linear versus Network Scheduling: A Critical Path Comparison', *Journal of Construction Engineering & Management*, 133(7), pp. 483–491.

Kemmer, S. L., Heineck, L. F. M. and Alves, T. D. C. L. (2008) 'Using the Line of Balance for Production System Design', *Proc. 16th Annual Conference of the International Group for Lean Construction*, Manchester, UK, pp. 299–308.

Kenley, R. (2004) 'Project Micro-Management: Practical Site Planning and Management of Work Flow', *Proc. 12th Annual Conference of the International Group for Lean Construction*, Helsingør, Denmark, pp. 3–5.

Kenley, R. and Seppänen, O. (2010) *Location-Based Management for Construction: Planning, Scheduling and Control*. London, UK: Spon.

Kleinfeld, I. H. (1976) 'Manpower Use in High-Rise Residential Construction', *Journal of the Construction Division*, 102(2), pp. 379–383.

Knapp, S., Charron, R. and Howell, G. (2006) 'Phase Planning Today', *Proc. 14th Annual Conference of the International Group for Lean Construction*, Santiago, Chile, Chile, pp. 431–441.

Koskela, L. (2000) 'An Exploration Towards a Production Theory and Its Application to Construction', PhD Thesis, VTT Technical Research Centre of Finland. Available at: www.vtt.fi/inf/pdf/publications/2000/P408.pdf

Koskela, L. and Ballard, G. (2003) 'What Should We Require from a Production System in Construction', in *ASCE Construction Research Congress*, Wind of Change: Integration and Innovation, Honolulu, Hawai, USA: American Society of Civil Engineers, pp. 19–21.

Koskela, L. and Ballard, G. (2006) 'Should Project Management be based on Theories of Economics or Production?' *Building Research & Information*, 34(2), pp. 154–163.

Koskela, L. and Howell, G. (2002) 'The Underlying Theory of Project Management is Obsolete', *Proc. PMI Research Conference*, PMI, pp. 293–302.

Koskela, L., Howell, G., Pikas, E. and Dave, B. (2014) 'If CPM Is So Bad, Why Have We Been Using It So Long', *Proc. 22nd Annual Conference of the International Group for Lean Construction*, Oslo, Norway, pp. 27–37.

Laufer, A. and Tucker, R. L. (1988) 'Competence and Timing Dilemma in Construction Planning', *Construction Management and Economics*, 6(4), pp. 339–355.

Linnik, M., Berghede, K. and Ballard, G. (2013) 'An Experiment in Takt-Time Planning Applied to Non-Repetitive Worked', *Proc. 21st Annual Conference of the International Group for Lean Construction*, Fortaleza, Brazil.

Lucko, G., Alves, T. D. C. L. and Angelim, V. L. (2014) 'Challenges and Opportunities for Productivity Improvement Studies in Linear, Repetitive, and Location-Based Scheduling', *Construction Management and Economics*, 32(6), pp. 575–594.

Lucko, G. and Gattei, G. (2016) 'Line-of-Balance Against Linear Scheduling: Critical Comparison', *Proceedings of the Institution of Civil Engineers - Management, Procurement and Law*, 169(1), pp. 26–44.

Lumsden, P. (1968) *The Line-of-Balance Method*. Oxford, UK: Pergamon Press.

Marchwinski, C. and Shook, J. (2003) *Lean Lexicon: A Graphical Glossary for Lean Thinkers*. Cambridge, MA: Lean Enterprise Institute.

Mattila, K. G. and Abraham, D. M. (1998) 'Linear Scheduling: Past Research Efforts and Future Directions', *Engineering, Construction and Architectural Management*, 5(3), pp. 294–303.

Mendez, J. R. and Heineck, L. F. M. (1998) 'Preplanning Method for Multi-Storey Building Construction Using Line of Balanceed', *Proc. 6th Annual Conference of the International Group for Lean Construction*, Guarujá, Brazil.

Meredith, J. R. and Shafer, S. M. (2009) *Operations Management for MBAs*. Hoboken, NJ: John Wiley & Sons.

Milberg, C. (2007) 'Tolerance Considerations in Work Structuring', *Proc. 15th Annual Conference of the International Group for Lean Construction*, East Lansing, MI, USA, pp. 233–243.

Mohr, W. E. (1979) *Project Management and Control (in the building industry)*, 2nd edn. Parkville, VIC: 5th, University of Melbourne, Department of Architecture and Building.

Mota, B. P., Mota, R. R. and Alves, T. D. C. L. (2008) 'Implementing Lean Construction Concepts in a Residential Project', *Proc. 16th Annual Conference of the International Group for Lean Construction*, Manchester, UK, pp. 251–257.

Moura, R., Monteiro, J. M. F. and Heineck, L. F. M. (2014) 'Line of Balance – Is It a Synthesis of Lean Production Principles as Applied to Site Programming of Works?', *Proc. 22nd Annual Conference of the International Group for Lean Construction*, Oslo, Norway, pp. 703–714.

O'Brien, J. J. (1975) 'VPM Scheduling for High Rise Buildings', *Journal of the Construction Division*, 101(4), pp. 895–905.

O'Brien, J. J., Kreitzberg, F. C. and Mikes, W. F. (1985) 'Network Scheduling Variations for Repetitive Work', *Journal of Construction Engineering and Management*, 111(2), pp. 105–116.

Office of Naval Material. (1962) *Line of Balance Technology: A Graphic Method of Industrial Programming*. NAVEXOS P1851 (Rev. 4–62), Reviewed and Approved 17. April Office of Naval Material. Washington, DC, USA: Department of the Navy.

Olivieri, H., Granja, A. D. and Picchi, F. A. (2016) 'Planejamento tradicional, Location-Based Management System e Last Planner System: Um Modelo Integrado', *Ambiente Construído*, 16, pp. 265–283.

Peer, S. (1974) 'Network Analysis and Construction Planning', *Journal of the Construction Division*, 100, p. 203.

Schoderbek, P. P. and Digman, L. A. (1967) 'Third Generation PERT/LOB', *Harvard Business Review*, 45(5), pp. 100–110.

Schramm, F. K. (2004) *O projeto do sistema de produção na gestão de empreendimentos habitacionais de interesse social*. Porto Alegre, Brazil: Federal University of Rio Grande do Sul.

Schramm, F. K., Costa, D. B. and Formoso, C. T. (2004) 'The Design of Production Systems for Low-Income Housing Projects', *Proc. 12th Annual Conference of the International Group for Lean Construction*, Helsingør, Denmark.

Schramm, F. K. and Formoso, C. T. (2007) 'Using Visual Interactive Simulation to Improve Decision making in Production System Design', *Proc. 15th Annual Conference of the International Group for Lean Construction*, East Lansing, MI, USA, pp. 357–366.

Schramm, F. K., Rodrigues, A. A. and Formoso, C. T. (2006) 'The Role of Production System Design in the Management of Complex Projects', *Proc. 14th Annual Conference of the International Group for Lean Construction*, Santiago, Chile, pp. 227–239.

Schramm, F. K., Silveira, G. L., Paez, H., Mesa, H., Formoso, C. T. and Echeverry, D. (2008) 'Using Discrete-Event Simulation to Support Decision-Makers in Production System Design and Operations', *Proc. 16th Annual Conference of the International Group for Lean Construction*, Manchester, UK, pp. 131–141.

Schramm, F. K., Tillmann, P. A., Berr, L. R. and Formoso, C. T. (2009) 'Redesigning the Production System to Increase Flexibility in House Building Projects', *Proc. 17th Annual Conference of the International Group for Lean Construction*, Taipei, Taiwan, pp. 347–358.

Selinger, S. (1973) 'Method for Construction Process Planning Based on Organisation Requirements', *PhD Thesis, Faculty of Civil Engineering, Haifa, Technion-Israel Institute of Technology*.

Seppänen, O. (2014) 'A Comparison of Takt Time and LBMS Planning Methods', *Proc. 22nd Annual Conference of the International Group for Lean Construction*, Oslo, Norway.

Seppänen, O., Ballard, G. and Pesonen, S. (2010) 'The Combination of Last Planner System and Location-Based Management System', *Lean Construction Journal*, pp. 46–54.

Su, Y. and Lucko, G. (2015) 'Comparison and Renaissance of Classic Line-of-Balance and Linear Schedule Concepts for Construction Industry', *Organization, Technology & Management in Construction*, 7(2), pp. 1315–1329.

Tommelein, I. D. (2017) 'Collaborative Takt Time Planning of Non-Repetitive Work', *Proc. 25th Annual Conference of the International Group for Lean Construction*, Heraklion, Greece, pp. 745–752.

Tsao, C. C. (2005) *Use of Work Structuring to Increase Performance of Project-Based Production Systems*. Berkeley, CA, USA: University of California.

Tsao, C. C. Y., Draper, J. and Howell, G. A. (2014) 'An Overview, Analysis, and Faciliation Tips for Simulations That Support and Simulate Pull Planning', *Proc. 22nd Annual Conference of the International Group for Lean Construction*, Oslo, Norway.

Tsao, C. C. Y., Tommelein, I. D., Swanlund, E. S. and Howell, G. A. (2004) 'Work Structuring to Achieve Integrated Product-Process Design', *Journal of Construction Engineering and Management*, 130(6), pp. 780–789.

Valente, C. P., Montenegro, G. A., Brito, F. L., Biotto, C. N. and Mota, B. P. (2014) 'Guidelines for Developing a Line of Balance for Non-Repetitive Areas (Common Areas) at a Vertical Residental Building', *Proc. 22nd Annual Conference of the International Group for Lean Construction*, Oslo, Norway, pp. 763–774.

Yamin, R. A. and Harmelink, D. J. (2001) 'Comparison of Linear Scheduling Model (LSM) and Critical Path Method (CPM)', *Journal of Construction Engineering & Management*, 127(5), pp. 374–381.

Yassine, T., Bacha, M. B. S., Fayek, F. and Hamzeh, F. (2014) 'Implementing Takt-Time Planning in Construction to Improve Work Flow', *Proc. 22nd Annual Conference of the International Group for Lean Construction*, Oslo, Norway.

16 The Last Planner® System as an approach for coping with the complexity of construction projects

Tarcisio Abreu Saurin and John Rooke

16.1 Introduction

In several scientific disciplines, complexity science (CS) is gaining traction as a result of the perceived limitations of the traditional scientific method for studying human organisation. The traditional approach tends to be reductionist and neglects contextual conditions in the search for generalisable findings. For example, in healthcare the use of CS has been advocated as both a theoretical and practical framework for addressing implementation difficulties of protocols of care, which often do not work in practice as anticipated by laboratory studies (Braithwaite et al., 2017).

CS offers an alternative paradigm by stressing the modelling of interactions between agents, which cannot be fully controlled, but at best influenced according to the desired direction. It suggests that system attributes such as diversity, evolution, and adaptive capacity can be key assets (Cilliers, 1998). CS also recognises that some types of variability may be beneficial and even essential for the survival of a system (Hollnagel, 2014).

In the construction industry, the use of CS as a lens for making sense and managing projects has been a topic of academic interest over the recent decades (Gidado, 1996; Bertelsen and Koskela, 2005; Ballard and Tommelein, 2012). Various tools for dealing with complexity have been developed, such as the Last Planner® System (LPS) (Ballard and Tommelein, 2012), the Design Structure Matrix, and Net Present Value analysis (Pikas et al., 2015b), and operationalised versions of the Transformation-Flow-Value (TFV) theory (Bertelsen and Bonke, 2011; Pikas et al., 2015a). These efforts have been motivated by the perceived strong implications of complexity for project management. Increased complexity tends to increase the time and cost of projects, requires more sophisticated planning, coordination, and control, and can thus be a criterion in the selection of a suitable project procurement arrangement (Baccarini, 1996).

To some extent, the complexity of construction projects is due to the inherent features of the industry, such as the dependency on environmental conditions (e.g. weather and soil conditions), and to the long duration of projects, which makes them more exposed to uncertainty from the environment, and changes in client requirements. On the other hand, managerial choices may amplify complexity by having a large number of subcontractors, or neglecting the control of basic waste, for example. As such, construction projects might be framed as complex *socio-technical* systems (CSS), which are a sub-set of the broader family of complex *adaptive* systems. Since many complex adaptive systems are not socio-technical (e.g. ecosystems, which may interact with construction projects), this distinction is necessary.

Based on a complexity framework, this chapter has two main objectives: (i) to discuss how the LPS has addressed complexity; and (ii) to identify opportunities for further theoretical and practical innovation. The emphasis on LPS is due to its prominent role as the main practice for the operationalisation of Lean Construction, and the resulting greater availability of theoretical and empirical evidence to support the discussion. A previous analysis of the compatibility between the LPS and complexity was conducted by Ballard and Tommelein (2012). These authors argue that Lean management practices in general, and the LPS, are better suited to complex projects than traditional methods. However, their study did not take a systematic and explicit CS approach, which hinders the exploration of the multidimensional nature of the complexity construct (Soliman, Saurin, and Anzanello, 2018).

The remainder of the chapter is organised into six sections. In the next section, the main attributes of CSS are described. The following section presents an overview of how complexity thinking has been addressed over time in the Lean Construction community. Following, what the complexity attributes look like in construction is discussed. The next section presents guidelines for the management of CSS, followed by a discussion of the relationships between the said guidelines and the LPS. The last section sets an agenda for future research.

16.2 What is complexity?

Given the multidisciplinary interest in complexity, slightly different definitions of the concept abound in the literature. In a review of complexity definitions in operations management, Jacobs and Swink (2011) found it to be described as a state manifested by the multiplicity, diversity, and interrelatedness of system elements. Walker et al. (2010) group complexity definitions into three main types: (i) the attribute view, which assumes that complex systems in general share some interrelated attributes, such as a large number of elements, emergent properties, nonlinear dynamics, feedback loops, and adaptive behaviour; (ii) complexity as a quantitative measure, which usually reflects one or more of the complexity attributes; and (iii) complexity as emergent phenomena.

The attribute view is commonly used in the management and social sciences literature, and is adopted as a basis in this chapter. The attributes are usually divided into two main categories: those that represent structural properties of the system, such as the number and diversity of parts; and those that represent functional characteristics of the system, such as resilience (El Maraghy et al., 2014).

The attribute view accounts for several dimensions of both structural and functional complexity, which is useful when envisioning the use of management guidelines for coping with complexity. Indeed, certain guidelines may be more suitable to cope with certain complexity attributes. Furthermore, the attribute view can partially account for the other two views, while the opposite is likely to be more difficult. The attributes can be quantified, and emergent phenomena are some of the attributes often cited in literature (Righi and Saurin, 2015).

Both in construction management and other disciplines, complex systems thinking is hardly operationalised and is often used only retrospectively. This means that the explicit use of complexity thinking is often limited to its adoption as a lens for explaining past events, with accompanying improvement recommendations. Of course, interventions in CSS may take complexity into account implicitly. For instance, Leykum et al. (2007) concluded, based on a systematic literature review, that better reported outcomes when treating patients with type II diabetes were obtained when the intervention leveraged complexity attributes in design and implementation.

It is also important to differentiate complex from complicated systems. According to Dekker et al. (2013), complicated systems are ultimately knowable and controllable, as they afford an exhaustive description. Although complicated systems may have thousands of parts, they follow a set of predefined rules (e.g. a computer software), and interactions are mostly linear. Order in complicated systems is achieved by figuring out one best method to operate them, and there is a clear boundary where the system ends and its environment begins (Dekker et al., 2013). The operational functioning of a building and its technical components is an example of a complicated system. By contrast, CSS cannot be fully described and modelled, and their behaviour is unpredictable to some extent (Dekker et al., 2013). In this sense, the operational and administrative functions performed on a construction site form a CSS. One implication of this distinction is that, presumably, there can be precise measures of complication, but not of complexity.

16.3 Complexity thinking in Lean Construction

The Lean Construction community has long been interested in construction as a CSS. Koskela (1992) can be read as critiquing the conventional management thinking on the grounds that it does not address the true complexity of production processes. The LPS represents the practical development of a planning system designed to cope with this complexity as it occurs on construction projects (Ballard and Howell, 1994a, 1994b, 1998). The system draws on Koskela's seminal 1992 paper, as well as the systems thinking of Sanvido's (1984) and Laufer and Howell's (1993) work on planning, in the ongoing development of a model of distributed planning and production control.

However, it is not until Bertelsen (2002), where it is suggested that complexity thinking may provide a better theoretical explanation of LPS, that there is an explicit attempt to apply the lessons of complex systems thinking to the development of Lean Construction theory. Subsequently, Bertelsen (2003a) identified three complex subsystems in construction – the project production system; the procurement system; and the social system of the construction site – thus theorising the importance in construction project management of organising over planning (see also, Johnston and Brennan, 1996; Koskela and Howell, 2002). This initial attempt to identify project subsystems is followed by an analysis of the complex nature of clients (Bertelsen and Emmitt, 2005).

A number of attempts have also been made to analyse the whole industry as a single system of complex flows. Taking this approach, Bertelsen and Sacks (2007) search for the overarching rules that guide the industry's activity. They identified at least two, namely profit maximisation and queueing theory, suggesting that these two interact to induce negative outcomes. In a similar vein, but with a greater emphasis on improvement, Saurin et al. (2013a) suggest six guidelines for the analysis and improvement of complex Lean systems, some consequences of which are developed in the present chapter (see section: guidelines for coping with complexity, below). Saurin (2016) introduces a complexity-derived technique, the Functional Resonance Analysis Method or FRAM (Hollnagel, 2012), as a means of understanding institutional obstacles to production in relation to construction health and safety management.

In contrast to these 'whole system' approaches, attempts have been made to drill further down into Lean Construction concepts. Thus, it has been suggested that Shingo's (1985) concept of operations can in some contexts be substituted for Transformation in the TFV theory (Rooke et al., 2007; Bertelsen, 2017), facilitating an analysis of the activity that drives flows. Complexity thinking has also been used as a lens for exploring the trade-off between slack and efficiency, which may arise in construction projects as part of Lean Construction implementation. Saurin (2017) suggests that CS offers a stronger theoretical background for analysing the said trade-off, while Lean offers a more practical approach. The connections between these diverse approaches aforementioned, their integration into mainstream Lean Construction theory, and their relevance to LPS remain to be fully spelt out.

16.4 Attributes of complexity in construction

The attribute view of complexity is common in construction management (e.g. Baccarini, 1996; Dao et al., 2016), and the attributes are similar to those mentioned in other disciplines. Bertelsen (2003b) seeks to identify the various complexity attributes of construction projects. Table 16.1 presents key attributes proposed by Cilliers (1998) and examples in

Table 16.1 Attributes of complexity and examples in construction

Attributes	*Examples*
Large number of elements	Large number of workers, transportation equipment, construction materials, subcontractors, regulations, designers, clients.
Diversity of elements	Subcontractors with different organisational cultures and levels of expertise, different profiles of clients in the same project, customisation possibilities offered to clients, different contractual arrangements with designers and subcontractors.
Dynamic interactions	Formal and informal exchange of information between project members, flow of people and materials in the construction site, interactions with designers, regulators, clients.
Nonlinear interactions	A small variation in the output of an operation (e.g. a design error) can cause large disruptions (e.g. demolition of built areas for reworking it).
Interactions with immediate neighbours	Crews tend to interact mostly with those that are physically proximal, without having an overview of the broader impacts of their actions.
Couplings	The major construction stages (e.g. earthworks, foundations, structure, masonry, roofing, etc.) must follow a rigid sequencing, and in this sense they are linear and tightly coupled. This implies that there is little or no margin of manoeuvre for alternative sequencing. However, organisational arrangements in the construction supply chain tend to be loosely coupled (Dubois and Gadde, 2002).
Feedback loops	Too much pressure for compliance with quality standards makes workers feel anxious about reporting deviances. In turn, this hinders learning on the real causes of defects, and encourages oversimplified assumptions that non-compliance is the root cause. As a result, management may increase compliance pressure, thus forming a closed feedback loop.
Openness	Construction sites are subject to external variability such as changes in the weather, changing regulations, and the macroeconomic situation of the country/region. There may also be interactions with other construction sites from the same or from another company, e.g. a same subcontractor reduces the size of its crew in a less important project, and transfers the workers to another, more important, project.
Path dependence	The way the system is currently managed is due to previous experiences and challenges overcome in the past. For example, a company may have a preference for certain construction technologies and contractual arrangements because they worked in the past.

construction. Saurin et al. (2013) presented a description of what some of these attributes look like in a refurbishment project.

Ballard and Tommelein (2012) argue that there is a need for developing the means for assessing the complexity of projects before they are started. This would support decision-making regarding the extent to which Lean is necessary in project management, as well as the nature of the Lean practices that should be applied. In their view, '*the greater the complexity the more the Lean methods are needed*'. There are many proposals for the quantification of the complexity of construction projects, usually in terms of estimating the intensity of some core complexity attributes and the resulting calculation of complexity scores (e.g. Bosch-Rekveldt et al., 2011; Dao et al., 2016). These concerns with measurement are motivated by the need for setting more precise cost and time estimates for projects, as well as by the need for defining the level and type of planning and management efforts (Baccarini, 1996).

16.5 Guidelines for coping with complexity

Bertelsen and Koskela (2005) propose four strategies for dealing with complexity in construction: creating slack; reducing complexity through modularisation and standardisation; codification through the implementation of LPS; and improving improvisation skills. Another attempt to bring a more pragmatic application of CS to bear has been the use of the concept of path dependency to understand, to plan and initiate change in construction organisations (Morrey et al., 2012).

Based on a literature review, Saurin, Rooke, and Koskela (2013) identified six guidelines for coping with complexity. Table 16.2 presents the conceptual association between the guidelines and 14 principles of the Toyota Production System, proposed by Liker (2004). Liker's principles and the guidelines are conceptually consistent with each other. Saurin et al. (2013) presented a description of what these guidelines look like in a refurbishment project, while Righi and Saurin (2015) discuss the use of the guidelines in an emergency department. It is worth noting that contingency is a core characteristic of a socio-technical system design (Clegg, 2000) and, as such, the mentioned guidelines are context-dependent and their use can trigger undesired interactions. For example, even in a complex socio-technical system, privacy may sometimes take priority over visual control (Bernstein, 2012), and slack may introduce more parts and interactions into the system, thus creating new error possibilities (Perrow, 1984).

The operationalisation of the guidelines may benefit from the use of modelling approaches fit to complexity. These approaches may support the understanding of the dynamics of CSS over time, and thus offer insight into leverage points and risk assessment of Lean interventions. For instance, social network analysis can be a powerful means for modelling social interactions, which tend to be supported by Lean Construction (Wehbe, Al Hattab, and Hamzeh, 2016). The use of this approach offers plenty of opportunities for operationalising the guideline 'encourage diverse perspectives when making decisions'. Similarly, other approaches can support the analysis of the impacts of apparently commonsense changes in the work system, and thus being useful for the guideline 'monitor unintended consequences of improvements and small changes'. For this purpose, it is possible to mention three approaches that have been used in the context of Lean Construction: system dynamics (Mota, Viana, and Isatto, 2010), agent-based -modelling (Hattab and M. Hamzeh, 2016), and the Functional Resonance Analysis and Modelling method (Hollnagel, 2012; Saurin, 2016).

Table 16.2 Guidelines for coping with complexity (adapted from Saurin et al., 2013)

Guidelines	*Dimensions of the guidelines*	*Liker's principles*
Give visibility to processes and outcomes	Systems should be intuitive, so as to reduce imaginary complexity. Visibility should be given to informal work practices, which may encompass either useful innovations or latent hazards that over time may be taken for granted as part of normal work. However, privacy may be important for adapting and innovating, and these may be hindered if visual management is used as a means for enforcing the use of ineffective rules.	P2: Create continuous process flow to bring problems to the surface. P5: Build a culture of stopping to fix problems, to get quality right first time. P7: Use visual control so no problems are hidden.
Encourage diversity of perspectives when making decisions	Diversity of perspectives may help to tackle uncertainty. Agents involved in decision-making should hold complementary skills. Some requirements for the implementation of this guideline are: high levels of trust, reduction of power differentials, and identification of apt decision-makers.	P13: Make decisions slowly by consensus, thoroughly considering all options (implement decisions rapidly).
Monitor unintended consequences of improvements and small changes	The impacts of small changes and improvements may be large in CSS, due to nonlinear interactions. Improvements and small changes interact between themselves, and this poses opportunities for unintended consequences. As small changes happen all the time, they offer frequent opportunities for reflection on practice. Small changes and improvements may be either non-intentional or intentionally self-initiated by the organisation (e.g. through kaizen) as well as originated from external sources (e.g. a client changes its order).	P7. P6: Standardised tasks are the foundation for continuous improvement and employee empowerment. P12: Go and see for yourself to thoroughly understand the situations.
Design slack	Slack means spare resources, of any sort, which can be called on in times of need. Also, slack must cope with variability, thus contributing to the resilience of CSS. Since variability is normal in these systems, slack also tends to	P3: Use pull systems to avoid overproduction. P4: Level out the workload.

(*Continued*)

Table 16.2 (Cont.)

Guidelines	*Dimensions of the guidelines*	*Liker's principles*
	be useful in everyday work, rather than only in extreme situations. Slack may take a number of forms, such as redundant equipment, underutilised space, excess of labour, generous time margins. Slack may have side effects such as contributing to maintain problems hidden, and disguising small changes.	
Monitor and understand the gap between work-as-imagined and work-as-done	It is impossible for standardised operating procedures to cover all situations. CS regards procedures as dynamic, local, and situated constructions, which need adaptation in the face of variability. This is in contrast with the traditional view of procedures as 'devised by experts (management) to guard against the errors and mistakes of fallible human operators at the sharp end, who are more limited than the experts in their competence' (Hale and Borys, 2013). Procedures may be of different types (e.g. goal oriented, action oriented) and, for all types, the gap between them and practice should be monitored.	P6, P12
Create an environment that supports resilience	All the previously mentioned guidelines support resilient performance. As complexity cannot be fully eliminated, agents must have the skills to adapt to it (i.e. resilience skills). Resilience skills are defined as individual and team skills of any type necessary to fill in the gaps of procedures, in order to maintain safe and efficient operations during both expected and unexpected situations. The use of resilience skills requires organisational support, such as granting authority to people to self-organise as well as the provision of training.	P1: Base your management decisions on a long-term philosophy, even at the expense of short term goals. P8: Use only reliable, thoroughly tested technology. P9: Grow leaders who thoroughly understand the work, live the philosophy and teach it to others. P10: Develop exceptional people and teams who follow your company's philosophy. P11: Respect your extended network of partners and suppliers by challenging them and helping them improve. P14: Become a learning organisation through relentless reflection and continuous improvement.

Another aspect related to the guideline focused on unintended consequences refers to the need for evaluating the impacts of Lean Construction in the long run and on broader systems, which are not limited to the intervention site. Indeed, interventions in one part of CSS reverberate over the whole system, and waste removed from one subsystem may reappear elsewhere. Nevertheless, in theory, if waste cannot be completely eliminated, it could be transferred to a less critical subsystem, with more slack and less safety hazards (Soliman and Saurin, 2017).

16.6 The complexity guidelines and the Last Planner®

In this section, an analysis is made of the compatibility between the six guidelines for coping with complexity and the LPS.

16.6.1 Encourage diversity of perspectives when making decisions

Planning meetings carried out as part of the LPS implementation usually involve several stakeholders, such as production managers, site managers, safety experts, subcontractors, and client representatives. This mixed composition provides cognitive diversity, encourages the sharing of information, facilitates the identification of conflicting requirements, and the design of reliable plans.

However, power and influence might be unequally shared between last planners (i.e. people who serve as the coupling point between planning and production, according to Aslesen and Tommelein, 2016) undermining collaborative work and leading to unreliable promises. Thus, there seems to be an opportunity for the development of decision-making tools that may support the last planners by providing data that can be used to balance unequal power structures. Furthermore, even where power and status differentials are not such as to distort communication, it may still be difficult to achieve a shared understanding of desired value and the means necessary to achieve it. The identification of 'common understanding' as a necessary project flow provides a useful starting point for investigating this phenomenon (Pasquire and Ebbs, 2017).

16.6.2 Monitor unintended consequences of improvements and small changes

The application of the LPS involves a number of decisions on production control, either with small (e.g. definition of the number of workers composing a crew) or large implications for site operations (e.g. sequencing of tasks). Due to the cyclical and dispersed nature of the LPS, decisions are frequent and based on the most up-to-date information from the locus of production. This provides opportunities for learning based on feedback. Since the planning and corresponding feedback cycles are short (e.g. weekly), the LPS allows for the early identification of unintended consequences of decisions before they escalate and cause greater damage. The idea of having last planners at the front line is that conditions are constantly monitored and significant changes are immediately dealt with at the appropriate level of the organisation, which is often at the point of production. In this respect, organisational support through visual management is useful for making the last planners and other stakeholders aware of changes, which otherwise could go unnoticed.

Furthermore, personal commitments between interdependent last planners are an essential part of the LPS, as the promises made by them define which tasks can be scheduled (Ballard and Tommelein, 2012). When making promises, the last planners should follow the should-can-do heuristic by first considering what *should* be done under ideal conditions, then appreciating what *can* realistically be done, and thus defining what they *will* do (Ballard and Howell, 1998). Regardless of this collaborative character, a CS perspective raises the question of whether the said heuristic is not too linear. In fact, the last planners are unlikely to have all the necessary information available, in a timely manner, when making promises.

An opportunity for improving the LPS exists in the possibility of developing means for making more visible, in real time, the interactions between work packages. This implies the need for greater levels of process transparency and computational support, as well as a functional model of site operations that can be easily updated on a regular basis. Innovations in this direction have recently been made through the systematic application of visual tools to site and project management (Brady et al., 2018).

16.6.3 Design slack

In Lean Construction, the use of the term 'buffer' is more common than slack, and has been addressed by many studies. Alves and Tommelein (2004) define buffers as resource cushions (e.g. money, time, materials, space), used to protect processes against variation and resource starvation. Russell et al. (2012) acknowledge four main types of buffers in construction: inventory, time, capacity, and plans.

In fact, many day-to-day planning decisions made in the scope of LPS offer opportunities for the design of those and other types of slack resources. For instance, the completion of delayed work packages may require the deployment of slack resources in terms of time and labour. Furthermore, the LPS has an explicit mechanism for the design of slack, namely the planning of a backlog of work packages. These packages have no pending constraints and they should be carried out if the scheduled packages need to be cancelled for any reason. This is an example of using plans as slack.

Nevertheless, the use of the guideline design slack presents opportunities for the development of Lean Construction and the LPS. In fact, the design of slack involves several decisions such as the definition of how much slack is enough, where the slack resources should be located, the nature of the resources that should compose slack, and how to make the status of slack resources visible to the interested parties. These decisions have not been explicitly framed as a 'slack' problem in Lean Construction, which also tends to be limited to the buffer types previously mentioned. A broader taxonomy of slack resources and the development of methods for their analysis and explicit design (Saurin and Werle, 2017) may trigger innovative uses of this concept in Lean Construction.

16.6.4 Monitor and understand the gap between work-as-imagined and work-as-done

Work-as-imagined is what designers, managers, regulators, and authorities believe happens, or should happen. Work-as-done is what actually happens in the workplace, as a result of adjustment to current conditions, constraints, and the context of the moment (Hollnagel, 2014). From a CS perspective, monitoring the gap is important because it sheds light on the real reasons for achieving the expected outcomes, which otherwise may be simplistically assumed by managers as a result of perfect planning and full compliance with it. The nature of the LPS as a distributed planning system tends to narrow the gap between work-as-imagined and work-as-done, by locating the planning process as close to the site of production as possible.

In fact, the LPS supports the application of this guideline by delegating responsibility for defining work-as-imagined to the last planner (at least partially), who initiates and directly supervises that work. Since this last planner is present most of the time at the *gemba* and he is well versed in the relevant trade, he is aware of work-as-done and can immediately check it against work-as-imagined.

The LPS also has a well-known mechanism for checking the adherence between work-as-imagined and work-as-done, namely the percentage of completed work packages (PPC) indicator. Nevertheless, this check is normally limited to the verification of achieved goals. The question arises as to *how* the satisfaction of goals was assessed. There are reports that some managers do not account for the quality of the work package when assessing its completion (Leão, Formoso, and Isatto, 2014).

The imagined *when* might also not be precisely checked. For instance, a work package scheduled to occur from Monday to Wednesday may in reality be finished on Friday. This delay may be neglected by the PPC indicator, which is commonly monitored on a weekly basis. Also, the *who* question is not usually monitored, which could be important for the analysis of productivity and quality. It is worth noting here that the relevant *who* does not refer to specific individuals. Rather, it is concerned with the professional trades involved in the task and the corresponding number of workers. Overall, there is an opportunity for the development of more robust means for comparing work-as-imagined and work-as-done.

16.6.5 Give visibility to processes and outcomes

The concept of a last planner allows the visibility of work-as-done through the principle of going to the *gemba*, in order to see what the situation looks like on the spot, rather than relying on indirect sources of information (Liker, 2004). Visibility is enhanced under LPS because the number of eyes is increased dramatically according to the number of last planners that are included in the management team. As such, the LPS does not imply that visibility must be available to a central planner, which is consistent with the complexity principle of decentralised control (Perrow, 1984).

In addition, visual management is a central tenet of Lean and, although take-up in construction has been slow, there is a growing body of knowledge on this topic (Tezel, Koskela, and Tzortzopoulos, 2016). From an operational perspective, there are many ways of incorporating visual management on the construction site, as well as in the documents that comprise the plans themselves, such as by using visual aids, and physical and computer-based mock-ups (Tezel, Koskela, and Tzortzopoulos, 2016). As a drawback, visibility can be hindered when last planners lack observation and management skills to understand and mitigate the downstream implications of processes and outcomes. The identification of personality traits of last planners, which could favour the development of such skills, has also been an underexplored topic. For instance, in the context of Lean production, Camuffo and Gerli (2018) concluded that humility is a desirable trait for implementing the principle of going to the *gemba*.

16.6.6 Create an environment that supports resilience

Although resilience is an inherent property of CSS, it can be either supported or hindered by organisational design and development. Resilience is that quality of a complex system that prevents it from breaking down. It is, thus, the vital key to the design of effective and robust complex systems. In fact, the LPS principle of planning in greater detail as the time for action approaches (Ballard and Tommelein, 2012) offers the opportunity to adjust the forecasts to the conditions on the spot. This adjustment across the planning levels is an example of resilience of the production planning system itself, which also supports the resilience of the production teams at the front line.

Furthermore, the previously discussed guidelines support resilience and, therefore, the LPS also plays a role in developing resilience through the implementation of the guidelines. For instance, the monitoring of the gap between work-as-done and work-as-imagined can provide measures of the amplitude and frequency of performance adjustments, and thus offering insights into why and how resilience manifests. In turn, the monitoring of unintended consequences of improvements and changes helps to track how variability propagates throughout the system, and thus it sheds light on how resilient performance is triggered as well as on how there may be a chain of resilient actions. The diversity of perspectives when making decisions contributes to resilience by reducing uncertainty in terms of when and how to adjust performance. As for the design of slack, it can provide alternative ways of achieving the goals and it makes processes loosely coupled, therefore providing time for the exploration of innovative solutions for adjusting performance. Visual management can be useful for the real-time monitoring of the availability of slack resources, and therefore can also support resilience.

In addition, the concept of resilient skills is relevant for the discussion of the role played by the LPS regarding this guideline. Resilience skills are 'skills of any type necessary to adjust performance, in order to maintain safe and efficient operations during both expected and unexpected situations' (Saurin et al., 2014, p. 30). These skills are particularly useful for the achievement of shared understanding and the learning of collaborative methods. However, the said skills are not usually present in construction organisations, and they cannot be learnt by individuals in isolation (see Chapter 5 by Rooke).

The language action perspective, which has been used to analyse interactions in the LPS implementation (Macomber and Howell, 2003), provides one such set of skills. This perspective postulates that the coordination of action happens with language acts such as requests, offers, promises, assertions, assessments, and declarations (Parrish, 2014), and provides support for the development of skills associated with making and sustaining sound commitments. These play a key role when adjusting plans to the real conditions on the construction site.

In this respect, an opportunity for improvement relates to the development of innovative teaching games, which explicitly support the development of communication and planning skills under the language-action paradigm. These skills can be interpreted as a sub-set of broader resilience skills. To this end, the games should explore the constraints (e.g. resource scarcity) and contextual factors (e.g. time pressure, unexpected delays due to inclement weather) that trigger the need for being resilient and making reliable promises and requests.

16.7 Conclusions

In this chapter, we provided a complex systems analysis of LPS, concluding that it is theoretically consistent with complexity thinking and offers practical ways of operationalising the guidelines for coping with complexity. However, problems have been shown to arise with the implementation of the LPS in practice.

In order to better understand and address these implementation problems, it is necessary to consider the LPS as a subsystem of the overall complex socio-technical system. The LPS subsystem will either thrive or diminish within its broader socio-technical system, depending on the degree of support available to it from other subsystems. Other competing subsystems, such as claims planning (Rooke, Seymour, and Fellows, 2003), or conventional top-down project management practices (Koskela and Howell, 2002), may restrict the LPS implementation. Overall, in spite of the theoretical alignment, a systematic analysis of the LPS through the lens of complexity indicated that there are gaps to be explored both in theory and practice, which are summarised in Table 16.3.

Table 16.3 Summary of the Last Planner® analysis from the perspective of the guidelines for coping with complexity.

Complexity Guideline	*Last Planner® strength*	*Opportunities for further research and recommendations for improvement*
Encourage diversity of perspectives when making decisions	Weekly and look-ahead planning meetings, participation of several stakeholders in these meetings.	To develop pedagogic tools and organisational mechanisms to support the last planners when applying the should-can-do heuristic. This can include techniques for training managers to recognise and combat the distortions to decision-making processes that arise out of power differences.
Monitor unintended consequences of improvements and small changes	Early identification of unintended consequences, due to short and frequent planning and control cycles.	To develop means for making more visible to (all) planners, in real time, the interactions between work packages and between these and other project flows.
Design slack	Many everyday planning decisions offer opportunities for the design of slack. There is also an explicit mechanism for it, namely the planning of a backlog of work packages.	To develop a broader taxonomy of slack resources and the development of methods for their analysis and explicit design, including the distinction between slack and waste.
Monitor and understand gap between work-as-imagined and work-as-done	LPS initiates and directly supervises the PPC measure.	To complement the goal-oriented nature of PPC with an approach that accounts for the quality of the product. This would help to achieve a more complete integration of value theory into the operation of the system.
Give visibility to processes and outcomes	The number of eyes (i.e. visibility) is increased dramatically the more last planners are included in the management team. Planning takes place in the *gemba*, mitigating the need for a central 'all-seeing' point of control.	To develop last planners' skills for the appreciation and management of the downstream effects of processes and outcomes.
Create an environment that supports resilience	The last planner is at the front line, and thus ideally placed to compensate for emerging variability. In parallel, the overall system of planning and control makes adjustments in the plans across the weekly, look-ahead and master planning levels.	To develop pedagogic and organisational approaches to the development of resilience skills in individuals and teams. Greater emphasis on people development, organisational support, and the elimination or re-design of interactions with competing subsystems.

An underlying assumption of these proposals refers to the need for providing better organisational support to the LPS as a system and to the last planner as an individual. Regardless of the key role of the last planner, their actual needs have been often neglected and their skills overestimated. An important step in this direction is Aslesen and Tommelein's (2016) study, which sheds light on four main types of behavioural patterns of last planners. A further, more general research avenue is the investigation of the extent to which other Lean Construction practices (e.g. Lean implementation of BIM) are aligned to CS.

References

Alves, T. and Tommelein, I.D. (2004) 'Simulation of buffering and batching practices in the interface detailing-fabrication-installation of HVAC ductwork', *Proc. 12th annual conference of the international group for lean construction*, Elsinore, Denmark. Available at: www.iglc.net.

Aslesen, S. and Tommelein, I.D. (2016) 'What "makes" the last planner? A typology of behavioral patterns of last planners', *Proc. 24th annual conference of the international group for lean construction*, Boston, MA, USA, pp. 43–52. Available at: www.iglc.net.

Baccarini, D. (1996) 'The concept of project complexity – a review', *International Journal of Project Management*, 14(4), pp. 201–204.

Ballard, G. and Howell, G. (1994a) 'Implementing lean construction: stabilizing work flow', Presented at the 2nd annual conference of the international group for lean construction, Santiago, Chile. Available in Alarcon, L. (ed.) (1997), Lean Construction, A.A. Balkema, Rotterdam, The Netherlands.

Ballard, G. and Howell, G. (1994b) 'Implementing lean construction: improving downstream performance', *Presented at the 2nd annual conference of the international group for lean construction*, Santiago, Chile. Available in Alarcon, L. (ed.) (1997), Lean Construction, A.A. Balkema, Rotterdam, The Netherlands.

Ballard, G. and Howell, G. (1998) 'Shielding production: essential step in production control', *Journal of Construction Engineering and Management*, 124 (1), pp. 11–17.

Ballard, G. and Tommelein, I. (2012) 'Lean management methods for complex projects', *Engineering Project Organization Journal*, *2*(1–2), pp. 85–96.

Bernstein, E. (2012) 'The transparency paradox: a role for privacy in organizational learning and operational control', *Administrative Science Quarterly*, 57(2), pp. 181–216.

Bertelsen, S. (2002) 'Bridging the gaps – towards a comprehensive understanding of lean construction', *Proc. 10th annual conference of the international group for lean construction*, Gramado, Brazil, pp. 23–35. Available at: www.iglc.net

Bertelsen, S. (2003a) 'Complexity – construction in a new perspective', *Proc. 11th annual conference of the international group for lean construction*, Virginia, USA. www.iglc.net

Bertelsen, S. (2003b) 'Construction as a complex system', *Proc. 11th annual conference of the international group for lean construction*, Virginia, USA. www.iglc.net

Bertelsen, S. (2017) *The unruly project: a new understanding of its nature and management.* Sven Bertelsen aps, Copenhagen.

Bertelsen, S. and Bonke, S. (2011) 'Transformation-flow-value as a strategic tool in project production' In: *Proceedings of the 19th annual conference of the international group for lean construction*, Lima, Peru. Available at: www.iglc.net

Bertelsen, S. and Emmitt, S. (2005) 'The client as a complex system', *Proc. 13th annual conference of the international group for lean construction*, Sydney, Australia. www.iglc.net/conferences/2005/papers/

Bertelsen, S. and Koskela, L. (2005) 'Approaches to managing complexity in project management', *Proc. 13th annual conference of the international group for lean construction*, Sydney, Australia, pp. 65–71. Available at: www.iglc.net

Bertelsen, S. and Sacks, R. (2007) 'Towards a new understanding of the construction industry and the nature of its production', *Proc. 15th annual conference of the international group for lean construction*, Sydney, Australia. Available at: www.iglc.net

Bosch-Rekveldt, M., Jongkind, Y., Mooi, H., Bakker, H. and Verbraeck, A. (2011) 'Grasping project complexity in large engineering projects: the TOE (Technical, Organizational and Environmental) framework', *International Journal of Project Management*, 29, pp. 728–739.

Brady, D., Tzortzopolous, P., Rooke, J., Formoso, C. and Tezel, A. (2018) 'Improving transparency in construction management: a visual planning and control model', *Engineering, Construction and Architectural Management*. Available at: http://eprints.hud.ac.uk/id/eprint/33542/

Braithwaite, J., Churruca, K., Ellis, L., Long, J., Clay-Williams, R., Damen, N., Herkes, J., Pomare, C. and Ludlow, K. (2017) *Complexity science in healthcare: aspirations, approaches, applications and accomplishments: a white paper*. Australian Institute of Health Innovation, Macquarie University, Sydney.

Camuffo, A. and Gerli, F. (2018) 'Modeling management behaviors in lean production environments', *International Journal of Operations and Production Management*, 38(2), pp. 403–423.

Cilliers, P. (1998) *Complexity and postmodernism: understanding complex systems*. Routledge, London, UK.

Clegg, C. (2000) 'Sociotechnical principles for system design', *Applied Ergonomics*, 31(5), pp. 463–477.

Dao, B., Kermanshachi, S., Shane, J. and Anderson, S. (2016) 'Project complexity assessment and management tool', *Procedia Engineering*, 145, pp. 491–496.

Dekker, S., Bergström, J., Amer-Wåhlin, I. and Cilliers P. (2013) 'Complicated, complex, and compliant: best practice in obstetrics', *Cognition, Technology and Work*, 15(2), pp. 189–195.

Dubois, A., & Gadde, L. E. (2002). The construction industry as a loosely coupled system: implications for productivity and innovation. *Construction Management & Economics*, 20(7), pp. 621–631.

El Maraghy, H., Al Geddawy, T., Samy, S. and Espinoza, V. (2014) 'A model for assessing the layout structural complexity of manufacturing systems', *Journal of Manufacturing Systems*, 33(1), pp. 51–64.

Gidado, K.I. (1996) 'Project complexity: the focal point of construction production planning', *Construction Management and Economics*, 14(3), pp. 213–225.

Al Hattab and M. Hamzeh, F. (2016) 'Analyzing design workflow: an agent-based modelling approach', *Procedia Engineering, 164*, pp. 510–517.

Hale, A., & Borys, D. (2013). Working to rule, or working safely? Part 1: A state of the art review. *Safety Science*, 55, pp. 207–221.

Hollnagel, E. (2012) *FRAM: the functional resonance analysis method. Modelling complex socio-technical systems*. Ashgate, Farnham, UK.

Hollnagel, E. (2014) *Safety-I and Safety-II: the past and the future of safety management*. Ashgate, Burlington, VT.

Jacobs, M.A. and Swink, M. (2011) 'Product portfolio architectural complexity and operational performance: incorporating the roles of learning and fixed assets', *Journal of Operations Management*, 29(7–8), pp. 677–691.

Johnston, R.B. and Brennan, M. (1996) 'Planning or organizing: the implications of theories of activity for management of operations', *Omega-International Journal of Management Science*, 24(4), pp. 367–384.

Koskela, L. (1992) *Application of the New Production Philosophy to Construction*, Technical Report Number 72, Centre for Integrated Facility Engineering, Stanford University, Stanford, CA. Available at: http://cife.stanford.edu/node/491

Koskela, L. and Howell, G. (2002) 'The underlying theory of project management is obsolete', *Proc. PMI research conference*, Project Management Institute, pp. 293–302.

Laufer, A. and Howell, G.A. (1993) 'Construction planning: revising the paradigm', *Proc. PMI Research Conference, project management institute.*

Leão, C.F., Formoso, C.T. and Isatto, E.L. (2014) 'Integrating production and quality control with the support of information technology', *Proc. 22nd annual conference of the international group for lean construction*, Oslo, Norway, pp. 847–858. Available at: www.iglc.net

Leykum, L.K., Pugh, J., Lawrence, V., Parchman, M., Noël, P.H., Cornell, J. and McDaniel, R.R. (2007) 'Organizational interventions employing principles of complexity science have improved outcomes for patients with Type II diabetes', *Implementation Science*, 2, p. 28. Available at (DOI) :10.1186/1748-5908-2-28.

Liker, J. (2004) *The toyota way: 14 Management principles from the world's greatest manufacturer.* McGraw-Hill, New York.

Macomber, H. and Howell, G.A. (2003) 'Linguistic action: contributing to the theory of lean construction', *Proc. 11th annual conference of the international group for lean construction*, Blacksburg, VA, pp 1–10. Available at: www.iglc.net

Morrey, N., Pasquire, C., Dainty, A. And Thomson, D. (2012) 'Path dependency to path creation: enabling strategic lean implementation', *Proc. 20th annual conference of the international group for lean construction*, San Diego, USA. Available at: www.iglc.net

Mota, B.P., Viana, D.D. and Isatto, E.L. (2010) 'Simulating the last planner® with systems dynamics', *Proc. 18th annual conference of the international group for lean construction*, Haifa, Israel, pp. 487–496. Available at: www.iglc.net

Parrish, K. (2014) 'Towards a language-action paradigm – experiences of a trade contractor', *Proc. 22nd annual conference of the international group for lean construction*, Oslo, Norway, pp. 1169–1179. Available at: www.iglc.net

Pasquire, C. and Ebbs, P. (2017) 'Shared understanding: the machine code of the social in a socio-technical system', *Proc. 25th annual conference of the international group for lean construction*, Heraklion, Greece, pp. 365–372. Available at: www.iglc.net

Perrow, C. (1984) *Normal Accidents: living with high-risk technologies.* Princeton University Press, Princeton, NJ.

Pikas, E., Koskela, L., Dave, B. and Liias, R. (2015a) 'Case study on design management: inefficiencies and possible remedies', *Proc. 23rd annual conference of the international group for lean construction*, Perth, Australia, pp. 547–557. Available at: www.iglc.net

Pikas, E., Koskela, L., Thalfeldt, M., Dave, B. and Kurnitski, J. (2015b) 'Complexity in designing energy efficient buildings: towards understanding decision networks in design', *Proc. 23rd annual conference of the international group for lean construction*, Perth, Australia. Available at: www.iglc.net

Righi, A.W. and Saurin, T.A. (2015) 'Complex socio-technical systems: characterization and management guidelines', *Applied Ergonomics*, *50*, pp. 19–30.

Rooke, J., Koskela, L., Bertelsen, S. and Henrich, G. (2007) 'Centred flows: a lean approach to decision making and organisation', *Proc. 15th annual conference of the international group for lean construction*, Michigan State University, East Lansing. Available at: www.iglc.net

Rooke, J., Seymour, D. and Fellows, R. (2003) 'The claims culture: a taxonomy of industry attitudes', *Construction Management and Economics*, 21(2), pp. 167–174.

Russell, M., Howell, G., Hsiang, S. And Liu, M. (2012) 'Causes of time buffer in construction project task durations', *Proc. 20th annual conference of the international group for lean construction*, San Diego, USA. Available at: www.iglc.net

Sanvido, V.E. (1984) *A framework for designing and analyzing management and control systems to improve the productivity of construction projects* (No. SUCE-TR-282).

Saurin, T.A. (2016) 'The FRAM as a tool for modelling variability propagation in lean construction',. *Proc. 24th annual conference of the international group for lean construction*, Boston, USA. Available at: www.iglc.net

Saurin, T.A. (2017) 'Removing waste while preserving slack: the lean and complexity perspectives', *Proc. 25th annual conference of the international group for lean construction*, Heraklion, Greece. Available at: www.iglc.net

Saurin, T.A., Rooke, J. and Koskela, L. (2013) 'A complex systems theory perspective of lean production', *International Journal of Production Research*, 51(19), pp. 5824–5838.

Saurin, T.A., Rooke, J., Koskela, L. and Kemmer, S. (2013) 'Guidelines for the management of complex socio-technical systems: an exploratory study of a refurbishment project', *Proc. 21st annual conference of the international group for lean construction*, Fortaleza, Brazil, pp. 13–22. Available at: www.iglc.net

Saurin, T.A., Wachs, P., Righi, A. and Henriqson, E. (2014) 'The design of scenario-based-training from the resilience engineering perspective', *Accident Analysis and Prevention*, 68, pp. 30–41.

Saurin, T.A. and Werle, N.B. (2017) 'A framework for the analysis of slack in socio-technical systems', *Reliability Engineering and Systems Safety*, 167, pp. 439–451.

Shingo, S. (1985) *A revolution in manufacturing: the SMED system*. A.P. Dillon (tr.), Productivity Press, Portland Ore, OR.

Soliman, M. and Saurin, T.A. (2017) 'Lean production and the complexity of socio-technical systems: a systematic literature review', *Journal of Manufacturing Systems*, 45, pp. 135–148.

Soliman, M., Saurin, T.A. and Anzanello, M. (2018) 'The impacts of lean production on the complexity of socio-technical systems', *International Journal of Production Economics*, 197, pp. 342–357.

Tezel, A., Koskela, L. and Tzortzopoulos, P. (2016) 'Visual management in production management: a literature synthesis', *Journal of Manufacturing Technology Management*, 27(6), pp. 766–799.

Walker, G.H., Stanton, N.A., Salmon, P.M., Jenkins, D.P. and Rafferty, L. (2010) 'Translating concepts of complexity to the field of ergonomics', *Ergonomics*, 53(10), pp. 1175–1186.

Wehbe, F., Al Hattab, M. and Hamzeh, F. (2016) 'Exploring associations between resilience and construction safety performance in safety networks', *Safety Science*, 82, pp. 338–351.

17 Framework to manage project structural complexity

Representation of the Lean Project Delivery System using a Multi-Domain-Matrix

Audrey Bascoul, Iris D. Tommelein and Stanislaus Tuholski

17.1 Timeline

The following timeline identifies papers in the Lean Construction literature that defined key milestones regarding complexity as it pertains to construction project delivery (Table 17.1). These and additional references, selected from the conference proceedings of the International Group for Lean Construction and from the *Lean Construction Journal*, are cited in this chapter.

17.2 Introduction

Koskela's (1992) groundbreaking report explored the causation between the adoption of a specific production theory (i.e., the conversion model of production) and waste induced by project complexity. He suggested that project complexity is induced by: (1) a peculiarity of construction, namely site production, which requires that the flow of teams is coordinated and (2) the lack of consideration of the Flow dimension of production in the conversion model. Koskela (1992) pointed out that the widely accepted conversion model may be sufficient for managing simple and certain projects, but is insufficient for managing complex projects: 'Only later, as the conversion model has been applied to more complex production, have the problems surfaced clearly' (ibid., p. 15).

The reason why the conversion model is insufficient for managing complex projects is that it neglects the flow dimension, which is an integral part of construction production. Koskela (1992) wrote that flows constitute a 'complexity problem', which results from a peculiarity of construction, namely that its projects are rooted in the ground and thus require site production. Site production refers to construction production taking place at the location of final assembly of the product, with the product being work-in-progress for some extended time. The complexity problem stems from the lack of coordination between 'the spatial flow of work stations (teams), which, in turn, is exacerbated by the adoption of a conversion view of production.'

The lack of management of project complexity has consequences. For example, it 'increases the costs beyond the sum of the costs of individual parts or steps' (Koskela, 1992, p. 21), and it decreases reliability as 'complex systems are inherently less reliable than simple systems. Also, the human ability to deal with complexity is bounded and easily exceeded' (ibid., p. 21).

Table 17.1 Key milestones in the Lean Construction literature about construction project complexity

Reference	*Description*
Koskela (1992)	Identifies a relationship between the adoption of the conversion model of production and project complexity, and the need to simultaneously address Transformation (conversion), Flow, and Value in project delivery.
Koskela et al. (1997)	Introduce the Design Structure Matrix (DSM) for Lean design management.
Howell (1999)	Points out that the new production philosophy is 'suited for complex, uncertain, and quick projects'. Defines: 'Complexity is defined by the number of pieces or activities that can interact.'
Ballard (2000)	Introduces the Lean Project Delivery System™ (LPDS) schematic.
Koskela (2000)	Characterises project complexity as 'hierarchical', with 'emerging properties' (similar to Simon's (1962) definition of structural complexity).
Bertelsen (2003a)	Formulates the hypothesis that 'construction (….) must be perceived as a complex system, operating on the edge of chaos'.
Ballard (2008)	Presents the current version of the Lean Project Delivery System™ schematic.
Biton and Howell (2013)	Suggest that 'the theoretical base of Lean Construction can be enhanced considering complexity within the Cynefin framework'.
Steinhaeusser et al. (2013)	Align Lean thinking with the Viable Systems Model (VSM) developed by Beer (1972) in the field of management cybernetics.

This chapter builds on Koskela's and other contributions from the Lean Construction community to identify how Lean thinking offers a comprehensive approach for managing project complexity. It presents a framework that integrates the Lean Project Delivery System™ (LPDS) with a Multi-Domain-Matrix (MDM) as a means to visualise and model structural complexity on projects.

This chapter is organised as follows. The first section explains how the Lean community built upon Koskela's work as the foundation to develop principles and methods for managing structural complexity. The second section introduces the LPDS-MDM framework. The third section tests the framework on a facility plant upgrade project. The discussion section and conclusion of this chapter offer some direction on further research into complexity management on projects. Related, the interested reader can refer to Saurin and Rooke's chapter (Chapter 16), discussing socio-technical complexity and analysing its alignment with the Last Planner® System.

17.3 Complexity in the Lean Construction literature

Koskela's (1992) report constitutes a milestone in the understanding of project complexity, because it is the first to explore the causation between the adoption of a specific production theory (i.e. conversion model of production) and waste associated with project complexity. While the report did not formally define project complexity, it suggested that project complexity stems from interdependences between project teams, in other words, flows of project teams need to be coordinated. In the Lean literature, Howell (1999) is the first author to venture an attempt at defining complexity (structural complexity): 'Complexity is defined by the number of pieces or activities that can interact.' Koskela (2000) builds on Simon's (1962) definition of complex systems, by stating in a footnote (vi p. 49) of his dissertation:

> Since the Second World War, analytical reductionism has been strongly criticised by the systems movement. It is argued that there exist, at certain levels of complexity, properties which are emergent at that level, *and* which cannot be reduced for explanation at lower levels. The idea is that architecture of complexity is hierarchical.

Koskela (1992) first touches on the notion of emergence of properties and second emphasises the fact that structural complexity may be revealed at the level at which it is looked at. This resonates with Klir (1985) and Espejo and Reyes (2011), in that complexity is in the eye of the beholder. Alves and Tsao (2007) reinforce the idea of emergence, stating that 'central to this concept is the idea that the interaction between all parts of a system will result in outcomes that differ from the sum of the outcomes of each individual part'.

However, the Lean community did not stop at the notions of hierarchy and emergence to articulate its understanding of project complexity. After all, Lean theory was deemed specifically well-suited for complex, uncertain, and quick projects (Howell, 1999; Ballard and Tommelein, 2012), which makes the quest for understanding complexity critical to its mission. In this, Bertelsen (2001, 2003a, 2003b)), Bertelsen and Koskela (2003), and Vrijhoef and Tong (2004) stressed the importance of understanding construction projects from a complexity perspective. In addition, the Lean community has pointed out that some complexity is self-inflicted and can be mitigated through, e.g., the use of simulation and appropriate buffering (Lee, Pena-Mora, and Park, 2003; Bertelsen and Koskela, 2005; Saurin, 2017), better design of supply systems (Arbulu and Ballard, 2004), early stakeholder involvement in the design phase (Vrijhoef and Tong, 2004), or product standardisation (Tommelein, 2006). Furthermore, visual management (Lindfors, 2000; Viana et al., 2014) can be a low-hanging fruit when one sets out to start managing project complexity.

Bertelsen (2003a) suggested that chaos theory be used to understand project complexity. Researchers and practitioners also reported the applicability and relevance of management cybernetics, and specifically the Viable Systems Model (VSM) to expand Lean theory (Gregory, 2007; Herrmann et al., 2008; Dominici and Palumbo, 2010; Steinhaeusser et al., 2013, 2014; Schmidt et al., 2014). Since site production is part of the 'complexity problem' (using Koskela's words), prefabrication (Larsson and Simonsson, 2012) and modularisation (Bertelsen, 2005) offer opportunities to manage complexity pertaining to site production, although they involve a different type of complexity (Höök and Stehn, 2005). Furthermore, Biton and Howell (2013) invited the Lean community to situate complexity in the Cynefin framework that had been articulated by Kurtz and Snowden (2003).

17.4 Research method

This chapter stems from a PhD research project titled Managing Project Structural Complexity by Integrating Facility Management in Planning, Designing, and Execution of High-End Facility Upgrades (Bascoul, 2017), in which the researcher followed a design science approach to understand and characterise project structural complexity as well as develop and test solutions to manage project complexity in high-end facility upgrades.

This chapter focuses on the Lean Project Delivery System™ (LPDS) – Multi-Domain-Matrix (MDM) framework, which was one of the solutions developed. The solution was developed and tested through case-study research on a cooling tower replacement project.

The case-study research took place in two phases: (1) an exploratory phase and (2) an explanatory phase. The exploratory phase lasted from January 2017 to mid-April 2017 and the explanatory phase from mid-April 2017 to October 2017.

This chapter focuses on the solution developed in the explanatory phase. The researchers collected data from January 2017 to December 2017, attending weekly construction progress meetings, and observing field work. The researchers had access to all emails exchanged for that project from April 2017 to December 2017. Furthermore, researchers had access to all project documentation on the shared network of the organisation, which allowed them to retrace the project history preceding January 2017. Informal interviews were conducted, since two of them were mobilised at the owner's site, full time. Researchers also had access to the commissioning agent's online documents repository.

The framework presented in this chapter was developed during the explanatory phase of the case-study research, during the analysis of the project breakdown associated to the cooling tower selection. The framework was presented to the team involved in the project, and refined with feedback from the project team through informal interviews.

The next section discusses the development of a new representation of LPDS to highlight the potential of Lean Construction for managing project complexity.

17.5 LPDS-MDM framework

The framework presented in this section is based on the LPDS representation and MDM. On the one hand, the LPDS representation was a starting point to develop the framework for two reasons. First, the large Research and Development (R&D) organisation giving access to the case study presented here, had an interest in Lean Construction (including the LPDS), due to its fitness for complex and uncertain projects. Second, the LPDS representation already captures some structural complexity (e.g. the edges linking the project phases show that they are interdependent). Conveying the idea of structural complexity in the LPDS was intentional, as Ballard (2000) writes: 'The LPDS will be developed as a philosophy, a set of interdependent functions.'

The schematic representation of the LPDS (Ballard, 2000, 2008) is composed of 11 modules. These are grouped in five triads (Figure 17.1): 'project definition', 'Lean design', 'Lean supply', 'Lean assembly', and 'use', and supported by the production control module, the work structure module, and the learning loops (i.e. post-occupancy evaluations).

On the other hand, the representation of structural complexity using MDM, an extension of the Design Structure Matrix (DSM), informed the development of the framework for three reasons. First, the DSM methodology is widely adopted to model structural complexity in new product development. Second, new product development and the planning of construction projects are very similar. Third, successful applications of DSM in the AEC industry are numerous.

Multi-domain matrices were developed under a Dependency and Structure Modelling methodology also called Design Structure Matrix methodology. This methodology uses matrices to manage the design of complex engineering systems (e.g. Maurer, 2007) by

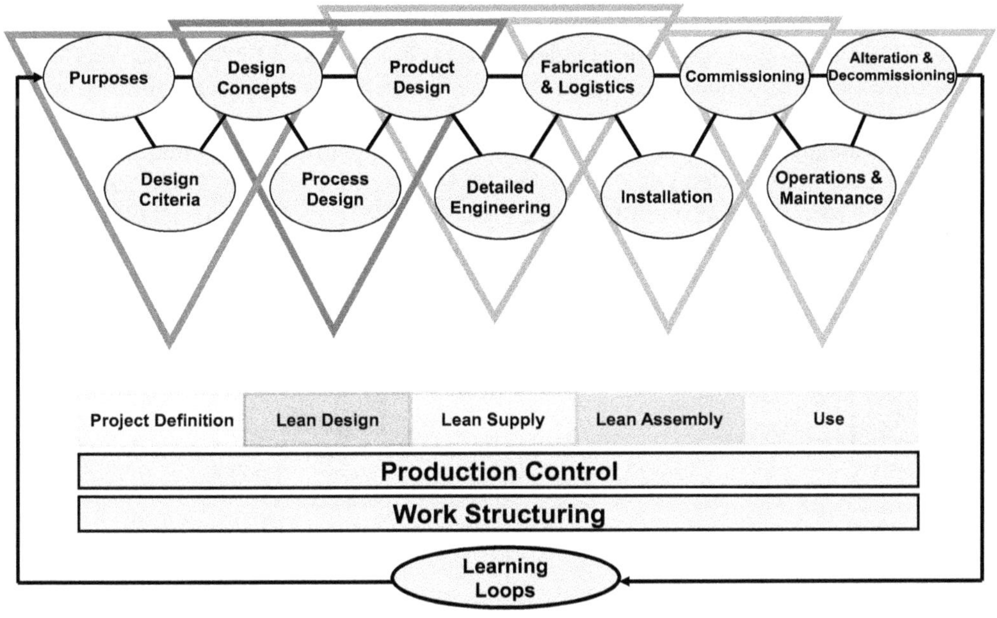

Figure 17.1 The LPDS schematic (adapted from Ballard, 2008)

modelling and analysing the dependences between the elements that compose the system. When elements of the matrix belong to the same domain, the matrix is called an intra-domain matrix or DSM; when they belong to two domains, the matrix is called an inter-domain matrix or Domain Mapping Matrices (DMM). A larger matrix called Multi-Domain-Matrix (MDM) can be created by assembling DSMs and DMMs.

The Lean community has successfully applied the DSM methodology to various domains. Furtmeier and Tommelein (2010) applied the MDM methodology to a plumbing installation process. Hickethier, Tommelein and Gehbauer (2012) used MDM to visualise information flows and reduce rework in design. Zegarra and Alarcón (2015) built upon the DSM methodology to offer a production control tool for team coordination. Pikas et al. (2015) and Khalife et al. (2018) used the DSM methodology to reduce waste in design.

Tuholski (2008) draws on the LPDS representation to propose the LPDS-MDM framework. This chapter augments Tuholski's (2008) LPDS-MDM framework by making three refinements. First, Tuholski's framework does not extend to a fine enough level of granularity to convey projects' structural complexity: 'How complex or simple a structure is, depends critically upon the way in which we describe it' (Simon, 1962). That is, the level of granularity chosen by Tuholski does not convey effectively how interdependent the triads are (i.e. 'project definition', 'Lean design', 'Lean supply', 'Lean assembly', and 'use'). A correction would be to stop at a recursion level of 'decomposition' that is fit for looking at (a certain degree of) project complexity. Second, because Tuholski's framework stops at a too high level of granularity, interpretation of the LPDS-MDM framework using the Transformation-Flow-Value (TFV) model of production presents some limitations. For example, Tuholski (2008) writes that Lean design, Lean supply, and Lean assembly pertain to the Transformation view of production. However, the three triads should be managed with the Flow and

Value views in mind as well. Third, Tuholski's framework does not show the learning feedback loop that links the last triad (product use) to the first triad (project definition).

Given these, classifying triads per production model as proposed by Tuholski (2008) does not seem suitable for understanding and managing structural complexity. Instead, seeing each triad as the combination of the three views of production seems more relevant.

Correspondingly, this chapter presents the revised LPDS-MDM framework (Figure 17.2). This framework goes one level deeper into the hierarchic structure of the LPDS schematic: it shows the 11 modules grouped in five triads. Each module corresponds to a DSM domain. Relationships between elements across modules are captured in the 55 DMMs (shaded in Figure 17.2) below the diagonal.

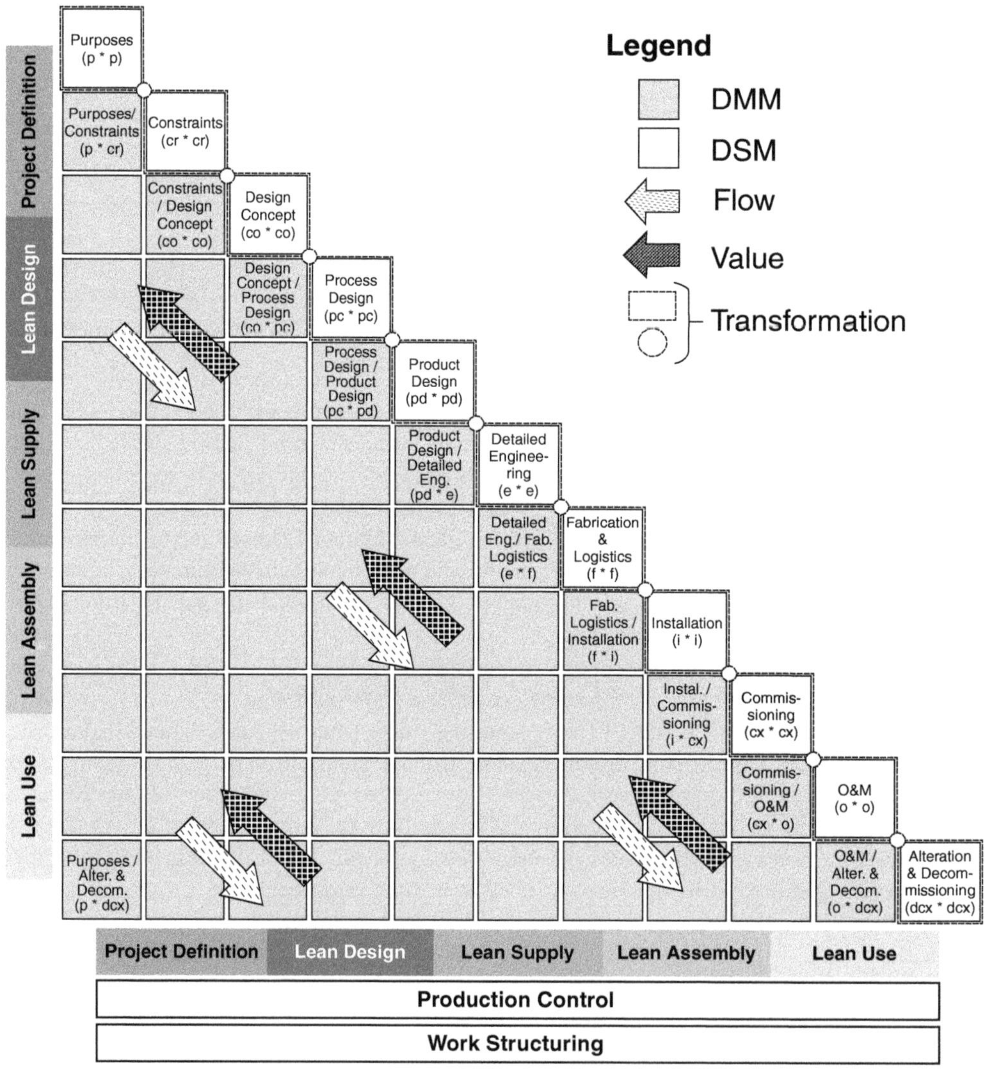

Figure 17.2 Revised LPDS-MDM framework

Figure 17.2 also captures the TFV perspectives. The Transformation view of production is captured along the diagonal of the MDM. Traditional project management focuses on the handoffs from one domain to the next. The Flow view of production encompassing work-, information-, and material-flows is represented in the lower diagonal part of the DMMs. For example, the interdependencies between project purposes (left-most white DSM in Figure 17.1) and design constraints (second-from-left DSM) and how these evolve with time could be captured in the p * cr DMM. The Value view of production is captured by the feedback loop linking DSMs and DMMs in the use triad to those in the project definition triad. In a case study, we will describe how failure to manage interdependences in the DSMs and MDMs hampered project performance with respect to the cooling tower selection.

17.6 Case study

17.6.1 Project overview

In the case study, the owner, a large Research and Development (R&D) campus decided to upgrade two wooden cooling towers, whose poor conditions posed reliability and safety concerns. The towers cool down the Low Conductivity Water (LCW) that feeds a Critical R&D facility and the Treated Water (TRW) that feeds multiple other buildings. In total, 13 buildings rely on the cooling towers to operate. Furthermore, the owner is taking advantage of this necessary upgrade to replace other aging equipment in the plant as part of this Design-Bid-Build (DBB) project. Project team members joined the project at different points in time and used traditional PM methods to plan, schedule, and coordinate work.

The planned phasing of the project aimed for the replacement of cooling tower 1 during the winter shutdown period called 'Phase 1' and operating with one cooling tower during a short transition period. The replacement of cooling tower 2 was scheduled to happen shortly after, during 'Phase 2'. However, during the commissioning of the first cooling tower, the lack of cooling capacity raised concerns among the project participants. The team spent the three following months diagnosing the problem, which contributed to delaying the start of the second phase of the project.

Before illustrating the use of the LPDS-MDM framework on the cooling tower selection, the next paragraph describes challenges faced in facility plant upgrades. It offers background on the complexity associated with planning the work.

17.6.2 Challenges in facility plant upgrades

Plants are a particular type of facility. Unlike a hospital, residential housing, or an office building, a facility plant delivers value by supporting the operations of other buildings, in this case reliable water flow to support equipment cooling.

A first challenge in facility plant upgrades lies in minimising the duration and severity of service interruption, because shutdowns affect a multitude of customers. Yet, the extent of those impacts is difficult to predict for at least two reasons. First, the site's as-builts are outdated, which is a recurrent problem in large organizations that have expanded incrementally over the years. Second, the location and condition of underground network components is unknown or uncertain. In addition to minimising service interruption, the construction team must coordinate with the serviced buildings' occupants and give them advance notice of shutdown time windows. However, forecasting shutdown time windows months in advance requires a highly reliable construction schedule. A schedule is reliable when the construction team consistently hits the target milestones spelled out in it, including the end date of the project. Close coordination of the design outputs with construction sequencing, along with work structuring and production control can increase schedule reliability. A procurement process that supports supply chain alignment also increases project performance.

A second challenge in facility plant upgrades lies in the densely occupied construction space, which constrains the sequence of activities. The challenge is exemplified in this case. The plant's footprint is approximately 30 metres (100 feet) by 11 metres (35 feet). Its two floors and roof are loaded with mechanical and electrical equipment, with unoccupied space accounting for less than 20% of the total building footprint.

For example, the second floor had some unoccupied space between the gridlines 3–4: B-C (Figure 17.3), while the roof did not have any in its existing condition (Figure 17.5). Therefore, the installation of new equipment could take place only after its installation location was freed up. In some cases, the footprint of a new piece of equipment – called 'product space' in Riley, and Sanvido (1995) – was the same as the piece of equipment it was replacing: e.g. the heat exchangers in 4–6:B-C (Figure 17.3) were being replaced with new heat exchangers in 4–6:B-C (Figure 17.4). In other cases, the product space of a new piece of equipment was different from the piece of equipment it was replacing: e.g., the new cooling tower 1 in 3–4:B-C (Figure 17.6) was actually replacing existing cooling tower 1 in 4–5:B-C (Figure 17.5).

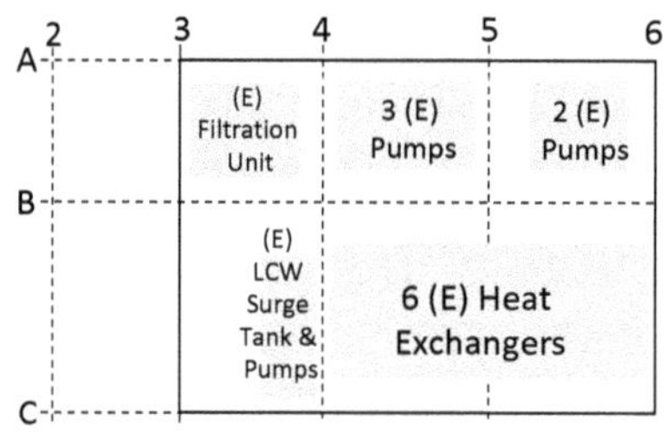

Figure 17.3 Existing (E) equipment layout on second floor (adapted from Bascoul, Tuholski, and Tommelein, 2017)

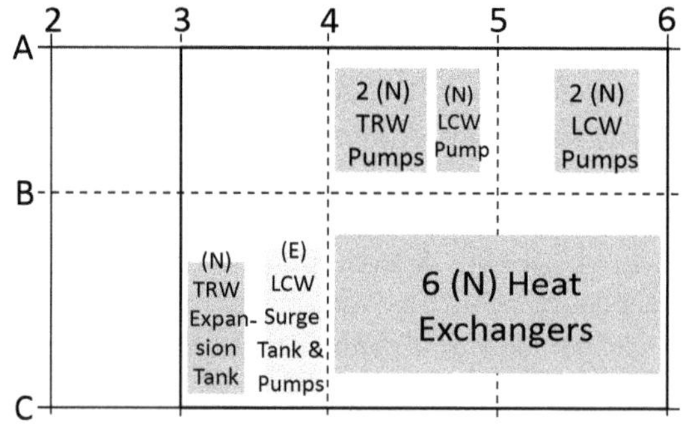

Figure 17.4 New (N) equipment layout on second floor (adapted from Bascoul, Tuholski, and Tommelein, 2017)

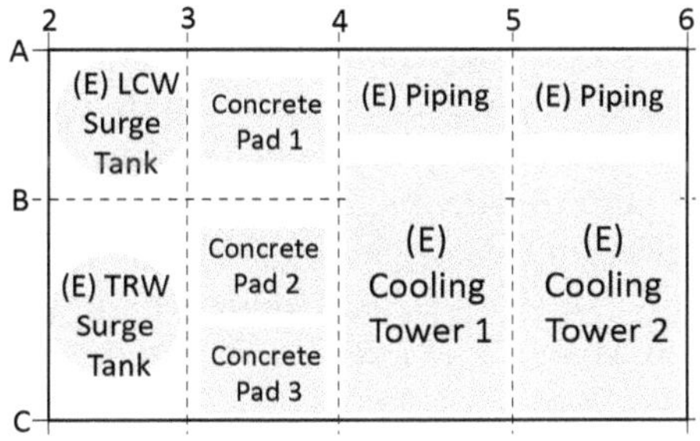

Figure 17.5 Existing (E) equipment layout on roof (adapted from Bascoul, Tuholski, and Tommelein, 2017)

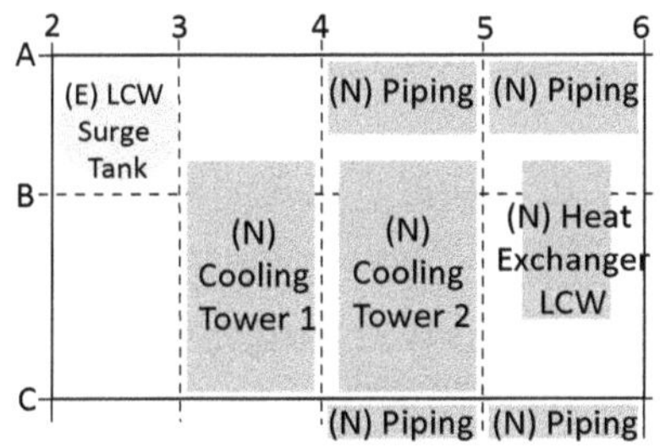

Figure 17.6 New (N) equipment layout on roof (adapted from Bascoul, Tuholski, and Tommelein, 2017)

The uncertain/unknown existing conditions, the density of the equipment, the interdependence of building, and use schedules make planning the work complex. In fact, aspects of structural complexity in high-end facility upgrades can be classified as follows: (1) customer, (2) product, (3) process, (4) organisational, and (5) market. This classification builds upon Maurer's (2007) for new product development. Planning the work falls within process complexity.

Process complexity encountered in this case study is further explained in Bascoul, Tuholski, and Tommelein (2017), in which the DSM methodology is used to capture negative design iterations and make recommendations for minimising those. In this case study, the researchers used activity-based DSMs to compare: two preconstruction schedules, one prepared by the Engineer and one by the general contractor (GC), and observed work flow. The significant differences in the three DSM sizes and iteration loops show that the Engineer and the GC failed to consider iteration in design/construction activities. As a result, they created unrealistic schedules. This research conveyed the benefits of using DSM to manage design efforts including iteration, thus improving the reliability of the schedule and efficiency of construction.

Product complexity is the focus of the next paragraphs. It is illustrated with the cooling tower selection.

17.6.3 Project timeline

The Engineer initially served as a consultant during March 2016. During that time, the owner, the Engineer, and the commissioning agent jointly developed a document called 'Current Facility Requirements' (CFR) (**purposes**) (Figure 17.7).

The CFR describes the performance requirements of the facility 'where practical and known'. The project team selected the cooling tower March 30, 2016 (**detailed engineering**). They finalised the CFR mid-April 2016. The owner contracted with the Engineer for the design of the project mid-March. The Engineer started the design right after the cooling tower selection.

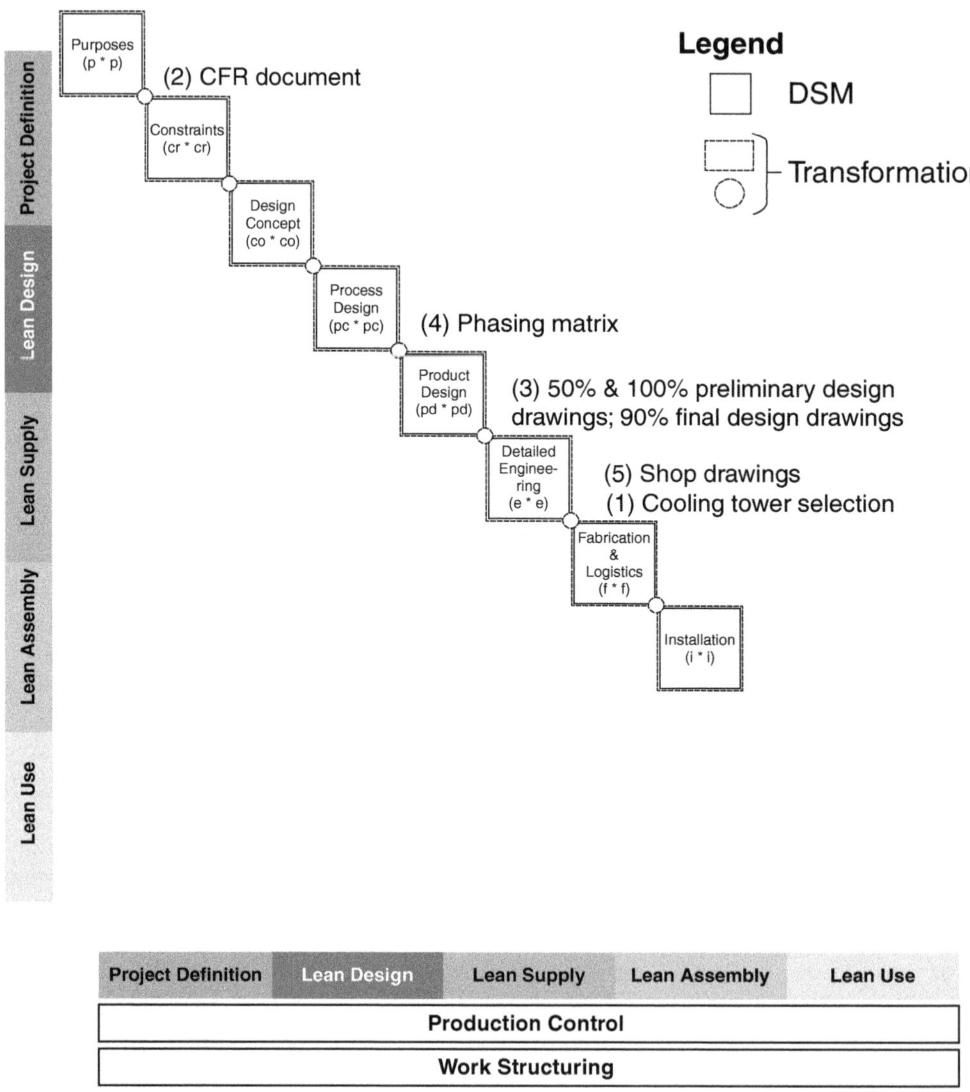

Figure 17.7 Actual project timeline

From the documents reviewed and interviews with project team members, the project team did not use a formal decision-making system for the cooling tower selection. Nonetheless, the researchers identified six steps that led to the final selection: (1) determine factors, (2) select alternatives, (3) narrow down the set of alternatives, (4) list advantages and disadvantages, (5) express preference, and (6) meet to select the final cooling tower model.

The Engineer submitted the 50% preliminary design April 29, 2016, and the 100% preliminary design two months later June 30, 2016 (Figure 17.7). The Engineer submitted the 90% final design drawings August 19, 2016 (**product design**). The owner selected the GC late August 2016 but the Construction Manager (CM)/GC contract was issued early December 2016. From August to December, the GC provided CM services upon the owner's request. When CM services were needed, the owner made the request through a letter describing the CM work scope and giving a lump sum. Eventually, the Engineer submitted the 100% final design drawings (issued for bid) September 23, 2016.

The Engineer issued the phasing matrix to demonstrate the feasibility of the proposed design August 26, 2016 (**process design**) (Figure 17.7). The goal of the phasing was to complete as much work as possible during the critical R&D facility's shutdown windows (**constraints**). As of September 2016, the facility's maintenance schedule showed a no LCW flow period from January 6, 2017 to January 19, 2017, and a no TRW and no LCW temperature control from January 6, 2017 to February 12, 2017. Following these no flow periods, the facility's maintenance schedule showed two-day maintenance windows every other week. The GC started to release shop drawings in October 2016 (**fabrication and logistics**).

The owner sent a letter to the GC to allow it to proceed with work installation December 13, 2016 (**installation**) (Figure 17.7). The commissioning of the first cooling tower took place March 12, 2017 (**commissioning**).

17.6.4 Problem encountered

During commissioning, the commissioning agent and the owner's FM (including the building energy manager) reported that the new cooling tower lacked cooling capacity. This raised concerns among the project participants. The team spent the three following months diagnosing the problem, which postponed the start of the second phase of the project. The team even turned off B2 chillers and B43 compressors to reduce the TRW load and prioritise the LCW load, namely the critical R&D facility, during the diagnosis time due to the apparent lack of cooling capacity. This severely affected the operations in B2 as its laboratory operations require the chillers to be on (experiments must be run at a specific temperature). When the temporary solution (at the expense of B2 and B43) was insufficient to reach the required LCW supply temperature and stability, the team injected city water directly into the system.

The project team brought up multiple reasons to account for the lack of cooling capacity. Among those reasons, they questioned the cooling tower capability. The next paragraphs use the 5-Whys to cast light on why the team questioned the cooling tower selection. 5-Whys is a problem-solving technique that consists of asking why at least five times successively in order to get to the root cause of the problem (Ohno, 1988).

W1. Why was the cooling tower selection questioned during phase 1?

The cooling tower selection was questioned following the start-up of the first cooling tower, because the team did not expect the experienced behaviour of the cooling tower, as attested by this email:

> Today May 1, 2017 the critical R&D facility had to go offline during user operations due to high LCW temperatures. This is a pretty significant failure of this system and it's supposed to get worse tomorrow. Here's a picture of the current conditions. Note that the LCW LOAD calculation is inaccurate because it uses a static supply temperature of 73°F (22.8°C) for the calculation and we are far above that.

The 'significant failure of this system' stems from the fact that the project team expected more cooling capacity even in the unusual warm conditions on that day. The LCW temperature hit 75.3°F (24°C) with only 80 short tons (72.6 metric tons) on the TRW system. This first answer brings the following question:

W2. Why was the behaviour of the cooling tower unexpected?

The behaviour of the cooling tower was unexpected, because the project team believed that the installed cooling tower would be able to meet the demand during the transition phase. The transition phase refers to the period of time during which only the new cooling tower installed operates.

W3. Why did the team think that the installed cooling tower would be able to meet the demand during the transition phase?

From the reviewed project documentation, it seems that the reliability of a single cooling tower during the 'transition phase' was never discussed during the cooling tower selection, which lacked a formal decision-making process. Reliability never appeared as a factor for selection. At the time, the design of the project had barely started, and the phasing of the project had not even been thought through. In fact, project documentation and memos show that the project team had not eliminated the use of a temporary cooling tower during the transition phase either. The temporary cooling tower would have allowed some flexibility in the execution of the plan. Selecting the cooling tower model before exploring alternatives such as the use of a temporary cooling tower can be summarised by determining the means and methods (the 'how') before defining the end (the 'what').

Furthermore, the lack of shared language led to confusion in terms of what 'capacity' (in US tons) meant, what risks were associated with each 'option', and the consequences of choosing a 'wet-bulb temperature' of 65°F (18.3°C) against 67°F (19.4°C) for example. In fact, the project team did not seem to fully understand the interdependences between the design parameters and trade-off between size and cooling capacity. This third level of root causes identification brings the two following questions: (1) why was the team unaware of (i) the interdependences between the design parameters and (ii) the trade-off between size and cooling capacity? and (2) why was the cooling tower selected before having a better understanding of what the project phasing would be? The next two paragraphs answer question (1) in W4.1 and W4.2 and the fifth level of 5-why analysis stemming from question (1) in W5.

W4.1. Why was the team unaware of (i) the interdependences between the design parameters and (ii) the trade-off between size and cooling capacity?

The team was unaware of the interdependences between the design parameters, because either understanding them was not part of their scope, or they did not have time to look at them, or they did not have the mechanical expertise themselves to understand the system. Furthermore, the interdependences were not communicated or made visible to them. The same applies to the trade-off between size and cooling capacity. This fourth level of root cause identification brings the following question:

W5. Why were interdependences between design parameters not made visible to project team members?

Making interdependences explicit and modelling them is not common practice in engineering for two reasons. First, the AEC industry simply lacks awareness about the importance of understanding and managing structural complexity in engineering design. Second, the AEC industry lacks knowledge about the use of tools such as DSM to model interdependences.

W4.2 Why was the cooling tower selected before phasing the project?

First, the organisational complexity induced by the project funding mechanism encouraged the project team to use the money allocated for that purpose before the end of the fiscal year. This way, the team avoided running the risk of not having the same amount of money allocated in the following year in case of delayed cooling tower selection/purchase. Second, the project team bought the cooling tower ahead of time to mitigate the impact of unreliable procurement time of the cooling tower through a time buffer. As a result, the project team did not fully assess whether a single cooling tower would be able to handle all the demand during the transition phase.

The LPDS-MDM framework (Figure 17.8) shows how traditional Project Management (PM) methods tend to have project teams focus on contract deliverables (e.g., programming documents, design drawings, shop drawings, installation milestones) rather than on collaboratively managing the project's structural complexity.

In retrospect, how could the team have managed project structural complexity?

17.6.5 Recommended approach

Figures 17.9 to 17.11 show how the project team could have used Lean principles and methods to manage project structural complexity. The following recommendations are based on the LPDS: (1) integrate FM early, (2) use Choosing-By-Advantages (CBA) to make sound decisions, and (3) make decisions at the last responsible moment.

17.7 Discussion

The LPDS-MDM framework as presented depicts interdependences between domains (e.g., project purpose and process design, product design and installation) that were overlooked in this project delivered using DBB. For example, a project purpose was to

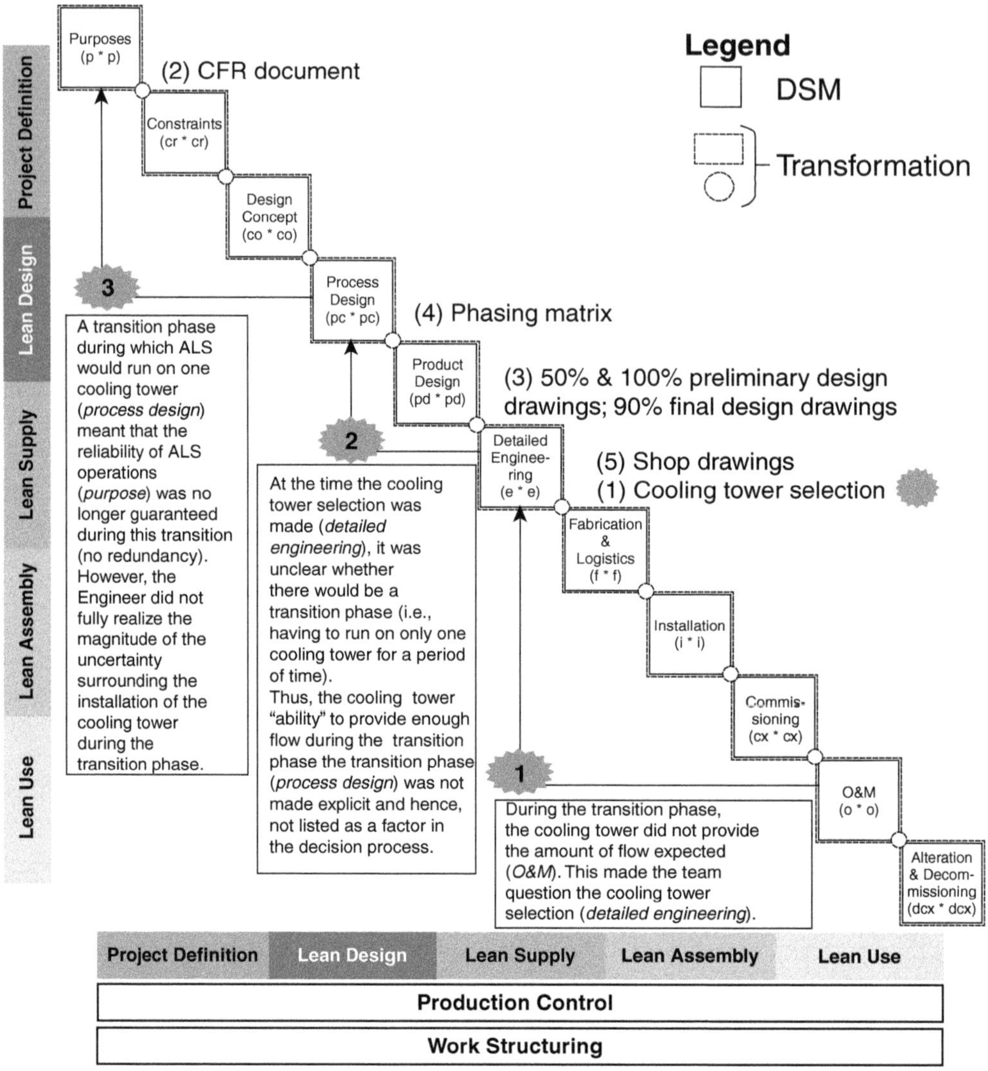

Figure 17.8 Analysis of problem encountered using LPDS-MDM framework

minimise disruptions to the facility's operations, but the project phasing did not take into account design iterations and uncertainty around existing conditions. This interdependence between project purpose and process design was overlooked. The product design showed a steel structure (supporting the cooling towers) anchored to the roof, but the location of the rebar in the roof was unknown at the time, which made the installation of the steel structure unsafe with the initial design. This interdependence between product design and installation was overlooked.

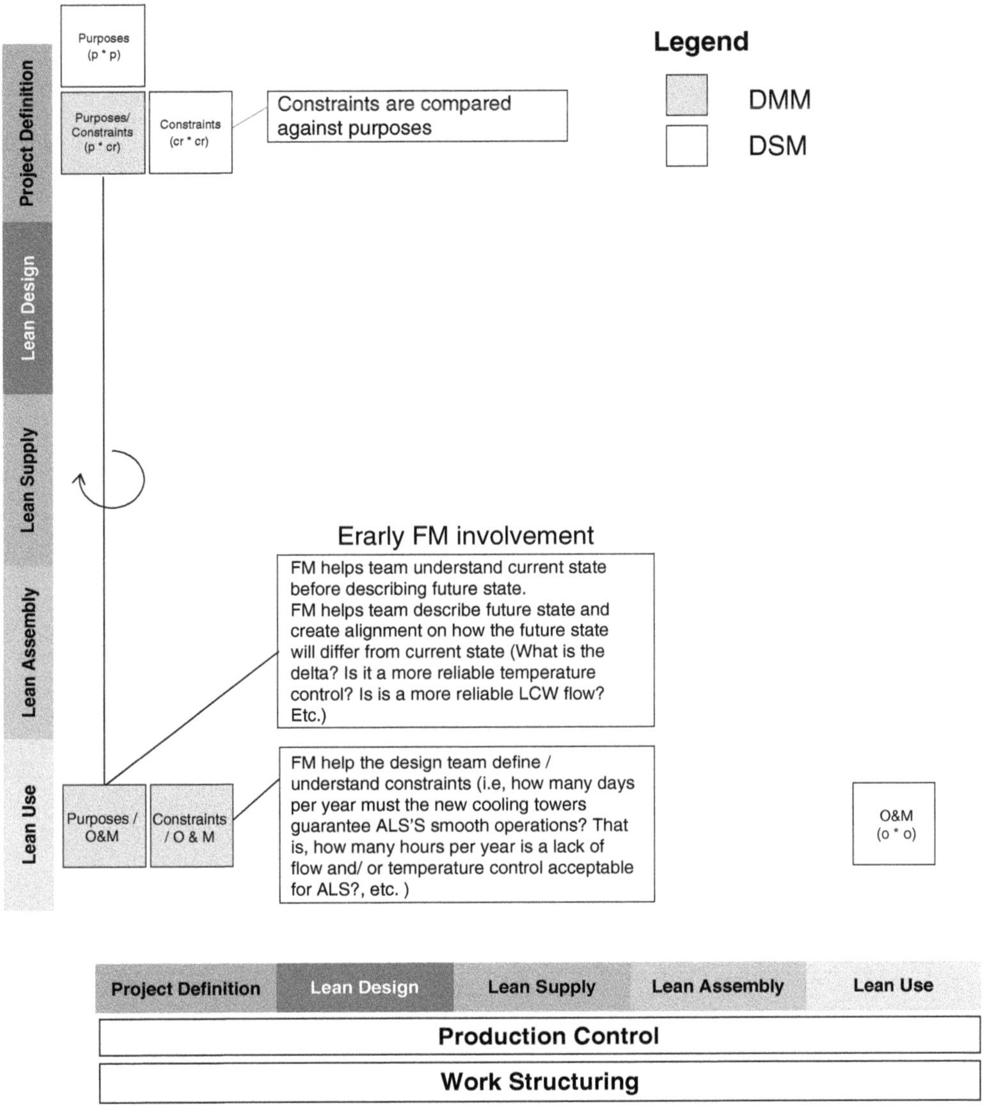

Figure 17.9 Proposed Approach to Manage Structural Complexity in Cooling Tower Selection (1/3)

Because the LPDS-MDM framework shows all areas where potential dependencies may occur (through 55 DMMs), it helps the team consider dependencies that the team would not anticipate without the use of such framework. The use of the framework seems promising to help teams look for and learn to see dependencies. Learning to see dependencies is a first step to managing them. After capturing dependencies in the

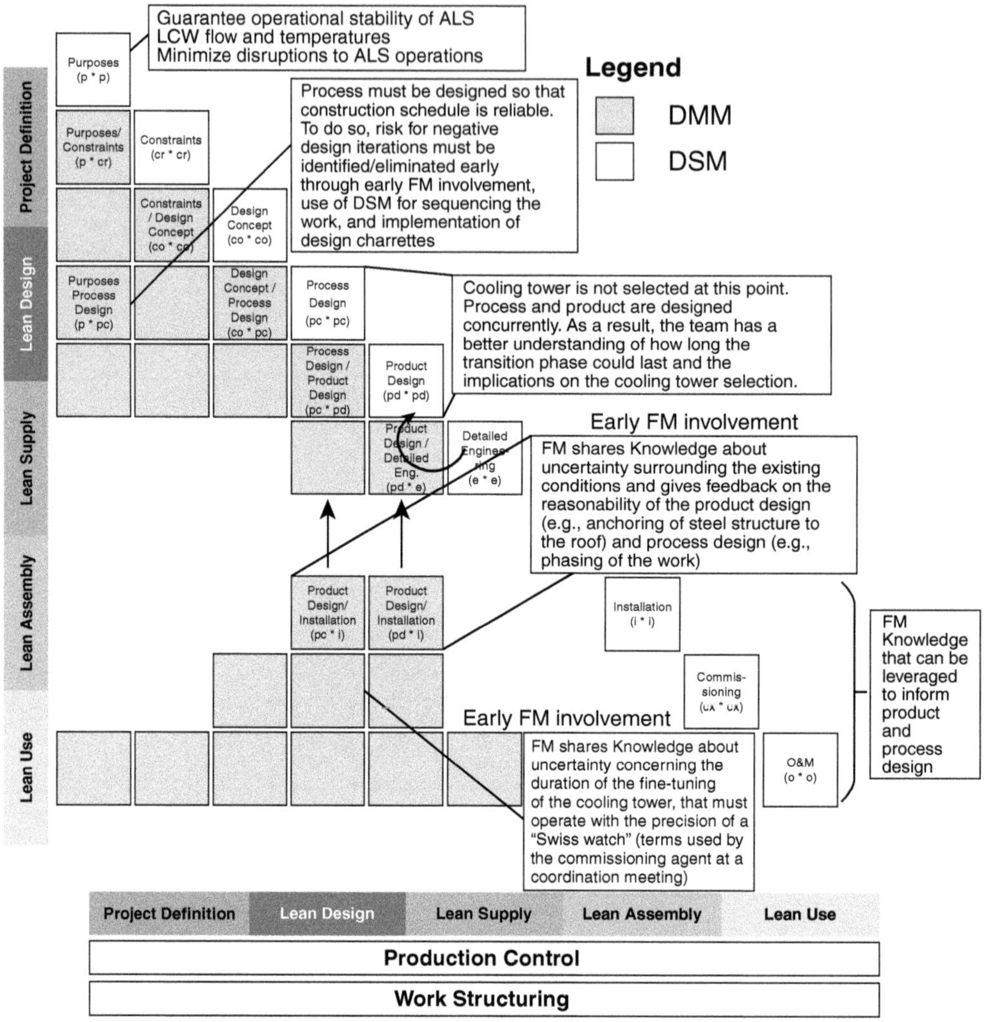

Figure 17.10 Proposed Approach to Manage Structural Complexity in Cooling Tower Selection (2/3)

LPDS-MDM framework, teams can refer to the LPDS to identify Lean principles and methods to manage them. By revealing those interdependencies between elements within the same and across different domains, project managers will be in a better position to manage the complexities they face.

Since this framework was applied after-the-fact on the project, future research should attempt to use the framework at the beginning of the project, and before project breakdowns occur.

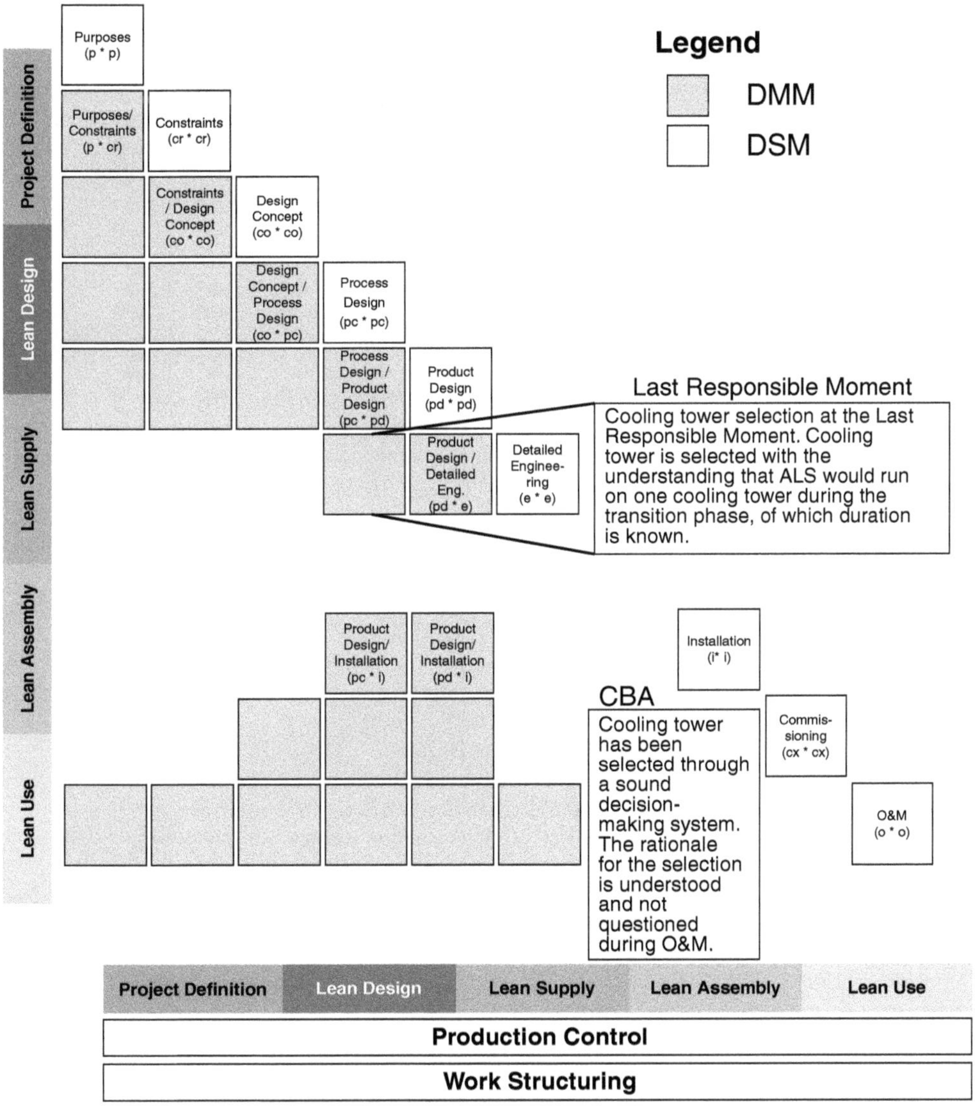

Figure 17.11 Proposed Approach to Manage Structural Complexity in Cooling Tower Selection (3/3)

17.8 Conclusion

Different levels of complexity exist in projects, with some of these differences lying in the eyes of the observers. Effective management of complexity involves first acknowledging this characteristic and second, visualising and recognising such complexity.

Regarding recognition, the differentiation between self-inflicted versus innate complexity of the project production system is critical to guide efforts for managing complexity.

Making processes and products simpler is a means to reduce self-inflicted complexity. For example, in complex projects where each part is unique and different from any other part, standardisation may reduce some excessive complexity.

Further research could investigate how one may recognise different degrees of complexity and reveal whether complexity is innate to the project delivery system or self-inflicted. It might shed light on which aspects of complexity are value-adding vs. waste. Clearly, the Lean Construction community has a long path ahead to understand and harness complexity.

References

Alves, T. and Tsao, C.C.Y. (2007) 'Lean construction – 2000 to 2006', *Lean Construction Journal*, 3 (1), pp. 46–70.

Arbulu, R. and Ballard, G. (2004) 'Lean supply systems in construction', *Proceedings of 12th Annual Conference of the International Group for Lean Construction*, Helsingør, Denmark. Available at: www.iglc.net.

Ballard, G. (2000) 'Lean project delivery system', in *White Paper*. Vol. *8*. Lean Construction Institute, Arlington, VA, pp. 1–6.

Ballard, G. (2008) 'The lean project delivery system: an update', *Lean Construction Journal*, pp. 1–19.

Ballard, G. and Tommelein, I. (2012) 'Lean management methods for complex projects', *Engineering Project Organization Journal*, 2 (1–2), pp. 85–96.

Bascoul, A., Tuholski, S., and Tommelein, I.D. (2017) 'Use of DSM to capture unplanned design iterations on a facility plant upgrade project', *19th Int'l DSM Conf.*, Espoo, Finland, 11–13.

Bascoul, A.M. (2017) *Managing Project Structural Complexity by Integrating Facility Management in Planning, Designing, and Execution of High-End Facility Upgrades*. PhD Thesis, Dept. Civ. & Env. Engrg., Univ. of California, Berkeley, 265.

Beer, S. (1972). *Brain of the firm*. New York: Herder & Herder.

Bertelsen, S. (2001) 'Lean construction as an integrated production', *Proceedings of 9th Annual Conference of the International Group for Lean Construction*, Singapore. Available at: www.iglc.net.

Bertelsen, S. (2003a) 'Complexity - a new way of understanding construction', *Proceedings of 11th Annual Conference of the International Group for Lean Construction*, Virginia, USA. Available at: www.iglc.net.

Bertelsen, S. (2003b) 'Construction as a complex system', *Proceedings of 11th Annual Conference of the International Group for Lean Construction*, Virginia, USA. Available at: www.iglc.net.

Bertelsen, S. (2005) 'Modularization — a third approach to making construction lean?', *Proceedings of 13th Annual Conference of the International Group for Lean Construction*, Sydney, Australia. Available at: www.iglc.net.

Bertelsen, S. and Koskela, L. (2003) 'Avoiding and managing chaos in projects', *Proceedings of 11th Annual Conference of the International Group for Lean Construction*, Virginia, USA. Available at: www.iglc.net.

Bertelsen, S. and Koskela, L. (2005) 'Approaches to managing complexity in project production', *Proceedings of 13th Annual Conference of the International Group for Lean Construction*, Sydney, Australia. Available at: www.iglc.net.

Biton, N. and Howell, G. (2013) 'The journey of lean construction theory review and reinterpretation', *Proceedings of 21st Annual Conference of the International Group for Lean Construction*, Fortaleza, Brazil. Available at: www.iglc.net.

Dominici, G. and Palumbo, F. (2010) 'A viable system view of the Japanese Lean production system', *World Complexity Science Academy 1st Conf. Systemic Sci. & the Key Global Challenges of Our Times*, WCSA, Bologna, Italy, 1–35.

Espejo, R. and Reyes, A. (2011) *Organizational Systems: Managing Complexity with the Viable System Model*. Berlin, Germany, Vol. 264. Springer & Science Media.

Furtmeier, F.A., and Tommelein, I.D. (2010) 'Explorative application of the Multi-Domain Matrix methodology in lean design', *Proceedings 18th Annual Conference of the International Group for Lean Construction*, Haifa, Israel. Available at: www.iglc.net.

Gregory, A.J. (2007) 'Target setting, Lean systems and viable systems: A systems perspective on control and performance measurement', *Journal of the Operational Research Society*, 58 (11), pp. 1503–1517.

Herrmann, C., Bergmann, L., Halubek, P., and Thiede, S. (2008) 'Lean production system design from the perspective of the viable systems model', *41st CIRP Conf. Manuf. Systems*, Tokyo, Japan, 309–314.

Hickethier, G., Tommelein, I.D., And Gehbauer, F. (2012) 'Reducing rework in design by comparing structural complexity using a Multi-Domain Matrix', *Proceedings of 20th Annual Conference of the International Group for Lean Construction*, San Diego, CA, USA. Available at: www.iglc.net.

Höök, M., and Stehn, L. (2005) 'Connecting Lean Construction to prefabrication complexity in Swedish volume element housing', *Proceedings of 13th Annual Conference of the International Group for Lean Construction*, Sydney, Australia. Available at: www.iglc.net.

Howell, G. (1999) 'What is Lean Construction – 1999', *Proceedings of 7th Annual Conference of the International Group for Lean Construction*, Berkeley, CA, USA. Available at: www.iglc.net.

Khalife, S., Mneymneh, B.E., Tawbe, A., Chatila, M.H., and Hamzeh, F. (2018) 'Employing simulation to study the role of Design Structure Matrix in reducing waste in design', *Proceedings of 26th Annual Conference of the International Group for Lean Construction*, Chennai, India. Available at: www.iglc.net.

Klir, G.J. (1985) 'Complexity: Some general observations', *Systems Research*, 2 (2), pp. 131–140.

Koskela, L. (1992) *Application of the New Production Philosophy to Construction*. Center for Integrated Facility Engineering, (CIFE), Department of Civil Engineering, Stanford University. Technical Report number: 72.

Koskela, L. (2000) *An exploration towards a production theory and its application to construction*. PhD Thesis, VTT Technical Research Centre of Finland, VTT Publication 408. Available at: www.vtt.fi/inf/pdf/publications/2000/P408.pdf

Koskela, L., Ballard, G., and Tanhuanpää, V. (1997). "Towards Lean Design Management." *Proceedings of 5th Annual Conference of the International Group for Lean Construction*, Gold Coast, Australia, July 16–17, 10.

Kurtz, C.F. and Snowden, D.J. (2003) 'The new dynamics of strategy: Sense-making in a complex and complicated world', *IBM System Journal*, 42 (3), pp. 462–483.

Larsson, J., And Simonsson, P. (2012) 'Decreasing complexity of the on-site construction process using prefabrication: A case study', *Proceedings of 20th Annual Conference of the International Group for Lean Construction*, San Diego, CA, USA. Available at: www.iglc.net.

Lee, S.H., Pena-Mora, F., and Park, M. (2003) 'Reliability and stability buffering approach in concurrent design and construction projects', *Proceedings 11th Annual Conference of the International Group for Lean Construction*, Virginia, USA. Available at: www.iglc.net.

Lindfors, C.T. (2000) 'Value chain management in construction: Controlling the housebuilding process', *8th Annual Conference of the Interational Group for Lean Construction*, Brighton, UK. Available at: www.iglc.net.

Maurer, M. (2007) Structural awareness in complex product design. PhD Thesis, Technische Universität München (TUM), Munich, Germany, 259.

Ohno, T. (1988) *Toyota Production System: Beyond Large-Scale Production*. Portland, OR, USA, Productivity Press.

Pikas, E., Koskela, L., Thalfeldt, M., Dave, B., and Kurnitski, J. (2015) 'Complexity in designing energy efficient buildings: Towards understanding decision networks in design', *Proceedings of 23rd Annual Conference of the International Group for Lean Construction*, Perth, Australia, 29–31 Jul, 213–222.

Riley, D.R., Sanvido, V.E. (1995) "Patterns of construction-space use in multistory buildings', ASCE', *Journal of Construction Engineering and Management*, 121 (4), pp. 464–473.

Saurin, T.A. (2017) 'Removing waste while preserving slack: The Lean and complexity perspectives', *25th Annual Conference of the International Group for Lean Construction*, Heraklion, Greece. Available at: www.iglc.net.

Schmidt, M.T., Elezi, F., Tommelein, I.D., Berghede, K. & Lindemann, U. (2014) 'Supporting Organizational Design Towards Lean With the Viable System Model', *Proceedings of 22nd Annual Conference of the International Group for Lean Construction*, Oslo, Norway. Available at: www.iglc.net.

Simon, H.A. (1962) 'The architecture of complexity', *Proceedings of the American Philosophical Society*, 106 (6), pp. 467–482.

Steinhaeusser, T., Elezi, F., Tommelein, I.D., and Lindemann, U. (2013) 'Management cybernetics as a theoretical basis for lean thinking', *Proceedings of 21st Annual Conference of the International Group for Lean Construction*, Fortaleza, Brazil. Available at: www.iglc.net.

Steinhaeusser, T., Elezi, F., Tommelein, I.D., and Lindemann, U. (2014) 'Management cybernetics as a theoretical basis for lean construction thinking', *Lean Construction Journal*, pp. 01–14.

Tommelein, I.D. (2006) 'Process Benefits from Use of Standard Products - Simulation Experiments using the Pipe Spool Model', *Proceedings of 14th Annual Conference of the International Group for Lean Construction*, Santiago, Chile. Available at: www.iglc.net.

Tuholski, S.J. (2008) *Transformation, Flow, and Value Constellations in AEC Projects*, PhD Thesis, Civil and Env. Eng. Dept., Univ. of Calif., Berkeley, 276.

Viana, D., Formoso, C.T., Wesz, J., and Tzortzopoulos, P. (2014) 'The role of visual management in collaborative integrated planning and control for engineer-to-order building systems', *Proceedings of 22nd Annual Conference of the International Group for Lean Construction*, Oslo, Norway. Available at: www.iglc.net.

Vrijhoef, R. and Tong, M.K. (2004) 'Understanding construction as a complex and dynamic system: An adaptive network approach', *Proceedings of 12th Annual Conference of the International Group for Lean Construction*, Helsingør, Denmark. Available at: www.iglc.net.

Zegarra, O. and Alarcón, L.F. (2015) 'A production control tool for coordination of teams, meetings and managerial processes', *Proceedings of 23rd Annual Conference of the International Group for Lean Construction*, Perth, Australia. Available at: www.iglc.net.

18 Uncertainty management

A development area for Lean Construction[1]

Ole Jonny Klakegg, Olav Torp, Bo Terje Kalsaas, Trond Bølviken and Gøril Hannås

18.1 Introduction

In this chapter, we present the knowledge field uncertainty management and thereafter relate it to Lean Construction and the Last Planner System (LPS) of production planning and control (Ballard, 2000; Ballard and Tommelein, 2016). The starting point in our approach is that uncertainty involves both risk and opportunities, where risk is the threat of negative outcomes (potential downsides) and opportunities are the potential for creating added value for stakeholders (potential upsides) (Hillson, 2004).

Uncertainty management is an important area of expertise within Project Management (PMI, 2013), yet uncertainty management has been scarcely discussed in the area of Lean Construction. However, the mechanisms in the LPS are strongly oriented towards reducing uncertainty and managing variation (Howell, 2012). In uncertainty management, it is important to be able to anticipate how a project can be affected by uncertainty and to choose strategies suitable for handling the uncertainties that a project meets during project and production planning and execution.

Good practice includes balancing risks and opportunities, and making sure future potential events are systematically identified and acted upon. An uncertainty management plan should be prepared in the project (Hillson, 2004). Project management needs to identify and analyse uncertainties, and plan measures for reducing risk and exploiting opportunities. The aims of Lean Construction and uncertainty management are compatible, and therefore purposeful practices should complement each other and make up a stronger whole. Here, uncertainty management in a project is seen in the context of Lean Construction in order to shed light on key dimensions of uncertainty management from a Lean perspective.

Uncertainty management encompasses managing both decision-making and execution, risk and opportunities, in a proactive, balanced manner. The first part of the chapter deals with concepts, while the second deals with how uncertainty is present in various phases of a project. This is followed by looking at how uncertainty is managed, including risk and opportunity matrices and uncertainty registers. The chapter then discusses how uncertainty management and LPS can be integrated. The chapter ends with some observations on learning and future development.

18.2 Conceptual framework for uncertainty management in construction projects

The focus of uncertainty management is a person's inability to predict or anticipate future events, including where numerous outcomes are possible, the scope of those outcomes (i.e. how great the consequences of a possible event will be), as well as the extent to which they will have positive or negative effects in relation to a project's goal (Hillson, 2004). Two words dominate discussions in this field: risk and uncertainty.

In the literature, the terms risk and uncertainty are used in different implicit and explicit meanings. Hillson (2004) uses the terms in a different way than the one presented in the introduction, describing risk as aleatoric[2] and uncertainty as epistemic.[3] He defines the terms risk and uncertainty as follows:

- Aleatoric – risk is a potential event where a set of possible outcomes are known, and the probability of each outcome can be measured or estimated, although the precise outcome of the event in an individual case is not known in advance;
- Epistemic – uncertainty is related to a lack of information about possible outcomes, including both their nature and associated probabilities. An uncertainty is thus an unknown event with an unknown set of possible outcomes.

The concepts of risk and uncertainty are used in different ways in different disciplines. Both relate to some extent to future events and the outcomes or consequences the events may have. Both of these concepts are used in project environments, despite different perceptions about what is most appropriate (Johansen et al., 2016). In simple terms, the relationship between the concepts can be linked to probability. Probability is a measure of uncertainty, while uncertainties exist without (specifying) probabilities (Aven and Renn, 2009).

Roughly, the literature on uncertainty in projects can be divided into two integrated areas, each with its own natural standpoint on or understanding of the phenomenon (Hillson, 2004):

- Planning and decision-making: uncertainty comes from lack of information
- Management and control: uncertainty is an aspect of the world we have to live with.

A much-used definition of uncertainty was developed by the organizational theorist Jay R. Galbraith (1977, p. 5):

> Uncertainty is the difference between the amount of information required to perform a task and the amount of information already possessed by the organisation.

This understanding of uncertainty as lack of necessary knowledge is today rather common (Austeng et al., 2005a; Vatn, 2015). Uncertainty can also be defined as a lack of information about either parameters that are characteristic of the object being modelled/analysed (Kolltveit and Reve, 2002), or about the effects of the parameters of the object being analysed/modelled (Telenor, 2005). This understanding is fundamental in handling decision-making under uncertainty, and characteristic for early phases of project development, although also present at later project stages.

The picture is complicated by the question of whether uncertainty and risk can be understood as being subjective or objective phenomena, i.e. whether uncertainty exists independent of how it is understood, or that uncertainty is affected by interpretation, contingent on an individual's knowledge and experience, as well as attitudes to risk (Aven and Renn, 2009). These different understandings affect how uncertainty is managed in practice, with respect to both analytical approaches and management methods (Johansen et al., 2016).

So, does all uncertainty vanish if we have all the information, knowledge and know-how relating to a subject/project? Such a situation is only theoretically possible. In reality, something unexpected/unknown could still occur. Despite the fact that experience provides a lot of information and knowledge about a project, which reduce the uncertainty internal to the project, we must also account for uncertainties in the project's sphere that could affect the result and that are outside the project's control (Christensen and Kreiner, 1991b; Karlsen, 1998; Jordanger, 2005). This could be a political decision that goes a completely different way to the one anticipated, a key stakeholder going bankrupt during the project, or an unexpected event, such as an earthquake, flood or fire, that affects the situation. In addition to information, knowledge and know-how, uncertainty thus consists of something unknown/unpredictable, as well as the fact that the world and society surrounding the project are changing/developing. An expanded definition of uncertainty is thus (Andersen, 2005, p. 130):

> Uncertainty – lack of information, knowledge and control over a relevant circumstance.

The concept of uncertainty is often used in relation to decision-making and planning. In order to arrive at balanced decisions, we want to identify, describe and preferably quantify all relevant uncertainties, for example regarding cost estimates or schedules. When moving into the area where management and control are more prominent, other definitions dominate and the concept of risk is often used. The Project Management Institute defines the concept of risk as follows (PMI, 2013, p. 310):

> an uncertain event or condition that, if it occurs, has a positive or negative effect on one or more project objectives such as scope, schedule, cost, or quality.

ISO 31000 (2009) and ISO Guide 73 (2009), in principle, define risk in a similar way. It is not difficult to see that this definition differs from, but also has similarities to, the definition based on uncertainty as a lack of information, expertise and control. One important aspect of this definition of risk is that it can be both positive and negative. The PMI and ISO have a different way of using the word risk from the authors of this chapter. While PMI and ISO say that risk could be both positive and negative, we say that risk is the negative impacts of uncertainty, and opportunity is the positive impacts.

Risk can be said to be a combination of the probability of an event occurring and the consequences of that event (ISO 16085, 2006). This link was first formulated by the economists Keynes (1921) and Knight (1921), separately. The Norwegian Planning and Building Act, Norwegian literature and Norwegian standards generally define risk as something negative, which can be formulated as follows (NS 5814, 2008, Requirements for risk assessment):

> Risk is associated with the probability and consequences of undesirable events occurring that represent a risk to life and health or economic assets.

The concept of risk as it is defined here is most clearly expressed as an undesirable event that can have consequences in the form of negative effects on a project's key objectives, for example costs, time, quality, reputation or health, safety and environment (HSE). A well-developed conceptual framework and systematics for risk and risk management exist within the specialist fields of safety, reliability and HSE. The requirement for overview analyses, known as Risk and Vulnerability Analyses (RVAs), for development plans stipulated in the Norwegian Planning and Building Act is based on this definition. Uncertainty in the form of a lack of information, knowledge and control can affect a project in both positive and negative ways. Risk describes the negative outcomes of uncertainty.

Risk for one party can be an opportunity for another. What is counted as a risk and the scale of the risk will be different for different stakeholders, and depend on the stakeholder's viewpoint on the project, but also on the working climate within the project, as well as the contract models used to regulate incentives and risk allocation between the parties. This will be discussed later.

The concept of opportunities is used for the positive outcomes of uncertainty (Hillson, 2004). Opportunities are thus linked to the probability and consequences of desirable events that can have a positive effect on the results of a project in the form of costs, time, quality reputation or HSE. Rationality predicts that humans want to have opportunities, but also want to avoid risk. If focusing on risk only, the ability to exploit the opportunities can be lost, resulting in cost overruns due to high cost of prevention. If focusing on opportunities only, cost can be underestimated, positive effects overestimated and unwanted events can occur which threaten, for example, HSE and reputation. This explains why balanced analysis and decision-making is important in project planning and execution.

Ballard and Vaagen (2017) discuss the possibility of concealing opportunities to turn risks into opportunities, which can be done by building alternative pathways into project master schedules, enabling the project team to move forward once an uncertain event happens or doesn't happen. An example is to exaggerate structural dimensioning when the loads are uncertain. Ballard and Vaagen (2017) call this strategy hedging. According to Webster's *New World Dictionary*, College Edition, published in 1968 by the World Publishing Company, to hedge is to try to avoid or lessen a loss by making counterbalancing investments. In the realm of building design, the need for hedging may occur when the selection of a component is delayed and uncertain, which, among several alternatives, will be chosen. Each alternative may have different design prerequisites, e.g. different structural support, controls, piping, etc. Investing in those different design prerequisites will have a cost, which can be calculated and compared to the cost of waiting until the selection is made.

Ballard and Vaagen (2017) make a distinction between uncertainty expressed through variability and uncertainty expressed through single events. The former is often also referred to as estimate uncertainty and is used to describe uncertainty in cost and time estimates. Variability/estimate uncertainty means that we do not know for sure what things will cost, or how long things will take, because we lack information about many parameters that may affect the outcome.

Figure 18.1 shows an allocation function, for example for the costs associated with a cost item of a project, and illustrates in a simplified manner the link between uncertainty and risk/opportunities. It shows why it is not possible to describe the big picture based on the concept of risk only. In order to describe the big picture, an infinite number of different events must be defined and described to cover the entire event space. The only way of doing this is to describe a continuous function. Such event space is described using the

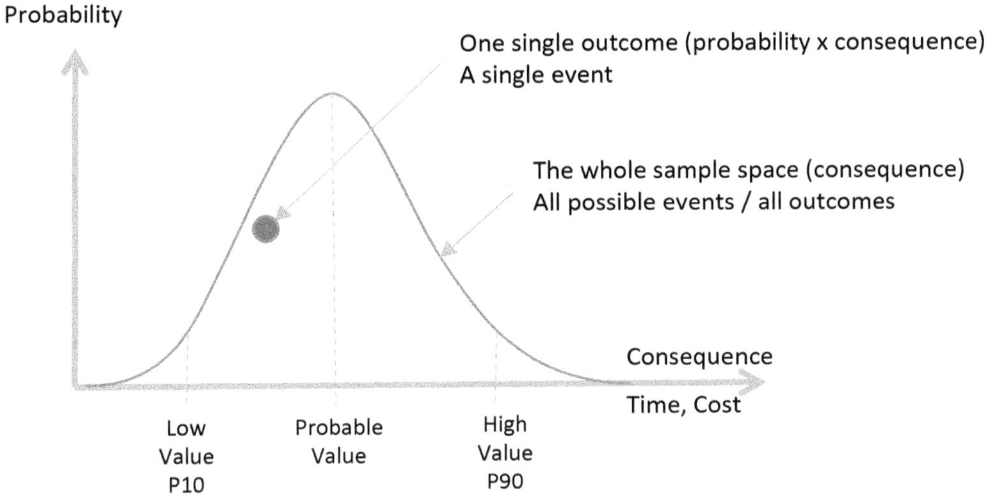

Figure 18.1 Event space and individual outcomes for each element

concept of uncertainty. On the other hand, the concept of risk is a powerful tool for identifying, quantifying and taking measures regarding single events. Uncertainty, risk and opportunities are all important concepts in the field of project management.

Wikström and Gustafsson (1999) write about creative chaos, which we can relate to uncertainty as an aspect of the world we have to live with regarding management and control (Hillson, 2004). Wikström and Gustafson base their argument on a comparison of production at a paper factory and at a shipyard, and ask how it can be possible to act rationally in an environment that hinders rational planning. They make the point that one must, largely, accept that, rather than seeing uncertainty and unpredictability as flaws in our information, one should see them as fundamental aspects of the phenomena we are trying to manage.

Wikström and Gustafsson (1999) mention a talk given by Karl Popper in 1988 entitled 'A world of propensities' in which he argues that the classic assumption of a deterministic world must be replaced by another in which the future always contains genuine uncertainty. Popper does not talk about uncertainty being due to lack of information, rather that all properties of phenomena and their propensities to provide certain effects are partly determined by the situation/context that exists.[4] One example of this is production in the paper factory, which is relatively stable and takes place in a relatively controlled environment, while there is more uncertainty in shipbuilding. Christensen and Kreiner (1991a) share this understanding and also regard uncertainty as a source of learning and opportunities.

This reasoning is also close to the principles underlying the LPS (Ballard, 2000; Ballard, Hammond, and Nickerson, 2009), although this has not been made completely explicit by Ballard in philosophical terms. In the LPS, for example, the traditional detailed production planning at an early stage in the critical path method (CPM) is skipped in favour of planning in greater detail as we approach production, and there are principles related to collaborate planning, making promises and learning that can be

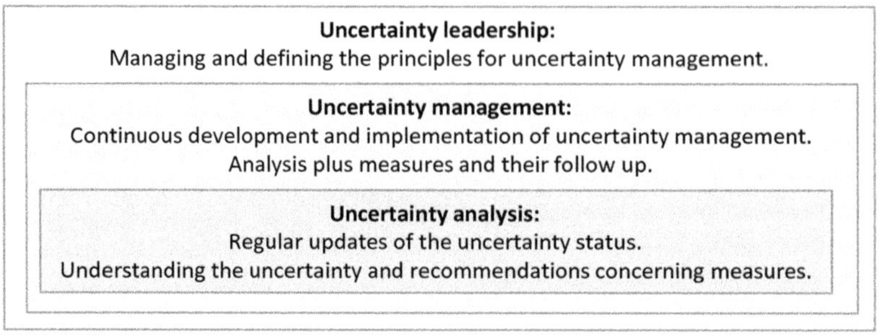

Figure 18.2 From uncertainty analysis to uncertainty leadership (adapted from Austeng et al., 2005b)

interpreted as being in line with Popper's argument to not regard the world as deterministic. Moreover, the aspect of 'making promises' in the LPS across tradesmen from different companies can be regarded as a risk-sharing mechanism on the shop floor level for progress in production.

Figure 18.2 illustrates a system-based model of the link between the concepts of uncertainty leadership, uncertainty management and uncertainty analysis, and what the individual concepts encompass (Austeng et al., 2005b).

Uncertainty leadership means accepting the consequences of uncertainty existing and, based on this, carrying out assessments, decisions and measures. The leadership team's duties in this context are numerous, for example (Austeng et al., 2005b):

- Culture building;
- Organising;
- Decision-making;
- Implementing effective uncertainty management.

The system perspective of uncertainty leadership is about developing and implementing the systematic management of uncertainty, opportunities and risk at all levels (Austeng et al., 2005b). This is an important part of uncertainty leadership. Nonetheless, other aspects of uncertainty leadership perhaps have just as great an impact on the result.

As shown in Figure 18.2, uncertainty is not something that should only be addressed at a project level. Just as important, it should be included in the company's or organisation's management philosophy. Uncertainty leadership is about establishing a culture for managing uncertainty. Culture means the particular way in which people work in the form of roles, duties, responsibilities, methods, acceptance of behaviour and habits in the organisation's environment (ISO 16085, 2006).

Uncertainty management is typically described as activities for identifying, estimating and controlling costs and revenues associated with uncertainties. The purpose is to minimise losses and maximise earnings. In order to do this, uncertainty should be identified, analysed and managed (PMI, 2013), and together this constitutes uncertainty management. Uncertainty management encompasses more than uncertainty analysis. Analysis is an important part of planning, but measures also have to be implemented in order to produce results. Commissioning, monitoring and control of activities are all part of management.

What the analyses have in common, whether they fall under the category uncertainty analysis, risk analysis or opportunity analysis, is that they are tools for planning. They are more or less formal mental exercises that are conducted prior to the start of an activity. It is not certain that the analyses will provide accurate or correct answers, but thinking through the challenges in advance better equips us to deal with what might occur. The benefit will be obvious when, during the execution, we are able to deal with the uncertainty, control and prevent the risk and exploit the opportunities. However, doing this successfully requires one more factor in addition to analysis: awareness on the part of managers and those who will actually execute the activities. The analyses can raise such awareness.

18.3 Uncertainty in the different stages of construction projects

The perspective in this chapter is primarily based on the project management's challenges when it comes to managing uncertainty in different stages of a construction project. Thus, the focus is on managing uncertainty at a construction project level. This affects those uncertainties that are relevant. Uncertainties that are not associated with the project, for example at a customer or company level, are not explicitly included at the project level. These uncertainties are relevant at, for example, a portfolio level. Examples of such uncertainties could be uncertainty linked to the reputation of a company, or financial uncertainty that affects more the executing company than the construction project directly, assuming that we are not talking about financial crises. Some of these uncertainties are grasped in the term pooled interdependency (Thompson, 1967/2003; Kalsaas and Sacks, 2011). Moreover, Christensen and Kreiner (1991a) question whether external interest groups impose uncertainty to projects, and differentiate between operational efficiency (internal to the project) and output efficiency, where also external interests are accounted for.

18.3.1 The initial phase

The initial phase basically involves understanding why a project exists, and the project's objectives. The users' needs, the owners' investment capacity and willingness, and the market's ability to identify reasonable solutions are all examples of challenges that are more or less characterised by uncertainty. A supplier who encounters a project at this stage should use resources for understanding these factors. It is dangerous to believe that the simple goal statements that are often presented in management documents or project management plans tell us anything exact about the future. Owners and users will go through a learning process during a project (Boyd and Chinyio, 2006) and can easily change their minds or change their interpretations of what is needed.

A building (or whatever object that is constructed) is not built for its intrinsic value. Its value derives from what it can be used for; the properties, capacities and functions it has (see also Chapter 6 by Tillmann and Miron). The owner is investing money in producing a building that appeals to its users. The building should either be well suited to the needs of a specific user, or it should be easy to sell or lease out, and thereby generate profitability for the owner in the long term. Contractors or suppliers who want to offer more valuable contributions to the customer than their competitors have two options. Either be able to understand the users' and owners' needs and requirements so well that they can offer a better solution than their competitors, or they must have such good control of their production system that they can deliver as good a solution as their competitors, but at a lower price. Strategy literature discusses this distinction well (Porter, 1996).

Good work in a project's initial phase is heavily dependent on the collection of information and the making of realistic assumptions about the future. This stage of the project life cycle makes the greatest demands on the stakeholders' ability to identify uncertainties in an often unclear and complex volume of information collected from a wide range of sources. Since many key decisions have not yet been made, numerous assumptions will have to be made about the future, assumptions about a very uncertain future, and which could change many times. Analysis models that are used at this stage are therefore correspondingly rough and simple. Systematically collecting the right information at an adequate and general level is an art. This philosophy is related to the underlying philosophy of the LPS, in which one of the underlying principles is to plan at a steadily more detailed level as we get closer to production (Ballard, Hammond, and Nickerson, 2009). There is very little evidence in the literature that the LPS can be used very early in the project.

The challenges in the initial phase are linked to Lean via the focus on recognising what provides value for the client (owner and users). Koskela's (2000) Transformation-Flow-Value (TVF) theory is central in Lean Construction. Lean Construction theory says little about uncertainty as a phenomenon and concept, but a lot about the importance of understanding what contributes to stakeholders' value creation, including framework conditions for better cooperation through concepts for sharing risk and opportunities between stakeholders, or formulated as alignment of commercial interests, conferring the concept of Integrated Project Delivery (IPD) (e.g. Matthews and Howell, 2005; Howell, 2013; Kalsaas, Una Obiose Nwajei, and Bydall, 2018) (see also Chapter 9 by Alvez and Lichtig).

The initial phase concludes with a decision to fund the project and start production planning. Such a decision is often based on an investment analysis in which the investment and future costs and revenues are compared in a present discounted value calculation.

18.3.2 Project and production planning

Project and production planning start with the client's decision to execute the project. This decision is normally based on, among other factors, a cost estimate and a schedule.

In design-bid-build contracts, the contractor starts production planning with a tender calculation. When pricing a tender, it is normal for the contractor to calculate his own costs before pricing risk separately and then adding a profit margin. The price of risk entails a premium in the cost calculation for various known and unknown factors that may affect the costs during the project period. How the contractor prices the risk premium depends on, among other things, the contract model and the allocation of risk between the client and the contractor (Baloi and Price, 2003).

The tender calculation is also based on a schedule. Both the cost estimate and schedule will contain numerous uncertainties. Therefore, some contractors also conduct a quantitative uncertainty analysis (Drevland, 2013) as part of their tender calculation. This means they will price the uncertainties based on probability calculations in order to make strategic assessments based on the risk and opportunities inherent in the project. When the contractor gets the job, the tender calculation is transferred to a production estimate, which provides the basis for managing and executing the project.

A much-used method for scheduling is the CPM (Kelley and Walker, 1959). The main focus in CPM is on the activities that are time critical (are on the critical path). This enables us to maintain an overview of the activities that are crucial to the progress of the construction project. While CPM can be classified as deterministic, the programme evaluation and review technique (PERT) can be considered stochastic[5] (Kenley and Seppänen, 2010). Similar stochastic models can also be found for cost planning (Drevland, 2013; Torp, Drevland, and Austeng, 2015). The key factors in PERT are probability-based estimates of how long it will take to execute work packages as a form of uncertainty assessment. PERT only includes uncertainties on the critical path. Other more advanced methods also consider paths other than the critical one. In the critical chain method (Goldratt, 1997) the main focus is on identifying uncertainties with respect to duration, using the bottlenecks and time buffers to manage the project.

Although uncertainty management is only to a limited degree explicitly mentioned in the descriptions of Lean Construction, it is implicitly present in Lean Construction in general and more specific in the LPS (Ballard, 2000) (see Chapter 3 by Ballard). As tasks move from the highest and most general plan level (Master Schedule) to the lowest and most specific (Weekly Work Plan), uncertainty is reduced as constraints are removed and the tasks get more and more specified. The tasks are planned in more detail the closer to production one gets (Ballard, Hammond, and Nickerson, 2009). Howell (2012) focuses on uncertainty regarding what should be built (the goal) and how it should be built (the method) and discusses the use of contingencies. He asserts that 'the Last Planner System (LPS) reduces uncertainty by improving the predictability of workflow on a project, in effect reducing the uncertainty caused by the way work is managed.' In other words: LPS emphasises reducing production uncertainty through better planning and production control, which is integrated.

The focus on reducing unintentional variation is central in Lean thinking. This also applies to Lean Construction and the LPS (Koskela, 2000; Ballard, 2000). We can, for example, view the effort to make work packages sound in lookahead planning as a measure aimed at reducing variation and uncertainty. To 'make ready' is another phrase for making work packages sound. Variation in processes and procedures during the execution of construction projects increases the risk of cost and time overruns and quality defects. Lack of, or poor, routines and processes for how projects should be executed can result in activities being executed differently depending on who executes them and how the task is defined. This creates random process variation, generating risk of delays, poor quality and extra costs.

18.3.3 Uncertainty management during execution

The execution phase in construction projects is dominated by resource-demanding, complex and, to some extent, challenging technical operations. Uncertainties must also be analysed and assessed systematically during execution in order to obtain the information necessary to make correct decisions, plans and management. Production in essence is controlling processes that are more or less characterised by variability.

In a Lean perspective, uncertainty and risk management can be linked to the management of variation in planned processes. Process variation is a driver and an underlying cause of the risk of waste linked to, for example, duplicating work, troubleshooting, delays and unnecessary waiting periods. There is, therefore, a connection between process variations due to work not being executed in accordance with a specific process or procedure and the risk of errors and defects in a project.

If, for example, drawings and other specifications that are sent from a technical department to a purchasing department in a large business organisation contain errors or defects due to poor or non-existent routines for documentation and control points, this creates the potential for variation in the way a purchase order is handled and sent to an external supplier. Sometimes this will work out okay, other times it will not. It depends on chance, for example whether and when the nonconformity is discovered, the potential consequences of the nonconformity, who is at work and handling the matter, and how the nonconformity is rectified.

The greater the volume being handled through varying execution processes, the greater the consequences of system errors and defects. For example, if a high number of purchase orders has to be handled via processes that are not to some degree standardised and documented, a small fraction of errors could result in big overall costs for the business. Correspondingly, a weakness in safety routines could potentially have major consequences in areas where the company has a low or zero tolerance for errors.

18.3.4 Uncertainty management at different management levels

Table 18.1 shows how different types of analysis relate to each other at a management level in a project, and from which perspectives they are typically executed. All of these management levels must function throughout the entire lifetime of the project. The type of analyses we actually choose to conduct depends on the nature of the project, the situation surrounding the project, the expertise of those involved and the owner's formal requirements. Typically, the purpose of uncertainty analyses conducted in overarching management is to obtain an overview of the uncertainties associated with the big picture in, for example, costs and progress. These analyses are usually different to those uncertainty analyses that are conducted as part of tactical management, which focus more on managing specific uncertainties well.

Table 18.1 Examples of management level and analysis types

Management level	*Analysis perspective*	*Analysis type*
Overarching management (Owner management)	Overarching analysis (Decisions concerning size of budget, buffers, management system)	Uncertainty analysis
Tactical management (Project management)	Comprehensive analysis (Clarifications of scope, resource allocation, priorities, management)	Risk and opportunity analysis
Operative management (Task management)	Secondary analysis on specific tasks (Concrete measures, safety, follow-up of management system)	RVA (Risk and vulnerability, failure mode and effects analysis) SJA (safe job analysis)

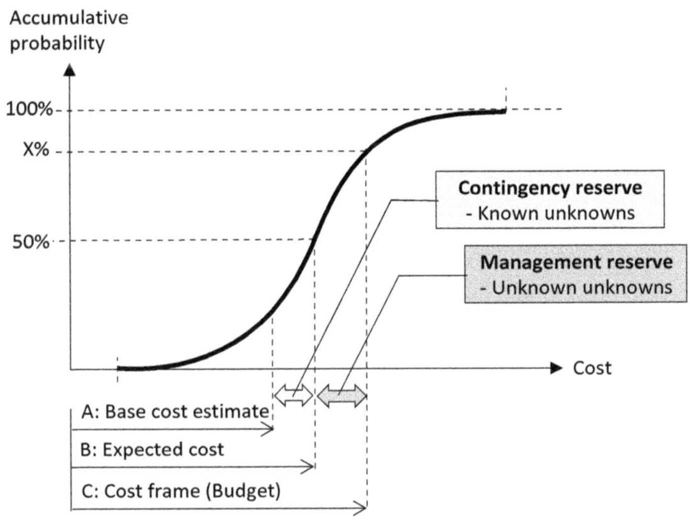

Figure 18.3 Accumulative distribution – example analysis result, strategic level

Overarching management is often associated with project owner management and decisions in the initial phase. Figure 18.3 shows an illustrative example. In this context, tactical management can be linked to the project manager's planning and the contractor's review of the project prior to submitting a tender – and planning immediately after winning the contract, e.g. producing a milestone plan. The operative analyses are more closely linked to the contractor's daily activities on the construction site. Project planning, management, and control on the construction site come under operative management.

As illustrated in Figure 18.3, planning and decision-making on a strategic level includes the purposeful dimensioning of buffers to handle uncertainties. Known unknowns are expected to happen, we just do not know exactly when or how much. The contingency reserve should cover the consequences and is managed by project management. Unknown unknowns are unpredictable consequences of changes in the project environment. The owner/client manage the corresponding management reserve as a buffer against budget overspend. The owner needs to decide what level of probability (X%) against overspend is accepted – this decides the size of the buffer.

Based on the assessments and analyses, we can choose not to carry out an activity if it is too risky, or to implement risk-reducing measures that mean that the activity can be safely carried out after all. Examples of analyses that belong in this category are RVA, FMEA (failure mode and effect analysis) and SJA.

18.4 Current status in uncertainty management

In the following, we look at how to make the analyses work as part of a systematic management system.

18.4.1 Analytical processes

An important contribution to reducing process variation in project execution is implementing good processes and routines, with a well-defined allocation of roles and responsibilities (PMI, 2013). The introduction of digital systems that both facilitate and force the adoption by companies of a more process-based approach to production are technical means that can have positive effects in reducing and managing project risk. We will not explicitly go into these issues, although they are with us when looking closer into uncertainty management.

The literature contains many process models that describe uncertainty management, for example Chapman and Ward (2003), PMI (2013), Austeng et al. (2005b), Torp, Karlsen, and Johansen (2007), Raz and Hillson (2005). Most of the models are similar and based on the same steps, but with slightly differing levels of detail. Based on Torp, Karlsen, and Johansen (2007), uncertainty management can be described by the model shown in Figure 18.4.

A project's uncertainty management process must be planned. One important starting point for this planning is the organisation's guidelines and policy for how uncertainty leadership should be exercised and how uncertainty management should be implemented in an individual project. This should be described in the organisation's project model or project management system. This has to be incorporated into the steering document or project management plan for each project. Some contractors will also produce a quality plan that sets out the framework for this.

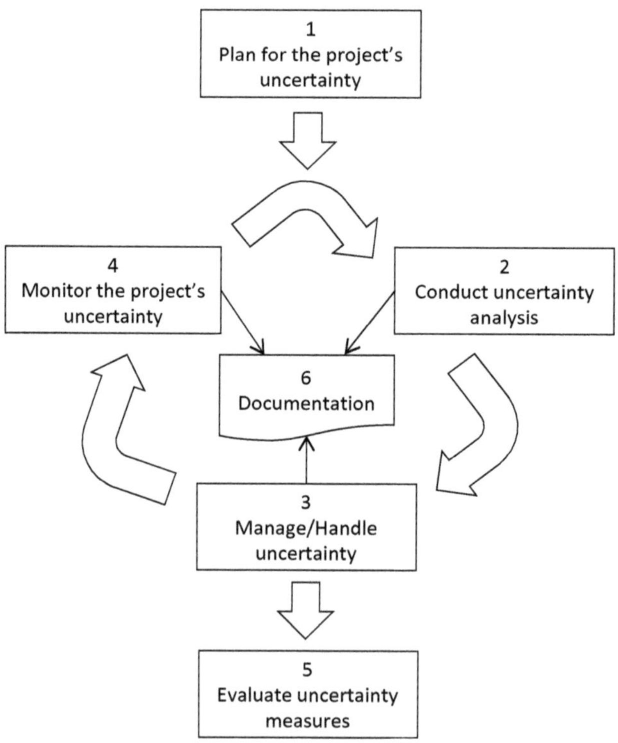

Figure 18.4 A process for uncertainty management (adapted from Torp, Karlsen, and Johansen, 2007)

This initial activity is intended to specify who will perform uncertainty management, define the roles, define the process that must be used for uncertainty management, specify the resources required to implement the process, and define how uncertainties and the management of these should be communicated and coordinated between a project's stakeholders (ISO 16085, 2006). The activity must be carried out at the start of a project and repeated when there is a need to adjust the plan.

The activity should result in a plan for uncertainty management. The six Ws framework has been developed as an aid for this stage of the process (Chapman and Ward, 1997):

- Who? Clarify which roles the various parties will have in the process;
- Why? Clarify why uncertainty analyses/management will be carried out;
- What? Clarify what will be analysed and managed;
- Which way? Clarify how and with which methods the uncertainty management will be carried out;
- Wherewith? Clarify which resources will be allocated to the uncertainty management;
- When? Clarify when the uncertainty management must be carried out.

One key element of the uncertainty management system is an uncertainty management plan, including a plan for the uncertainty analyses. Also, a plan for each uncertainty analysis is important. In literature, we find many recipes for good analysis processes. A typical example is the step-by-step process described by Klakegg (1993). Torp and Klakegg (2016) give an illustrative example, while Austeng et al. (2005b) provide a broader overview of the topic. The main features of these process guidelines are:

- The process is a group task, not to be performed by any individual alone;
- A process leader is a supervisor facilitating the process, not a boss;
- The group must be broadly composed: it must have sufficient expertise, experience, breadth of backgrounds, gender, ages, etc;
- Participants should come both from the project (they know the project) and from outside the project (they can come up with new and different ideas);
- The process should start with defining a goal for the process and end with a conclusion in line with this.

The process for uncertainty analyses could be simplified to the three steps shown in Table 18.2.

The first step can be done very simply in most cases and may be seen as an analysis in its own right. A single brainstorming session is usually effective. Sometimes, simple aids are used such as post-it notes or boards. In some cases, we can use special IT tools for qualitative analysis. Other possible methods that can be supplemental to a brainstorming session are stakeholder analyses and SWOT analyses. The key to a good process is a capable process leader who can get the group to open up and share their thoughts about uncertainties, both opportunities and risks, and work systematically. It is appropriate to try to identify as many points as possible. This is because it is impossible to know in advance which thoughts are important – beyond what is already in the participants' minds. Often, identifying the most important points has proved to require both time and inspiration from what others say. This is about learning, which we will come back to.

Table 18.2 Simple process model for analyses

1 Identify uncertainties (risks/opportunities)	**Qualitative analysis**: Group process (brainstorming) to put into words what is uncertain, can change or could happen.
2 Estimate input (data/analyse/results)	**Quantitative analysis:** Quantify input data (e.g. costs), analyse using IT tools to calculate the consequences.
3 Conclude (define measures/follow-up)	**Recommendations**: Discuss the consequences of the findings and formulate what should be done, when (set deadlines) and by whom (responsibilities).

Step number 2 is the quantitative part. Here, we have to put figures on estimate and event uncertainties. In the major uncertainty analyses conducted to underpin overarching decisions, this is the most salient part of the analysis process. The point is to reveal the uncertainty in a cost estimate or in a schedule, not to make a detailed estimation. This is important, especially in the initial phase. The analysis will often, but does not necessarily need to, include complicated mathematical models and software. In operational analyses, it is often useful to invite those directly involved to take part in the discussion without expressing anything mathematically. The connections may be so clear that it is possible to see the magnitude of the uncertainty (or risk, depending on the purpose of the analysis) and reach conclusions concerning measures and follow-up without calculations.

The third step involves taking the consequences of what the two previous steps revealed. First, a number of potential uncertainties were identified, then their consequences were assessed and analysed. We have now come to the important question: how could the uncertainties best be handled? As discussed above, this is a question of identifying measures that can be implemented to reduce risk and exploit opportunities (within the space for action) and measures that can help us live with the uncertainties (build up the buffers we need and establish a plan B that can be used if required). The results produced by the analysis are incorporated into the project's planning. One important point here is to see the big picture and the connections in everything that is identified. Be aware of relationships that interact and that can produce reinforcing combination effects, and think things through so that measures are not implemented that counteract each other. It is not given that all proposals from the analysis should be implemented. The analysis is input for the project management or the person responsible for the relevant task. The people responsible must consider what they want to include in the plan.

18.4.2 Analytical tools for uncertainty and risk management

As shown earlier, overarching analyses mean that an uncertainty analysis focused on estimate uncertainty and uncertainty drivers is used to 'get the big picture'. By combining a set of items (see Figure 18.1) that make up a model of the project, we arrive at an overall result for the total cost or time (see Figure 18.3). The model is usually analysed with a software tool (Monte Carlo simulation). Uncertainty drivers (or 'cost drivers' in costs analyses and 'progress drivers' in time analyses) represent internal or external effects on a project or solution; forces that can shift assumptions in the plan, or people's perceptions, ambiguities, or simply areas where we lack information or knowledge. These are identified in the analysis's qualitative step. Event uncertainty can also be included in this analysis, but usually this is more relevant at the tactical or operational level.

This type of analysis normally provides probability distributions for the project's total costs (Figure 18.3), as well as uncertainty profiles that show what items or drivers add the most uncertainty to the current situation (Drevland, 2013). This information provides a necessary basis for determining the client's management regime for the project, especially with respect to the right dimensioning of buffers and provisions, and clarifying the conditions that will trigger the use of these buffers. The uncertainty profile is also useful for defining means to manage uncertainty and to prioritise between alternatives. For the contractor, the shift to probability distribution can say something about how likely it is that he will be able to do the work at a specific cost, which is useful information in, for example, the tender phase.

The uncertainty profile is a useful aid when it comes to prioritising resources. Lichtenberg (2000) demonstrates how engineers and others tend to devote all their time and attention to the areas they know best and have the most information about. This is of little use with respect to uncertainty. The uncertainty analysis will show where we have too little information and knowledge. This is where we should focus our attention. This will provide input for the project management regarding the need for more information and better planning. In the tender phase, the contractor can use this to assess where opportunities lie and where there are risks that should be reflected in the tender price.

So far, on a strategic level the analytical approach has been based on an epistemic worldview (uncertainty is lack of information). When shifting towards more operational levels, we will find methods and tools based on an aleatoric worldview (risk and opportunities are facts of life and must be managed as such). This approach is generally based on analysing discrete events – so-called 'event uncertainty'. This type of analysis can be used by both the client and the contractor. Event uncertainty is often analysed using probability/consequence matrices (see the example in Figure 18.5). The group assesses the individual events they envisage could occur. The group then assesses the probability and consequences of these potential events occurring. The categories are specified in advance, so the group only needs to consider to which category an event belongs.

Determining categories for probability and consequences is no simple task. The example of probability in Figure 18.4 is 'typical' for construction projects. It is based on the following logic: events with a very high probability of occurring (more than 50%) must be treated as actual, not hypothetical, events. Something must be done about them. Human rationality is limited, and we have little chance of differentiating between low probabilities. Only statistics can tell us about probabilities at such low levels as, for example, 1% and lower, but we seldom have such statistics when it comes to building and construction projects. Between these two extremes, we choose intervals found appropriate in relation to acceptable risk and the need for measures. In Figure 18.5, five categories have been chosen. The categories are set out in order to ensure that the different situations receive the right attention from the perspective of the risk owner. The risk owner is the person or organisation who has the decision-making responsibility when it comes to taking risks and who will bear the consequences should

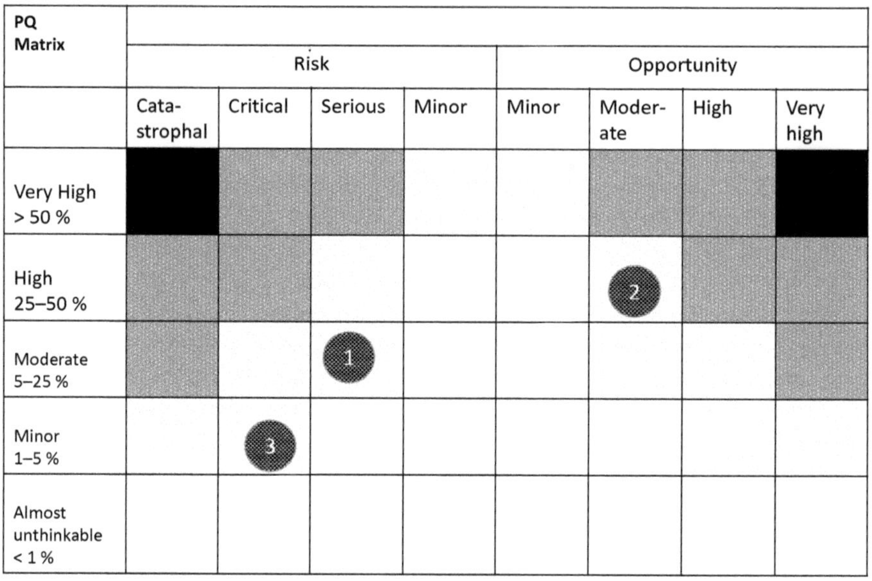

Figure 18.5 Example probability and consequence matrix

the event occur. Correspondingly, we must think through the consequence categories. What is a minor or major consequence depending on who we are (role and responsibilities), how economically robust the project or business is (how large a loss we can withstand or will accept) and the framework conditions we are working under (can we get more money, how exposed are we).

Example: Event number 1 has potentially negative consequences (risk). It is considered to be moderately probable (has between a 5% and 25% probability of occurring) and its consequences would be serious. In this example, the event thus ends up in the light grey with dotted shading category – it must be closely monitored and measures to counter the event must be implemented. Event number 2 has positive consequences (opportunity). Its probability is low (between 1% and 5%) and consequences moderate, and thus it ends up in a white zone. The project management should consider whether it is possible to increase the probability in order to cash in on this opportunity; what is required to trigger it or monitor it in case the situation changes.

This example also shows why we must consider the concept of manageability in uncertainty management. Manageability expresses the degree to which we can affect the probability of and/or outcome of uncertain factors (Austeng et al., 2005a). Low manageability indicates that implementing the measure can only improve the information about the factor, not reduce its probability or consequences. Manageability is medium if the measures are able to affect either the probability or consequences of the event occurring. High manageability is if the measures can affect both the probability and the consequences (Statoil, 2005). Probability, consequences and manageability are important perspectives when it comes to managing uncertainty, but they are not the only perspectives to be considered. In addition, factors such as resource use, urgency (measures must be implemented immediately), legality (the extent to which we are allowed/obliged to implement) and moral obligations, and our ethical room for manoeuvre (is it acceptable to do/not do this) need to be considered. Therefore, we have to see things from different perspectives and cannot manage uncertainty based on the matrix in Figure 18.5 alone.

18.4.3 Strategies for handling/managing uncertainty

Once we have identified the uncertainties and analysed the consequences in the form of cost, time or other consequences, it is time to choose a strategy to deal with them.

The purpose is to decide whether the uncertainty (and the balance between risk and opportunities) is acceptable to the stakeholders. If not, measures must be initiated. These can be measures for reducing risk or increasing opportunities. Risk should be reduced to a level acceptable to the individual party. Here, the strategy could be to reduce the total risk exposure for everyone or to allocate it in some other way between the parties. Correspondingly, as with risk, we must look at the opportunities – the sum can be increased for everyone, or it can be allocated in some other way between the parties. Note that risk and opportunities are often closely connected, like two sides of the coin.

A number of strategies for managing uncertainty are available. The strategy with the greatest probability of being effective in relation to, for example, cost (could also be other parameters like time, safety, etc.) should be chosen for each uncertainty. Thereafter, we should develop specifications for implementing the strategy. Regarding risk, strategies such as Avoid, Mitigate, Transfer to others or Accept can be chosen (PMI, 2013).

What may be a risk for some actors can represent an opportunity for others in construction projects. Earlier in the chapter, we pointed out that risk and opportunities are two sides of the same coin (a development can have both positive and negative consequences). Situations occur where one party wins and the other party loses. An example: An unclear contract in a design-bid-build project results in a dispute between the client and the contractor. The conclusion is that the client has to give a change order. For the client, this represents a risk of costs rising. For the contractor, it represents an opportunity to charge extra and thereby increase the revenue.

18.4.4 Monitoring and managing a project's uncertainty

Planning is an important and integral part of the management of construction projects; but, in the final step of the uncertainty management process, we exit planning mode and transition to operational management mode. Operational management is about controlling and taking care of the activities and events when and where they occur. It builds on the earlier stages of the process and exploits the knowledge that has been gained through the previous steps. While each analysis represents a snapshot, operational management represents the continuous focus on the day-to-day activities, their preconditions, interconnections and consequences. Operational management should never take its eyes off the uncertainties!

All uncertainties – both estimate and event uncertainties, both risks and opportunities that are identified in analyses or through continuous reporting from the activities, are registered in a log, often referred to as the project's uncertainty register. Figure 18.6 shows an example. The first version of the uncertainty register is normally developed from the results of the uncertainty analysis in the initial phase. The uncertainty register should contain as much information about all of the uncertainties (description of the

ID	Uncertainty	Specification of uncertainty	P	C	R/M	Measure	Person	Deadline
1	Design development	Challenging design development	1	1	1	Follow up design review	Håkon	01.12.2017
2	Ground conditions	Ground pollution/old fill is discovered	1	2	2	Prepare environmental measures	Even	01.02.2017
3	Schedule	Delay in pipeline engineering	1	2	2	Follow up status meeting with pipes	Kari	01.01.2017
		Delayed delivery of foundation bolts	1	2	2	Early ordering establish storeroom	Håkon	01.01.2017
		Delayed delivery of steel	1	1	1	Send early enquiry	Kari	01.01.2017
4	Market	High market prices for project parts	2	2	4	Follow up qualified suppliers	Anders	01.03.2017
		Low market prices for project parts	1	2	2	Follow up qualified suppliers	Anders	01.03.2017
5	Project organisation and management	Late distribution of basis for design and build contract	2	2	4	Follow up closely in relation to distribution of time	Nils	01.01.2017
		Well-qualified project organization	2	3	6	Active recruitment of well-qualified people	Håkon	01.01.2017

	Risk: Low
	Risk: Moderate
	Opportunity: Low
	Opportunity: Moderate

Figure 18.6 Example uncertainty register

event, causes, uncertainty costs, possible measures, consequences of possible measures, responsibilities, deadlines, etc.) as you deem appropriate. The minimum information is: identification of opportunity/risk, who is responsible for the task to which the event is linked (and thereby for the uncertainty it entails), which measures, or follow-up, have been decided on in relation to the risk, and who is responsible for this follow-up (person responsible for the measures). Besides this, afterwards you should also register how things went and what could be learnt from the situation. In principle, the risk matrix in Figure 18.5 and the uncertainty register in Figure 18.6 should be updated continuously, although in practice periodic updating is the norm. We recommend to integrate focus on uncertainty, risks and opportunities into the project's meeting structure and work processes so that uncertainty management does not become something alongside or in addition to the other management activities. For example, risk and opportunities should be considered at the start of each phase, such as building shell, airtight building, interior fittings, etc.

Based on the uncertainty register, a priority list of the most important uncertainties is established. The priority list is often also called the focus list, the uncertainty profile or the top ten list. The priority list should include the 5 to 15 uncertainties, with the largest impact on the project objectives. The goal of the priority list is to establish a strong focus on the few important uncertainties and, thereby, avoiding important and less important topics in the uncertainty register from getting the same attention.

Based on, among others, ISO 16085 (2006), it is fair to say that the human aspect of uncertainty management can be described as good leadership. Management must ensure that the culture and individuals' attitudes underpin the desired strategy. When it comes to safety risk, an important part is to get everyone to understand their responsibility not to carry out operations without planning and safety measures, and not accepting others doing so either. Managers must ensure that there is openness about risk factors, not in order to look for scapegoats, but because this is an important factor in ensuring safe working conditions and project success. The means are the same as those in good project management: communicating the goals (understanding priorities, cooperation), good communication (enthusiasm, clarity), clear responsibilities (delegation of authority, trust) and presence (good role model). This mindset generally corresponds with the ideas concerning involvement and empowerment that we find in Lean Construction and the LPS.

Following up uncertainty factors in the project properly will provide the management with good management information and enable them to implement the necessary measures at an early stage (Norwegian Government Agency for Financial Management, 2005). It is equally important to follow up the uncertainty management processes and assess whether or not they are working as they should. This learning is essential with respect to improving uncertainty management.

18.5 Integrating uncertainty management and the Last Planner System

As we see it, increased and explicit focus on uncertainty has something to add to LPS. As described earlier, LPS and LPS-inspired planning and control focuses heavily on reducing uncertainty, although the term uncertainty is only, to a limited degree, explicit and prominent in LPS. Plans become increasingly detailed the closer we get to execution and there is a heavy focus on eliminating constraints before the tasks are included in the weekly work plan. Those constraints, which are also denoted the seven preconditions (Koskela, 1999) or seven flows (Bertelsen et al., 2006), impose uncertainty/contingency on the production plan.

The various trades and disciplines collaborate in planning and make and receive mutually binding promises. Nonconformities between what is planned and actually done is measured (using PPC) in order to learn from errors and for the purposes of continuous improvement (Ballard, Hammond, and Nickerson, 2009). LPS can be regarded as a learning system beyond PPC in which the collaborative planning sessions are central learning arenas (Kalsaas, 2012). This corresponds well with the philosophy in the successive principle (Lichtenberg, 2000), which emphasises that it is crucial to focus on uncertain issues instead of focusing on what we already know and control. The meeting structure and involvement across disciplines in LPS reduces the contextual uncertainty and increases transparency across the disciplines. This contributes to reduced risk and, at the same time, opens up for grasping opportunities.

For uncertainty management to function, a culture of collaboration it vital. We must ensure that the signals and knowledge about what can happen, or is happening, are identified and shared among the participants in the project. There must, therefore, be no doubt that such information is not only acceptable, but wanted, and that the information will be taken seriously and acted upon. The uncertainty management processes should not be used to apportion blame or to 'get' those responsible. We must make the effort to create an attitude of trust and sharing. Accepting uncertainty means accepting that there are limits to what we know and can control.

Our experience is that common practice, also in companies using LPS, is to approach uncertainty analyses and management as something additional to, and separate from, production planning and control. We argue that uncertainty management should be a completely natural and integrated part of planning and control in a project. Integrating uncertainty management and LPS could, as we see it, create an even more effective system to manage the project-based production in construction.

As previously described, uncertainty management starts above the project level and before the project is staffed. The first uncertainty analyses can be done by the client, by the client's principal, by consultants, or by contractors prior to project start-up. This means that project management can receive information about uncertainty at the start of the project. This could for example be:

1. Uncertainty analyses or risk analyses prepared by the client as a basis for making early decisions. The risk picture may be subject to negotiations between client and contractor before a contract allocating the uncertainties between the parties is agreed;
2. A design-build contractor will want to be involved in the project at an early stage in order to reduce uncertainty by influencing constructability and cost aspects before the tender is submitted or the contract is signed. A design and build contract tender is often based on comprehensive uncertainty assessments, with the uncertainties being priced by the design and build contractor and subcontractors as a basis for the tender. Much of the basis for calculating uncertain costs is at this level provided by a comparison of historic cost figures with current estimates and tenders from suppliers;
3. The tender can be based on decisions concerning construction methods, for example the use of concrete elements or in-situ casted concrete. A decision may also have been made that the concrete elements will be bought from specific suppliers, based on an uncertainty analysis that must be followed up by the project.

Uncertainties identified in the project development and tender phases can provide a basis for diligence, monitoring and follow-up in the project. These uncertainties should give inputs to the main schedule, and the main schedule should give inputs to the uncertainty management on the lower levels of planning. Moreover, a design-build contractor will continue to work on optimising his project budget within the constraints set by the contract, client expectations, earlier decisions and other preconditions, for example laws, regulations and standards. An example is how steel reinforcement drawings could be changed to achieve improved constructability and then cost and time with equal quality. Another example is the construction of an underground garage covered with a waterproof membrane below terrain. The risk for leakages is significant and the consequences for repairing very costly. This can be analysed in relation to possible causes, measures for preventing the probability of punctures, and the measures to be taken if a leak should occur.

Wehbe and Hamzeh (2013) propose a framework for integrating FMEA as an addition to constraints removal in the look-ahead planning in LPS. Aslesen et al. (2013) propose to extend LPS to include safety risk management by including safety risk considerations on each of the plan levels of LPS. The argument is that accidents and injuries on the construction site are deeply intertwined with the workflow that LPS seeks to control.

Torp et al. (2018) build further on the reasoning of Aslesen et al. (2013) and propose to use the plan and meeting structure of LPS as the basis for an integration of LPS and uncertainty management. They do, however, propose that the concept of safety risk discussed by Aslesen et al. (2013) is extended to the broader notion of uncertainty. The main reason for this is that, whereas safety is only about preventing negative outcomes (risks) in the form of accidents, uncertainty, whether related to costs, progress, quality, corporate reputation or health, also includes the possibility of positive outcomes. Torp et al. (2018) propose that a focus list should be established at the level of the master schedule. The list should not only consider schedule uncertainties (uncertainty in the workflow and the flows of resources feeding the workflow) but give an overall picture of the top uncertainties for the project. The uncertainties on the list should, if possible, be estimated in monetary values. Both risks and opportunities should be included. The relevant uncertainties in the focus list on the level of the master schedule should be transferred down, concretised and translated to the phase schedule. In addition, new uncertainties could be identified at this planning level and added to the uncertainty register (and if they are important enough, to the focus list). Next would be to translate and operationalise the focus list to the look-ahead level. Again, the focus list should be translated, specified and presented in a way that makes it possible for larger parts of the organisation to take part in the mitigation and elimination of risks and the realisation of opportunities. Moving to the level of the weekly work plans, the translation process should be carried out once more. This way, a separate focus list is established at all plan levels. The list should build on the corresponding list on the higher level, but re-formulating (translating) the uncertainties if needed in order to make them relevant, understandable and manageable at the level in question. Through this top-down process, the entire organisation could be mobilised in a better way to manage uncertainty.

In the above-described process, there would be different ways of handling and transforming uncertainties. This must be seen in relation to the organisational level that

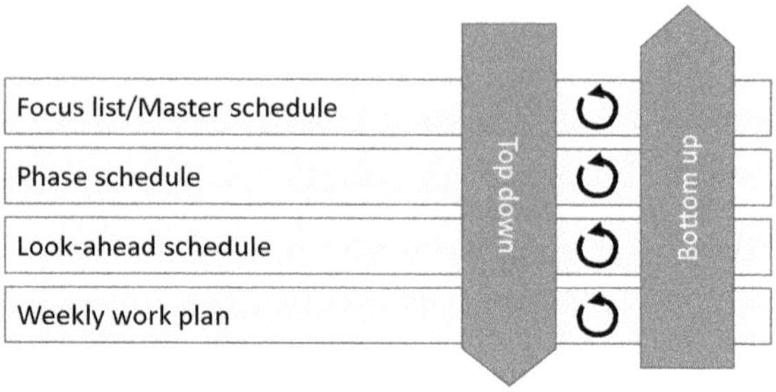

Figure 18.7 Local handling of uncertainty integrated into LPS (adapted from Torp et al., 2018)

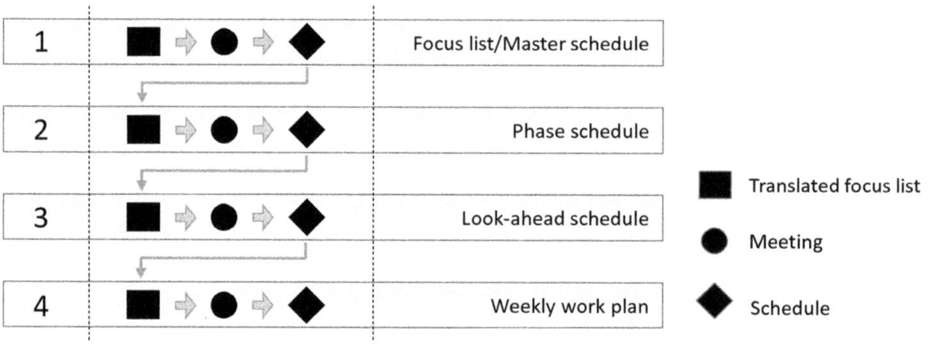

Figure 18.8 Top-down translation of uncertainties (adapted from Torp et al., 2018)

'owns' the different plans. Some uncertainties could be handled on the level on which they are identified, and not transferred down to the underlying levels. Others could or should be handled outside LPS, while some are translated and transferred down to the next plan level. Uncertainties can also be identified on the operational level. Some of these might be handled on the operational level, others might be required to be handled at a higher organisational level. The proposed approach is shown in Figures 18.7 and 18.8 below.

As pointed out by Torp et al. (2018), there are some important issues that must be assessed and tested before concluding whether their proposal will work or not in practice:

- Is there a risk that loading even more tasks onto LPS may result in dilution?
- Which specific questions concerning uncertainty should be asked at the different planning levels and how should the answers be documented, communicated and followed up?
- Will the translation of the focus list to the different plan levels in LPS function in practice?
- Is LPS equally suitable for managing all types of uncertainty, or are there specific types of uncertainty that demand other approaches?

18.6 Conclusion: learn from history – grasp the opportunities

This chapter presents a review of the concepts of uncertainty and risk in construction projects. By uncertainty, we understand the totality of risks and opportunities, where risk is the threat of negative events (potential downsides) and opportunities are the potentials for creating added value for stakeholders (potential upsides). We have explained that uncertainties originate from lack of information when it comes to decisions and planning, and that uncertainty is a feature of the world (a feature we must live with) as far as management and control are concerned.

Lean Construction and uncertainty management are related, but different concepts. We have argued that an increased and explicit focus on uncertainty has something to add to the LPS, and that the plan and meeting structure of LPS has something to add to uncertainty management. We have also discussed how the two could be integrated and have presented important issues that must be assessed and tested before concluding whether or not the presented approach will work in practice.

Uncertainty management can be described as the implementation of processes that make us able to 'have the benefit of hindsight in advance'. Such a statement is very similar to the statement that the LPS has both 'headlights and rear lights', while traditional planning only has 'rear lights' (Kalsaas, 2017). Headlights are, then, the handling of upstream uncertainties (the seven flows).

Continuous improvement and learning are important aspects of a project, including the management of uncertainty. ISO 16,085 states that the purpose of evaluating the uncertainty management process is to provide stakeholders with feedback about the quality of the uncertainty management process.

As stated in the title of this chapter, we firmly believe that uncertainty management has something to add and, therefore, should become a development area for Lean Construction. The authors believe that this chapter will contribute to Lean Construction grasping this opportunity.

Notes

1 The chapter is a revised version of an article by the authors previously published in Norwegian (Klakegg et al., 2017).
2 Aleatoric: random (Oxford Dictionary).
3 Epistemic: from Greek *epistêmê* (Oxford Dictionary), which means knowledge, insight or realisation.
4 Popper was rooted in an aleatoric view of the world and not an epistemic one (see the discussion above).
5 An event is described as stochastic if it occurs with a known or predictable frequency or probability, but without one being able to say precisely when it will occur.

References

Andersen E.S. (2005) *Prosjektledelse – Et Organisasjonsperspektiv*. Oslo, Norway: NKI-forlaget.

Aslesen, S., Sandberg, E., Stake, S. and Bølviken, T. (2013) 'Integrating safety analyses in production planning and control – A proposal', *Proc. 21st Annual Conference of the International Group for Lean Construction*, Fortaleza, Brazil. Available at: www.iglc.net.

Austeng, K., Midtbø, J.T., Jordanger, I., Magnus Sen, O.M. and Torp, O. (2005a) *Usikkerhetsanalyse – Kontekst og grunnlag*. Trondheim, Norway: Department of Civil and Transport Engineering, NTNU. Concept report number 10.

Austeng, K., Torp, O., Midtbø, J.T., Helland, V. and Jordanger, I. (2005b) *Usikkerhetsanalyse – Metode.* Trondheim, Norway: Department of Civil and Transport Engineering, NTNU. Concept report number 12.

Aven, T. and Renn, O. (2009) 'On risk defined as an event where the outcome is uncertain', *Journal of Risk Research, 12*(1), pp. 1–11.

Ballard, G. (2000) *The Last Planner System of Production Control.* PhD Thesis, The University of Birmingham, Birmingham, UK Available at: http://leanconstruction.org.uk/media/docs/ballard2000-dissertation.pdf.

Ballard, G., Hammond, J. and Nickerson, R. (2009) 'Production control principles', *Proc. 17th Annual Conference of the International Group for Lean Construction,* Taipei, Taiwan. Available at: www.iglc.net.

Ballard, G. and Tommelein, I. (2016) 'Current process benchmark for the last planner system', *Lean Construction Journal, 89,* pp.57–89.

Ballard, G. and Vaagen, H. (2017) 'Project flexibility and lean construction', *Proc. 25th Annual Conference of the International Group for Lean Construction,* Heraklion, Greece, pp. 589–596.

Baloi, D. and Price, A.D.F. (2003) 'Modelling global risk factors affecting construction cost performance', *International Journal of Project Management, 21*(4), pp. 261–269. https://doi.org/10.1016/S0263-7863(02)00017-0.

Bertelsen, S., Koskela, L., Henrich, G. and Rooke, J. (2006) 'Critical flow – Towards a construction flow theory', *Proc. 14th Annual Conference of the International Group for Lean Construction,* Santiago, Chile, pp. 31–40. Available at: www.iglc.net.

Boyd, D. and Chinyio, E. (2006) *Understanding the Construction Client.* Oxford, UK: Blackwell Publishing.

Chapman, C. and Ward, S. (1997) *Project Risk Management.* Chichester, UK: John Wiley & Sons.

Chapman, C. and Ward, S. (2003) *Project Risk Management.* Chichester, UK: John Wiley & Sons.

Christensen, S. and Kreiner, K. (1991a) *Prosjektledelse I Løst Koblede Systemer: Ledelse Og Læring I En Ufuldkommen Verden.* Copenhagen, Denmark: Jurist- og Økonomforbundets Forlag.

Christensen, S. and Kreiner, K. (1991b) *Prosjektledelse Under Usikkerhet.* Oslo, Norway: Universitetsforlaget.

Drevland, F. (2013) *Kostnadsestimering Under Usikkerhet.* Trondheim, Norway: Department of Civil and Transport Engineering, NTNU. Concept topic booklet number 4.

Galbraith, J.R. (1977) *Designing Complex Organizations.* Reading, MA: Addison-Wesley Longman.

Goldratt, E.M. (1997) *Critical Chain.* Great Barrington, MA: North River Press.

Hillson, D. (2004) *Effective Opportunity Management For Projects: Exploiting Positive Risk.* New York: Marcel Dekker.

Howell, G. (2012) 'Uncertainty and contingency: Implications for managing projects', *Proc. 20th Annual Conference of the International Group for Lean Construction,* San Diego, CA, USA. Available at: www.iglc.net.

Howell, G. (2013) 'Uncertainty, organizational structure & collaboration: Questions for research', *Proc. 21st Annual Conference of the International Group of Lean Construction,* Fortaleza, Brazil. Available at: www.iglc.net.

ISO 16085. (2006) *Systems and Software Engineering – Life Cycle Processes – Risk Management.* Geneva, Switzerland: International Organization for Standardization, Final Draft.

ISO 31000. (2009) *Risk Management – Principles and Guidelines.* Geneva, Switzerland: International Organization for Standardization.

ISO Guide 73. (2009) *Risk Management – Vocabulary.* Geneva, Switzerland: International Organization for Standardization.

Johansen, A., Eik-Andresen, P., Dypvik Land- Mark, A., Ekambaram, A. and Rolstadås, A. (2016) 'Value of uncertainty: The lost opportunities in large projects', *Administrative Sciences, 6*(3), p. 11.

Jordanger, I. (2005) *Positiv Usikkerhet Og Økt Verdiskaping.* Trondheim, Norway: Department of Civil and Transport Engineering, NTNU. Concept report no. 14.

Kalsaas, B.T. (2012) 'The last planner system style of planning: Its basis in learning theory', *Journal of Engineering, Project, and Production Management, 2*(2), pp. 88–100.

Kalsaas, B.T. (2017) 'Last planner – et system for planlegging og styring', in Kalsaas, B.T. (ed.) *Lean Construction. Forstå og forbedre prosjektbasert produksjon.* Bergen, Norway: Fagbokforlaget, pp. 35–60.

Kalsaas, B.T. and Sacks, R. (2011) 'Conceptualization of interdependency and coordination between construction tasks', *Proc. 19th Annual Conference of the International Group for Lean Construction*, Lima, Peru. Available at: www.iglc.net.

Kalsaas, B.T., Una Obiose Nwajei, U.O. and Bydall, C. (2018) 'A critical perspective on integrated project delivery applied in a norwegian hospital project', *Proc. 9th Annual Conference for Engineering, Production and Construction Management* (EPPM), Cape Town, South Africa.

Karlsen, J.T. (1998) *Mestring Av Omgivelsesusikkerhet: En Emirisk Studie Av Prosjekter.* PhD thesis, Trondheim, Norway: NTNU.

Kelley, J.E. and Walker, M.R. (1959) 'Critical-path planning and scheduling', *Proc. Eastern Joint Computer Conference.* Boston, MA, USA.

Kenley, R. and Seppänen, O. (2010) *Location based Management for Construction: Planning, Scheduling and Control.* London, UK: Spon Press.

Keynes, J.M. (1921) *A Treatise on Probability.* London, UK: Macmillan.

Klakegg O.J. (1993) *Trinnvisprosessen.* Trondheim, Norway: Department of Building and Construction Engineering, NTNU.

Knight, F.H. (1921) *Risk, Uncertainty and Profit.* New York: Hart, Schaffner and Marx.

Kolltveit, B. and Reve, T. (2002) *Prosjekt – Organisering, Ledelse Og Gjennomføring.* Oslo, Norway: Universitetsforlaget.

Koskela, L. (1999) 'Management of production in construction: A theoretical view', *Proc. 7th Annual Conference of the International Group for Lean Construction.* Berkeley, CA, USA, pp 241–252. Available at: www.iglc.net.

Koskela, L. (2000) *An Exploration Towards a Production Theory and Its Application to Construction.* PhD Thesis, VTT Technical Research Centre of Finland. Available at: www.vtt.fi/inf/pdf/publications/2000/P408.pdf.

Lichtenberg, S. (2000) *Proactive Management of Uncertainty using the Successive Principle: A Practical Way to Manage Opportunities and Risks.* Lyngby, Denmark: Polyteknisk Press.

Matthews, O. and Howell, G.A. (2005) 'Integrated project delivery an example of relational contracting', *Lean Construction Journal, 2*(1), pp. 46–61.

Norwegian Government Agency for Financial Management. (2005) *Risikostyring i Staten. Håndtering av risiko i mål- Og Resultatstyringen.* Oslo, Norway: Norwegian Government Agency for Financial Management.

NS 5814. (2008) *Krav til Risikovurderinger.* Oslo: Standards Norway.

PMI. (2013) *A Guide to the Project Management Body of Knowledge.* 5th edition. Newtown Square, PA: Project Management Institute.

Porter, M.E. (1996) 'What is strategy?', *Harvard Business Review* (November–December), pp. 61–78.

Raz, T. and Hillson, D. (2005) 'A comparative review of risk management standards', *Risk Management: An International Journal, 7*(4), pp. 53–66.

Statoil (2005) *Uncertainty Management in Projects, WD0622* – internal document. Stavanger, Norway: Statoil.

Telenor. (2005) *Platform for Risk Management at Telenor.* Oslo: Company Internal.

Thompson, J.D. (1967/2003) *Organizations in Action. Social Science Bases of Administrative Theory.* New York: McGrawhill. [First published in 1967].

Torp O., Aslesen S., Bølviken T., Fritzsønn L.P., Haagensen Å., Lombardo S. and Saltveit T. (2018) 'Is integration of uncertainty management and the last planner system a good idea?', *Proc. 26th Annual Conference of the International Group for Lean Construction*, Chennai, India. Available at: www.iglc.net.

Torp O., Drevland F. and Austeng K. (2015) *Prosess for kostnadsestimering under usikkerhet. Concept topic booklet.* Trondheim, Norway: Department of Civil and Transport Engineering, NTNU.

Torp O., Karlsen J.T. and Johansen A. (2007) *Teori, Kunnskapsgrunnlag Og Rammeverk Innen Usikkerhetsstyring Av Prosjekter.* Trondheim, Norway: Norwegian Centre of Project Management, NTNU.

Torp, O. and Klakegg, O.J. (2016) 'Challenges in cost estimation under uncertainty - A case study of the decommissioning of barsebäck nuclear power plant', *Administrative Sciences*, *6*(4), p. 14.

Vatn, J. (2015) *Project Risk Management.* Trondheim, Norway: Department of Production and Quality Engineering, NTNU.

Wehbe, F.A. and Hamzeh, F.R. (2013) 'Failure mode and effect analysis as a tool for risk management in construction planning', *Proc. 21st Annual Conference of the International Group of Lean Construction*, Fortaleza, Brazil. Available at: www.iglc.net.

Wikström, K. and Gustafsson, C. (1999) 'Et skapande kaos', *Projektperspektiv, Åbo: Åbo Akademi*, *2*, pp.5–10.

19 The evolution of Lean Construction education at US-based universities

Zofia K. Rybkowski, Lincoln H. Forbes and Cynthia C. Y. Tsao

19.1 Introduction: the context of Lean education

Lean Construction pioneer, consultant, and educator Hal Macomber claims that very few contractors and design firms are truly Lean because most are operating from only partial experience. Like the ancient Indian parable of the five blind men touching only part of an elephant, where each one erroneously comes to believe the animal has the shape and form of a snake (trunk), rope (tail), tree-trunk (leg), fan (ears), spear (tusk), or wall (side), few companies grasp the fuller picture of Lean (Schmaltz, 2003; Hal Macomber, *Personal Communication*, November 22, 2017).

While there are many definitions of Lean, this chapter defines Lean as *reducing waste and adding value using continuous improvement in a culture of respect* (Rybkowski and Forbes, 2016). This description suggests there are at least four critical conceptual parts to the Lean 'elephant' (e.g. waste, value, continuous improvement, and respect); if any one of the four components is missing, an organisation is not leveraging all critical facets of Lean.

Understanding the full size and shape of the Lean animal therefore requires that Lean education be both broad and deep. Being able to exercise Lean thought means not only nurturing an understanding of Lean concepts and principles, but also developing an ability to generate new processes while applying existing ones. Educational specialists reference the importance of teachers engaging students at all levels of Bloom's Taxonomy pyramid (Bloom et al., 1956; see Figure 19.1). While remembering/recalling information is certainly foundational to the educational process, it is not sufficient; more advanced forms of learning such as applying, analysing, and creating are also necessary.

To explore the immediate impact of implementing Lean principles with OAEC practitioners, Lean pioneers Greg Howell and Glenn Ballard began experimenting with serious games and simulations in the 1980s (Tsao and Howell, 2015). Serious games are distinguished from simple 'gaming' in that the primary aim is educational for the former, i.e. to *learn* through entertainment rather than to be purely entertained (Wouters, Tabbers, and Paas, 2007). Serious games facilitate participant learning through trial and error without risking interference with actual practice. Organisations wishing to inculcate Lean thinking use serious games and simulations to teach concepts such as Lean processes, supply chain management, sustainable production, logistics, capacity planning, etc. (Pourabdollahian, Taisch, and Kerga, 2012).

Like scientific experiments, serious games often include a control group, and they sequentially modify a single variable that leads to measurable improvements – all while engaging participants in an enjoyable state of play. Psychologist Csikszentmihalyi described *flow* as the state in which an individual is deeply immersed in an activity,

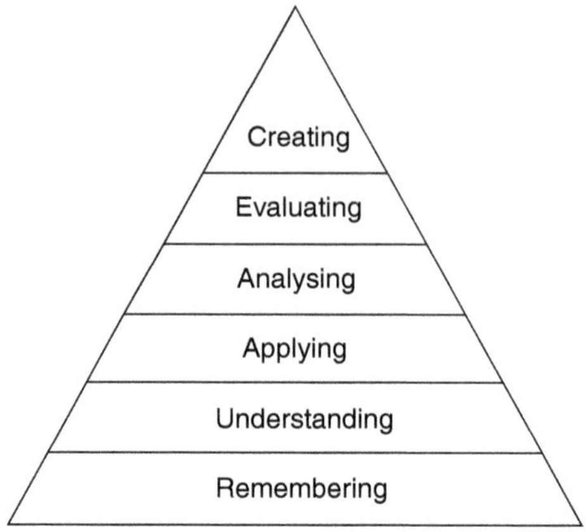

Figure 19.1 Bloom's taxonomy of educational learning objectives

resulting in a sense of contentment with little awareness of the passage of time (Csikszentmihalyi, 1990, 2002). The most successful simulations immerse players in a state of flow while imparting the lessons intended to be conveyed (Figure 19.2).

Figure 19.2 Students at Texas A&M University develop both a mathematical and intuitive understanding of fundamental Lean principles such as pull, one-piece flow, and multi-skilling, by playing serious games and simulations. The game shown here by Visionary Products Inc. (2008) is often called the 'Lego® Airplane Game' by Lean educators (photo by Z. Rybkowski)

Academics, consultants, and industry practitioners use serious games and simulations, but they are not a panacea. Like all scientific experiments, they must be validated internally and externally. *Internal validation* means an experiment does indeed measure what researchers believe it is measuring, and *external validation* means the behavioural outcomes predicted by experimental results are actually applicable to conditions external to that experiment (e.g. in the 'real world'; Jackson, 2012).

In addition to internal and external validation, participants must be able to make a cognitive connection between the Lean principles illustrated in the simulations and ways in which those principles can be applied to actual construction projects. Neeraj et al. (2016) attempted to forge a connection between Lean simulations and their on-site manifestations by linking to published project case studies the principles that the simulations illustrate. A number of commonly played Lean simulations are itemised in Table 19.1.

Table 19.1 Examples of serious games and simulations played by educators to teach Lean principles and Lean Construction processes

Games that demonstrate need for Lean		
Repairman Game	*Activity:* Participants are asked to brainstorm at least 10 reasons a repairman cannot reach his daily quota. *What game reveals*: We tend to blame the worker for system underperformance. *The Lean response*: Instead, identify actual root causes of poor system performance in order to generate effective countermeasures.	(Mossman, 2013; Seddon, 2017)
Deming's Red Bead Experiment	*Activity*: Participants are asked to dip a dimpled paddle into a bin of red and white beads and minimise the number of red beads they collect. *What game reveals*: We tend to use proverbial sticks and carrots to improve worker performance when the underlying system may be the actual root cause of underperformance. *The Lean response*: Managers should collaborate together with workers to improve the system, with their help.	(Lean Simulations, 2017a)
Games that demonstrate impact of company culture/focus		
Silent Squares	*Activity*: Participants are given various pieces of a square and asked to assemble them. *What game reveals*: We tend to see 'winning' as a zero-sum game. *The Lean response*: When we instead focus outside ourselves toward an overall goal, performance improves.	(Lean Construction Institute (LCI), 2017a)
Helium Stick/Magic Stick	*Activity*: Participants stand in two lines, balancing a stick between them, using	(Discovery Village, 2017)

(Continued)

	a single finger each and asked to lower the stick as a team. *What game reveals*: Individual wills can undermine a common goal. *The Lean response*: Focusing on a collective goal will improve performance.	
Win-As-Much-As-You-Can	*Activity*: Participants engage in a game of numbers to maximise their winnings. *What the game reveals*: People tend to see winning on an individual level, and as a zero-sum game. *The Lean response*: Players win as much as possible when they instead focus on overall outcomes.	(Tsao et al., 2013)
Red-Black Game (*Variants:* Maroon-White Game/Red-Green Game)	*Activity*: Teams are asked to maximise their points during a number of rounds of play. *What the game reveals*: Most individuals see gains only at the personal level. *The Lean response*: When individuals strategise collectively and optimise the whole over the parts, they find they will observe greater gains over the long term.	(CSBSJ, 2012; Smith and Rybkowski, 2013)
Marshmallow Challenge	*Activity*: Participants are asked to work in teams to build a tower of a specified height made from dried spaghetti and masking tape with a marshmallow on top. *What the game reveals*: Some team members struggle for dominance and are unable to complete the tower. *The Lean response*: Teams that learn to collaborate effectively are able to complete the tower.	(Wujec, 2010)
Ball Game	*Activity*: Participants are asked to pass a ball between them in multiple rounds, further reducing time to complete the process during subsequent rounds. *What the game reveals*: Conventional thinking of how to pass a ball makes improvement difficult. *The Lean response*: Teams permitted to collectively brainstorm can generate multiple effective solutions.	(Gloger, 2008)
Games that demonstrate mechanics of general Lean principles		
Parade of Trades ('Dice Game')	*Activity*: Participants arranged as a project team assume roles of sequential trades and toss 'trick' die. Each team unknowingly holds die with identical averages, but with different number ranges. Teams record inventories, overcapacity, and final durations. *What the game reveals*: Variability increases overall time and inventory.	(Tommelein and Riley, 1999; Lean Construction Institute (LCI), 2017b)

(*Continued*)

Table 19.1 (Cont.)

	The Lean response: Reduce variability per activity and performance improves; game often serves as an introduction to the Last Planner® System of Production Control.	
Lego® Airplane Game	*Activity*: Participants work along an assembly line to create Lego® airplanes. *What the game reveals*: Push systems and batching lead to greater WIP (Work-in-Process), fewer completed airplanes and a greater number of errors. *The Lean response*: A pull system and one-piece flow lead to reduced WIP, a greater number of completed airplanes, reduced cycle time, and fewer errors. A cross-training/multiskilling round of the games leads to even greater improvement.	(Visionary Products Inc., 2008)
LEAPCON simulation	*Activity*: Participants assume roles as clients, general contractor, and subcontractors, to build high-rise apartment buildings using Lego® blocks. *What the game reveals*: Traditional management practices lead to underperformance. *The Lean response*: Multiple rounds of the simulation demonstrate that pull-scheduling, single-piece flow, work-restructuring, and multi-skilling, lead to increased throughput, reduced cycle delivery time for apartments, and improved cash flow.	(Sacks, Esquenazi, and Goldin, 2007)
Make-a-Card simulation *(This is a lighter weight, lower cost version of the Lego® Airplane Game)*	*Activity*: Participants work along an assembly line to create cards with sticker dots and specified alterations. *What the game reveals*: Push systems and batching lead to greater WIP (Work-in-Process), fewer completed dot-cards and greater number of errors. *The Lean response*: A pull system and one-piece flow lead to reduced WIP, reduced cycle time, a greater number of completed dot-cards, and fewer errors.	(Lean Construction Institute, 2017c)
Lean Cups Game *(This is a lighter weight, lower cost version of the Lego® Airplane Game)*	*Activity*: Participants work along an assembly line to create a specified arrangement of paper cups. *What the game reveals*: Push systems and batching lead to greater WIP (Work-in-Process), fewer completed cup-assemblies and greater number of errors. *The Lean response*: A pull system and one-piece flow lead to reduced WIP, reduced cycle time, a greater number of completed cup-assemblies, and fewer errors.	(Lean Simulations, 2017b)

(Continued)

Table 19.1 (Cont.)

Light Fixture Simulation *(This is a lighter weight, lower cost version of the Lego® Airplane Game)*	*Activity*: Participants work along an assembly line to create origami-style light fixtures from paper. *What the game reveals*: Push systems and batching lead to greater WIP (Work-in-Process), fewer completed paper light fixtures and greater number of errors. *The Lean response*: A pull system and one-piece flow lead to reduced WIP, reduced cycle time, a greater number of completed paper light fixtures, and fewer errors.	(Tariq Abelhamid, presented at the Lean Academic Forum, May 15, 2010)
5S Numbers Game	*Activity*: Participants cross out numbers on game sheets during various rounds as directed by a facilitator. *What the game reveals*: Disorder of existing systems wastes time and creates frustration. *The Lean response*: Implementing the 5Ss (e.g. Sort, Set in Order, Shine, Standardise, and Sustain) saves time and improves worker morale.	(Superteams, 2016)
Games that demonstrate tools specific to Lean Construction/Lean-IPD		
DPR Pull-planning block tower simulation/ Turner variation	*Activity*: Trade partners schedule the assembly of a block tower using coloured post-it notes. *What the game reveals*: Push planning with a sole scheduler makes adjustments difficult. *The Lean response*: Pull planning done by those responsible for the actual work improves performance.	DPR Construction, 2019; Zettel, 2014; Tsao, Draper, and Howell, 2014)
Villego® Last Planner® Simulation	*Activity*: Trade partners schedule the assembly of a Lego® house using coloured post-it notes. *What the game reveals*: Push planning with a sole scheduler lengthens time and makes adjustments difficult. *The Lean response*: Pull planning done by those responsible for the actual work improves performance. This is a thorough introduction to the Last Planner® System of Production Control.	(Villego, 2017)
Target Value Design Simulation	*Activity*: Teams build a free-standing 2-ft tall tower, no more than 2 inches out-of-plumb with a marshmallow on top from drinking straws, coffee stirrers, dried spaghetti, bamboo skewers, and masking tape. During the first round, they are unaware of unit costs. Between the first	(Rybkowski et al., 2016)

(*Continued*)

Table 19.1 (Cont.)

	and second rounds, unit costs are revealed. *What the game reveals*: When teams design unaware of costs, those costs can be higher than necessary. *The Lean response*: When participants know value criteria and unit costs, they can often still fulfil a client's needs, but at a greatly reduced cost.

Although serious games are regarded as a hallmark of Lean education, no known course on Lean consists solely of simulations. Instead, serious games are typically embedded into a *structured framework*, so they offer a deep dive into specific concepts at strategic moments within the course.

While there are many ways to structure a Lean course due to the fact that there are numerous instructors who teach Lean, some examples of structured frameworks taken from interviews and informal discussions about Lean course offerings include:

- *Factory Physics* and its application to construction (Hopp and Spearman, 2001);
- Eleven (11) principles from *Technical Report #72* (Koskela, 1992);
- Fourteen (14) principles from *The Toyota Way* (Liker, 2003);
- *Lean history and theory* from manufacturing to construction (Taylor, 1947; Spriegel, Myers, and Irwin, 1953; Gilbreth and Gilbreth Carey, 1963; Ohno, 1988; Koskela, 1992; Ballard, 2000; Liker, 2003; Deming (Dawson-Pick, 2004); etc.);
- *Modern Construction* text (Forbes and Ahmed, 2011); and
- Course modules from the Associated General Contractors (AGC, 2017).

One example of a framework around which to structure Lean courses is the Toyota Way's House of Lean shown in Liker (2003, p. 33), or the 'kaizen stairway' which graphically depicts Lean principles to reduce waste and add value, using continuous improvement in a culture of respect (see Figure 19.3).

19.1.1 The need for Lean: setting the stage

Similar to introducing Lean concepts to OAEC industry practitioners, teaching Lean at universities needs to first establish the *need* for Lean. To students with limited industry experience (e.g. some have yet to work on an actual construction site), Lean principles seem to represent obvious 'common sense'. However, as Greg Howell often noted, 'Unfortunately, the sense is not common.'

When one of the co-authors of this chapter started teaching Lean concepts to a mixed group of upper-level undergraduates and masters-level construction science graduate students, various methods were used to heighten awareness of the challenges typically faced during project delivery. For example, teams of students were assigned to interview 14 representative stakeholders about specific problems they encountered during a typical workday. These stakeholders included building owners, architects, structural engineers, mechanical engineers, contractors, specialty contractors, permitting agents, inspectors, vendors,

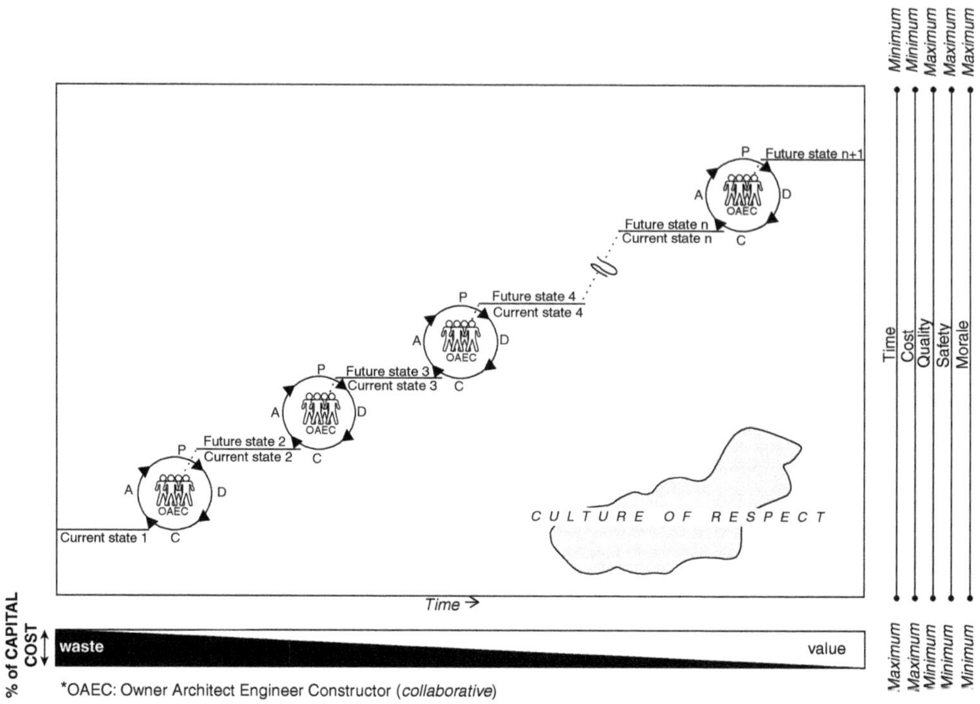

Figure 19.3 The 'kaizen stairway' represents a series of continuous improvements (Δ) where, as waste is eliminated and replaced with value, there is a corresponding improvement in time, cost, quality, safety, and stakeholder morale on a project. These five benefits are inscribed into the roof pediment of Liker's (2003) House of Lean (adapted from Rybkowski and Kahler, 2014)

financiers, insurers, attorneys, bonding agencies, and union hall representatives. Students frequently expressed surprise about the difficulties these practitioners shared during their conversations. Practitioners' 'deltas' (that is, suggestions for improvement) would eventually form the basis for development of the students' final projects – specifically, the invention, testing, and reporting of unique and innovative Lean simulations designed to address specific challenges mentioned during the stakeholder interviews (Rybkowski et al., 2011, 2012, 2016; Rybkowski and Kahler, 2014; Bhaidani et al., 2016; Bhatt et al., 2016).

Similarly, early in a course, students can learn about how Lean philosophy, methods, and tools help address organisational problems. Two exercises that assist in setting the stage in this regard include: (1) Deming's Red Bead Experiment (Lean Simulations, 2017a), and (2) the Repairman Exercise (Mossman, 2013; Seddon, 2017). In the Red Bead Experiment, a facilitator assumes the role of a manager who instructs his or her 'employees' (i.e. audience volunteers) to randomly dip a dimpled paddle into a bin of red and white beads, but to avoid picking up the red beads because they represent an organisation's problems. The facilitator offers proverbial sticks and carrots to motivate the participants, including threats of firing and promises of trips to Hawaii. However, despite their best efforts, players are unable to satisfy their manager's demands because there is a disproportionately large number of red beads in the bin. The purpose of Deming's exercise is to illustrate that the problem of poor performance often rests not with the employees, but with a system and its operational procedures that make it impossible for even the most diligent of employees to succeed.

Like the Red Bead Experiment, the Repairman Exercise helps students grasp some of the challenges that await them as future managers in the construction industry. During the Repairman Exercise, a facilitator invites the audience to brainstorm the reasons a house-call repairman may be unable to meet his or her manager's target to complete an average of eight service visits per day. The facilitator captures the reasons identified by the audience on a displayed flip chart or computer projection. After listing at least 10 to 20 challenges (e.g. traffic, the homeowner is not in, key tools are not available on the truck, bad weather, etc.), the facilitator invites participants to successively revisit each item on the list and ask: 'Is this the fault of the individual repairman or the system?' Inevitably, 80 to 90% of the fault lies with the system; yet, the repairman is the one who is more often blamed. Blaming the worker for a problem that is not of his or her own making not only creates poor morale, it also does not solve the problem. Rather, we need to find ways to fix the system instead of blaming the workers. Thus, the intent of both Deming's Red Bead Experiment and the Repairman Game is to set the stage for what is to follow in a Lean course. Once they engage in stakeholder interviews and play either or both of these games, students tend to become more receptive to learning about Lean philosophy, methods, and tools.

19.1.2 University-based Lean education

To help start a more formal conversation about approaches to Lean teaching, Tsao, Alves, and Mitropoulos (2012) described academic Lean Construction syllabi at three universities, including an overview of course characteristics, grading metrics, readings assigned (both required and optional), and simulations played. Later, Tsao et al. (2013) expanded the initial study to describe Lean teaching approaches at seven universities (Table 19.2a). These initial papers and subsequent work describe strategies used by various faculty members, as well as facilitation tips for some of the most popular simulations used in Lean education (Tsao, Alves, and Mitropoulos, 2012; Tsao et al., 2013; Tsao, Draper, and Howell, 2014; Tsao and Howell, 2015). In addition to contacting US-based faculty members who they already knew were teaching Lean, the co-authors of this chapter requested referrals to find additional faculty members to help expand Table 19.2a. Then the co-authors asked the five faculty members to fill in a spreadsheet with existing categories and to add to these categories as needed to help expand Table 19.2a. Table 19.2b reflects Tsao et al.'s (2013) expanded inventory that includes the Lean Construction course content at these five additional universities (Rybkowski, Forbes, and Tsao, 2018).

It is clear from Tables 19.2a and 19.2b that, of the 12 university faculty members surveyed, the required reading assignments for most Lean Construction courses in academia include International Group for Lean Construction (IGLC) papers (9 out of 12), journal papers (7 out of 12), Koskela (1992; 6 out of 12), and Liker (2003; 6 out of 12). From Tables 19.3a and 19.3b, it is also clear that, of the 12 universities surveyed, the most frequently played simulations include the Parade of Trades (all 12); the Lego® Airplane Game (or its lower cost variants – the Cup Game, Light Fixture Game, or Make-A-Card) (all 12); and Silent Squares (7 out of 12). This suggests that faculty consider these works to be seminal to the understanding of Lean thought. Also, by studying Tables 19.2b and 19.3b, it is clear that additional material has been added to Lean course curriculum in the five years since the Tsao et al. (2013) paper was published with respect to 'Grading', 'Readings', and 'Simulations', reflecting the continuous improvement nature of Lean Construction thought and Lean teaching.

Table 19.2a Overview of seven introductory university-level courses on Lean Construction published in the original table by Tsao et al. (2013), with respect to format, grading, and readings

Overview *Instructor*	*U. Cincinnati* *Tsao*	*Arizona State* *Mitropoulos*	*San Diego St* *Alves*	*S. Illinois U.* *Azambuja*	*Amer. U. Beir.* *Hamzeh*	*Ill. Inst. Tech.* *Menches*	*Texas A&M* *Rybkowski*
Undergrad/Grad	Both	Graduate	Graduate	Undergrad	Both	Grad	Both
Required/Elective	Both	Required	Required	Elective	Elective	Required	Elective
Enrollment	10 to 26	8 to 24	8 to 23	10 to 25	20 to 22	40	10 to 25
Semester/Quarter	Quarter	Semester	Semester	Semester	Semester	Semester	Semester
Weeks	10 of 10	8 of 16	15 of 15	16 of 16	16 of 16	16 of 16	16 of 16
Started	2005	2004	2009	2010	2011	2012	2011
Ended	2008	2010	Continuing	Continuing	Continuing	Continuing	Continuing
Night/Day	Day	Night	Night	Day	Day	Night	Day
Guest lectures	1 to 2	0	5	0	1 to 2	0	2 to 3
GRADING							
Assignments	X	X	X	X	X	X	X
Contribution	X	X	X	X	X	X	X
Discussion forums					X		
Exams	X	X	X	X	X	X	X
Field trip	Toyota						Toyota
Reflection papers	X				X		
Simulations	X	X	X	X	X		X
Team projects	X	X	X	X	X	X	X
Additional: Video presentation							

OVERVIEW *Instructor*	*U. Cincinnati* *Tsao*	*Arizona State* *Mitropoulos*	*San Diego St* *Alves*	*S. Illinois U.* *Azambuja*	*Amer. U. Beir.* *Hamzeh*	*Ill. Inst. Tech.* *Menches*	*Texas A&M* *Rybkowski*
READINGS							
Ballard 2000							Required
Factory Physics			Recommend				
Gilbreths 1963							Required
Goldratt 1992	Required	Required					Ch 13 require d
IGLC papers		Required	Required	Required	Required		Required
Journal papers				Required	Required		Required
Koskela 1992	Required	Required	Required				Required
LCI white papers			Required				
Liker 2003	Required	Required	Recommend	Required	Required	Required	Required
Oglesby 1989		Required					
Taylor							Required
Womack 1991	Required	Required				Recommend	
Additional:							
AGC Textbook							
Aker, 2nd ed.							
Forbes & Ahmed 2010							
Martinez							
Schmaltz 2013							

Table 19.2b Additional Lean programmes and coursework published five years after publication of Table 19.2a, with respect to format, grading and readings (Rybkowski, Forbes, and Tsao, 2018)

OVERVIEW *Instructor*	*N Carolina St* *Liu*	*Virginia Tech* *Muir*	*Colorado St* *Senior*	*Michigan St* *Abdelhamid*	*Pittsburg St* *Levens*
Undergrad/Grad	Grad	Undergrad	Grad	Grad	Both
Required/Elective	Elective	Elective	Elective	Elective	Elective
Enrollment	20 to 30	10 to 25	8	10	10 to 15
Semester/Quarter	Semester	Semester	Semester	Semester	Semester
Weeks	15	15	16	15	16 of 16
Started	2008	2018 (antic.)	1999	2001	2016
Ended	Continuing	New	Continuing	Continuing	Continuing
Night/Day	Day	Day	Evening	Evening	Day
Guest lectures	1	2 to 3	3 to 4	1 to 2	None
GRADING					
Assignments	X	X	X	X	X
Contribution	NA	X	X	X	X
Discussion forums	NA			X	
Exams	X	X	X	X	
Field trip	X	X		X	LCI Kansas City
Reflection papers	X	X	X	X	
Simulations	X	X	X	X	X
Team projects	X	X	X	X	X
Additional:					
Video presentation			X		

OVERVIEW *Instructor*	*N Carolina St* *Liu*	*Virginia Tech* *Muir*	*Colorado St* *Senior*	*Michigan St* *Abdelhamid*	*Pittsburg St* *Levens*
READINGS					
Ballard 2000	Required	Required	Required	Optional	
Factory Physics				Optional	
Gilbreths 1963				Optional	
Goldratt 1992	Required		Required	Optional	
IGLC papers	Required	Required	Required	Required	
Journal papers	Required	Required	Required	Required	
Koskela 1992	Required		Required	Optional	
LCI white papers	Required	Required		Optional	
Liker 2003				Optional	
Oglesby 1989	Required		Required	Required	
Taylor			Required	Optional	
Womack 1991	Required			Optional	
Additional:					
AGC Textbook				Optional	Required
Aker, 2nd ed.				Optional	Optional
Forbes & Ahmed 2010				Required	
Martinez				Required	
Schmaltz 2013				Required	

Table 19.3a Overview of seven introductory university-level courses on Lean Construction published in the original table by Tsao et al. (2013), with respect to simulations

OVERVIEW *Instructor*	*U. Cincinnati* *Tsao*	*U. Cincinnati* *Mitropoulos*	*San Diego St* *Alves*	*S. Illinois U.* *Azambuja*	*Amer. U. Beir.* *Hamzeh*	*Ill. Inst. Tech.* *Menches*	*Texas A&M* *Rybkowski*
SIMULATIONS							
5S Game							X
Airplane Game	X		X	X	X		X
Cocktail Napkin							X
Cups Game		X				X	
Delta Design	X						
Deming s Red-Bead							X
Helium Stick	X				X		
Leapcon				X		X	
Magic Tarp	X						
Maroon-White					Variant		X
Origami Game	X						
Parade Game	X	X	X	X	X	X	X
Radioactive Popcorn				X			
Silent Squares			X	X	X	X	
TVD Game							X
Win As Much As	X				X	X	
Additional:							
Ball Game							
DPR Block Tower							
Gemba Walk							
Last Planner (AGC)							
Leadership Styles							
Lego Hotel/Tower							
Light Fixtures							
Make-a-Card							
Marshmallow Challenge							
NASA Survive/ Moon							
No./Task Switching							
Oops							
Original Dice Game							
Prison Door Case							
Repairman							
Villego							

Table 19.3b Additional Lean programmes and coursework published five years after publication of Table 19.3a, with respect to simulations (Rybkowski, Forbes, and Tsao, 2018)

OVERVIEW *Instructor*	*N Carolina St* *Liu*	*Virginia Tech* *Muir*	*Colorado St* *Senior*	*Michigan St* *Abdelhamid*	*Pittsburg St* *Levens*
SIMULATIONS					
5S Game				X	
Airplane Game	X	X		X	
Cocktail Napkin					
Cups Game					
Delta Design				X	
Deming s Red-Bead			X	X	
Helium Stick	X				
Leapcon				X	
Magic Tarp					
Maroon-White				X	
Origami Game					
Parade Game	X	X	X	X	X
Radioactive Popcorn					
Silent Squares	X	X		X	
TVD Game				X	
Win As Much As			X	X	
Additional:					
Ball Game				X	X
DPR Block Tower		X			X
Gemba Walk		X			
Last Planner (AGC)				X	
Leadership Styles				X	X
Lego Hotel/Tower				X	X
Light Fixtures			X	X	
Make-a-Card				Variant	X
Marshmallow Challenge				X	
NASA Survive/ Moon				Variant	X
No./Task Switching				X	
Oops	X				
Original Dice Game				X	
Prison Door Case				X	
Repairman					
Villego	X			X	X

The two most commonly played simulations demonstrate that faculty consider the following concepts critical to Lean thought: reduction of variability (illustrated by Parade of Trades); and management of materials that includes pull, one-piece flow, and balanced workflow (illustrated by Lego® Airplane Game and its variants). Silent Squares is the third most commonly played simulation. This may be due to the fact that it is an easier simulation to set up, and it helps faculty demonstrate the value of optimising the whole over the parts (that is, prioritising global optimisation over local optimisation), which is a critical Lean principle.

Interestingly, Joe Levens' course at Pittsburg State University in Pittsburg, KS, represents a significant departure from previous university courses on Lean Construction because it follows the standard Associated General Contractors' (AGC) textbook on the topic and – with the exception of the optional web-based audio book by Paul Akers (2014) – the course is not supplemented with additional required readings. One primary aim of the course is to prepare students to sit for the AGC certificate examination after completing the course, if they so choose. Pittsburg State students and industry representatives alike take the course together in the same university classroom. The site of the course is fluid as well; Pittsburg State students are equally permitted to earn university credits by sitting in AGC courses whenever they are offered in the AGC of Kansas' Wichita office, especially when Levens is serving as the local AGC Lean course instructor (Joe Levens, *Personal Communication*, November 21, 2017).

19.1.3 The US academic Lean knowledge engine

A key generator of many US-based Lean academic courses resides at the University of California, Berkeley. The UC Berkeley curriculum represents an evolved philosophical amalgamation from the work of Lean Construction pioneers Iris Tommelein and Glenn Ballard. Tommelein and Ballard started collaborating and explicitly researching Lean Construction in the late 1990s (e.g. Tommelein and Ballard, 1997) and formally teaching Lean Construction as a standalone course in 1998 (Iris Tommelein, *personal communication*, August 8, 2018). The faculty who teach at US-based universities listed in Tables 19.2a and 19.2b and 19.3a and 19.3b, and who are also former students of Tommelein and Ballard, include: Cynthia Tsao, formerly at the University of Cincinnati and now at Wentworth Institute of Technology, Thais Alves at San Diego State University, Farook Hamzeh formerly at Colorado State University and the American University of Beirut, and now at the University of Alberta, Zofia Rybkowski at Texas A&M University, and Min Liu at North Carolina State University.

For these faculty members, a direct lineage to the UC Berkeley research programme is clear as they earned their doctoral degrees in part through engagement in Lean Construction research. This is interesting because it demonstrates how the diffusion of Lean Construction philosophy into industry has been facilitated by the education of PhD students who mature into Lean thinkers; upon graduation, these former students relocate to become faculty in other universities, where they spread the principles of Lean through courses they then teach. It would be an interesting study unto itself to determine whether Lean Construction philosophy could have been so rapidly disseminated without the participation of academic researchers and their doctoral students.

Additional diaspora not included in Tables 19.2a and 19.2b and 19.3a and 19.3b include Yong-Woo Kim and Hyun Woo 'Chris' Lee, both at the University of Washington, and Kristen Parrish at Arizona State University. Each of these faculty members, in turn, instruct graduate and undergraduate students who join the OAEC industry following graduation. James Choo and Colin Milberg, also former doctoral students of Tommelein and Ballard, transfer Lean principles directly to OAEC industry practitioners through consulting and AGC's Lean certificate programme (as in the case of Milberg). Former doctoral student Paz Arroyo worked as faculty at Pontificia Universidad Católica de Chile and now serves as quality leader at DPR Construction. Similarly, Patricia Tillman completed a case study at UC Berkeley as part of her PhD research and later worked as a postdoctoral scholar at UC Berkeley; she now serves as senior Lean manager at the University of California, San Francisco. In other words, while not all doctoral students who research Lean Construction for dissertation topics become traditional academics, many assume a similar role as full-time educators ('Lean champions') within companies – a role that did not previously exist.

As of fall semester 2017, Tommelein's first graduate-level course at UC Berkeley, *Lean Construction Concepts and Methods*, covered the following topics: the House of Toyota, TFV (Transformation-Flow-Value) theory, project production, Lean Project Delivery System (LPDS), work structuring, waste, value stream mapping, design of construction operations, the Last Planner™ System of Production Control, line of balance scheduling, takt planning, standardisation, project definition, Target Value Design/Delivery (TVD), mistake proofing (poka-yoke), plan-do-check-act (PDCA) cycle, and Choosing-by-Advantages (CBA).

Illustrative research papers supplemented theoretical lectures, as did guest lecturers, gemba walks at Lean companies, and hands-on simulations such as the airplane game and Villego. Tommelein's second graduate-level course at UC Berkeley, *Lean Construction and Supply Chain Management*, further extends the systems-engineering perspective on project-production systems to include suppliers of goods and services. Among other topics, this course teaches discrete-event simulation as a tool to model and analyse system impacts of interdependence and variation. Over the years, Iris Tommelein's course – as well as colleague Glenn Ballard's – has served as a generative wellspring for the courses that UC Berkeley's doctoral research students teach once they accept faculty positions at other universities across the US and around the world. Tommelein's course is being singled out here from other courses described in Tables 19.2a and 19.2b and Tables 19.3a and 19.3b because of UC Berkeley's seminal role – and because the breadth and depth of its content is too great to contain within the scope of the inventory tables.

19.2 Academic course frameworks

Several university faculty commented about the conceptual frameworks they employed. For example, Tariq Abdelhamid of Michigan State University noted,

> The conceptual framework I use for the course has always been structured around the premise that Lean Construction is a set of principles and methods that significantly and continuously change what and how we build. I am very keen on emphasising that Lean Construction has been incorrectly assumed to be related only to the construction phase of a project, while Lean Construction is really Lean in the construction industry, with all of the industrys' different providers (owners, architects, engineers, constructors, suppliers, regulators, etc.) considered benefactors of what it has to offer.
>
> (Tariq Abdelhamid, *Personal Communication*, January 5, 2018)

This comment vividly reflects many of the challenges associated with teaching a university-level Lean course, i.e. (1) the reality that the discipline is constantly evolving, as evidenced by the ever-expanding Tables 19.2a and 19.2b and Tables 19.3a and 19.3b, and (2) the fact that, to maximise success, a culture of Lean must permeate not only that of the contractor, but also that of the associated stakeholders as well.

Getting all project stakeholders to value and support Lean principles is a common challenge, and understanding this hurdle helps explain why the AGC chose to standardise an otherwise ever-changing, continuously improving, body of knowledge. This also helps explain why pioneering legal attorneys such as Will Lichtig took it upon themselves to develop some of the earliest relational contracts for construction, such as the Integrated Form of Agreement (IFOA), the Integrated Project Delivery contracts for the American Institute of Architects (AIA), and Consensus Docs for AGC. It also explains why Tariq Abdelhamid introduces students to relational contracts as part of his Lean Construction course at Michigan State. Since the force of law sides with written contracts, and many Lean projects are now operating under Lean IPD contracts (or similar) (see Chapter 9 by Alves and Lichtig for a detailed discussion on IPD), the impact of Lean Construction for improving project outcomes has greater potential than earlier, voluntary (i.e. not legally enforceable) collaborative OAEC efforts such as 'partnering'.

19.2.1 Serious games and simulations

Serious games and simulations have traditionally played a critical role in teaching the principles of Lean Construction to newcomers (Tsao and Howell, 2015), and they have become a hallmark of Lean coursework and understanding whether OAEC projects have successful outcomes or not. The benefits of pull, one-piece flow, and multi-skilling are not always intuitively obvious, so simulations provide miniature, controlled experiments that create opportunities for an 'aha' moment among players. The benefits of different Lean principles can be demonstrated through successive rounds of play, and metrics collected during play can often help reinforce key takeaways.

Again, some manifestations of these Lean principles on a construction site include takt planning (pull, on-piece flow, and cross training, as illustrated by the Lego® Airplane Game) (Frandson, Berghede, and Tommelein, 2013); Last Planner™ System of Production Control (reduced variability, as illustrated by the Parade-of-Trades Game); and IPD legal contracts (optimising the whole over the parts, as illustrated by the Red-Black Game). Relatively recent newcomers to the Lean Construction simulation scene include Villego™ (Villego® Last Planner® Simulation, 2017) and the DPR block tower game (King, 2011; Tsao, Draper, and Howell, 2014; Zettel, 2014), which illustrate Last Planner™ and pull planning, respectively, and the eponymous 5S Numbers game (Superteams, 2016).

Some simulations have taken on different manifestations, driven in part by the need to either save on material cost (e.g. Legos can be expensive), or weight and volume for Lean Construction educators who need to travel to various off-campus locations to offer Lean workshops for education or research purposes (Legos are also heavy and consume valuable luggage space). Consequently, simulations such as the Lego Airplane Game have spawned multiple, lighter-weight simulation variations such as the 'Make-a-Card Simulation' from LCI (LCI, 2017), the Lean Cups Game (Lean Simulations, 2017b), or the Light Fixture Simulation (Tariq Abdelhamid, presented at the Lean Academic Forum, May 15, 2010). The LEAPCON simulation presents some similar concepts as the Lego Airplane Game (Sacks, Esquenazi, and Goldin, 2007), and it is more appropriate to the OAEC industry as it requires participants to construct buildings, rather than airplanes, out of Legos.

A more recent addition to the Lean Construction game scene is the TVD simulation, which asks participant teams to collaboratively design and build a 2-foot-tall free-standing tower that is no more than two inches out of plumb, calculate its cost, and then redesign the tower, to reduce the total cost by 20% while maintaining key values, as defined by the 'customer' (Rybkowski et al., 2016). Simulations such as the Magic Stick (Discovery Village, 2017), Win-as-much-as-you-can (Tsao et al., 2013), or the Red-Black Game and its variants (e.g. Maroon-White Game; Red-Green Game; College of Saint Benedict and Saint John's University, 2012; Smith and Rybkowski, 2013) teach the importance of optimising the whole over the parts. These are a few of the ever-evolving array of simulations that are used not only to illustrate Lean principles, but to create buy-in among those who will be implementing Lean.

19.3 Need for research on Lean education

Despite observed successes in Lean Construction education, a number of university educators have also observed confusion about what Lean is and what it is not. In 2012, Naney, Goser, and Azambuja (2012) argued that Lean principles need be more rapidly disseminated in order to reach a tipping point among stakeholders in the construction industry, and that one barrier to acceptance is the diffuse nature of Lean concepts. Rybkowski, Abdelhamid, and Forbes (2013) reported a similar concern when her team distributed 'cocktail napkins' to both pioneers and novices of Lean Construction and asked each to sketch his or her personal definition of Lean Construction – an exercise that mimics real-world scenarios where practitioners try to explain to OAEC stakeholders the value of implementing Lean. From the sketches, the researchers discovered considerable variability in perceptions of the core concepts of Lean.

Additionally, some educators have found partial adoption of Lean tools within companies, or adoption of Lean tools without also implementing the softer skills such as creating a culture of respect for people or a continuous improvement process that is inclusive (Liker, 2003; Santorella, 2011). The problem with partial implementation of Lean is that it leads some practitioners to believe that Lean 'doesn't work', or they resent its implementation since misunderstanding of concepts such as 'just-in-time', for example, have led some individuals to mistakenly believe that Lean gives licence to implement last-minute directives to their stakeholder partners (Lean Construction pioneers instead prefer the term 'last responsible moment' to avoid this misunderstanding).

For these reasons, it would be helpful to determine the extent to which an evolving Lean canon – such as that developed by the AGC – is helping to offer a more clear and comprehensive understanding of Lean Construction. Many students who are studying Lean Construction at universities do not become academics, but rather obtain positions in industry following graduation. It can be argued, therefore, that the real test of the effectiveness of Lean Construction education via university courses is the extent to which these students can implement Lean principles on job sites, or at least convince their superiors that Lean is worth a try. Rigorous longitudinal research on this topic is currently missing and is sorely needed.

19.4 Conclusion

Lean Construction education – both at universities and in practice – is arguably rooted in Lean production theory. The aim of this chapter has been to capture a snapshot of Lean Construction education by surveying a sample of US-based university course materials. The sample reveals a constant evolution of academic resources that include additional readings and simulations. This chapter's inventory of university curricula suggests there is observable growth with respect to grading, assigned readings, and simulations played. A number of industry practices have been influenced by Lean theory developed by academics – a process which reinforces the importance of understanding how knowledge is being created, transferred, and formalised within institutions of higher education.

At the time of publication of the Tsao et al. (2013) paper, the Associated General Contractor's Lean certification course was still under development. Five years later – by the time the Rybkowski, Forbes, and Tsao (2018) paper was published – not only is the AGC course well developed and offered to practitioners throughout the US, we are observing its emergence as a fully credited offering for at least one US-based university. We observe that graduating students seeking to enhance their credentials in a competitive job market through industry-recognised certifications are increasingly requesting coursework that can do this. Nevertheless, while these potentially positive signs suggest that Lean Construction is 'going mainstream', systematic research is needed to determine how effective formal academic coursework is in helping students actually implement Lean on jobsites following graduation. Ironically, while the AGC course and its offering in formal education is potentially good news, a very real concern also arises, i.e. How might standardisation of curricula impact the *continuous improvement* process in Lean thought itself, which is so fundamental to Lean? Future research is needed to investigate these concerns.

Regardless of questions and concerns, it is encouraging to observe that the number of Lean courses offered at US-based universities is on the rise. This increase in Lean Construction education opportunities can help grow a workforce of multiple stakeholders committed to implementing Lean on projects not only because they are aware of the benefits of Lean project delivery, but also because they understand *why* Lean improves project performance. As we intuitively suspect from experience, knowing 'the why' makes 'the what' so much easier to implement.

References

Akers, P.A. (2014) *2-Second Lean: How to grow people and build a lean culture*, 3rd edition. Available at: www.youtube.com/watch?v=zbL4LwmWAy0 (Accessed: 27th Nov 2017).

Associated General Contractors (AGC) (2017) 'Lean Construction Education Program'. Available at: www.agc.org/learn/education-training/lean-construction-education-program (Accessed: 26th Nov).

Ballard, H.G. (2000) *Last Planner System of Production Control.* PhD Thesis. The University of Birmingham, Birmingham, UK.

Bhaidani, N., Rybkowski, Z., Smith, J., Choudhury, I. and Hill, R. (2016) 'Percent Planned Complete: Development and Testing of a Simulation to Increase Reliability in Scheduling', *Proc. 24th Annual Conference of the International Group for Lean Construction*, Boston, MA, USA.

Bhatt, Y., Rybkowski, Z.K., Kalantar, N. and Fernandez-Solis, J.L. (2016) 'Trainathon lean simulation game: Determining perceptions of the value of training among construction stakeholders', *Proc. 24th Annual Conference of the International Group for Lean Construction*, Boston, MA, USA.

Bloom, B.S., Engelhart, M.D., Furst, E.J., Hill, W.H. and Krathwohl, D.R. (1956) *Taxonomy of educational objectives*, Handbook 1: Cognitive domain. Longman, New York, USA.

College of Saint Benedict and Saint John'sUniversity (2012) 'The Red/BlackGame.'. Available at: www.cs.csbsju.edu/~lziegler/redblack.html (Accessed: 1st April).

Csikszentmihalyi, M. (1990) *Flow: The Psychology of Optimal Experience.* Harper and Row, New York, USA.

Csikszentmihalyi, M. (2002) *Flow: The Psychology of Happiness.* Random House, London, UK.

Dawson-Pick, D., producer. (2004) *W. Edwards Deming: Prophet Unheard* (film). BBC Education & Training, Films for the Humanities, Princeton, NJ, USA. 30 min.

Discovery Village (2017) 'Helium Stick'. Available at: www.discoveryvillage.in/activity/helium-stick/ (Accessed: 27th Nov 2017).

DPR Construction (2019). DPR Pull-Planning Game. Available at: https://randdmarketing.four51 ordercloud.com/DPR/product/DPR-2431(Accessed: 15 December 2019).

Forbes, L.H. and Ahmed, S.M. (2011) *Modern Construction: Lean Project Delivery and Integrated Practices.* CRC Press/Taylor & Francis Group, Boca Raton, FL, USA. p. 490.

Frandson, A., Berghede, K. and Tommelein, I.D. (2013) 'Takt Time Planning for Construction of Exterior Cladding', *Proc. 21st Annual Conference of the International Group for Lean Construction*, Fortaleza, Brazil.

Gilbreth F.B., Jr. and Gilbreth Carey, E. (1963) *Cheaper by the Dozen*. Thomas Y. Crowell Co., New York, USA.

Gloger, B. (2008). *Ball Point Game: Feel the scrum flow* www.borisgloger.com/wp-content/uploads/Publikationen/Tools/Ball_Point_Game.pdf (Accessed: 10 December 2019).

Hopp, W.J. and Spearman, M.L. (2001) *Factory Physics: Foundations of Manufacturing Management*. Irwin/McGraw-Hill, Boston, MA.

Jackson, C. (2012) 'Internal and external validity in experimental games: A social reality check', *European Journal of Development Research*, 24(1), pp. 71–88.

Koskela, L. (1992) *Application of the New Production Philosophy to Construction*, Center for Integrated Facility Engineering, (CIFE), Department of Civil Engineering, Stanford University. Technical Report number: 72.

Lean Construction Institute (LCI) (2017a) *Silent Squares Simulation*. Available at: www.leanconstruction.org/learning/game-simulations/silent-squares-simulation/(Accessed: 15 December 2019).

Lean Construction Institute (LCI) (2017b) *Parade of Trades Simulation*. Available at: www.leanconstruction.org/learning/game-simulations/parade-of-trades-simulation/(Accessed: 15 December 2019).

Lean Construction Institute (LCI) (2017c) *Make-a-Card Simulation*. Available at: www.leanconstruction.org/learning/game-simulations/make-a-card-simulation/(Accessed: 26 November 2019).

Lean Simulations (2017a) 'Red Bead Experiment'. Available at: www.leansimulations.org/2010/09/red-bead-experiment.html (Accessed: 27th Nov).

Lean Simulations. (2017b) 'The Lean cups game – What's a summer without a few cold ones?'. Available at: www.leansimulations.org/2011/06/lean-cups-game-whats-summer-without-few.html (Accessed: 27 November 2019).

Liker, J. (2003) *The Toyota Way: 14 Management Principles from the World's Greatest Manufacturer*. McGraw-Hill, New York, USA.

Mossman, A. (2013) '*The Repairman Game*', PowerPoint slides, The Change Business Ltd. (unpublished work).

Naney, D., Goser, C. and Azambuja, M. (2012) 'Accelerating the adoption of lean thinking in the construction industry', *Proc. 20th Annual Conference of the International Group for Lean Construction*, San Diego, CA, USA, pp. 291–300.

Neeraj, A., Rybkowski, Z., Fernandez-Solis, J.L., Hill, R., Tsao, C., Seed, B. and Heinemeier, D. (2016) 'Framework Linking Lean Simulation Principles to their Application on Construction Projects', *Proc. 24th Annual Conference of the International Group for Lean Construction*, Boston, MA, USA.

Ohno, T. (1988) *Toyota Production System: Beyond Large-Scale Production*. Productivity Press, Cambridge, MA, USA.

Pourabdollahian, B., Taisch, M. and Kerga, E. (2012) 'Serious games in manufacturing education', *Procedia Computer Science*, 15, pp. 256–265.

Rybkowski, Z.K., Abdelhamid, T.S. and Forbes, L.H. (2013) 'On the back of a cocktail napkin: An exploration of graphic definitions of lean construction', *Proc. 21st Annual Conference of the International Group for Lean Construction*, Fortaleza, Brazil.

Rybkowski, Z.K. and Forbes, L. (2016) 'Chapter 8: Lean Construction', in *Handbook of Construction Management: Scope, Schedule, and Cost Control*. ed.. (A. Razzak Rumane), Boca Raton, FL, CRC Press, Taylor & Francis Group, pp. 677–729.

Rybkowski, Z.K., Forbes, L.H. and Tsao, C.Y. (2018) 'The evolution of lean construction education (Part 1 of 2): At US-based universities', *Proc. 26th Annual Conference of the International. Group for Lean Construction*, Chennai, India, pp. 1013–1023.

Rybkowski, Z.K. and Kahler, D. (2014) 'Collective kaizen and standardization: the development and testing of a new lean simulation', *Proc. 22nd Annual Conference of the International. Group for Lean Construction*, Oslo, Norway.

Rybkowski, Z.K., Munankami, M., Gottipati, U., Lavy, S. and Fernández-Solis, J. (2011) 'Impact of cost constraints on aesthetic ranking following Target Value Design exercises', *Proc. 19th Annual Conference of the International. Group for Lean Construction*, Lima, Peru.

Rybkowski, Z.K., Munankami, M., Shepley, M.M. and Fernández-Solis, J.L. (2016) 'Development and testing of a lean simulation to illustrate key principles of Target Value Design: A first run study', *Proc. 24th Annual Conference of the International Group for Lean Construction*, Boston, MA, USA.

Rybkowski, Z.K., Munankami, M., Smith, J. and Kulkarni, A. (2012) 'Survey instrument to facilitate continuous improvement of lean teaching materials: a first-run study', *Proc. 20th Annual Conference of the International Group for Lean Construction*, San Diego, CA, USA.

Sacks, R., Esquenazi, A. and Goldin, M. (2007) 'LEAPCON: Simulation of Lean Construction of High-Rise Apartment Buildings', *JCEM, ASCE*, 133(7), pp. 529–539.

Santorella, G. (2011) *Lean Culture for the construction Industry: Building Responsible and Committed Project Teams*. CRC Press: Taylor & Francis Group, Boca Raton, FL, USA.

Schmaltz, D.A. (2003) *The Blind Men and the Elephant: Mastering Project Work*. Berrett- Koehler Publishers, Inc., San Francisco, CA, USA.

Seddon, J. (2017) 'The washing machine repairman'. Available at: www.vanguard-method.com (Accessed: 28th Nov 2017).

Smith, J.P. and Rybkowski, Z.K. (2013) 'The Maroon and White Game: A simulation of trust and long-term gains and losses', *Proc. 21st Annual Conference of the International Group for Lean Construction*, Fortaleza, Brazil.

Spriegel, W.R. and Myers, C.E., eds. (1953) *The Writings of the Gilbreths*. Richard D. Irwin, Inc., Homewood, IL, USA.

Superteams. (2016) *The 5S Numbers Game. Available at*: www.superteams.com/new-site/wp- content/uploads/2016/04/SuperTeams5SGameHandout.pdf (Accessed: 26th Nov 2017).

Taylor, F.W. (1947) *The Principles of Scientific Management*. W. W. Norton, New York, USA.

Tommelein, I.D. and Riley, D.R. (1999) 'Parade Game: Impact of workflow variability on trade performance,' *Journal of Construction Engineering & Management*, 125 (5), pp. 304–310.

Tsao, C., Azambuja, M., Hamzeh, F., Menches, C. and Rybkowski, Z.K. (2013) 'Teaching lean construction: perspectives on theory and practice', *Proc. 21st Annual Conference of the International Group for Lean Construction*, Fortaleza, Brazil.

Tsao, C.C.Y., Alves, T. da C.L. and Mitropoulos, P. (2012) 'Different perspectives on teaching lean construction', *Proc. 20th Annual Conference of the International Group for Lean Construction*, San Diego, CA, USA.

Tsao, C.C.Y., Draper, J. and Howell, G.A. (2014) 'An overview, analysis, and facilitation tips for simulations that support and simulate pull planning', *Proc. 22nd Annual Conference of the International Group for Lean Construction*, Oslo, Norway.

Tsao, C.C.Y. and Howell, G.A. (2015) 'A history of the use of simulations, the parade of trades, and the pull planning technique in lean construction learning and development', White Paper, March 03.

Villego® Last Planner® Simulation (2017) Available at: www.villego.com/ (Accessed: 26th Nov 2017).

Wouters, P., Tabbers. H.K. and Paas, F. (2007) 'Interactivity in video-based models', *Educational Psychology Review*, 19(3), pp. 327–342.

Visionary Products Inc. (2008). 'Lean zone production methodologies: A cellularmanufacturing simulation for 6 to 8 participants.' www.visionaryproducts.biz/index.cfm?action=view&pdfid=232. (February 28 2008).

Wujec, T. (2010). *Build a Tower, Build a Team* (TED) www.ted.com/talks/tom_wujec_build_a_tower_build_a_team?language=en (Accessed: 10 December 2019).

Zettel, G. (2014) Block Tower A-B-C.pptx. Turner Construction PowerPoint slides for amending the DPR Pull Planning Game, Anaheim, CA, USA. January.

20 Challenges and opportunities for early project collaboration

Danilo Gomes and Patricia Tzortzopoulos

20.1 Introduction

This chapter addresses early project collaboration in the context of Lean Construction. It is well known that collaboration can be negatively impacted by misunderstandings between project stakeholders with different ideas of what constitutes collaboration (Gray, 2004). In this context, the influence of the project organisation on the way project stakeholders make sense of the situation is significant, and as such mediating structures generate a new context of social interactions (Kocaturk et al., 2012). Early project collaboration involves interactions to create new objects that facilitate reflective practices and boundary-crossing between specialised disciplines (Carlile, 2004).

The chapter outlines early project collaboration, discusses how collaboration has been conceptualised in Lean Construction research, and how collaboration has been interpreted in the development of Lean contracts, systems and approaches. Based on a literature review, the discussion proposes ways to address misunderstandings around the concept of collaboration in early project interactions.

20.2 Early project collaboration: significance and challenges

The early engagement of project stakeholders from all construction disciplines is needed to optimise early project activities and reduce the need for compromise and changes in later tasks (Macmillan et al., 2001; Dong, Kleinsmann and Deken, 2013; Paranandi, 2014). Consequently, a multidisciplinary approach has been assumed as an inherent aspect of early project collaboration (AIA, 1995; RIBA, 2013).

In the context of Lean, the *Project Definition* phase in the Lean Project Delivery System (Figure 20.1) suggests that early project activities involve collective decision-making to generate a concept design based on the determination and translation of requirements (i.e. individuals' *purposes*) into performance criteria (i.e. *design criteria*) for both product and process (Ballard, 2008).

Poor performance in early project activities can result in negative environmental impacts, cost overruns and client dissatisfaction (Egan, 1998; Hsu and Liu, 2000; Qin, Wright, and Jordanov, 2000; Macmillan et al., 2001; Bertelsen, 2003; Zimina, Ballard, and Pasquire, 2012). Stakeholders need to deal with a high level of uncertainty and vagueness in design requirements and constraints (Qin, Wright, and Jordanov, 2000; Hsu and Liu, 2000). Furthermore, the diversity of commercial interests and traditional practices usually leads to poor integration between different disciplines, affecting the decision-making process and

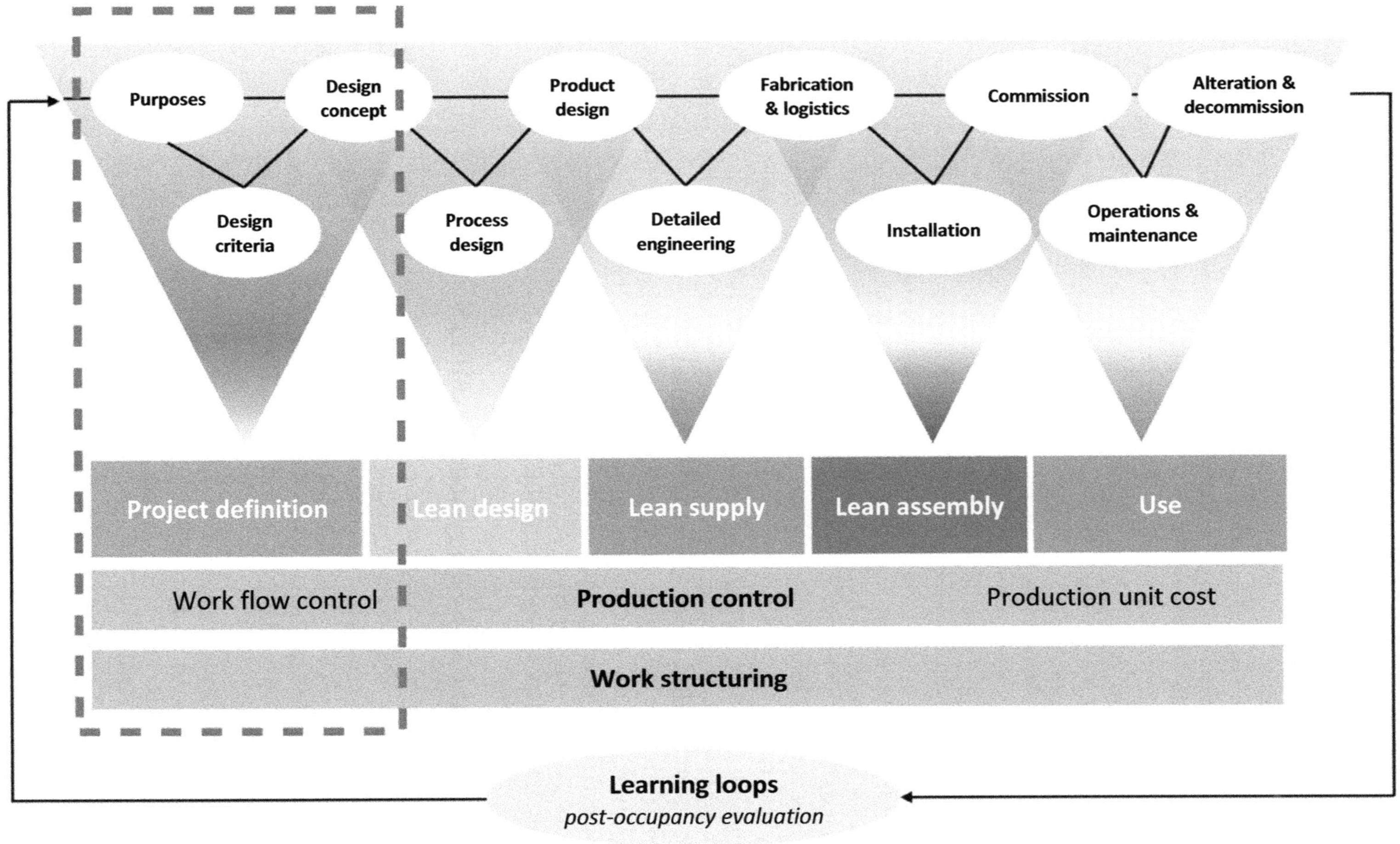

Figure 20.1 Early project stage at the Lean Project Delivery System (Ballard, 2008)

pushing problem-solving downstream (Pikas et al., 2015; Adamu, Emmitt, and Soetanto, 2015). Poor early project collaboration resulting from the late engagement of design participants leads to lack of or delayed input from participants in coupled tasks, in which the iteration needed has to start with incomplete information (Koskela, 2000). Furthermore, early decisions or intentions are sometimes not taken into account in later tasks, leading to continuous changes in design objectives or criteria (Koskela, 2000).

The ad hoc, unplanned and intuitive nature of participants' behaviours at early project activities means that the coordination of activities is challenging (Cross and Cross, 1995; Cross, 2011), and can result in a sense of disorganised behaviour among project stakeholders (Macmillan et al., 2001). The challenge is that these tacit iterations are not easy to track, affecting the understanding of the reasoning behind decision-making (Cross and Cross, 1995).

Project stakeholders engaged in collaborative interactions tend to reveal early on their different solution agendas and different ways of working (Brereton et al., 1996). Nonetheless, misunderstandings happen due to different languages, diverse disciplinary standards and wrong assumptions between design disciplines (Parrish et al., 2008). Thus, different ways of understanding and approaching the activity, along with diverse business purposes, leads to poor collaboration in terms of structure and focus, resulting in frustration within design teams (Valkenburg, 1998; Kvan, 2000; Macmillan et al., 2001).

There are also issues when participants in a construction project do not discuss what they understand as *collaboration* and use the term to refer to activities that may vary in intent and degree of participation (Kvan, 2000). Previous research identified that the term *collaboration* has been used widely, carrying a variety of meanings (D'Amour et al., 2005; Schöttle, Haghsheno, and Gehbauer, 2014; Poirier, Forgues, and Staub-French, 2016), with no consistent understanding of what collaboration is (Gray, 1989). Such diversity of understandings seems to be reinforced by the traditional configuration of construction maintaining tight boundaries around specialised functions of practice (e.g. business, design and construction), embodied within different professional disciplines (Forgues, Koskela, and Lejeune, 2009).

20.3 Collaboration in the context of the TFV theory

Koskela (1992, 2000) explored how the lack of conceptual clarity around the notion of *production* leads to collaboration problems. The author discussed that construction has either denied the existence of a theory of production or borrowed partial theories from other contexts (Koskela, 2000). Accordingly, traditional theories describing production in construction are based in the concept of transformation, and, consequently, can only provide a partial understanding, leaving certain practices without a sound theoretical foundation (Koskela, 2000).

Moreover, design and construction are understood differently, as design production is based on information as opposed to the manipulation of physical objects in the construction stage (Koskela, 2000). As a result, collaboration problems emerge especially in the interface between design and construction, affected by the temporary nature of project organisations and the lack of prior interactions between stakeholders (Latham, 1994; Koskela, 2000). For example, lack of constructability is usually seen as a consequence of poor collaboration, in which construction alternatives and constraints were not sufficiently considered during design (Koskela, 2000).

Koskela (1992, 2000) proposed the TFV theory as an integration of the conventional *Transformation* with *Flow* and *Value Generation*, providing a more comprehensive view of production (Koskela, 2000). TFV suggests that production should be conceptualised not only as a transformation of inputs into outputs, but also as a flow composed of transformation, inspection, moving and waiting, as well as a process where value for the customer is created through the fulfilment of requirements (Koskela, 2000).

The TFV theory breaks down construction projects into three complementary management functions, namely, contract management, responsible for setting up the production system, process management, responsible for getting this production system to operate to produce the project and value management, responsible for ensuring that the value delivered fulfils the client's requirements (Bertelsen and Koskela, 2002). Such re-conceptualisation led researchers to question the underlying theory of project management, finding deficiencies in the way the nature of work in projects is understood (Koskela and Howell, 2002).

According to the TFV theory, the design of production systems should focus on eliminating barriers for collaboration by creating an alliance with the customers (Koskela, 2000). Thus, Lean practices stress the importance of *planned collaboration* among different disciplines, supporting optimisation to increase value generation (Koskela, 2000). As such, collaboration is envisioned as the positive result of improved flow and value, emerging from new production system structures.

TFV provides a theoretical foundation that aligns with methods such as Concurrent Engineering, allowing the modelling and systematic management of flow and value in construction projects (Koskela, 2000). TFV also provides grounds for the development and implementation of approaches that depend on collaborative interactions between project participants (e.g. the Last Planner System of Production Planning and control, see Chapter 3 by Ballard), and technological solutions for collaboration in construction (e.g. BIM approaches, see Chapter 4 by Dave and Sacks), allowing the collective production of *value* and avoiding *waste* (Koskela, 2000; Maia, Lima and De Paula Barros Neto et al., 2011; Biton and Howell, 2013).

The main principles proposed in the TFV theory are further explored through practical applications, in terms of *contracts*, *systems* and *approaches* within a Lean Construction context, presented in the following section.

20.4 Collaboration in Lean Construction

This section briefly presents some Lean developments, emphasising how they support collaboration.

20.4.1 Collaborative contracts

New procurement routes have been considered as one of the most important changes necessary to allow more collaboration among project stakeholders in construction (Egan, 1998; Koskela, Ballard, and Howell, 2003; Bertelsen, 2003; Parrish et al., 2008; Mossman, Ballard, and Pasquire, 2011; Zimina, Ballard and Pasquire, 2012; Alves and Shah, 2018). This starts from the assumption that construction projects generally involve coalitions of representatives from different organisations with different cultures and ways of working (Forgues and Koskela, 2009).

In traditional contracting, risk is allocated to individual contract parties (Koskela, Howell, and Lichtig, 2006). Such distributed approach to risk does not help in responding to the emerging risks in the project, coming from a web of dynamic and evolving relationships among project stakeholders (Koskela et al., 2007). According to Koskela, Howell, and Lichtig (2006), traditional forms of contract establish the '*rules of the game*' that are focused in determining predictable flows within and between the organisations and usually do not consider problem-solving activities at the project level.

Such approach to contracting is embedded in current bodies of knowledge and training curricula for construction professions, in a way that prescriptive tools and processes are a representation and manifestation of the professional silos and constitute significant barriers for specialists' interactions (Forgues and Koskela, 2009).

Some Lean initiatives propose the use of relational contracts, which are designed to focus the attention of project participants on the interdependencies among them. They do so by binding the project deliverables to collaborative activities, compelling the project team to work together, enabling risk and profit to be shared across the project team. One example of this type of contract is the Integrated Project Delivery (IPD), which suggests a single prime contract between the project team and the owner (Lichtig, 2005; Ashcraft, 2012; Alves and Shah, 2018). Thus, an IPD contract should cover three basic project domains: project organisation, commercial terms and operating system (Alves and Shah, 2018). For details on IPD, please refer to Chapter 9 by Alvez and Lichtig.

20.4.2 Collaborative systems

In addition to new contractual approaches, Lean initiatives developed new systems, in which collaboration is realised as a relationship (Schöttle, Haghsheno, and Gehbauer, 2014). Initiatives such as the Lean Project Delivery System (LPDS) and Target Value Delivery (TVD) have focused on the design of new contexts using new media to support collective decision-making, aiming to achieve better value creation (Ballard, 2000; Mossman, Ballard, and Pasquire, 2011; Zimina, Ballard, and Pasquire, 2012; Alves, Lichtig, and Rybkowski, 2017).

The LPDS and TVD were envisioned, respectively, as new operating system and delivery process (Mossman, Ballard, and Pasquire, 2011; Pasquire, 2012). In TVD the team treats the value articulated by the client system as design criteria, that need to be managed through the whole life cycle (Mossman, Ballard, and Pasquire, 2011; Zimina, Ballard, and Pasquire, 2012). In combination, both LPDS and TVD propose the creation of a structured process to support joint work from early project stages (Whelton, Ballard, and Tommelein, 2001; Silveira and Alves, 2018). A detailed description of TVD is presented in Chapter 8 by Ballard.

Furthermore, the Last Planner System (LPS) was proposed as a planning and control system focused on workflow, coordinating the interdependencies between tasks and resources in a time frame (Ballard, 2000; Spitler, 2014 – see Chapter 3 by Ballard for details on the LPS). One assumption in the LPS is that collaboration will be the result of a complex network of promises, operating in situated and dynamic conditions (Ballard and Howell, 2003). Hence, task management is based on commitments to accomplish distributed responsibilities. Notably, this method creates involvement and ownership in the process (Ballard, 2000; Fundli and Drevland, 2014). LPS was initially developed for construction planning and control, but there are examples of its use in design. One example of the adaptation of the LPS to design is the *Collaborative Design Management* case presented by Fundli and Drevland (2014). In this model, a start-up meeting seems to be key for establishing the relationship between the project participants (Fundli and Drevland, 2014).

Consequently, the adoption of such systems implies that collaborative design processes should be supported by specific methods of data collection and structured meetings, in order to align requirements, criteria and concepts in group decision-making (Ballard, 2000; Alves, Lichtig, and Rybkowski, 2017). Thus, it is expected that management practices help engage multiple stakeholders involved in resolving the purposes of a project (Whelton, Pennanen, and Ballard, 2004; Silveira and Alves, 2018). This would be possible through a process that seeks early and frequent feedback to establish up to date information about needs and values between the stakeholders (Whelton, Pennanen, and Ballard, 2004). In this case, someone in the project assuming the role of the workplace planner is responsible to produce this set of information obtained from focused dialogue with the stakeholder groups (Whelton, Pennanen, and Ballard, 2004).

20.4.3 Collaborative approaches

There are initiatives attempting to formalise early project interactions to support rational methods of decision-making based on the consideration of design alternatives (Whelton, Ballard, and Tommelein, 2001; Parrish et al., 2008). In such initiatives, design is considered a collaborative decision-making process, which is more than just the exchange of information that involves the cognitive aspects of communication (Whelton, Ballard, and Tommelein, 2001). According to the same authors, formal structures for the design process should provide more transparency to the design rationale, allowing designers to negotiate conflicts between design constraints with more clarity.

For example, the use of Set-based Design (SBD) is argued to provide the necessary change in the structure of conversations between team participants, encouraging the consideration of multiple design alternatives (Parrish et al., 2008). SBD starts with mapping the design spaces to define what decisions need to be made and establishing the available design options (Parrish et al., 2008). The set of alternatives or range of values are identified in this process (Ballard, 2000). According to Ballard (2000), this map of the design space defines boundaries, in which all design contributors are free to develop their work. Integration occurs through intersections, which means that the team members need to work on solutions within the intersections of sets, where the interface dimensions, for example, need to be based on shared values (Ballard, 2000).

Choosing by Advantages (CBA) is a decision-making approach that considers advantages of alternatives and supports comparisons based on these advantages (Suhr, 1999 as cited in Parrish and Tommelein, 2009). In comparing alternatives, participants have to establish factors, which are dependent on their ability to discern unique advantages of the alternatives (Parrish and Tommelein, 2009). Factors are determined by a list of attributes, which must reflect facts wherever possible, postponing value judgement, making the process more transparent and defensible (Parrish and Tommelein, 2009). Choosing by Advantages is discussed in Chapter 10 by Arroyo.

In addition, Intensive Big Room is suggested as a co-creation method and structured context of co-working to facilitate multidisciplinary decision-making (Alhava, Laine, and Kiviniemi, 2015). The focus is to use co-location as an environment or as a structure to support collective problem-solving. In this case, a dedicated facility, physical or virtual, is used as shared workspace for frequent meetings with project participants (Alhava, Laine, and Kiviniemi, 2015). According to the same authors, it creates a platform for problem-solving, in which the team can discuss alternatives and question assumptions in parallel, reducing latency in design and avoiding waste due to unnecessary rework.

20.5 Different interpretations of collaboration

It is possible to identify differences in the interpretation of collaboration on the above discussed Lean contracts, systems and approaches. It can be argued that the different ways in which collaboration is addressed relates to different understandings of what constitutes collaboration in construction. An exploration of the differences between these underlying concepts of collaboration is needed to provide more clarity around the idea of collaboration in Lean Construction. This section discusses such differences and explores the limitations of having divergent understandings of collaboration in a project.

20.5.1 Collaborative contracts: collaboration as the project organisation

New forms of contract, such as IPD, assume that collaboration is achieved through organisational changes, enabling the early involvement of stakeholders and supporting the alignment of commercial terms for project-level teams, creating a unified project culture (Parrish et al., 2008; Mossman, Ballard, and Pasquire, 2011).

Such organisational changes impinge changes in traditional systems of work, changing roles and responsibilities between participants. While procurement methods, seen as 'structural artefacts', can play an important role, they do not seem to be enough to provoke changes in attitude and behaviour (Zimina, Ballard, and Pasquire, 2012; Alves and Shah, 2018). Results from a study on the influence of new procurement forms in the performance of collaborative teams showed that the adoption of a relational contract was not sufficient to mitigate socio-cognitive barriers between the project stakeholders (Forgues and Koskela, 2009). According to Forgues and Koskela (2009), while new procurement routes can provide a better context for collaboration, it is possible to recognise fundamental limitations regarding project managers' and designers' ability to perform in such new situations.

20.5.2 Collaborative systems: collaboration as a project mechanism

It has been argued that Western management thinking has been dominated by *thing* metaphysics, which has led to deficient conceptualisations and the adoption of counterproductive methods in construction management (Koskela and Kagioglou, 2006). Hence, traditional approaches to project management have interpreted construction as an ordered and linear phenomenon, which is a misconception, leading to misunderstandings about the nature of construction projects (Bertelsen, 2003). Moreover, the *thing* metaphysics has hindered learning, understanding and implementation of process based approaches to construction management (Koskela and Kagioglou, 2006), which is a core aspect of the TFV theory.

The process paradigm assumes that a construction project is a complex, dynamic phenomenon in a complex and non-linear setting (Bertelsen, 2003); consequently, project management should be seen as a continuous and distributed activity, involving dealing with emergent changes while fit is maintained between different parts of the production system (Koskela et al., 2007).

While the TFV theory indicated a fundamental change from the '*thing*' paradigm to a '*process*' paradigm, the development of Lean applications in terms of systems revealed challenges related to the use of prescriptive approaches to early project collaboration. This is because one of the major consequences of the adoption of the '*process*' paradigm was the proliferation of many studies prescribing sets of management practices that are implicitly assumed to generate collaboration through a predetermined process (e.g. see the *Collaborative Design Management* proposed by Fundli and Drevland, 2014).

The main assumption embedded in the adoption of prescriptive design process structures is that they can support rational decision-making, providing more transparency to resolve conflicts and making it more efficient (Whelton, Ballard, and Tommelein, 2001). Consequently, it is believed that the existence of these processes, designed as *mechanisms* for the project system, is sufficient to produce collaborative outcomes.

This assumption is also present in the concept of Integrated Design, which was summarised by Forgues and Koskela (2009) as an approach that assembles, integrates and harnesses all the collective skills and capabilities of clients and stakeholders engaged in a supply chain. Similarly, concepts such as Co-design and Participatory Design have been used to refer to strategies and processes that put together the expertise from designers and the people that will be affected by the changes produced by the design (Sanders and Stappers, 2008).

However, the adoption of, e.g. Integrated Design failed to address the socio-technical problems affecting the collaborative performance of multidisciplinary teams in construction (Forgues and Koskela, 2009). While traditional project management models had arguably failed to create the necessary conversation to develop shared perspectives and common concerns (Howell et al., 2004), some of the Lean applications had also limited impact on early project collaboration due to its deterministic nature.

For example, it is possible to say that while the use of the Last Planner System can benefit interactions at construction planning and late design stages (e.g. detailed design), there is poor evidence of its applicability at the early project stages, or on successful implementations of Phase Scheduling (mid-term planning) in design. The challenge related to the adoption of LPS at the early project stages is that design problems are ill-defined and involve the generation of a product and a correspondent production system (even if this happens unconsciously); thus, there is no way of dissociating planning from doing.

Moreover, in principle, the LPS is based on the assumption that there is belief and trust between stakeholders. This is made clear in the phase scheduling, where the focus is on describing specific goals in terms of plans and handoffs between stakeholders to achieve those goals (Ballard, 2000), without necessarily sharing the reasoning process that supports the definitions of the goals.

It could be argued that building interdependencies in a temporal/spatial planning activity creates awareness, but not necessarily understanding, about the process across the project team. This means that team members may know the operational sequence, but not necessarily understand why all the procedures were determined in a particular way. Consequently, this can limit team members' ability to contribute in the decision-making, and therefore limit collaboration, especially if this happens at early project stages.

20.5.3 Collaborative approaches: collaboration as socio-constructive interaction

Lean approaches focused to support early project stages seem to interpret collaboration as socio-constructive interaction. A socio-constructive perspective assumes that individuals develop subjective meanings of their experiences directed toward certain objects or things (Berger and Luckmann, 1966). Thus, social constructive interactions (e.g. conversations among individuals in the social space) generate collective meaning by articulating, producing and negotiating social objects, e.g. schemata and theories of collaboration (Gergen, 1985).

Stakeholders engaged at early stages use techniques to cope with the lack of input and progress with design activities even if this brings additional costs, added risks and reduced functionality (Koskela, 2000). Most of the time design decisions are made based on assumptions that will be checked later, but if they are found to be wrong it leads to rework (Koskela, 2000). Consequently, to avoid or decrease the level of rework, design solutions tend to be over-dimensioned to absorb all possible future decisions, which can be considered a sub-optimal solution (Koskela, 2000).

In Set-based Design, each project participant must understand what the project brief is asking, articulate the levels of detail and accuracy required to define alternatives, taking into account the values and constraints that emerge from each member (Parrish et al., 2008). In this case, the different disciplinary background of project participants may affect some basic assumptions made on the definition of abstract concepts in design (i.e. defining a set on the design space), which consequently affects the understanding of the levels of detail and accuracy required by the other party to contribute to that design decision (Parrish et al., 2008). This requires a certain ability from the project stakeholders to work upon incomplete information. Moreover, it requires the recognition that emergent misunderstandings are unpredictable and will change and evolve in different forms depending on the situation.

In order to overcome misunderstandings, team members need to interact over design semantics, which are justifications of design decisions and properties associated with performance and behaviour of design (Maher, Cicognani, and Simoff, 1996). This means that team members regularly need to engage in breakdown activities to describe the rationale behind certain design decisions. As concepts need to be built over time to turn the initial idea into something more robust, project team members are involved in a persuasion process to convince the others of the value of certain concepts (Cross and Cross, 1995; Cross, 2011). Consequently, the evolution of design content also depends on stakeholders' negotiation strategies (Brereton et al., 1996).

When these conflicts in understanding are identified and analysed at an early project stage, relational uncertainty can be reduced making it easier to reach compromise, avoiding escalation in the relationship (Vaaland, 2004). Choosing by Advantages seems to allow such collective reflective interactions.

In Choosing by Advantages, arguments are built upon data that is relevant to a particular decision, providing stronger support and less ambiguity (Arroyo, Ballard, and Tommelein, 2014). The emergent behaviour is that the affected party asks for clarification (Parrish et al., 2008), and as such the interdependent factors for decision-making are collectively understood.

Designers will usually try to build shared understanding about the design situation using graphic tools and verbal communication (Cross and Cross, 1995). Shared understanding in the context of collaborative design is seen as a mutual view amongst the team members on a relevant design topic and design activity (Valkenburg, 1998; Kleinsmann, 2006; Kleinsmann and Valkenburg, 2008). According to Smart (2011), shared understanding is more than sharing similar views, it implies similarity of understanding in relation to a particular phenomenon (i.e. goals, task, situation), involving the emergence of a collective ability to form expectations and predictions regarding future states, actions and events. Bittner and Leimeister (2013) suggest that shared understanding can be seen as a collective and dynamic ability to conceive and coordinate actions towards common goals or objectives of multiple agents in a project activity.

As a collective ability, shared understanding emerges through dialogue. For example, designers show skills to construct an effective argument to get their version of the consequences on a project situation accepted in a meeting (Lloyd and Busby, 2001). Using language mechanisms of engagement, exaggeration and imagery (i.e. metaphors), designers try to create situations of implied objectivity over common assumptions and/or about past experiences (Lloyd and Busby, 2001). In face-to-face interactions, these persuasion strategies will help designers reconcile their differences and moderate their commitments by paying attention and reflecting over assumptions shared by the project team (Brereton et al., 1996). In this case, individual technical expertise (i.e. understanding) seems to emerge embedded in a context of meanings, in which interactions would be generally related to the surfacing of negative information, the public negotiation of dilemmas and the resolution or dissolution of conflicting views (Schön, 1983), towards the construction of shared understanding.

Initial studies about the use of CBA suggest that these interactions could be related to the development of rhetorical skills (Arroyo, Ballard, and Tommelein, 2014; Koskela, 2015). For instance, the use of logical reasoning, to describe and summarise the advantages of alternatives (Arroyo, Ballard, and Tommelein, 2014), may be an example of the use of causal knowledge to persuade towards a suggested argument in the project. In this process, the group needs to set the importance scale, defining the advantage and assigning a degree of importance to advantages based on the multiple perspectives of the participants (Parrish and Tommelein, 2009). However, these studies also indicate that the nature of collective interactions involved in the discussions to decide the advantages of alternatives have not yet been sufficiently studied (Arroyo, Ballard, and Tommelein, 2014; Alhava, Laine, and Kiviniemi, 2015). Overall, there seems to be a need for further exploration on how to develop skills to comply with these new contextual components (e.g. procurement methods, project strategies, management systems) (Mossman, Ballard, and Pasquire, 2011; Zimina, Ballard, and Pasquire, 2012).

In summary, early project collaboration can be seen as a deeper sense-making activity that is not necessarily tied to any particular tool and which is usually related to a higher creative element of a project (Jonson, 2005). Early design is generally developed using a combination of tools (i.e. verbal and non-verbal languages: graphical visualisations and gestures) to build a dialogue between project stakeholders (Schön, 1983; Donn, 2014). In the social context of project teams, verbalisation is emphasised as a fundamental aspect to support collaborative interactions (Cross and Cross, 1995; Jonson, 2005), and Set-based Design and Choosing by Advantages seem to support this as they support the collective reflective nature of these interactions.

20.6 Discussion

The previous sections suggested that a major aspect of early project collaboration is the forging of connections between diverse discourses (Dorst, 2006). However, this can be hampered if there are differences in the way participants perceive and understand collaboration (Gray, 2004). Thus, it can be argued that individuals and organisations deeply involved with the adoption and implementation of Lean may encounter difficulties due to potential misunderstandings about the nature of collaboration.

Table 20.1 summarises how Lean contracts, systems and approaches differ on the way they conceive collaboration. Consequently, stakeholders may develop wrong expectations about each other's contributions.

It can be argued that these different ways of interpreting collaboration (Table 20.1) can be seen as ideal types shared by certain groups. Therefore, these different notions of collaboration tend to operate as competing appreciative systems,[1] framing the way project participants interact in the construction project. This happens because these conceptualisations differ fundamentally in ontological terms, and, consequently, actions and artefacts produced by one project participant can be seen as incoherent or even contradictory by another participant (Gomes and Tzortzopoulos, 2019).

Emergent misunderstandings at early projects can be considered dilemmas, as they usually involve contradictory ways of conceptualising collaboration. Reflection and awareness over conflicts of understanding may lead designers to understand the intractability of their dilemmas and to suggest alternative design decisions (Schön, 1983). Moreover, the resolution or dissolution of conflicting views emerging in early project interactions should also be treated as the negotiation of organisational dilemmas (Schön, 1983).

However, in order to overcome these conflicts of understanding we should not be searching for a definitive interpretation of collaboration, but instead we need to embrace the fact that these diverse interpretations can complement each other (Bardram, 1998).

Hence, to see collaboration solely as a *mechanism* (organisation and/or process) can be problematic. According to Gharajedaghi and Ackoff (1984), traditional conceptions of human organisation and project management can severely limit our ability to understand and manage collaborative interactions. More importantly, different notions of collaboration produce terminologies that sometimes are interpreted as equivalent, potentially leading to misunderstandings that can only be resolved through dialogue (Gharajedaghi and Ackoff, 1984).

The collaborative performance of a team member depends on how people interact with others, articulating shared concepts (Dong, Kleinsmann, and Deken, 2013). Thus, early project interactions involve the negotiation of shared perspectives, in which individuals' understanding of the project situation are conceived and externalised through artefacts, making the implicit frame more explicit, which then becomes a target of negotiation to establish a shared frame, resolving conflicts (Schön, 1983; Hey, Joyce, and Beckman, 2007).

Early project interactions can be interpreted as a sophisticated activity system, from an Activity Theory perspective (Forgues, Koskela, and Lejeune, 2009). Accordingly, human interactions have a dialectical nature, and they are always embedded within a sociocultural context of other humans and work activities (Engeström, 1987). This means that as different stakeholders may develop different perceptions over the same task, the natural evolution of an organisation to deliver a project involves the management of the individuals' diverse conceptions of the task, linked to the stability of the means and the object of work (Bardram, 1998). The dynamics of these interactions are caused by reflections on the means and object of work (i.e. artefacts), stimulated by a breakdown or by a deliberate shift of focus within the interactions (Bardram, 1998). Thus, the social construction of collaboration in the activity requires the resolution of misunderstandings

Table 20.1 Different interpretations of collaboration in the context of Lean Construction

Collaborative Contracts		*Collaborative Systems*		*Collaborative Approaches*	
Main idea:	**Collaboration as the project organisation**	**Main idea:**	**Collaboration as management strategies and/or process models**	**Main idea:**	**Collaboration as a socio-constructive interaction**
Key Principles:		**Key Principles:**		**Key Principles:**	
Collaboration involves:	coalitions of representatives from different organisations with different cultures and ways of working (Forgues & Koskela, 2009), leading to organisational changes.	Collaboration involves:	management, as a continuous and distributed activity, to maintain a fit between different parts of the production system (Koskela et al., 2007).	Collaboration involves:	the socio-construction of the activity, dealing with the undetermined nature of early project interactions (Schön, 1983).
How	Relational Contract models establish interdependencies among project participants, binding the project deliverables to collaborative activities, making risk and profit to be shared across the project team (Alves & Shah, 2018).	How	Management practices and strategies based on formal process structures that are expected to support rational decision-making, providing transparency to resolve conflicts and becoming more efficient (Whelton, Ballard & Tommelein, 2001).	How	Different disciplinary background of project participants affects the understanding of the levels of detail and accuracy required by the other party to contribute on that design decision (Parrish et al., 2008). Interactions involve forging of connections between diverse discourses (Dorst, 2006).
Examples of Lean Applications		**Examples of Lean Applications**		**Examples of Lean Applications**	
	Integrated Project Delivery		*Lean Project Delivery System* *Target Value Design* *Last Planner System* *Collaborative Design Management*		*Set-Based Design* *Choosing-by-Advantages* *Intensive Big Room*

(*Continued*)

Table 20.1 (Cont.)

Collaborative Contracts	*Collaborative Systems*	*Collaborative Approaches*
Consequences, Limitations and Challenges:	**Consequences, Limitations and Challenges:**	**Consequences, Limitations and Challenges:**
Leads to changing traditional systems of work, changing roles and responsibilities between participants.	Failed to address the socio-technical problems affecting the collaborative performance of multidisciplinary teams in construction (Forgues & Koskela, 2009).	Stakeholders are involved in a persuasion process to convince the others of the value of certain concepts (Cross and Cross, 1995; Cross, 2011). In order
Not enough to provoke changes in attitude and behaviour (Zimina, Ballard & Pasquire, 2012)	The adoption of LPS at Early Project Stages is challenging because there is no way of dissociating planning from doing at this stage.	To build shared understanding, stakeholders tend to use language mechanisms of engagement, exaggeration and imagery (i.e. metaphors) (Lloyd & Busby, 2001)
Not sufficient to mitigate socio-cognitive barriers between the project stakeholders (Forgues & Koskela, 2009).		As misunderstandings often happen over the consequences of certain "facts" (Lloyd and Busby, 2001), to overcome them, team members need to discuss design semantics, and the reasoning behind design decisions (Maher, Cicognani & Simoff, 1996).

about the task that, eventually, become re-embodied through implementation and routinisation (Bardram, 1998).

Therefore, collaboration is fundamentally dependent on individuals' ability to build a coherent interpretation of the purpose of the collective action. This is not to say that perceptions cannot be collectively constructed. Nevertheless, it suggests that participants that usually engage in working in a specific way can develop a particular understanding of the nature of collaboration. Hence, collective reflection leading to breakdown and the resolution of misunderstandings could be considered a way to remove conceptual barriers and build common ground between project participants (Coyne and Snodgrass, 1993). As suggested by Schrage (1995, p. 13):

> Collaboration is the process of shared creation: two or more individuals with complementary skills interacting to create shared understanding that none had previously possessed or could have come to on their own. Collaboration creates a shared meaning about a process, a product, or an event.

Shared creation depends on the notion of shared understanding, as a collective ability developed in the context of the project activity. Hence, collaboration as shared creation depends on the constructive dialogue between project stakeholders, allowing a dialectical process in which these individuals contribute in increasing the collective understanding on how to best proceed in the task.

20.7 Final thoughts

This chapter discusses how divergent interpretations about the nature of collaboration in construction may lead to misunderstandings among project stakeholders, especially at the early project stages. In order to overcome these misunderstandings, researchers and practitioners need to embrace the socio-constructive nature of these interactions, as misunderstandings about what represents collaboration in construction projects can emerge through the 'clash' of different concepts of collaboration embedded in Lean contracts, systems and approaches.

In this context, the development of the TFV theory represented a huge step towards a change of perception in construction allowing the realisation of a different concept of collaboration. However, previous research already pointed out a few limitations of Lean applications which have been based solely on a deterministic view of collaboration.

It can be argued that while some Lean contracts, systems and approaches (e.g. LPDS, TVD, LPS) seem to be an effective way of supporting collaborative interactions at later stages of the construction project (in terms of supporting collective interactions once the object of design is stable), they do not completely address the socio-constructive nature of collaborative interactions at the early project stage.

Early project collaboration should be interpreted as a dialectical activity, in which collaboration can be considered as a *social construct*. Consequently, prescriptive and deterministic strategies and approaches should be avoided. Alternatively, the creation and utilisation of project artefacts should emerge through collective reflection, building shared understanding among the project participants.

Moreover, this socio-constructive view of collaborative interactions allows different interpretations of what constitutes collaboration to coexist. In this case, project stakeholders, as activity systems, tend to work towards the resolution of misunderstandings, conceiving concerted actions as an outcome of collaboration.

Activity Theory can help improve the understanding of the socio-constructive and dialectical nature of collaborative interactions at early project stages, by supporting investigation into how different concepts of collaboration coexist. Therefore, further empirical research should be developed to better understand how project participants create shared understanding using diverse Lean applications in early project stages. There is a need to develop applied research into the socio-constructive aspect of Lean initiatives that may enable reflection on existing Lean approaches (e.g. procurement methods, organisational structures, design methods). Research should also be developed to explore how some of the Lean applications could include more socio-constructive practices (e.g. gamification of Lean). In addition, further developments may also require a reflection about the traditional ideas of what constitutes knowledge and professional skills, used as foundations for the current models of professional training in the construction industry.

Note

1 The notion of *Appreciative Systems* was initially proposed by Vickers (1965) as an epistemological model by which individuals make sense of what is known as the 'social process'. The concept of *appreciative systems* contends that appreciation is enacted by individuals' ability to select and perceive (part of) reality to make judgements about it, contributing to the idea stream, and leading to actions that become part of the events stream (Checkland, 1994).

References

Adamu, Z.A., Emmitt, S. and Soetanto, R. (2015) 'Social BIM: Co-creation with shared situational awareness', *Journal of Information Technology in Construction*, 20, pp. 230–252.

AIA. (1995) *AIA Document D200. Project Checklist.* Washington, DC, USA: The American Institute of Architects, 1735 New York Avenue, N. W.

Alhava, O., Laine, E. and Kiviniemi, A. (2015) 'Intensive big room process for co-creating value in legacy construction projects', *Journal of Information Technology in Construction*, 20(11), pp. 146–158.

Alves, T.C.L., Lichtig, W. and Rybkowski, Z.C. (2017) 'Implementing target value design: Tools and techniques to manage the process', *HERD: Health Environments Research and Design Journal*, 10(3), pp. 18–29.

Alves, T.C.L. and Shah, N. (2018) 'Analysis of construction contracts: Searching for collaboration', *2018 Construction Research Congress*, ASCE, New Orleans, LA, USA, pp. 148–157.

Arroyo, P., Ballard, G. and Tommelein, I.D. (2014) 'Choosing by advantages and rhetoric in building design: Relationship and potential synergies', *Proc. 22nd Annual Conference of the International Group for Lean Construction*, Oslo, Norway. Available at: www.iglc.net/Papers.

Ashcraft, H.W. (2012) *The IPD Framework*, San Francisco, CA, USA: Hanson Bridgett white paper.

Ballard, G. (2000) 'Positive vs. negative iteration in design', *Proc. 8th Annual Conference of the International Group for Lean Construction*, Brighton, UK. Available at: www.iglc.net/Papers.

Ballard, G. (2008) 'The lean project delivery system: An update', *Lean Construction Journal*, pp. 1–19.

Ballard, G. and Howell, G.A. (2003) 'An update on last planner', *Proc. 11th Annual Conference of the International Group for Lean Construction*, Virginia, USA. Available at: www.iglc.net/Papers.

Bardram, J. (1998) 'Designing for the dynamics of cooperative work activities', *Proc.1998 ACM conference on Computer supported cooperative work*, ACM, pp. 89–98.

Berger, P.L. and Luckmann, T. (1966) *The Social Construction of Reality: A Treatise in the Sociology of Knowledge*, London: Penguin UK.

Bertelsen (2003) 'Construction as a complex system', *Proc. 11th Annual Conference of the International Group for Lean Construction*, Virginia, USA. Available at: www.iglc.net/Papers.

Bertelsen, S. and Koskela, L. (2002) 'Managing the three aspects of production in construction', *Proc. 10th Annual Conference of the International Group for Lean Construction*, Gramado, Brazil. Available at: www.iglc.net/Papers.

Biton, N. and Howell, G. (2013) 'The journey of lean construction theory review and reinterpretation', *Proc. 21st Annual Conference of the International Group for Lean Construction*, Fortaleza, Brazil. Available at: www.iglc.net/Papers.

Bittner, E.A.C. and Leimeister, J.M. (2013) 'Why Shared Understanding Matters-Engineering a Collaboration Process for Shared Understanding to Improve Collaboration Effectiveness in Heterogeneous Teams', *46th Hawaii International Conference on System Sciences (HICSS)*, IEEE, pp. 106–114.

Brereton, M.F., Cannon, D.M., Mabogunje, A. and Leifer, L.J. (1996) 'Collaboration in design teams: How social interaction shapes the product', in N. Cross, H. Christiaans, and K. Dorst (eds), *Analysing Design Activity*, Chichester, UK: John Wiley & Sons, pp. 319–341.

Carlile, P.R. (2004) 'Transferring, translating, and transforming: An integrative framework for managing knowledge across boundaries', *Organization Science*, *15*(5), pp. 555–568.

Checkland, P. (1994) 'Systems theory and management thinking', *American Behavioural Scientist*, *38*(1), pp. 75–91.

Coyne, R. and Snodgrass, A. (1993) 'Cooperation and individualism in design', *Environment and Planning B: Planning and Design*, *20*(2), pp. 163–174.

Cross, N. (2011) *Design Thinking: Understanding How Designers Think and Work*, Oxford, UK: Berg Publishers.

Cross, N. and Cross, A.C. (1995) 'Observations of teamwork and social processes in design', *Design Studies*, *16*(2), pp. 143–170.

D'Amour, D., Ferrada-Videla, M., San Martin Rodriguez, L. and Beaulieu, M.D. (2005) 'The conceptual basis for interprofessional collaboration: Core concepts and theoretical frameworks', *Journal of Interprofessional Care*, *19*(1), pp. 116–131.

Dong, A. (2005) 'The latent semantic approach to studying design team communication', *Design Studies*, *26*(5), pp. 445–461.

Dong, A., Kleinsmann, M.S. and Deken, F. (2013) 'Investigating design cognition in the construction and enactment of team mental models', *Design Studies*, *34*(1), pp. 1–33.

Donn, M. (2014) 'BIM and the predesign process: Modelling the unknown', in K. M. Kensek and D. Noble (eds), *Building Information Modelling: BIM in Current and Future Practice*, Hoboken, NJ: John Wiley & Sons, pp. 143–155.

Dorst, K. (2006) 'Design problems and design paradoxes', *Design Issues*, *22*(3), pp. 4–17.

Egan, J. (1998) *Rethinking Construction, Construction Task Force Report for Department of the Environment, Transport and the Regions*, London, UK: HMSO.

Engeström, Y. (1987) *Learning by Expanding*, Helsinki, Finland: Orienta-Konsultit Oy.

Forgues, D. and Koskela, L. (2009) 'The influence of a collaborative procurement approach using integrated design in construction on project team performance', *International Journal of Managing Projects in Business*, *2*(3), pp. 370–385.

Forgues, D., Koskela, L.J. and Lejeune, A. (2009) 'Information technology as boundary object for transformational learning', *Journal of Information Technology in Construction*, *14*, pp. 48–58.

Fundli, I.S. and Drevland, F. (2014) 'Collaborative Design Management – A Case Study', *Proc. 22nd Annual Conference of the International Group for Lean Construction*, Oslo, Norway. Available at: www.iglc.net/Papers.

Gergen, K.J. (1985) 'The social constructionist movement in modern psychology', *American Psychologist*, *40*(3), pp. 266–275.

Gharajedaghi, J. and Ackoff, R.L. (1984) 'Mechanisms, organisms and social systems', *Strategic Management Journal*, *5*(3), pp. 289–300.

Gomes, D. and Tzortzopoulos, P. (2019) 'Metaphors of collaboration in construction', *Canadian Journal of Civil Engineering*, Available at. https://doi.org/10.1139/cjce-2018-0461

Gray, B. (1989) *Collaborating: Finding Common Ground for Multiparty Problems*, San Francisco, CA, USA: Jossey-Bass. pp. 360.

Gray, B. (2004) 'Strong opposition: Frame-based resistance to collaboration', *Journal of Community & Applied Social Psychology, 14*(3), pp. 166–176.

Hey, J.H., Joyce, C.K. and Beckman, S.L. (2007) 'Framing innovation: Negotiating shared frames during early design phases', *Journal of Design Research, 6*(1–2), pp. 79–99.

Howell, G., Macomber, H., Koskela, L. and Draper, J. (2004) 'Leadership and project management: time for a shift from Fayol to Flores', *Proc. 12th Annual Conference of the International Group for Lean Construction*, Helsingør, Denmark. Available at: www.iglc.net/Papers.

Hsu, W. and Liu, B. (2000) 'Conceptual design: Issues and challenges', *Computer-Aided Design, 14*(32), pp. 849–850.

Jonson, B. (2005) 'Design ideation: The conceptual sketch in the digital age', *Design Studies, 26*(6), pp. 613–624.

Kleinsmann, M. (2006) *Understanding Collaborative Design*. PhD Thesis, Delft University of Technology, Delft, The Netherlands.

Kleinsmann, M. and Valkenburg, R. (2008) 'Barriers and enablers for creating shared understanding in co-design projects', *Design Studies*, 29(4), pp. 369–386.

Kocaturk, T., Balbo, R., Medjdoub, B. and Veliz, A. (2012) "An innovative approach to technology mediated architecture design education', A framework for a web-based socio-cognitive eco-system', *eCAADe 30, CAAD Curriculum*, 1, pp. 55–66.

Koskela, L. (1992) *Application of the New Production Philosophy to Construction*, Center for Integrated Facility Engineering, Department of Civil Engineering, Stanford University. Technical Report number: 72.

Koskela, L. (2000) *An Exploration Towards a Production Theory and its Application to Construction*, PhD thesis. Julkaisija, Finland: Technical Research Centre of Finland, VTT.

Koskela, L., (2015) 'Where rhetoric and lean meet', *Proc. 23rd Annual Conference of the International Group for Lean Construction*, Perth, Australia. Available at: www.iglc.net/Papers.

Koskela, L., Ballard, G. and Howell, G. (2003) 'Achieving change in construction', *Proc. 11th Annual Conference of the International Group for Lean Construction*, Virginia, USA. Available at: www.iglc.net/Papers.

Koskela, L., Howell, G. and Lichtig, W. (2006) 'Contracts and production', *CIB W92 Symposium on Sustainability and Value through Construction Procurement*, December 2006, Salford Quays. (Unpublished).

Koskela, L., Rooke, J., Bertelsen, S. and Henrich, G. (2007) 'The TFV theory of production: New developments', *Proc. 15th Annual Conference of the International Group for Lean Construction*, East Lansing, MI, USA. Available at: www.iglc.net/Papers.

Koskela, L.J. and Howell, G. (2002) 'The underlying theory of project management is obsolete', *Proc. of the PMI Research Conference*, PMI.

Koskela, L.J. and Kagioglou, M. (2006) 'On the metaphysics of management', *Proc. 14th Annual Conference of the International Group for Lean Construction*, Santiago, Chile. Available at: www.iglc.net/Papers.

Kvan, T. (2000) 'Collaborative design: what is it?', *Automation in Construction*, 9(4), pp. 409–415.

Latham, M. (1994) *Constructing the Team: Joint Review of Procurement and Contractual Arrangements in the UK Construction Industry*. UK: Department of the Environment.

Lichtig, W.A. (2005) 'Sutter health: Developing a contracting model to support lean project delivery', *Lean Construction Journal*, 2(1), pp. 105–112.

Lloyd, P. and Busby, J. (2001) 'Softening up the facts: Engineers in design meetings', *Design Issues*, 17(3), pp. 67–82.

Macmillan, S., Steele, J., Austin, S., Kirby, P. and Spence, R. (2001) 'Development and verification of a generic framework for conceptual design', *Design Studies, 22*(2), pp. 169–191.

Maher, M.L., Cicognani, A. and Simoff, S. (1996) 'An experimental study of computer mediated collaborative design', *Proc. 5th Workshop on Enabling Technologies: Infrastructure for Collaborative Enterprises*, pp. 268–273.

Maia, S., Lima, M. and De Paula Barros Neto, J. (2011) 'A systemic approach to the concept of value and its effects on lean construction', *Proc. 19th Annual Conference of the International Group for Lean Construction*, Lima, Peru. Available at: www.iglc.net/Papers.

Mossman, A., Ballard, G. and Pasquire, C. (2011) 'The growing case for lean construction', *Construction Research and Innovation*, 2(4), pp. 30–34.

Paranandi, M. (2014) 'BIM as a catalyst to foster creativity through collaboration', in K.M. Kensek and D. Noble (eds), *Building Information Modeling: BIM in Current and Future Practice*, Hoboken, NJ: John Wiley & Sons, pp. 237–249.

Parrish, K. and Tommelein, I.D. (2009) 'Making design decisions using choosing by advantages', *Proc. 17th Annual Conference of the International Group for Lean Construction*, Taipei, Taiwan. Available at: www.iglc.net/Papers.

Parrish, K., Wong, J.M., Tommelein, I.D. and Stojadinovic, B. (2008) 'Set-based design: case study on innovative hospital design', *Proc. 16th Annual Conference of the International Group for Lean Construction*, Manchester, UK. Available at: www.iglc.net/Papers.

Pasquire, C. (2012) 'The 8th flow – common understanding', *Proc. 20th Annual Conference of the International Group for Lean Construction*, San Diego, CA, USA. Available at: www.iglc.net/Papers.

Pikas, E., Koskela, L., Dave, B. and Liias, R. (2015) 'Case study on design management: Inefficiencies and possible remedies', *Proc. 23rd Annual Conference of the International Group for Lean Construction*, Perth, Australia. Available at: www.iglc.net/Papers.

Poirier, E., Forgues, D. and Staub-French, S. (2016) 'Collaboration through innovation: Implications for expertise in the AEC sector', *Construction Management and Economics*, *34*(11), pp. 769–789.

Qin, S.F., Wright, D.K. and Jordanov, I.N. (2000) 'From on-line sketching to 2D and 3D geometry: A system based on fuzzy knowledge', *Computer-Aided Design*, *32*(14), pp. 851–866.

RIBA. (2013) *RIBA Plan of Work 2013*. Dale Sinclair (ed), 66 Portland Place, London, UK: RIBA.

Sanders, E.B.N. and Stappers, P.J. (2008) 'Co-creation and the new landscapes of design', *Co-design*, *4*(1), pp. 5–18.

Schön, D.A. (1983) *The Reflective Practitioner: How Professionals Think in Action*. New York: Basic Books.

Schöttle, A., Haghsheno, S. and Gehbauer, F. (2014) 'Defining cooperation and collaboration in the context of lean construction', *Proc. 22nd Annual Conference of the International Group for Lean Construction*, Oslo, Norway. Available at: www.iglc.net/Papers.

Schrage, M. (1995) *No More Teams! Mastering the Dynamics of Creative Collaboration*, ed New York, USA: Currency Doubleday. p. 239.

Silveira, S.S. and Alves, T.C.L. (2018) 'Target value design inspired practices to deliver sustainable buildings', *Buildings*, 8(9), pp. 116.

Smart, P.R. (2011) Understanding and Shared Understanding in Military Coalitions s.n. (In Press)', *University of Southampton Institutional Repository*. Available at: https://eprints.soton.ac.uk/267735/ [Accessed 01st October 2019].

Spitler, L.E. (2014) 'The effect of inter-team dynamics on the constructability of the BIM model', *Proc. 22nd Annual Conference of the International Group for Lean Construction*, Oslo, Norway. Available at: www.iglc.net/Papers.

Vaaland, T.I. (2004) 'Improving project collaboration: Start with the conflicts', *International Journal of Project Management, 22*(6), pp. 447–454.

Valkenburg, R.C. (1998) 'Shared understanding as a condition for team design', *Automation in Construction*, 7(2), pp. 111–121.

Vickers, G. (1965) *The Art of Judgment: A Study of Policy Making*, London: Chapman & Hall.

Whelton, M., Ballard, G. and Tommelein, I. (2001) 'Application of design rationale systems to project definition – Establishing a research project', *Proc. 9th Annual Conference of the International Group for Lean Construction*, Singapore, Singapore. Available at: www.iglc.net/Papers.

Whelton, M., Pennanen, A. and Ballard, G. (2004) 'Fostering collaboration and learning in project definition: a case study in workplace planning', *Proc. 12th Annual Conference of the International Group for Lean Construction*, Helsingør, Denmark. Available at: www.iglc.net/Papers.

Zimina, D., Ballard, G. and Pasquire, C. (2012) 'Target value design: Using collaboration and a lean approach to reduce construction cost', *Construction Management and Economics*, *30*(5), pp. 383–398.

Index

For Product Safety Concerns and Information please contact our EU
representative GPSR@taylorandfrancis.com
Taylor & Francis Verlag GmbH, Kaufingerstraße 24, 80331 München, Germany

www.ingramcontent.com/pod-product-compliance
Ingram Content Group UK Ltd.
Pitfield, Milton Keynes, MK11 3LW, UK
UKHW052217190926
475702UK00006B/1070
* 9 7 8 0 3 6 7 1 9 6 5 5 4 *